THERMOMECHANICS OF COMPOSITES UNDER HIGH TEMPERATURES

SOLID MECHANICS AND ITS APPLICATIONS
Volume 65

Aims and Scope of the Series

The fundamental questions arising in mechanics are: *Why?*, *How?*, and *How much?* The aim of this series is to provide lucid accounts written by authoritative researchers giving vision and insight in answering these questions on the subject of mechanics as it relates to solids.

The scope of the series covers the entire spectrum of solid mechanics. Thus it includes the foundation of mechanics; variational formulations; computational mechanics; statics, kinematics and dynamics of rigid and elastic bodies; vibrations of solids and structures; dynamical systems and chaos; the theories of elasticity, plasticity and viscoelasticity; composite materials; rods, beams, shells and membranes; structural control and stability; soils, rocks and geomechanics; fracture; tribology; experimental mechanics; biomechanics and machine design.

The median level of presentation is the first year graduate student. Some texts are monographs defining the current state of the field; others are accessible to final year undergraduates; but essentially the emphasis is on readability and clarity.

For a list of related mechanics titles, see final pages.

Thermomechanics of Composites under High Temperatures

by

YU. I. DIMITRIENKO

Bauman State Technical University,
Moscow, Russia

KLUWER ACADEMIC PUBLISHERS
DORDRECHT / BOSTON / LONDON

A C.I.P. Catalogue record for this book is available from the Library of Congress.

ISBN 978-90-481-5122-6

Published by Kluwer Academic Publishers,
P.O. Box 17, 3300 AA Dordrecht, The Netherlands

Sold and distributed in the North, Central and South America
by Kluwer Academic Publishers,
101 Philip Drive, Norwell, MA 02061, U.S.A.

In all other countries, sold and distributed
by Kluwer Academic Publishers
P.O. Box 322, 3300 AH Dordrecht, The Netherlands

Printed on acid-free paper

TABLE OF CONTENTS

Introduction ix

Nomenclature xiii

Chapter 1. High-Temperature Environment and Composite Materials **1**
- 1.1. Main Types of High-Temperature Effects on Composite Materials 1
- 1.2. Ablation Processes of Composites 19
- 1.3. Phenomena in Composite Materials under High Temperatures 22
- 1.4. A Physical Model of Ablative Composite 25

Chapter 2. General Equations of Multiphase Continuum Mechanics for Ablative Composites **27**
- 2.1. Conservation Laws 27
- 2.2. Constitutive Relations for Phases of Ablative Composites 32
- 2.3. Relations at the Phase Interface 38
- 2.4. Equation of Phase Transformation Rate 42
- 2.5. Infinitesimal Strains of Solid Phases 44

Chapter 3. Mathematical Model of Ablative Composites **53**
- 3.1. Main Assumptions 53
- 3.2. Statement of Thermomechanics Problem for Ablative Composites 57
- 3.3. Method of Asymptotic Averaging 59
- 3.4. Averaging of Processes in Ablative Composites 63
- 3.5. Analysis of Asymptotic Averaging Method 65
- 3.6. Statement of Problems for Composites with Ablative Matrix and Fibres 66

Chapter 4. Behaviour of Matrices at High Temperatures **73**
- 4.1. Varying Density of Matrices at High Temperatures 73
- 4.2. Effective Elastic Properties of Ablative Matrices at High Temperatures 79
- 4.3. Heat Expansion/Shrinkage of Matrices at High Temperatures 89
- 4.4. Strength Properties of Matrices at High Temperatures 92
- 4.5. Heat Conductivity and Heat Capacity of Matrices at High Temperatures 101
- 4.6. Gas-Permeability of Matrices at High Temperatures 103

Chapter 5. Reinforcing Fibres under High Temperatures **109**
5.1. Changing Phase Composition of Fibres under Heating 109
5.2. Heat Conductivity and Heat Capacity of Ablative Fibres 113
5.3. Varying Elastic Properties of Fibres under Heating 115
5.4. Heat Deformation of Fibres 117
5.5. Strength Properties of Fibres under High Temperatures 117
5.6. Short Fibres and Dispersed Particles 125

Chapter 6. Unidirectional Composites under High Temperatures **127**
6.1. Structural Model of Unidirectional Composites 127
6.2. Model of Microcomposite 130
6.3. Thermo-Elastic Characteristics and Heat Conductivity of Unidirectional Composites 135
6.4. Strength Properties of Unidirectional Composite under High Temperatures 138
6.5. Heat Expansion/Shrinkage 155

Chapter 7. Textile Ablative Composites Materials **157**
7.1. Model of a Structure of Ablative Textile Composite Material 157
7.2. Model of a Layer with Curved Threads 159
7.3. Constitutive Relations for Ablative Textile Composites 164
7.4. Thermo-Elastic Moduli and Heat Conductivity Coefficients of Textile Composites 165
7.5. Heat Deformations 170
7.6. Coefficients of Phase Interaction 174
7.7. Thermal Strength 174
7.8. Thermal Properties of Textile Composites 184
7.9. Gas Permeability 189

Chapter 8. Composites Reinforced by Dispersed Particles **191**
8.1. Model of the Composite 191
8.2. Thermo-Elastic Characteristics 192
8.3. Strength 196
8.4. Thermal Properties 198

Chapter 9. Phenomena in Composite Materials Caused by Gradient Heating **203**
9.1. Internal Heat-Mass-Transfer and Stresses in Ablative Composites under Gradient Heating 203
9.2. Plane Problems of Thermomechanics for Composites under High Temperatures 210

9.3. Heat Deformations, Stresses and Load-Bearing Capacity of a Composite Plate under Gradient Heating 213

Chapter 10. Linear Ablation of Composites 223
10.1. Main Types of Linear Ablation of Composites 223
10.2. Combustion Rate 224
10.3. Sublimation Rate 228
10.4. Thermomechanical Erosion Rate 231
10.5. Melting Rate 243
10.6. Comparison of Theoretical and Experimental Results 244
10.7. Heat Balance on Ablative Surface 250
10.8. Criteria of Efficiency of Composites 252

Chapter 11. Thermal Stresses in Composite Structures under High Temperatures 255
11.1. Axisymmetic Problems of Thermomechanics for Composites under High Temperatures 255
11.2. Thermal Stresses in Composite Structures of Heat-Energetical Systems 260
11.3. Thermal Stresses in Thermoprotective Structures under Gasodynamical Heating 273
11.4. Thermal Stresses in Thermoprotective Structures under Aerodynamical Heating 281
11.5. Thermal Stresses in Composites under Local Technological Heating 289

Chapter 12. Mechanics of Composite Thin-Walled Shells under High Temperatures 295
12.1. General Equations for Thin-Walled Ablative Shells under High Temperatures 296
12.2. Main Assumptions for Thin-Walled Ablative Shells 302
12.3. Peculiarities of the Theory of Composite Shells under High Temperatures 303
12.4. Cylindrical Composite Shells under High Temperatures 308
12.5. Failure of Composite Structures under High Temperatures 315

Appendix. Methods of Experimental Investigation of High-Temperature Properties of Composite Materials 323
A1. Determination of Density under Heating 323
A2. Determination of Thermal Characteristics under Heating 324
A3. Determination of Gas Permeability 329
A4. Determination of Heat Deformations under Heating 329

A5. Determination of Strength and Elastic Modulus of Composites under High Temperatures 330
A6. Gas-Dynamical Testing of Composites 334

References **337**

Subject Index **345**

INTRODUCTION

The thermomechanical properties of composites on polymer matrix at high temperatures are essentially different from those at normal temperatures. The main distinctions briefly consist in the following:

- at high temperatures there occurs an irreversible variation (degradation) of all mechanical and thermal properties of a material that usually has a complex non-linear character depending on time exposure under high temperature;
- there are complicated internal physico-chemical processes in a matrix and fibres under high temperatures called by the general notion of ablation; the internal physico-chemical and mechanical processes run differently in the matrix and fibres, and this leads to the appearance of considerable internal thermal stresses. Generally speaking, a composite under high temperatures can be considered as a multiphase system consisting of solid, gaseous and fluid phases interacting mechanically and chemically with each other.

There are three levels of temperature: normal, elevated and high. *Normal*, or *room* temperatures are 10 – 30 °C; *elevated* temperatures are 30 – 200 °C; *high* temperatures are those above 200 °C. However, the dividing line between elevated and high temperatures depends on the material involved; a temperature is called high for a particular composite material if, at this temperature, irreversible internal physico-chemical transformations occur in the matrix and/or fibres of the material.

The mechanical behaviour of composites at normal and elevated temperatures has already been studied in great detail. On the other hand, some experimental and theoretical experience has been accumulated on thermal and physico-chemical ablation processes of composite materials under high temperatures. However, problems of macro- and micromechanics of composites at high temperatures, especially under the ablation conditions, have not previously been investigated.

The purpose of the present book is to simulate the thermomechanical behaviour of composites at high temperatures, taking account of internal physico-chemical transformations which occur in their matrix and fibres under intense heating. The book is the first to present a systematized mathematical description of the complex internal mechanical and thermal processes in composites under high temperatures.

The problem is complex because it straddles many fields: thermal physics, thermal chemistry, solid mechanics, etc. Since concepts such as thermodecomposition and ablation are new to mechanicians calculating microstresses in composites, we define the principal ideas of physico-chemical transformations in Chapter 1.

In Chapters 2 and 3 we develop mechanico-mathematical models of internal structure of a composite under ablation conditions when different phase transformations occur, and we also give a mathematically rigorous representation of the basis of continuum mechanics applicable to ablative composites.

Chapters 4 - 8 are devoted to problems of micromechanics of composites under high temperatures. Calculation methods are given for determining the mechanical, strength and thermal properties for different types of matrix (Chapter 4) and fibres (Chapter 5) under high temperatures, in terms of the properties of the separate phases. We give models and calculation methods for mechanical and thermal properties of unidirectional composites (Chapter 6), textile composites (Chapter 7) and also composites reinforced by dispersed particles (Chapter 8) under their matrix and fibre ablation conditions. Practically all computed results are compared with corresponding experimental data on high-temperature behaviour of composites.

Chapters 9 - 12 of the book are devoted to modelling of the behaviour of composites under conditions of gradient heating. Chapter 9 describes phenomena in ablative composite structures caused by gradient heating. We investigate the important phenomenon as intrapore pressure of gaseous products of matrix thermodecomposition and its influence on thermal stresses in composite structures.

Chapter 10 develops methods for calculating the rate of surface (linear) ablation of a composite under high temperatures in terms of ablation rates of its matrix and fibres. We pay special attention to the thermomechanical erosion type of ablation.

Chapter 11 gives calculation methods for thermal stresses in several typical structures made of composites under high temperatures. Finally, Chapter 12 is devoted to calculation methods of thermal stresses in thin-walled high-temperature structures made of composites under conditions of gradient heating and ablation. It should be noted that investigations into the thermomechanical behaviour of composites under ablation conditions are necessary not only for cases when composites are deliberately used at high temperatures, for example for thermal-protective structures of re-entry vehicles, rocket fairings, rocket engines etc., but also for structures with a relatively low level of tempepatures, for example for non-metallic fuel containers, when we need to determine their behaviour under extreme conditions: fire, explosion etc.

Although all experimental data shown in the book concern polymer-matrix composites, the principal results of the thermomechanical simulation can be applied to other types of composites, based on ceramic and carbon matrices, as the physico-chemical and thermomechanical processes at high temperatures are similar, but the processes run at a higher level of temperature.

I would like to thank Professor B.E.Pobedria, Chair of Composite Mechanics at Moscow Lomonosov State University, for fruitful discussions and valuable advice on different problems in the book; Professor I.S.Epifanovskiy of NPO Machinostroenia, Russia, discussions with whom promoted the better understanding of the physical sense of thermal effects in composites; Professor V.S.Zarubin, Chair of Applied Mathematics at Moscow Bauman State Technical University, who read the manuscript carefully and introduced many valuable comments.

I am grateful to Professor G.M.L.Gladwell of the University of Waterloo, Canada, who edited the book and improved the English text.

I also thank my wife, Dr. Irina D.Dimitrienko, Department of Mechanics and Mathematics at Moscow Lomonosov State University, who translated the book into English and prepared the camera-ready copy.

I hope that the book proves to be useful for investigators, academic scientists and engineers designing structures of composites, and also for material science specialists developing advance performance thermostable materials.

Moscow, April 1998 **Yuriy Dimitrienko**

NOMENCLATURE

$a^{(0)}$, $a_f^{(0)}$ – functions which characterize changing elastic properties of polymer phase of matrix and amorphic phase of fibres under heating, respectively;
a_1, $\widetilde{a}_1$, $\widetilde{a}_f$ – functions which characterize changing strength and elastic properties of matrix and fibres under heating, respectively;
$\widetilde{a}_2$, $\widetilde{a}_3$, $\widetilde{a}_4$, $\widetilde{a}_5$ – functions which characterize changing elastic properties of unidirectional composite under heating;
$a_{\theta 1}$, $a_{\theta 2}$ – functions which characterize changing elastic properties of textile composite under heating;
$\mathbf{a}^{(\varkappa)}$, $a^{(\varkappa)}$ – tensors-constants characterizing a type of anisotropy and their invariants, respectively;
$b^{(0)}$ – functions which characterize changing heat conductivity of phases under heating;
b_1, $\widetilde{b}_f$ – functions which characterize changing heat conductivity of matrix and fibres under heating, respectively;
$B_{m1} \ldots B_{m4}$, $B_{f1} \ldots B_{f3}$ – concentrations of stresses in matrix and fibres of microcomposite;
$\overset{\circ}{B}_m$, $\overset{\circ}{B}_f$ – coefficients of concentrations of heat stresses in matrix and fibres, Pa^{-1};
$B_2^{(s)}$, $B_3^{(s)}$, $s = 1, 2$ – coefficients of concentrations of stresses in layers of textile composite;
c_i, c_g, $\bar{c}_g$ – heat capacities of phases, $\mathrm{J \cdot kg^{-1} \cdot K^{-1}}$;
$C_{\alpha\beta}$ – coefficients of elastic moduli tensor of composite, Pa;
$\mathbf{C}_i$ – elastic moduli tensors of phases, Pa;
d_f – diameter of monofibres, m;
d – diameter of a pore channel, m;
D, D_0, D_p, D_g – rates of motion of a phase interface and surfaces of phase transformations (Chapter 2, 3), $\mathrm{m \cdot s^{-1}}$;
D, D_e, D_i, D_S, D_C, D_m, D_{TC} – rates of surface ablation, external and internal thermomechanical erosion, sublimation, combustion, thermomechanical and thermochemical ablation of composite, respectively, $\mathrm{m \cdot s^{-1}}$;
D_{Me}, D_{Mi} – rates of external and internal thermomechanical erosion of a matrix, respectively, $\mathrm{m \cdot s^{-1}}$;
$D_\perp$, $D_\parallel$, D_u, D_u' – rates of external thermomechanical erosion of fibres and unidirectional composite in transverse and longitudinal directions, respectively, $\mathrm{m \cdot s^{-1}}$;
D_{e1}, D_{e2} – rates of external thermomechanical erosion of textile composite in different directions, $\mathrm{m \cdot s^{-1}}$;
E_i, E_m, E_f – elastic moduli of phases, matrix and fibres, respectively, Pa;
E_μ', E_μ – elastic moduli of microcomposite in longitudinal and transverse directions, respectively, Pa;

$\bar{E}_\alpha^{(s)}$, $\alpha = 1,2,3$ – elastic moduli of layers with curved fibres in textile composite, Pa;
E_α, $\alpha = 1,2,3$ – elastic moduli of textile composite, Pa;
$\mathbf{E}$ – metric tensor;
E_{Am}, E_{Af}, E_{Ak} – activation energy of matrix, fibre and k-th heterogeneous chemical reaction, respectively, $\mathrm{J \cdot kg^{-1}}$;
e_i, e_g – internal energies of phases, $\mathrm{J \cdot kg^{-1}}$;
Δe°, Δe_f° – heat of volumetric ablation of matrix and fibres, respectively, $\mathrm{J \cdot kg^{-1}}$;
$\mathbf{e}_i$ – basis vectors of Cartesian coordinate system;
$e_{\alpha\beta}$ – deformations of a shell;
f_m – coefficients of phase interactions in a matrix, $\mathrm{Pa^{-1}}$;
$f_{\mu kl}$, f'_μ, f_μ – components of a phase interaction tensor in microcomposite and its principal values, respectively, $\mathrm{Pa^{-1}}$;
$\bar{f}_\alpha^{(s)}$ – coefficeints of phase interactions in layers of textile composite, $\mathrm{Pa^{-1}}$;
$\mathbf{f}$, f_{kl}, f_α – tensor of phase interactions in composite, its components and principal values, $\mathrm{Pa^{-1}}$;
f_{ij}, f_Σ – functions of a shape of i-th and j-th phase interface and of ablation surface of composite, respectively;
$\mathbf{F}_i$, $\mathbf{F}_i^*$ – gradients of phase deformations and phase transformations, respectively;
$\mathbf{F}$ – tensor of stress functions, $\mathrm{Pa \cdot m^{-2}}$;
F – stress function, $\mathrm{Pa \cdot m^{-1}}$;
$\bar{G}$ – dimensionless rate of ablation;
G_i – shear moduli of phases, Pa;
G_b, G_m, G_f, G'_μ, G_μ – shear moduli of matrix, fibre and microcomposite in longitudinal and transverse directions, respectively, Pa;
$\bar{G}_{\alpha\beta}^{(s)}$ – shear moduli of layers of textile composite, Pa;
$G_{\alpha\beta}$ – shear moduli of textile composite, Pa;
h'_{uT}, h'_{uC}, h_{uT}, h_{uC}, h'_{uS} – functions which characterize changing strength of unidirectional composite in tension, compression and shear in longitudinal and transverse directions, respectively;
H_α – Lamé's parameters;
I_{eff} – effective enthalpy, $\mathrm{J \cdot kg^{-1}}$;
I_i, I_g – enthalpies of phases, $\mathrm{J \cdot kg^{-1}}$;
I_e – enthalpy of overrunning gas flow, $\mathrm{J \cdot kg^{-1}}$;
J, J_f – intensity of volumetric ablation of a matrix and fibre, respectively, $\mathrm{kg \cdot m^{-3} \cdot s^{-1}}$;
K – gas-permeability coefficient of a matrix, s;
K_{nm} – components of gas-permeability tensor of composite, s;
$\mathbf{k}_i$, k_i – heat conductivity tensors and coefficients of phases, $\mathrm{W \cdot m^{-1} \cdot K^{-1}}$;
k_m, k_f – heat conductivity coefficients of matrix and fibre, respectively, $\mathrm{W \cdot m^{-1} \cdot K^{-1}}$;
$k_{\mu kl}$, k_μ, k'_μ – components of heat conductivity tensor of microcomposite and its principal values, respectively, $\mathrm{W \cdot m^{-1} \cdot K^{-1}}$;
k_{ukl} – components of heat conductivity tensor of unidirectional composite, $\mathrm{W \cdot m^{-1} \cdot K^{-1}}$;
$\bar{k}_{\alpha\beta}^{(s)}$, $\bar{k}_\alpha$ – components of heat conductivity tensor of layers of textile composite and its principal values, $\mathrm{W \cdot m^{-1} \cdot K^{-1}}$;

$\mathbf{K}_i$ – tensor of kinetic energy of phase transformations, $\mathrm{m \cdot s^2 \cdot kg^{-1}}$;
L – length of monofibres, shell (Chapter 12);
l_0, l_μ, l_u – characteristic scales of nonhomogenity in composite, m;
$M_{\alpha\beta}$ – moments in a shell, N;
M_g, $M_{g\alpha}$ – moments of pore pressure and phase interactions in a shell, N;
m_E, m_p and m_k, m_h – ratio of elastic moduli and heat conductivity coefficients of polymer and amorphic phases and also pyrolytic and amorphic phases, respectively;
$\dot{m}$ – mass rate of ablation of composite, $\mathrm{kg \cdot m^{-2} \cdot s^{-1}}$;
$\dot{m}_j$ – mass rate of formation of j-th chemical component, $\mathrm{kg \cdot m^{-2} \cdot s^{-1}}$;
$\mathcal{M}$ – mass rate of phase transformation, $\mathrm{kg \cdot m^{-2} \cdot s^{-1}}$;
M – Mach number;
M_j – molecular mass of j-th chemical component;
$\mathbf{n}$, n_α – normal vector and its components, respectively;
p – gas pressure, Pa;
p_e – local pressure of a gas flow passing over a surface of composite, Pa;
p_Σ – pressure head of flow about a surface, Pa;
P_g, $P_{g\alpha}$ – force of pore pressure and phase interaction in a shell, respectively, $\mathrm{N \cdot m^{-1}}$;
Q_α – transverse forces in a shell, $\mathrm{N \cdot m^{-1}}$;
$Q^{(s)}_{\alpha\beta}$ – matrix of rotation;
$\mathbf{q}_i$ – vectors of heat flux in phases, $\mathrm{W \cdot m^{-2}}$;
q_e, q_e^0 – convective heat flux and heat flux to a 'cold wall', respectively, $\mathrm{W \cdot m^{-2}}$;
q_R, q_{Rw}, q_{bl}, q_w, q_{sub}, q_{cs}, q_{cg} – heat fluxes (see paragraph 9.7) on an ablative surface, $\mathrm{W \cdot m^{-2}}$;
q_α – curvilinear coordinates;
R – gas constant, $\mathrm{J \cdot kg^{-1} \cdot K^{-1}}$;
R_1, R_2 – radii of curvature, m;
r – radius, m;
$\mathbf{S}_{ei}$ – vectors of external surface forces, Pa;
s – length of a curved arc, m;
S_a, S_l, S_b, S_p – functions characterizing linear dimensions of phases;
$\mathbf{T}_i$, $\mathbf{T}'_i$ – tensors of elastic and viscous stresses of phases, Pa;
t – time, s;
t_i^* – times of phase formation (Chapters 2, 3);
t_1^*, t_2^*, t_{**} – times of mechanical fracture of composites, s;
$T_{\alpha\beta}$ – forces in a shell, $\mathrm{N \cdot m^{-1}}$;
$\mathbf{u}_i$ – displacement vectors of phases, m;
$u_\alpha^{(i)}$ – displacements of i-th layer in a multilayer shell, m;
$\mathbf{u}_e$, $\mathbf{u}_{ei}$ – displacement vectors of external surfaces of composite and its phases, respectively, m;
U_α – displacements of a shell, m;
U_k – mass rate of k-th chemical reaction, $\mathrm{kg \cdot m^{-3} \cdot s^{-1}}$;
$\mathbf{u}$, u_k, u_r, u_θ, u_z – displacement vector of composite and its coordinates in Cartesian and cylindrical coordinate systems, respectively, m;
V – domain occupied by a composite;
V_i, $\overset{\circ}{V}_i$ – domain occupied by i-th phase in actual and initial configurations, respectively;

V_ξ, $V_{\xi i}$ – domain of a periodicity cell and domains occupied by phases within the periodicity cell;
$\mathbf{v}_i$, $\mathbf{v}_g$ – velocity vectors of phases, $\mathrm{m \cdot s^{-1}}$;
W – deflection of a shell, m;
W_i^* – energy dissipation functions of phases, (Pa $\cdot$ s^{-1};
$\mathbf{x}$, $\overset{\circ}{x}$, x^k and x_k, $\overset{\circ}{x}{}^k$ – radius-vectors and coordinates in actual and initial configurations, m;
x_0 – characteristic dimension of domain V occupied by a composite, m;
$Y_{m\alpha}$, $Y_{u\alpha}$, $Y_{u\alpha}^{(s)}$ – invariants of stress tensors σ_{mkl}, σ_{ukl}, $\bar{\sigma}_{kl}^{(s)}$, respectively, Pa;
y_i – mass concentrations of chemical components in a chemical reaction;
y – moving coordinate;
z_α – damage parameter;
α_i, α_m, α_f – heat expansion coefficients of phases, matrix and fibres, respectively, K^{-1};
(α/c_p) – heat transfer coefficient, $\mathrm{kg \cdot m^{-2} \cdot s^{-1}}$;
$\boldsymbol{\beta}_i$, β_i – chemical shrinkage tensor of phases and its principal components, respectively;
β_{sh} – chemical shrinkage of a matrix;
β_{nm} – reduced coefficients of pliability, Pa^{-1};
γ_α – shears in a shell;
Γ, Γ_f – gasification coefficients of matrix and fibres, respectively;
δ_{kl} – Kronecker symbol;
$\boldsymbol{\varepsilon}_i$ – deformation tensors of phases;
ε_{mkl}, $\varepsilon_{\mu kl}$ – components of deformation tensors of matrix and microcomposite, respectively;
$\boldsymbol{\varepsilon}$, ε_{kl}, ε_r, ε_θ, ε_z, $\varepsilon_{r\theta}$, $\varepsilon_{\theta z}$, ε_{rz} – deformation tensor of composite and its components in Cartesian and cylindrical coordinate system;
$\overset{\circ}{\boldsymbol{\varepsilon}}_i$ – heat deformation tensor of phases;
$\overset{\circ}{\varepsilon}_m$, $\overset{\circ}{\varepsilon}_f$ – heat deformations of matrix and fibres, respectively;
$\overset{\circ}{\varepsilon}_{\mu kl}$, $\overset{\circ}{\varepsilon}{}'_\mu$, $\overset{\circ}{\varepsilon}_\mu$ – components of heat deformation tensor of microcomposite and its principal values;
$\overset{\circ}{\boldsymbol{\varepsilon}}$, $\overset{\circ}{\varepsilon}_{kl}$, $\overset{\circ}{\varepsilon}_r$, $\overset{\circ}{\varepsilon}_z$, $\overset{\circ}{\varepsilon}_\theta$ – heat deformation tensor of composite and its principal components in Cartesian and cylindrical coordinate systems, respectively;
η_i – entropies of phases, $\mathrm{J \cdot kg^{-1} \cdot K^{-1}}$;
θ_i, θ_g – temperatures of phases, K;
θ_i^* – temperature of formation of i-th phase, K;
θ_e – temperature of overrunning gas flow, K;
ϑ, $\vartheta_{\max}$ – angle of bending threads in textile composite and its maximum values;
$\varkappa$ – small parameter (see (3.4));
$\varkappa_{\theta 1}$, $\vartheta_{\theta 2}$ – functions which characterize changing heat conductivity of composite in different directions under heating;
λ_i, μ_i, ν_i – Lamé's parameters and Poisson's ratios of solid phases, respectively;
$\boldsymbol{\Lambda}_i$, $\boldsymbol{\Lambda}_i^*$ – Almancy's finite deformation tensor and tensor of phase transformations of i-th phase;
$\bar{\nu}_g$, $\bar{\mu}_g$ – viscosity coefficients of gas phase, Pa·s;
ν_m, ν_f, ν'_μ, ν_μ – Poisson's ratios of a matrix, fibre and microcomposite in longitudinal and transverse directions;

$\bar{\nu}_{\alpha\beta}^{(s)}$ – Poisson's ratios of layers of textile composite;
$\nu_{\alpha\beta}$ – Poisson's ratio of composite;
$\bar{\nu}_{ij}$ – stoichiometric coefficients;
$\boldsymbol{\xi}$ – vector of local coordinates, m;
ρ_i, ρ_g – densities of phases, $kg \cdot m^{-3}$;
ρ_m, ρ_f, ρ – densities of matrix, fibre and composite, respectively, $kg \cdot m^{-3}$;
$\boldsymbol{\sigma}_i$ – Cauchy's stress tensor of i-th phase, Pa;
σ_{mkl}, σ_{fkl} – components of stress tensors in matrix and fibres, Pa;
$\sigma_{\mu kl}$, σ_{ukl} – components of stress tensors in microcomposite and unidirectional composite, Pa;
$\bar{\sigma}_{kl}^{(s)}$ – components of a stress tensor in layers of textile composite, Pa;
$\boldsymbol{\sigma}$, σ_{kl}, σ_r, σ_θ, σ_z, $\sigma_{r\theta}$, σ_{rz}, $\sigma_{\theta z}$ – stress tensor in composite and their components in Cartesian and cylindrical coordinate systems, Pa;
σ_{mT}, σ_{mC}, σ_{mS}, σ_f – ultimate strengths of a matrix in tension, compression and shear and fibre in tension, respectively, Pa;
σ'_{uT}, σ_{uT}, σ'_{uC}, σ_{uC}, σ'_{uS}, σ_{uS} – ultimate strengths of unidirectional composite in tension, compression and shear in longitudinal and transverse directions, respectively, Pa;
$\sigma_{\alpha T}$, $\sigma_{\alpha C}$, $\sigma_{\alpha\beta S}$ – ultimate strengths of textile composite in tension, compression and shear in different directions, respectively, Pa;
$\boldsymbol{\chi}_i$ – tensor of Bowen's chemical potential of i-th phase, $J \cdot kg^{-1}$;
φ_i, φ_g – volumetric concentrations of phases;
ψ_i, ψ_g – Helmholtz free energies of phases, $J \cdot kg^{-1}$;
ζ – part of fibres laid in the warp direction per volume unit of textile composite.

Subscripts

a (amorphic) – parameters of amorphic phase of fibre;
b (binder) – parameters of polymer phase of a matrix;
bl (blow) – parameters of blowing in;
C (compression) – characteristics of compression;
c, com (combustion) – characteristics of combustion;
E (elastic) – indicator of elastic properties;
e (external) – parameters of actions on an external surface of composite;
f (fibre) – parameters of fibre;
f – characteristics of melting (Chapter 10);
g (gas) – parameters of gas phase;
h (heat) – indicator of heat properties;
i – index of a phase in composite $i = a, b, p, l, g$;
j – the same as the subscript i;
k, l, m, n, q, r – number of component of tensors and vectors (values are equal to 1, 2, 3);
k – indicator of a number of k-th chemical reaction;
l – parameters of crystalline phase of fibre;
m (matrix) – parameters of a matrix;
max (maximum) – maximum value of parameter;
p (pyrolysis) – parameters of pyrolysis phase of a matrix;
R (radiational) – characteristics of irradiation;
S (shear) – characteristics of shear;

s, sub (sublimation) – characteristics of sublimation;
T (tensile) – characteristics of tension;
u (unidirectional) – parameters of unidirectional composite;
w (wall) – parameters of composite on its external surface;
r, θ, z – indicators of components of tensors in a cylindrical coordinate system;
α, β, γ – number of a component of tensors and vectors (values: 1, 2, 3);
μ (micro) – parameters of microcomposite;
Σ – parameters of a separation surface;
ξ – parameters in a periodicity cell.

Superscripts

$'$ – reinforcing direction (Chapters 6, 7);
$'$ – derivative with respect to coordinate (Chapters 9, 10);
$*$ – parameters of critical states (phase transformations, fracture);
(s) – indicator of a layer with curved fibre in textile composite;
(0), (1) – parameters of 'zero' and 'first' approximations in the method of asymptotic expansions;
(i) – parameters of the i-th layer of a multilayer shell.

CHAPTER 1

HIGH-TEMPERATURE ENVIRONMENT AND COMPOSITE MATERIALS

1.1. Main Types of High-Temperature Effects on Composite Materials

The aim of the present work is to model the behaviour of composite materials under high temperatures.

First we present a schematic classification of the most wide-spread types of high-temperature effects on composite structures (Figure 1.1).

The principal types of high-temperature effects on composites are aerodynamical heating, gas-dynamical heating, heating in an energetic device, action of a fire and technological heating.

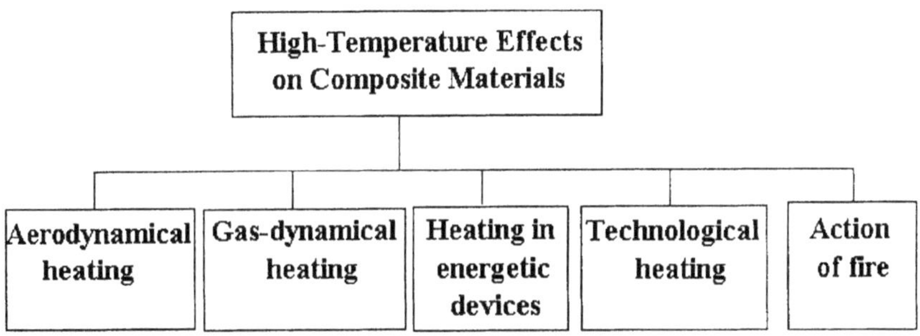

Figure 1.1. A schematic classification of the most wide-spread types of high-temperature effects on composite structures

1.1.1. AERODYNAMICAL HEATING

Types of Air-Space Structures made of Composites

We shall use the abbreviation ASV for any kind of air or space vehicle.

Aerodynamical heating is heating of composite materials employed as upper surfaces of ASV structures when they move in dense layers of the atmosphere with super– and hyper–sonic speeds.

Near a stagnation point of an aerospace structure the air is heated up to several thousands or even tens of thousands degrees; this is due to the deceleration of the

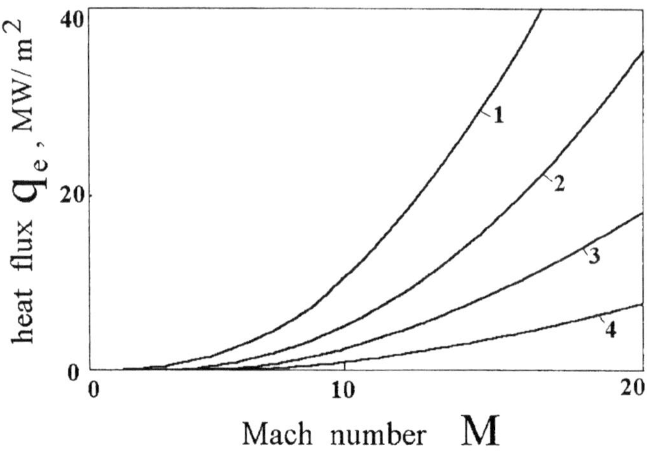

Figure 1.2. Typical dependences of heat flux q_e acting on the surface of ASV at the critical point on Mach number M for flights in the Earth's atmosphere at different heights H: 1 - 10 km, 2 - 20 km, 3 - 30 km and 4 - 40 km

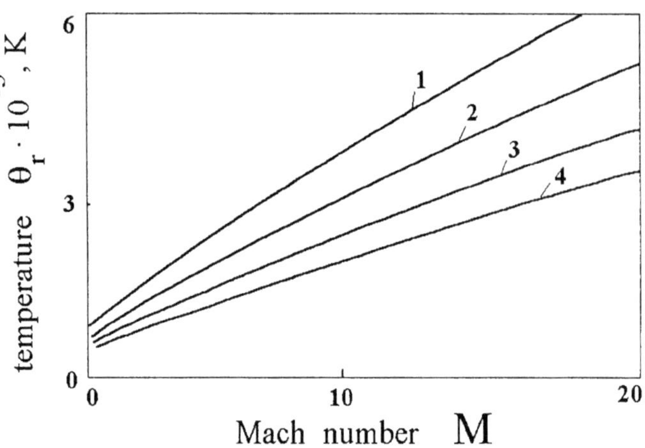

Figure 1.3. Typical dependences of equilibrium temperature θ_r at the surface of ASV on Mach number M for flights in the Earth's atmosphere at different heights H: 1 - 10 km, 2 - 20 km, 3 - 30 km and 4 - 40 km.

high speed flow passing over the body. This heating of the gas medium surrounding the ASV leads to the appearance of *convective heat fluxes* q_e on external surfaces of the vehicle. Figure 1.2 shows how the heat fluxes q_e for leading surfaces of ASV's depend on the vehicle speed; the result is expressed in terms of the Mach number M for different heights H within the Earth's atmosphere.

Figure 1.3 shows corresponding values of equilibrium temperatures θ_r on external surfaces of ASV's. The actual surface temperature θ_w is usually somewhat

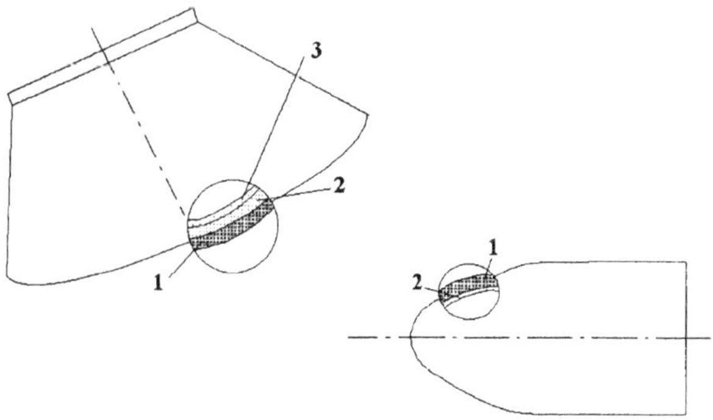

Figure 1.4. Typical schemes of application of composite thermal-protective materials in structures of re-entry vehicles: 1 - composite thermal-protective material; 2 - load-bearing shell; 3 - heat insulation

less than θ_r because the heat flux q_e is absorbed by heat conduction, sublimation and other thermochemical phenomena; nevertheless θ_w is still very high.

At present only the initial section of diagrams of Figures 1.2 and 1.3 corresponding to the Mach numbers $M \approx 2-3$ can be applied for flights of aircrafts with supersonic speeds. To develop advance hypersonic aircrafts for flights with $M \approx 6-10$ we need to design composite materials which could endure temperatures up to 1500 - 2000° C.

Another important field of application of composites is the thermal protection of ASV's at the stage of orbital injection (the *active section of the trajectory (AST)*) and especially at the stage of exiting the orbit onto the Earth's surface (the *passive section of the trajectory (PST)*) (Figure 1.4). Re-entry vehicles are decelerated when they enter the Earth's atmosphere from speeds of the order $M = 25-30$ down to zero; during this stage they undergo the action of considerable heat fluxes. Figure 1.5 shows the variation with time t of de-orbiting for the aerodynamic convective fluxes q_e for typical ASV's. Corresponding equilibrium temperatures θ_r are presented in Figure 1.6.

Figure 1.6 shows that characteristic times t of the action of high temperatures for re-entry vehicles are comparatively small and equal to 20 - 100 sec. Therefore, application of composite materials for these structures, as has been shown in practice, is especially effective as external thermal protection; this is due to the low heat conductivity and high heat of ablation of the composites.

Main Parameters of Aerodynamical Heating

The main characteristics of aerodynamical heating of composite structures, from the point of view of their heat strength, are the convective heat flux q_e supplied to the heated surface and the *local pressure* p_e of the gas flow passing over the structure surface. It is often necessary to take account of *pressure head* p_Σ of the gas flow moving along the composite surface Σ.

To determine q_e, p_e, p_Σ as functions of ASV flight speed (i.e. Mach number M), one can use models in which supersonic flow passes over the body [95]. Figure

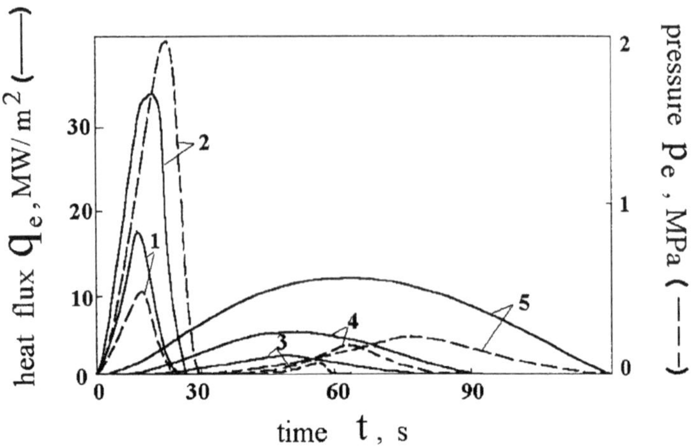

Figure 1.5. Typical dependences of convective heat flux q_e and flow pressure p_e acting on surfaces of re-entry vehicles of different types on time t of de-orbiting [66]:
1) middle-range ballistic missile;
2) long-range ballistic missile;
3) manned re-entry vehicle;
4) unmanned re-entry vehicle;
5) Space Shuttle

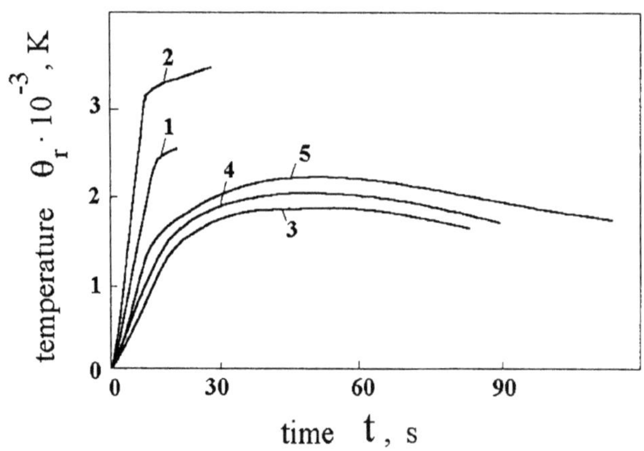

Figure 1.6. Typical distributions of equilibrium temperature θ_r on surfaces of different type vehicles re-entry onto the Earth *versus* time t of de-orbiting (symbols at curves are the same as in Figure 1.5)

1.7 exhibits typical passages over nose sections of spherical and conical shapes of ASV's; the structures were shown in Figure 1.4.

A shock wave of the flow appears ahead of the nose section of an ASV. Gas ahead of the shock wave is quiescent. Characteristics of the quiescent gas are pressure

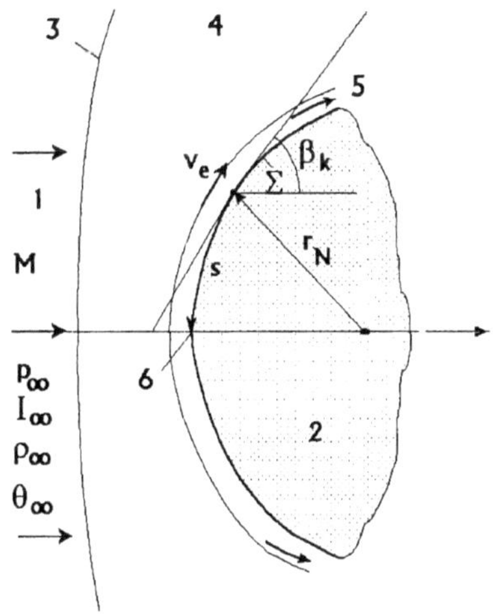

Figure 1.7. Typical passages over nose sections of ASV's with supersonic flow [95],
1 - field of quiescent flow;
2 - nose section passed over;
3 - shock wave;
4 - zone flowing behind the shock wave;
5 - boundary layer of gas moving along the heated composite surface Σ;
6 - point of stagnation of the flow

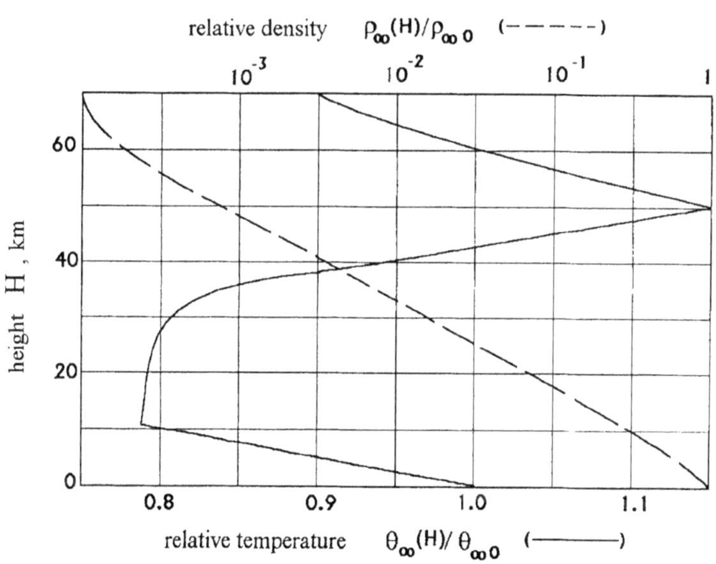

Figure 1.8. Dependences of standard parameters $\rho_\infty(H)/\rho_{\infty 0}$ and $\theta_\infty(H)/\theta_\infty$ on height H [9]

TABLE 1.1. Parameters of the Earth's and Mars's standard atmospheres [9, 94]

Parameters	Earth's atmosphere	Mars's atmosphere
Temperature θ_i, K, at height H_i, km		
0	288	210
11	217	195
25	217	175
35	247	155
50	331	140
80	185	140
95	185	140
β_0, km^{-1}	0.15	0.1
$\rho_{\infty 0}$, kg $\cdot$ m^{-3}	1.225	0.013

TABLE 1.2. Typical values of parameters of an ASV with a solid-propellant jet engine (see [109])

Parameters	m_0, kg	$\dot{m}$, kg $\cdot$ s^{-1}	S_M, m^2	θ_c, $^\circ$ C	p_c, MPa	P_{sp}, s
Values	$30 \cdot 10^3$	20	3... 10	3500	6... 10	250

p_∞, density ρ_∞, temperature θ_∞, enthalpy I_∞; they are defined only by properties of the standard atmosphere at the given flight height H of ASV. Dependences of $\rho_\infty(H)/\rho_{\infty 0}$ and $\theta_\infty(H)/\theta_{\infty 0}$ for the Earth's atmosphere are shown in Figure 1.8. These functions can be given in the form of approximate analytical relations:

$$\rho_\infty(H)/\rho_{\infty 0} = \exp(-\beta_0 H); \tag{1.1}$$

$$\theta_\infty(H) = \theta_{\infty i} + (\theta_{\infty i+1} - \theta_{\infty i})\left(\frac{H - H_i}{H_{i+1} - H_i}\right), \quad \text{if } H_i \leqslant H < H_{i+1},$$
$$i = 0, 1, 2, \dots,$$

where $\rho_{\infty 0}$ and $\theta_{\infty 0}$ are the values of density and temperature at height $H = 0$, and β_0, $\theta_{\infty i}$ and H_i are constants; their values for Earth's and Mars's atmospheres are shown in Table 1.1.

The enthalpy of quiescent gas flow I_∞ is determined as follows:

$$I_\infty = \bar{c}_{g\infty}\theta_\infty(H) = \frac{k_\infty}{k_\infty - 1}\frac{p_\infty(H)}{\rho_\infty(H)}, \tag{1.2}$$

where $k_\infty = \bar{c}_{g\infty}/c_{g\infty}$ is the ratio of gas heat capacities at constant pressure and constant volume in the quiescent state.

Immediately near the surface Σ of the nose section of an ASV there appears a boundary layer of gas passing over the surface (Figure 1.7).

Gas parameters within the boundary layer near the stagnation point (i.e. the point where the flow speed $\mathbf{v}_e = 0$, see Figure 1.7) are p_0' – pressure, ρ_0' – density, θ_0' – temperature and I_0' – enthalpy; they are connected to parameters of the quiescent flow by the relations [95]

$$p_0' = p_\infty \left(1 + \frac{k_\infty - 1}{2} M^2\right)^{k_\infty/k_\infty - 1}; \qquad I_0' = I_\infty \left(1 + \frac{k_\infty - 1}{2} M^2\right). \tag{1.3}$$

For an ASV with an axisymmetric shape and zero angle of attack, the local pressure p_e and *enthalpy* $\bar{I}_e$ *of the gas flow* in an arbitrary point of the surface Σ with coordinate s along meridian (Figure 1.7) are determined by the formulae

$$p_e = p_0' \sin^2 \beta_k(s) + p_\infty \cos^2 \beta_k(s);$$
$$\bar{I}_e = I_\infty \left(1 + r\frac{k_\infty - 1}{2} M^2\right). \tag{1.4}$$

Here r is the regeneration coefficient; $r = 0.84$ for laminar flow and $r = 0.89$ for a turbulent boundary layer; and $\beta_k(s)$ is the angle between the tangent plane to the surface Σ at the point with coordinate s, and the symmetry axis (Figure 1.7). The function $\beta_k(s)$ characterizes the shape of the surface Σ. For a nose surface Σ with a spherical shape we have

$$\beta_k(s) = \frac{\pi}{2} - \frac{s}{r_N}; \tag{1.5}$$

for a conical surface Σ $\beta_k(s) = \text{const}$, for a cylindrical surface Σ $\beta_k(s) = 0$.

Because it is decelerated, the overrunning flow is heated up to a temperature θ_0' at the critical point; this temperature can reach several thousands of degrees. The temperature θ_0', and also the temperature θ_e at an arbitrary point of the surface Σ passed over are connected to flow enthalpies I_0' and $\bar{I}_e$ by the relations

$$I_0' = \int_0^{\theta_0'} \bar{c}_g d\theta, \quad \bar{I}_e = \int_0^{\theta_e} \bar{c}_g d\theta, \tag{1.6}$$

where $\bar{c}_g$ and c_g are the heat capacities of the gas in current state at constant pressure and volume, respectively.

Convective heat flux q_e transmitted from the heated gas to the surface Σ is

$$q_e = \left(\frac{\alpha}{c_p}\right)(\bar{I}_e - I_w), \tag{1.7}$$

where $I_w = c_w\theta_w$ is the enthalpy of the composite at the surface Σ, and (α/c_p) is the *heat transfer coefficient*. Formula (1.7) is a consequence of the approximate solution of the gas momentum equations within the boundary layer based on the Reynolds analogy [95]. Many works (for example [70, 88, 95]) are devoted to evaluation of the heat transfer coefficient (α/c_p). For laminar flow one can use the approximate formula [94]

$$\left(\frac{\alpha}{c_p}\right) = \xi_e \left(\frac{p_e}{r_N + s}\right)^{0.5}, \qquad \xi_e = 4 \cdot 10^{-4}\ \mathrm{kg}^{0.5}/\mathrm{m}. \tag{1.8}$$

For turbulent flow the heat transfer coefficient can be determined as follows:

$$\left(\frac{\alpha}{c_p}\right) = \xi_t \frac{(p_e)^{0.5}}{(r_N + s)^{0.2}}, \qquad \xi_t = 3.8 \cdot 10^{-3}\ \mathrm{kg}^{0.5}/\mathrm{m}^{1.3}, \tag{1.9}$$

where r_N is the radius of the blunted nose section (see Figure 1.7).

Equations (1.7) - (1.9) show that near the stagnation point, the convective heat flux q_0 is determined in the following way:

$$q_0 = \left(\frac{\alpha}{c_p}\right)_0 (I_0' - I_w),$$

$$\left(\frac{\alpha}{c_p}\right)_0 = \xi_e \left(\frac{p_0'}{r_N}\right)^{0.5} \quad \text{for laminar regime,} \tag{1.10}$$

$$\left(\frac{\alpha}{c_p}\right)_0 = \xi_t \frac{(p_0')^{0.5}}{(r_N)^{0.2}} \quad \text{for turbulent regime.}$$

Besides q_e, one can also determine a *heat flux to the 'cold wall'* q_e^0:

$$q_e^0 = \left(\frac{\alpha}{c_p}\right) (\bar{I}_e - c_w \theta_0), \tag{1.11}$$

where $\theta_0 = 293$ K is the temperature of the 'cold' surface, and

$$q_e = q_e^0 - q_w, \qquad q_w = (\alpha/c_p) c_w (\theta_w - \theta_0), \tag{1.11a}$$

q_w is the heat flux carried off from the composite surface due to heat transfer.

Pressure head of the gas flow moving along the surface Σ, is determined with the help of the modified Newton model [9]:

$$p_\Sigma = \rho_e v_e^2 = p_0' \cos^2 \beta_k(s). \tag{1.12}$$

Equation (1.12) shows that $p_\Sigma = 0$ at the stagnation point, and the pressure head of the gas flow reaches its maximum on the cylindrical surface $p_\Sigma = p_0'$.

The parameters q_e, p_e, p_Σ of aerodynamical heating of the surface Σ not only depend on the surface coordinate s, but often also on time t. If we consider the problem of heating an ASV in flight, then these dependences are determined by solving the ballistic equations. When the ASV is in the active section of the trajectory (AST), these equations have the form [109]

$$\frac{dH}{dt} = V_\infty \sin\vartheta;$$

$$\frac{d\vartheta}{dt} = \frac{\alpha_A g_0}{(1-\dot{m}_0 t)V_\infty}\left(\dot{m}_0 P_{\rm sp} - \frac{c_x}{2}\frac{p_\infty(H)S_m}{m_0 g_0}M^2\right) - \frac{g_0}{V_\infty}\cos\vartheta;$$

$$\frac{dV_\infty}{dt} = \frac{g_0}{1-\dot{m}_0 t}\left(\dot{m}_0 P_{\rm sp} - \frac{c_x}{2}\frac{p_\infty(H)S_m}{m_0 g_0}M^2\right) - g_0; \qquad (1.13)$$

$$M = V_\infty(R\theta_\infty(H))^{-1/2}, \quad t > 0;$$

$$t = 0: \qquad H = 0, \quad V_\infty = 0, \quad \vartheta = \vartheta_0,$$

where the functions $p_\infty(H)$ and $\theta_\infty(H)$ are determined by (1.1), g_0 is the gravitational acceleration within the planet atmosphere, and m_0 (kg) is the initial mass of the ASV, $\dot{m}_0$ (s^{-1}) is the specific mass consumption of the propellant, S_m (m^2) is the area of the ASV midsection, $P_{\rm sp}$ (s) is the specific draft, c_x is the aerodynamical coefficient, ϑ is the angle between the velocity vector and the horizon, α_A is the attack angle, and V_∞ is the speed of the ASV. Typical values of parameters m_0, $\dot{m}$, S_m, $P_{\rm sp}$ are given in Table 1.2.

For an ASV at the PST stage, the ballistic equations are rewritten as follows [97, 109]:

$$\frac{dH}{dt} = -V_\infty \sin\vartheta; \qquad \frac{d\vartheta}{dt} = \frac{g_0}{V_\infty}\cos\vartheta;$$

$$\frac{dV_\infty}{dt} = -\frac{c_x p_\infty(H)S_m}{2m_0}M^2 + g_0\sin\vartheta; \qquad (1.14)$$

$$M = V_\infty(R\theta_\infty(H))^{-1/2}, \quad t > t_0;$$

$$t = t_0: \qquad H = H_{\rm bg}, \quad V = V_{\rm bg}, \quad \vartheta = \vartheta_{\rm bg},$$

where $H_{\rm bg}$, $V_{\rm bg}$ and $\vartheta_{\rm bg}$ are the height, speed and angle of the beginning of de-orbiting for the ASV.

On determining the flight height $H(t)$ and the Mach number $M(t)$ from equations (1.13) and (1.14) as functions of t, one can find values of $\rho_\infty(H)$, $\theta_\infty(H)$, $p_\infty(H)$, $I_\infty(H)$ by formulae (1.1); and then relations (1.4), (1.7), (1.12) allow values of $p_e(t)$, $q_e(t)$ and $p_\Sigma(t)$ to be defined as functions of time t.

Figure 1.9 shows graphs of functions $H(t)$, $M(t)$, $\bar{I}_e(t)$, and also $(\alpha/c_p)(t)$, $p_e(t)$, $q_e(t)$ and $p_\Sigma(t)$ depending on time t of de-orbiting the ASV within the Earth's atmosphere.

1.1.2. GAS-DYNAMICAL HEATING

Gas-dynamical heating signifies the heat effects of combustion products on composite materials employed for internal thermal protection of a combustion chamber.

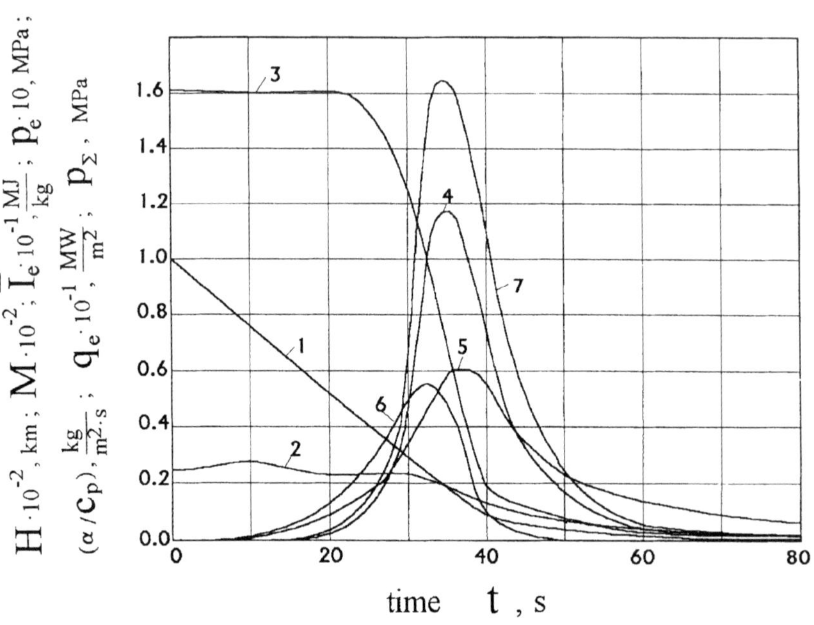

Figure 1.9. Dependences of parameters of a gas-dynamical flow (H - (1); M - (2); $\bar{I}_e$ - (3); p_e - (4); (α/c_p) - (5); q_e - (6); p_Σ - (7), acting on the lateral surface of the ASV on time t of de-orbiting onto the Earth

Figure 1.10 shows a typical scheme [7, 109] of a jet engine using a solid propellant which has elements made of composite materials.

Table 1.3 gives us characteristic temperature θ_e of hot gas within the combustion chamber, gas pressure p_e and propellant consumption $\dot{m}$ in the jet engine. In the jet engine there is a very high level of pressure p_e especially at a critical crossection (nozzle) at which there occurs a strong thermo-erosion of materials. Composites on ceramic, carbon and metallic matrices prove to be rather effective in withstanding this erosion.

The main parameters of gas-dynamical heating of composite materials are as follows: convective heat flux q_e from moving gaseous combustion products, gas pressure p_e on the composite surface Σ, pressure head p_Σ of gas passing over the internal surface of the nozzle Σ.

Convective heat flux q_e under gas-dynamical heating is determined by formula (1.7), in which the heat transfer coefficient (α/c_p) is evaluated as follows [9]:

$$\left(\frac{\alpha}{c_p}\right) = \xi_d \frac{\dot{m}^{0.8}}{r(x)^{1.8}}; \quad \xi_d = 10^{-3} \left(\frac{\text{kg}}{\text{m} \cdot \text{s}}\right)^{0.2}, \tag{1.15}$$

where $r(x)$ is the function of the surface shape Σ (the nozzle radius r can vary with the axial coordinate x_1, (see Figure 1.10)).

TABLE 1.3. Typical parameters of high-temperature effect on composite materials in different sections of a nozzle of a solid-propellant engine [7, 109]. (Sections correspond to Figure 1.10)

Parameters	Sections of nozzle		
	1	2	3
Gas temperature, θ_e, °C	3300	3000	2800
Gas pressure, p_e, MPa	6	4	3
Mach number, M	< 1	1	> 1
Heat transfer coefficient, (α/c_p), kg·m^{-2}·s^{-1}	0.4	2.8	0.2
Gas enthalpy, $\bar{I}_e$, MJ·kg^{-1}	10	10	9
Heat flux, q_e, MW·m^{-2}	3.5	2.5	1.8
Pressure head, p_Σ, MPa	0.4	4.0	0.2

Pressure p_c and flow enthalpy $\bar{I}_c$ distributions along the surface Σ are defined by the gas-dynamical relation [11]

$$\bar{I}_e = \bar{c}_g \theta_c \left(1 - \frac{k_\infty - 1}{k_\infty + 1} \lambda_\infty^2(x_1)\right);$$

$$p_e = p_c \left(1 - \frac{k_\infty - 1}{k_\infty + 1} \lambda_\infty^2(x_1)\right)^{k_\infty/(k_\infty - 1)}; \tag{1.16}$$

where parameter λ_∞ can be determined from the equation

$$\left(\frac{r_0}{r(x_1)}\right)^2 = \left(\frac{k_\infty + 1}{2}\right)^{1/(k_\infty - 1)} \lambda_\infty \left(1 - \frac{k_\infty - 1}{k_\infty + 1} \lambda_\infty^2(x_1)\right)^{1/(k_\infty - 1)}, \tag{1.17}$$

the approximate solution of which has the form

$$\lambda_\infty(x_1) = \frac{k_\infty + 1}{2} \left(1 - \left(1 - 2\left(\frac{2}{k_\infty + 1}\right)^{k_\infty/(k_\infty + 1)} \left(\frac{r_0}{r(x_1)}\right)^2\right)^{1/2}\right). \tag{1.18}$$

Here r_0 is the radius of the critical crosssection of the nozzle (Figure 1.10).

Distribution of pressure head p_Σ of gas flow passing over the nozzle surface Σ can be determined with the help of the modified Newton formula for calculating the pressure on an oblique area:

$$p_\Sigma = \frac{p_c}{2} \cos^2 \beta_k(x_1) \left(1 - \left(1 - \frac{k_\infty - 1}{k_\infty + 1} \lambda_\infty^2(x_1)\right)^{k_\infty/(k_\infty - 1)}\right), \tag{1.19}$$

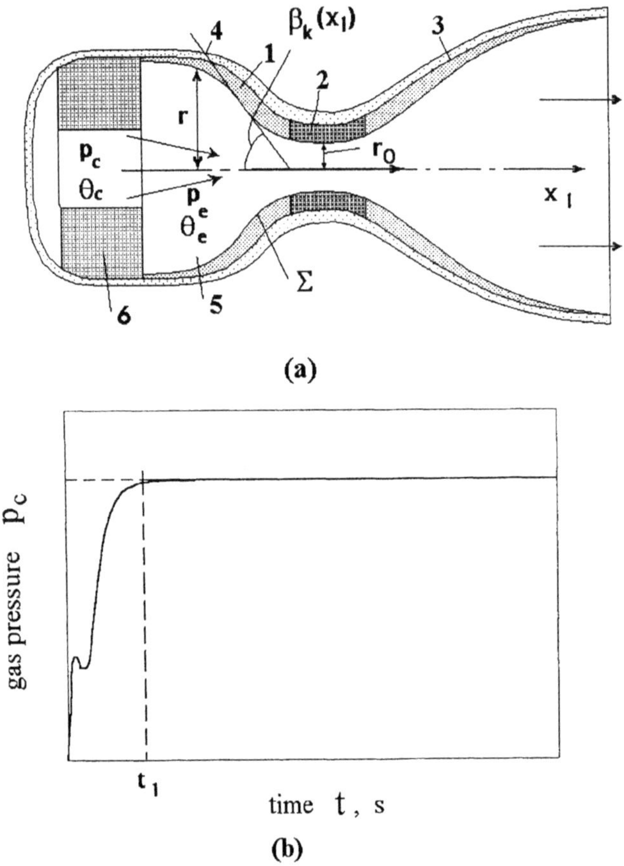

Figure 1.10. a) A scheme of application of composite thermal-protective materials in different sections (1, 2, 3) of a typical structure of a nozzle of a jet engine with solid propellant [7, 109]: 1 - entrance cone; 2 - nozzle insert; 3 - funnel; 4 - frame; 5 - combustion chamber; 6 - solid propellant;
b) Graph of changing gas pressure p in a combustion chamber of a solid-propellant engine [109], t_1 is the time of arrival at a stationary regime

where $\beta_k(x_1)$ is the angle of inclination of the tangential to the surface Σ (Figure 1.7).

Typical values of parameters p_e, $\bar{I}_e$, q_e, (α/c_p) and p_Σ are shown in Table 1.3.

Typical times t of heat-force action on composite materials at gas-dynamical heating are from several seconds to tens of seconds.

Heat-force action of combustion products on thermal-protective materials has, as a rule, a non-stationary character that is caused by such phenomena as acoustic pulsations of the gas flowing out from the combustion chamber, self-sustaining oscillations within the combustion chamber and others [9].

Typical regimes of changing averaged pressure $p_c(t)$ within the combustion chamber are shown in Figure 1.10,*b*. Temperature $\theta_c(t)$ changes in a similar way.

1.1.3 HEATING IN ENERGETIC SYSTEMS

Different energetic facilities (engines of aircrafts, machines, turbines of electric power stations etc.) use hot gas as a working element; therefore, structural materials applied in these systems are subject to high temperatures for long periods (from several minutes to tens of thousands of hours). Traditionally, in energetic facilities, heat-stable metallic alloys are applied. However, there are prospects for the application of composites in this field. It is anticipated that heat-stable lightweight polymer matrix composites will have important consequences for aircraft engines.

Figure 1.11 shows a schematic construction of a supersonic turbo-jet engine (STJE), which has some elements made of composite materials. Current temperatures for these structures in air are about 300° C. The temperatures of work surfaces and non-cooled blades of gas turbines at present are currently about 1100° C. Investigations are under way on the creation of working elements of turbines made of metallocomposites with high heat strength.

Figure 1.12 shows a typical flow over an inlet of STJE by the overrunning gas [11]. The flow regime usually has many jumps: a leading shock wave (oblique jump), and a set of jumps ahead of the leading edge of the external surface and within the channel of the inlet. The gas dynamics of such systems is rather complicated, and Figure 1.12 exhibits an example in which only two jumps (oblique and direct) are considered ahead of the leading edge of the external shell. There is a boundary layer behind the compaction jumps on the surface. Due to a complex geometrical configuration of the inlet, the direct jump usually interacts with the oblique jump and with the boundary layer of the central body (Figure 1.12).

The pressure p_e and temperature θ_e within the inlet channel can be calculated with the help of the theory of oblique jumps of compaction [11]:

$$p_e = p_1(2k_\infty M_1^2 - k_\infty + 1)(k_\infty + 1)^{-1}; \qquad \theta_e = \theta_1\left(1 + \frac{k_\infty - 1}{2} M_1^2\right); \tag{1.20}$$

where p_1, θ_1 and M_1 are the flow parameters behind the oblique jump (Figure 1.12) and are determined by the given value of the Mach number M:

$$p_1 = p_\infty(H)\left(2k_\infty M^2 \sin^2\beta_0 - k_\infty + 1\right)(k_\infty + 1)^{-1};$$

$$\theta_1 = \theta_\infty(H)\left(\frac{2}{M^2\sin^2\beta_0} + k_\infty - 1\right)\frac{2k_\infty M^2\sin^2\beta_0 - k_\infty + 1}{(k_\infty + 1)^2}; \tag{1.21}$$

$$M_1 = \frac{M\Big((k_\infty + 1)^2\cos^2\beta_0 + \sin^2\beta_0(k_\infty - 1 + 2M^{-2}\sin^{-2}\beta_0)^2\Big)^{1/2}}{(k-1)^{1/2}\Big(2k_\infty M^2\sin^2\beta_0 - \dfrac{2}{M^2\sin^2\beta_0} + \dfrac{4k_\infty}{(k_\infty - 1)} - k_\infty + 1\Big)^{1/2}}.$$

Here $\beta_0(M, \beta_k)$ is the inclination angle of the oblique jump (Figure 1.12) which depends on M and the angle β_k of the cone of the central body of the inlet; it is calculated from the equation

$$(k_\infty + 1)\tan(\beta_0 - \beta_k) = \tan\beta_0(k_\infty - 1 + 2M^{-2}\sin^2\beta_0). \tag{1.22}$$

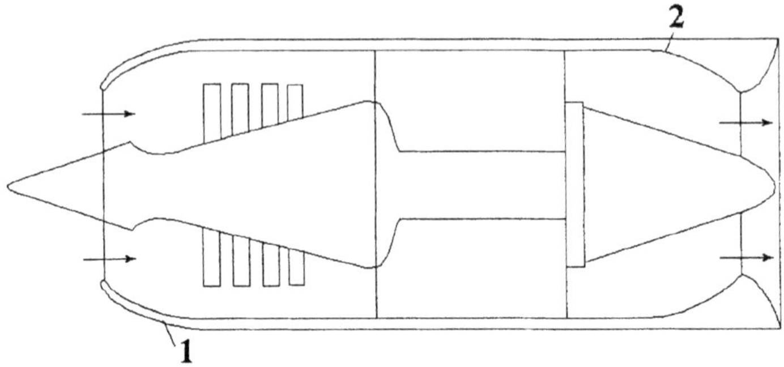

Figure 1.11. A scheme of a supersonic turbo-jet engine with elements made of high-temperature composite materials [11]: 1 - external shell of inlet; 2 - nozzle shell

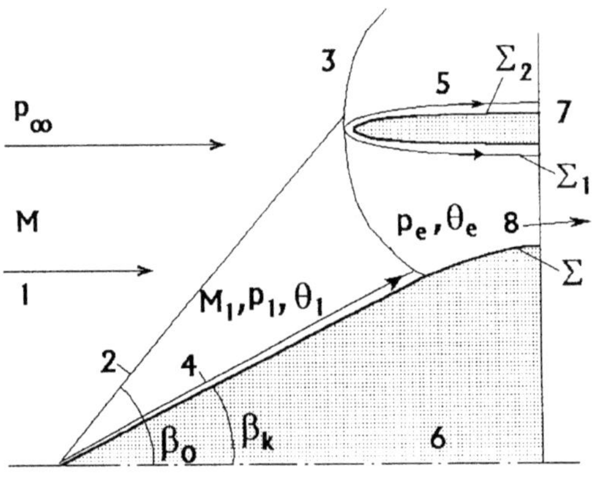

Figure 1.12. A scheme of gas flowing over an inlet of a supersonic turbo-jet engine (STJE) [11]: 1 - quiescent gas flow, 2 - shock wave (oblique jump), 3 - direct jump, 4 - boundary layer on the surface of the central body, 5 - boundary layer on the surface of the external surface of the inlet, 6 - central body of the inlet flowed over, 7 - external shell of the inlet flowed over, 8 - channel of the air-inlet

The heat transfer coefficient (α/c_p) in the inlet channel can be evaluated with the help of the theory of heat transfer in channels of jet engines nozzles:

$$\left(\frac{\alpha}{c_p}\right) = \xi_a \frac{p_e^{0.8}}{\theta_e^{0.4} r^{0.2}}, \qquad \xi_a = 0.6 \cdot 10^{-3} \, \frac{\mathrm{kg}^{0.2}\mathrm{s}^{0.8}\mathrm{K}^{0.4}}{\mathrm{m}}, \tag{1.23}$$

where r is the radius of the inlet channel crosssection.

Typical dependences of gas-dynamical flow parameters p_e, θ_e, (α/c_p) and q_e on the Mach number M are shown in Figure 1.13 for the case of $\beta_k = 15°$.

Due to constructional peculiarities of the STJE inlet, the pressure p_e in the channel is external for the central body (surface Σ in Figure 1.12) and internal for

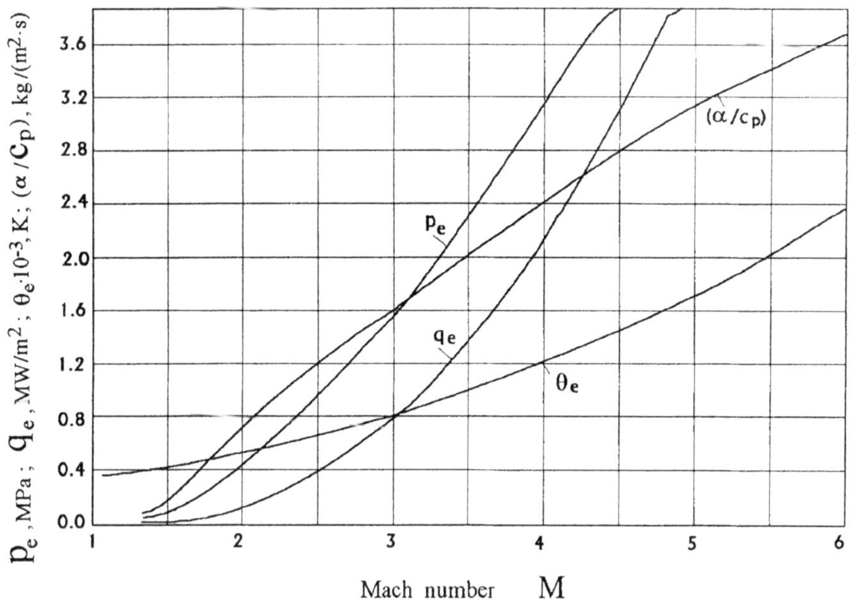

Figure 1.13. Parameters of gas-dynamical flow acting on the external surface of STJE inlet *versus* Mach number for angle $\beta_k = 15°$

the exterior shell of the inlet (surface Σ_1 (see Figure 1.12)). On the surface Σ_2 the flow pressure is considerably smaller than the one on the surfaces Σ_1 and Σ and can be assumed to be equal to

$$p_e \approx p_\infty \quad \text{at } \Sigma_2. \tag{1.24}$$

1.1.4. TECHNOLOGICAL HEATING

Technological heating is a directed action of heat fluxes related to technological processes on a structure made of composites. Here are some examples: heat cutting, drilling, material thermo-cleaving, or changing of properties due to burning out of gaseous components (pyrolytic technology, manufacturing technology of carbon–carbon materials). In heat cutting, the temperature level is usually high and can reach up to 2000° C; for pyrolytic technology it does not as a rule exceed 1000° C. The processes can be quasi-static or dynamic. For example, laser cutting of composites can be realized by continuous (quasi–static) laser radiation, or by impulse and impulse-periodic (dynamic) radiations [55].

The main parameter characterizing the technological heating process is the radiant heat flux q_R supplied to the composite surface Σ by radiation from a heat

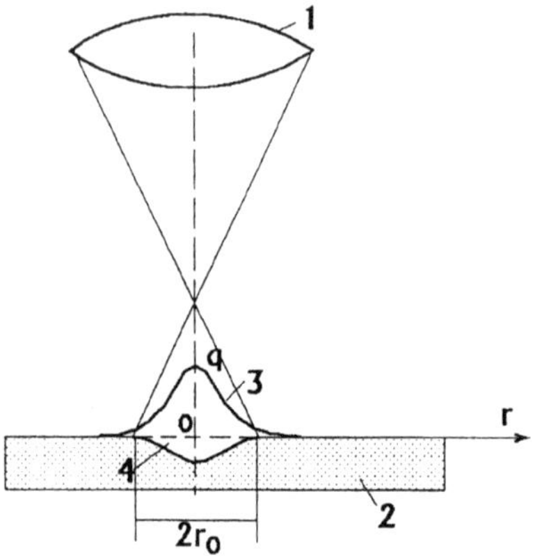

Figure 1.14. A scheme of technological heating of composite material by laser radiation: 1 - radiation source, 2 - composite material, 3 - heat flux, 4 - zone of evaporation of the material

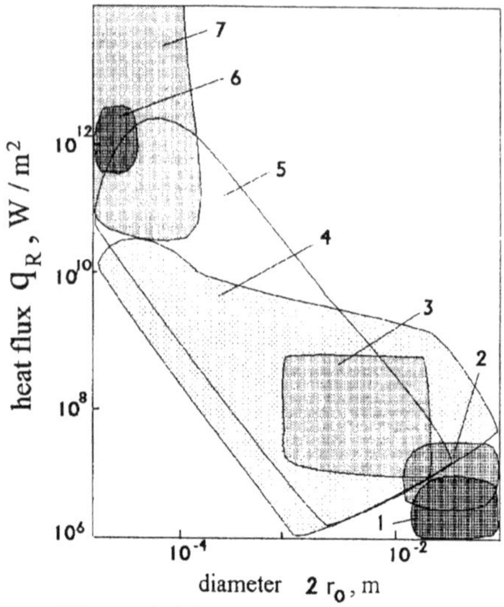

Figure 1.15. Typical values of radiant heat fluxes q_R from different heat sources [55]: 1 - gas flame, 2 - arc plasma, 3 - welding arc, 4 - electron beam, 5 - permanent laser, 6 - impulse-periodic laser, 7 - spark discharge.

source, for which the Gauss distribution is usually assumed [55]:

$$q_R = \frac{E_0 R_\Sigma}{2\pi r_0^2} \exp\left(-\frac{r^2}{r_0^2}\right). \tag{1.25}$$

Here E_0 is the power of incident radiation, r_0 is the radius of the radiation spot, R_Σ is the coefficient of radiation absorption by the composite surface and r is the radial coordinate (Figure 1.14).

The power of current sources of radiation can reach very high magnutudes, and Figure 1.15 shows typical domains of values of radiant heat fluxes q_R and diameters of the heating spot $2r_0$ for different technological heat sources.

One peculiarity of technological heating from laser sources is a localization of the heat flux supply, as magnitudes of the focusing spot r_0 are usually considerably smaller than the characteristic dimension x_0 of the whole composite structure treated.

1.1.5. ACTION OF FIRE

Action of fire most often occurs when composite structures are placed in a zone of fire flame (Figure 1.16). Alternatively, a source of flame action can be hot gas jets of heat-energetic systems, for example, jet engines (Figure 1.17). *Fire action* is characterized by the presence of a lighting flame, heat transfer from which occurs mainly by radiation and, to a lesser degree, by convection-conduction.

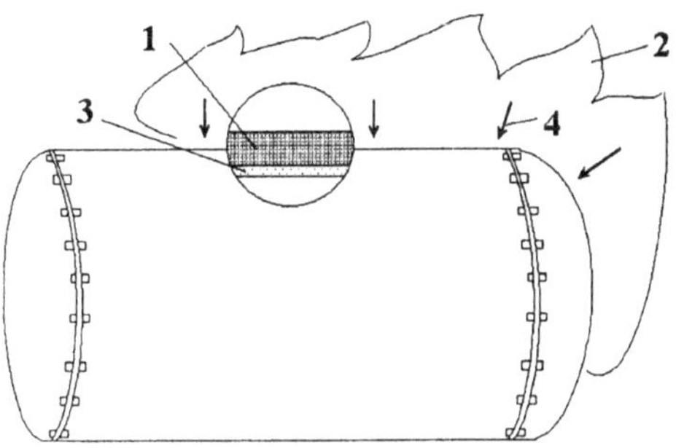

Figure 1.16. A scheme of a composite container (1) heated by flame (2), 3 - load-bearing shell; 4 - heat flux

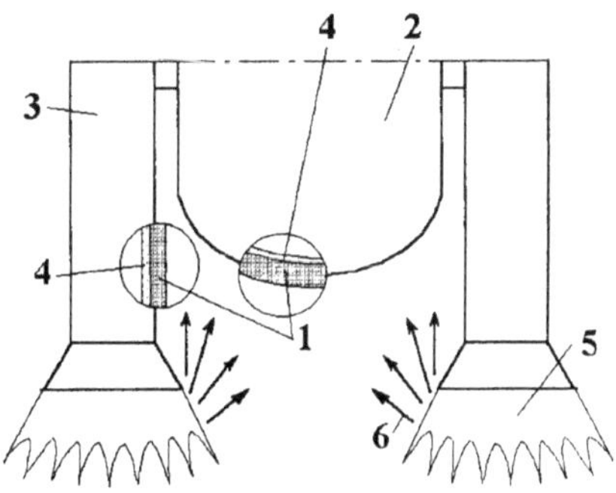

Figure 1.17. A scheme of radiational heat action of gas jets (5) of a jet engine (3) on composite thermal-protective materials (1) of air-space structure of the Space Shuttle type [52]; 2 - fuel tank; 4 - frame of the tank and engine; 6 - radiational heat flux

Radiant heat flux q_R incident onto a composite surface Σ from a flame source is defined, mainly, by flame temperature θ_c and distance h_0 from the flame center to the surface Σ:

$$q_R(r) = R_\Sigma \varepsilon_f \sigma_{SB} \theta_c^4 \frac{r_0^2}{h_0^2(1 + (r^2/h_0^2))^{3/2}}. \tag{1.26}$$

Here ε_f is the coefficient of heat radiation of the combustion products (typical values are $10^{-3} \ldots 10^{-2}$), σ_{SB} is the Stefan-Boltzman constant ($\sigma_{SB} = 5.6 \cdot 10^{-8}$ W/(m$^2 \cdot$ s^4)), r_0 is the radius of the flame source, r is the radial coordinate on the surface Σ. When the surface Σ is in immediate contact with a flame, one should assume $r_0 \approx h_0 \gg r$ in formula (1.26), and then

$$q_R = R_\Sigma \varepsilon_f \sigma_{SB} \theta_c^4. \tag{1.27}$$

The proportion q_{Rw} of the heat flux is carried off into the surroundings from the heated surface:

$$q_{Rw} = \varepsilon_w \sigma_{SB} \theta_w^4, \tag{1.27a}$$

where ε_w is the radiation coefficient, and θ_w is the temperature of the composite surface.

Due to growing requirements for safety of engineering plants, determination of a serviceability of composite structures in the fire zone is a very real problem.

TABLE 1.4. Typical values of heat fluxes q_R and combustion temperature of different substances

Source of flame formation	Products of solid propellant combustion	Products of wood combustion	Products of solid propellant combustion	Products of liquid jet fuel combustion
Temperature of combustion (flame) θ_c °,C	1000-1200	700-1000	2500-3300	2500-3000
Radiant heat flux q_R, $\mathrm{kW{\cdot}m^{-2}}$	6-10	2-6	150-350	150-240

Under the contact action of a flame, the temperature can reach 1500...2000° C for a time up to hundreds of seconds. Heating processes are usually quasistatic.

Table 1.4 shows typical temperatures θ_c of flames for some combustible substances and also radiant heat fluxes incident onto the composite surface Σ from them. When there is a direct contact of a flame and composite material, heat flux q_e is also transmitted by convection-conduction; for this case the heat transfer coefficient (α/c_p) can be calculated as in (1.9):

$$\left(\frac{\alpha}{c_p}\right) = \xi_d \frac{G^{0.8}}{r_0^{0.2}}. \tag{1.28}$$

Here G is the mass rate of combustion of fire source (for free combustion $G = \rho_e D_c$, where ρ_e is the density of the combustion source, D_c is the linear rate of combustion) and r_0 is the typical dimension of the flame contact zone.

To provide fire-explosion safety for containers made of composite materials and intended for transportation and preservation of fuels, agressive fluids etc. (Figure 1.16), it is necessary to conduct thorough investigations of the thermomechanical behaviour of large-scale structures made of the composites under the action of heat fluxes.

Radiant heating of composite thermal-protective materials by hot gas jets of jet engines is typical for bottom sections of ASV's (Figure 1.17). For example, for the Space Shuttle system, the level of radiational heat fluxes q_e on the bottom thermal protection of a fuel tank (Figure 1.17) can reach 100 - 300 kW/m^2, and the time of action is of order 100 s [52]. Typical values of q_R for several current jet engines using solid and liquid propellants are given in Table 1.4.

1.2. Ablation Processes of Composites

High temperatures lead to internal physico-chemical transformations in composites. One of the peculiarities of composite materials is that these high-temperature physico-chemical transformations are accompanied by ablation, i.e. by irreversible loss of mass.

Let us give a number of definitions.

1.2.1. CLASSIFICATION OF ABLATION PROCESSES

Ablation of a finite volume V of composite material means the process of removal of a substance from this volume; it may occur due to the action of temperature θ_e or mechanical loads $\mathbf{S}_e$. Removal of a substance means the loss of mass $m(t) = \rho(t)\overset{\circ}{V}(t)$ of the composite, where $\rho(t)$ is the composite density averaged over the undeformed volume $\overset{\circ}{V}(t)$.

Composite material is called *ablative*, even if the ablation process occurs in only one of its parts.

To say that a composite is ablative means that there is some process going on in the material under certain external actions. This concept is not a classification of materials, as the same composite may or maynot be ablative in accordance with external conditions which promote or prevent removal of substance from the composite.

For ablative material its mass m and, at least, one of the values ρ and $\overset{\circ}{V}$ depend on the time t of the thermo-force action.

Let us give the following classification of ablation processes (Figure 1.18).

We will say that in the finite volume of the composite the process of *surface (linear) ablation* occurs, if there is a removal of a substance from the surface $\Sigma(t)$ only due to the decrease of the volume $\overset{\circ}{V}(t)$ under thermo-force loading: $\overset{\circ}{V}(t) < \overset{\circ}{V}(0)$, and density $\rho(t)$ remains constant, $\rho(t) = \rho(0)$.

We will say that *volumetric ablation (thermodecomposition)* occurs, if removal of a substance from the composite under the thermo-force action is caused by physical-chemical transformations such as phase transformations or chemical reactions inside the material volume $\overset{\circ}{V}(t)$. For this case, the density decreases $\rho(t) < \rho(0)$, but the volume remains the same: $\overset{\circ}{V}(t) = \overset{\circ}{V}(0)$.

In volumetric ablation, in each elementary volume of the material, a part of the initial material becomes gaseous or liquid and then is moved away through pores into the surroundings due to filtration; the remainder passes into a new solid phase state called a *solid pyrolytic residue*.

Depending on the type of internal physico-chemical processes occurring in a material (purely thermal decomposition of the initial substance into gas and new solid phases, or formation of gas and solid phases due to internal chemical reactions with outer gas filtering through pores, for example oxidation), volumetric ablation

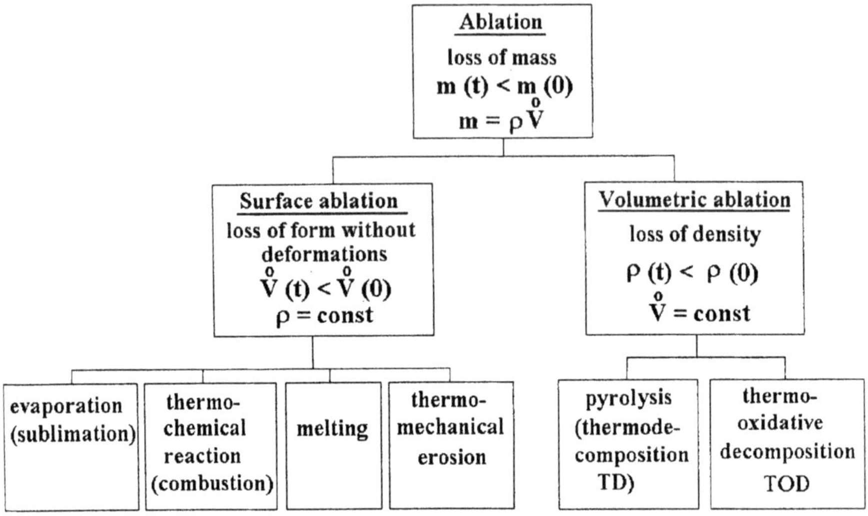

Figure 1.18 Classification of ablation processes in composites

can be subdivided into:

- pyrolytic *thermodecomposition* (TD);
- thermo-oxidative decomposition (TOD).

In surface ablation, the volume of composite material decreases due to physico-chemical transformations of all or some of the components of the material in a thin surface layer; initial solid phases pass into gaseous or liquid ones and then move away from the surface due to evaporation, or they are mechanically destroyed under action of external forces and then move away in the form of solid particles.

The surface ablation can be subdivided into four main types:

- evaporation (sublimation);
- chemical reactions with the surroundings (mainly, combustion);
- melting;
- thermomechanical erosion.

1.2.2. VOLUMETRIC ABLATION

Volumetric ablation occurs by *pyrolysis* if under the action only of heating in an inert medium, there are internal physico-chemical transformations of the following type in the material:

$$\text{solid phase} \xrightarrow{\theta,^{\circ}C} \text{gas} + \text{new solid phase}.$$

Pyrolysis is usually endothermal, a part of the heat flux supplied to the material is absorbed due to the endothermal effect of thermodecomposition. Gas generating in pyrolysis is moved away from the material into the environment through pores. This type of ablation takes place usually for a polymer matrix of composite material and also for organic reinforcing fibres.

The solid phase which appears in pyrolysis is, most often, carbon; this can form a mechanically strong frame, if the material remains in one piece; such materials are called *charring ablators.* The solid pyrolytic residue can also be in the form of dispersed particles (powder), then the ablative material loses its shape in the pyrolysed zone.

The new solid phase needs not be carbon, it can be, for example, silica dioxide if the polymer is silicon-organic. For this case the solid phase is called a *pyrolytic residue.* As a rule, pyrolytic residues have very low mechanical strength.

Volumetric ablation occurs according to the mechanism of *thermo-oxidative decomposition* (TOD) if internal physico-chemical transformations of the material components of the type:

$$\text{solid phase} + O_2 \xrightarrow{\theta,^{\circ}C} \text{gas} + \text{new solid phase}$$

take place in the material volume due to the combined action of external heating and reacting external oxygen filtering into the material through its pores.

Thermo-oxidative decomposition is usually endothermal, however, gases generated in heating can combine with oxygen and combust giving an exothermal effect. Rate of TOD exceeds, as a rule, a rate of TD. The decomposition process can be caused by other gases, besides oxygen, penetrating through pores into composites, for example by hydrogen [69]. This phenomenon is called hydrolysis.

This type of ablation is typical for polymer matrices and organic fibres and for carbon, boron and ceramic fibres in an oxidizing environment.

1.2.3. SURFACE ABLATION

Melting means ablation of material when there is a phase transformation of the following type on the material surface:

$$\text{solid phase} \xrightarrow{\theta,^{\circ}C} \text{liquid}$$

where liquid can transform into gas partially or completely.

Evaporation (sublimation) is the ablation process in which a phase transformation of components of a composite on its surface does not pass through a liquid phase:

$$\text{solid phase} \xrightarrow{\theta,^{\circ}C} \text{gas}$$

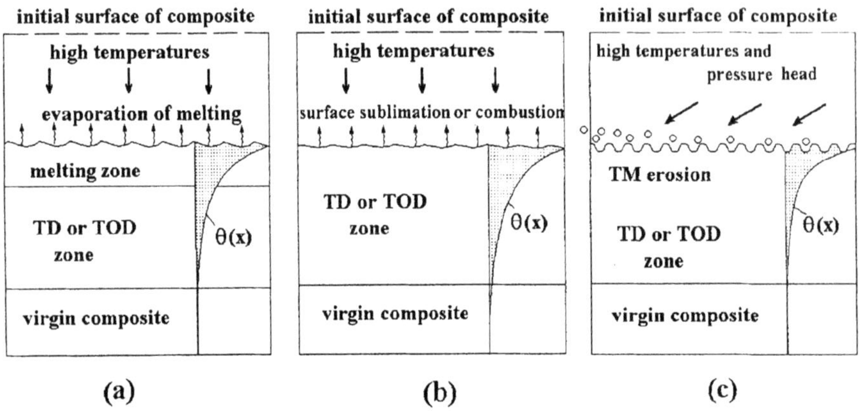

Figure 1.19 Four main types of surface ablation of composite materials under gradient heating: a) melting; b) sublimation and thermochemical reaction; c) thermomechanical erosion

Thermochemical reaction. This type of ablation takes place when there are physico-chemical transformations of solid material components of the following type

$$\text{solid phase} + \text{REG (or REL)} \xrightarrow{\theta,^{\circ}C} \text{gas}$$

on the surface Σ of the material due to the combined action of high temperatures and reacting environmental gas (REG) or reacting environmental liquid (REL). An example of such ablation is combustion of components of ablative material (for instance, carbon-plastics) under the action of hot oxygen-containing gas flows.

In *thermomechanical erosion*, external high-speed gas flow causes surface stresses to exceed the strength limit, leading to successive breaking away of particles of material from the surface.

Surface ablation of composite material occurs, as a rule, under nonuniform heating; the temperature gradient can reach 1000 K/mm and higher. In this case a part of material is in its normal state under normal or elevated temperatures, and not subject to ablation, while another part is subject to superhigh temperature up to 3000° C and consequent surface ablation. There may also be an intermediate part of the composite, which is subject to high temperatures, with accompanying a volumetric ablation process of the TD or the TOD type (Figure 1.19).

1.3. Phenomena in Composite Materials under High Temperatures

As has been noted, ablation processes of composite materials are processes of physical and chemical transformations occurring under the action of high temperatures.

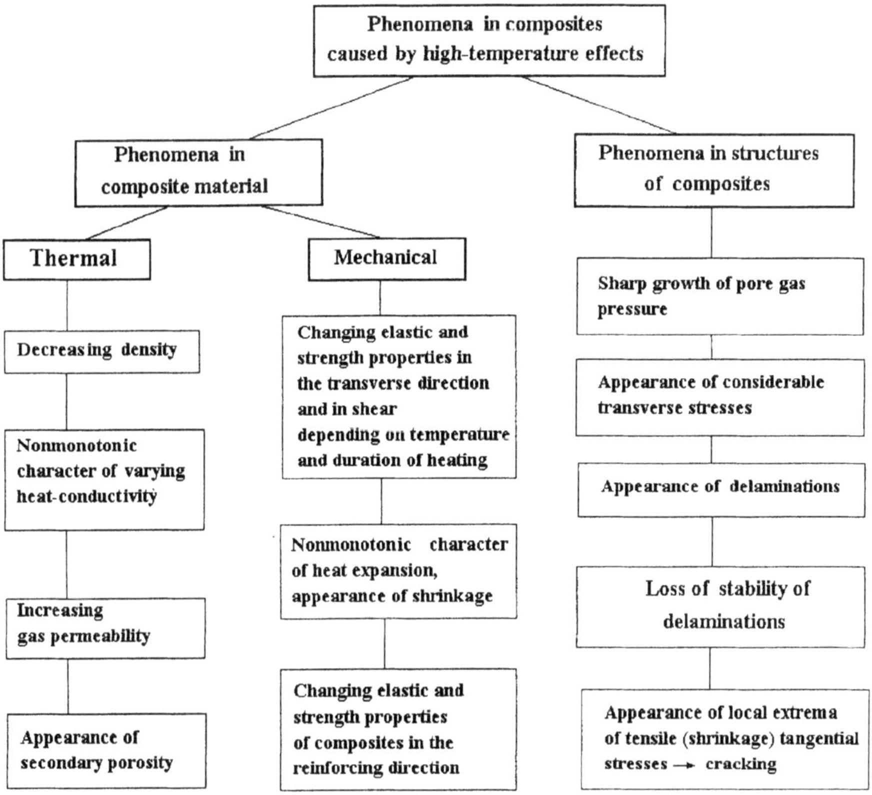

Figure 1.20. Phenomena in composites caused by high-temperature effects

Pyrolysis of a polymer matrix leads to formation of many pores filled with gaseous pyrolysis products. Besides, a new pyrolytic solid phase is formed; it is usually carbon.

Changing the physico-chemical structure of a material leads to changing macroscopic properties of the material and thus to changes in its structural behaviour.

Let us enumerate the principal phenomena caused by high-temperature effects in composite materials.

First of all, two types of such phenomena should be marked out (Figure 1.20):

- phenomena arising immediately in material independently of its application in structures (i.e. when its stress-strain state, temperature and filtrational fields are uniform);
- phenomena connected with a specific type of structure under gradient heating (when stress, temperature and pore pressure fields are nonuniform).

High-temperature effects change practically all properties of a material; we will

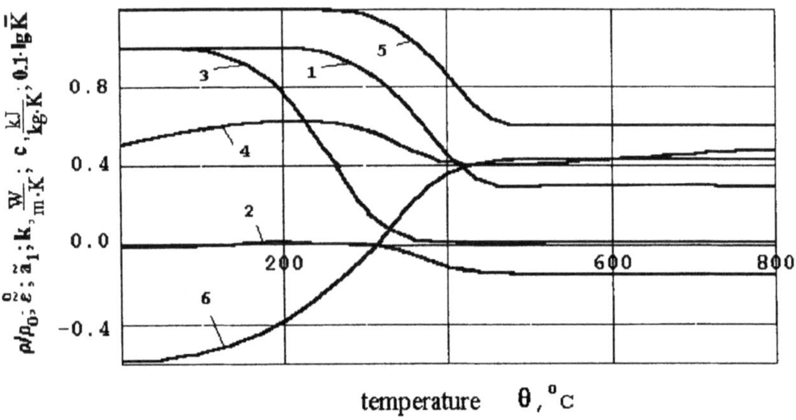

Figure 1.21. Dependences of heat and mechanical characteristics of epoxy composite on temperature θ of heating with rate 0.1 K/s: 1 - relative density ρ/ρ^0; 2 - heat deformation $\overset{0}{\varepsilon}$; 3 - function of changing elastic properties $\widetilde{a}_1$; 4 - heat conductivity k W/(m · K); 5 - heat capacity c, kJ/(kg · K); 6 - gas permeability $0.1 \lg \bar{K}$

consider the following two types of varying properties: thermal and mechanical.

Phenomena relating to varying thermal properties are as follows:

- decreasing composite density ρ in heating,
- nonmonotonic character of varying heat conductivity k,
- sharply increasing gas-permeability K,
- appearance of secondary porosity φ_g (porosity φ_g^0 formed after technological operations is usually called primary).

All these processes have an irreversible character. Figure 1.21 shows typical graphs of functions ρ, k and K depending on heating temperature θ for epoxy-glass composite.

Phenomena relating to varying mechanical properties under high temperatures are as follows:

- changing elastic and strength properties of composites in the direction transverse to reinforcing direction, and in shear, depending on temperature θ and duration of heating t; this concerns textiles, unidirectionally reinforced and laminated composites;
- changing elastic and strength properties of composites in the reinforcing direction for all types of composites (textile, unidirectional, laminated);
- nonmonotonic character of heat expansion $\overset{0}{\varepsilon}$, appearance of shrinkage at certain temperatures, etc.

Figure 1.21 shows graphs of the function $\overset{0}{\varepsilon}$ and the coefficient of lowering mechanical properties $\widetilde{a}_1$ for epoxy composite.

When composite materials are used as components of structures, stress, temperature and pore gas pressure fields, as a rule, prove to be nonuniform. High-temperatures redistribute these fields, and the following phenomena are observed in most structures:

- local extrema of pore pressure p of gaseous products of pyrolysis;
- tensile transverse stresses σ_{33} even in thin thermal-protective coatings; these can lead to delamination;
- local extrema of tensile (shrinkage) tangential stresses; these can lead to cracking.

Besides the above phenomena, surface ablation can occur in composite material, i.e. surface recession under the action of a large heat flux and erosion due to pressure head of gas jets. Usually, surface ablation occurs at temperatures exceeding $1000 - 1500^{\circ}$ C.

Finally we consider a physico-mathematical model of the internal structure of composites and of the processes occurring in the material under high temperatures. This model will allow us to describe the phenomena we have listed.

1.4. A Physical Model of Ablative Composite

Composite material under normal temperatures consists of two phases: reinforcing filler and *polymer phase* of matrix (in classic composite mechanics the fibre and matrix are usually called components of the composite), between which there exist only mechanical interactions. Under normal temperatures the geometrical configuration of the phases can greatly change only for the case of finite deformations (for example, for elastomer composites); for composite materials with small deformations (for glass-, carbon- and organo-plastics) it varies unconsiderably. Under high temperatures a polymer matrix ablates by pyrolysis and two new phases appear: *pyrolytic phase* containing solid products of decay (as a rule, carbon) and *gas phase* containing gaseous products of decomposition of the polymer (Figure 1.22); this gas phase occupies pores of the material.

Fillers can ablate too (for example, if they are organic fibres); then another solid phase appears in the composite, conventionally called a *'crystalline'* phase of the fibre, and pores filled with gas also appear in the fibre. The initial solid phase, the fibre was under normal temperature, is called *'amorphic'* in this case.

For organic fibres, the 'amorphic' phase is the initial polymer, and the 'crystalline' one is a pyrolytic residue, usually carbon.

For carbon fibres, the 'amorphic' phase is amorphic or crystalline carbon and no essentially new solid phase appears under heating in an oxidizing medium; the fibre goes completely into a gas phase. Similar processes occur with boron fibres in oxidizing surroundings.

For glass fibres, the 'amorphic' and 'crystalline' phases are the only physical states of the substance, and gas phase practically does not appear.

Besides mechanical interaction between the initial phases and new phases of matrix and filler, there is also a physico-chemical interaction: under the action of

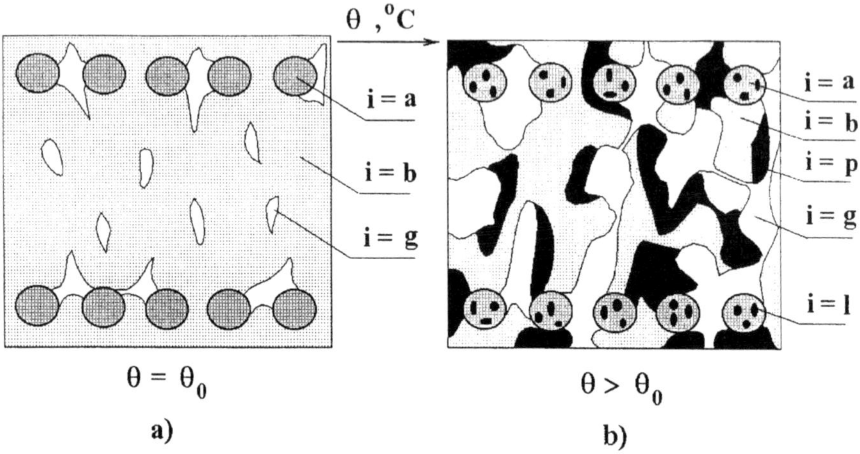

Figure 1.22. A physical model of an ablative composite describing the material structure changing under high-temperature heating (a) at the initial state; (b) after heating; where $i = a, b, p, g, l$ are the composite phases:
$i = a$ - reinforcing fibre at the 'amorphic' state;
$i = b$ - polymer phase of the matrix;
$i = p$ - pyrolytic phase (the solid residue of the matrix pyrolysis);
$i = g$ - pores and cracks filled in with gas phase;
$i = l$ - 'crystalline' phase of fibres

high temperatures, substances of 'old' solid phases pass gradually into 'new' phase states.

The geometrical shape of each the phase is a certain finite domain V_i. This domain is not static, i.e. it changes with time $V_i = V_i(t)$ under the action of high temperatures, even without the presence of deformations.

Figure 1.22 shows schematically a physical model of the change of an actual structure made of composite materials under high temperatures.

The actual process of structure-formation in a composite material under thermodecomposition is very complex, and its physical essence has not been yet described adequately. At first, as the temperature increases, nuclei of solid and gas phases appear. These have certain critical dimensions, as in the formation of critical centers in the melting of solid bodies, boiling liquids etc. Under further heating, the nuclei of gas and pyrolytic phases grow. The structure-formation scheme at TD suggested above and investigated below is the model of the process. The basis of the model is the assumption that all the phases (initial polymer, gas and new (pyrolytic and crystalline) phases) exist simultaneously at each time t in each elementary volume of the composite, and only the ratios of the volumetric parts of the phases change in the heating process. This assumption is used, sometimes in an implicit form, in most current models of polymer thermodecomposition. It will be shown below, that this assumption leads to a quite accurate description of thermomechanical phenomena in composites.

CHAPTER 2

GENERAL EQUATIONS OF MULTIPHASE CONTINUUM MECHANICS FOR ABLATIVE COMPOSITES

Models of multiphase media mechanics have been considered in many works [56, 60, 68, 83, 107 etc.]. In the present chapter general relationships of multiphase media mechanics are developed for ablative composites. These are the key points: derivation of expressions for a phase transformation rate with the help of thermodynamical equations at a phase interface; and determination of the way in which constitutive relations depend on the characteristics of the natural configurations of phases. In the first four paragraphs the general case of finite deformations of phases is considered using, in the main, the Euler (space) description given in works [22, 51, 56, 74, 80, 107 etc.]; and in the last paragraph the case of infinitesimal deformations of phases is analyzed.

2.1. Conservation Laws

2.1.1. MAIN CONCEPTS OF TENSOR ANALYSIS

Consider the motion of a continuum in the Cartesian coordinate system $O\mathbf{x}$ with basis vectors $\mathbf{e}_k$, $k = 1, 2, 3$. The geometrical location of the medium at the initial time $t = 0$ in the coordinate system is called the *initial configuration* $\overset{\circ}{V}$, and at time $t > 0$ – the *actual configuration* $V(t)$.

Let radius-vectors of a material point in the actual $V(t)$ and initial $\overset{\circ}{V}$ configurations be $\mathbf{x} = x^k\mathbf{e}_k$ and $\overset{\circ}{\mathbf{x}} = \overset{\circ}{x}{}^k\mathbf{e}_k$, respectively, where x^k and $\overset{\circ}{x}{}^k$ are the coordinates of the same material point M in configurations V and $\overset{\circ}{V}$. Introduce the *material coordinates* X^k individualizing the material point M. Motion of the medium is described by the equations $\mathbf{x} = \mathbf{x}(X^k, t)$; location of a material point with coordinates X^k in the initial configuration is determined by the equations

$$\overset{\circ}{\mathbf{x}} = \overset{\circ}{\mathbf{x}}(X^k).$$

Indices $k, s, p, r, \ldots$ here and below are indicators of components of tensors and take the values 1, 2, 3; summation from 1 to 3 is denoted by a repeated index.

Introduce vectors $\mathbf{R}_k$ of a *general local basis* and vectors $\mathbf{R}^k$ of a *coupling local basis* in the actual configuration, and vectors $\overset{\circ}{\mathbf{R}}_k$ and $\overset{\circ}{\mathbf{R}}{}^k$ in the initial configuration, respectively:

$$\mathbf{R}_k = \frac{\partial \mathbf{x}}{\partial X^k} = \frac{\partial x^s}{\partial X^k}\mathbf{e}_s, \qquad \overset{\circ}{\mathbf{R}}_k = \frac{\partial \overset{\circ}{\mathbf{x}}}{\partial X^k} = \frac{\partial \overset{\circ}{x}{}^s}{\partial X^k}\mathbf{e}_s,$$

$$\mathbf{R}^k = \frac{\partial X^k}{\partial x^s}\mathbf{e}_s, \qquad \overset{\circ}{\mathbf{R}}{}^k = \frac{\partial \overset{\circ}{X}{}^k}{\partial x^s}\mathbf{e}_s.$$

The concept of the *dyad* [90] of vectors $\mathbf{a}$ and $\mathbf{b}$ is introduced, and denoted by $\mathbf{a}\otimes\mathbf{b}$. With the help of dyads $\mathbf{R}^k\otimes\mathbf{R}^s$, $\mathbf{R}_k\otimes\mathbf{R}_s$, $\overset{\circ}{\mathbf{R}}{}^k\otimes\overset{\circ}{\mathbf{R}}{}^s$ of the vectors of the local basis and others, or of the Cartesian basis $\mathbf{e}^k\otimes\mathbf{e}^s$, we may form *dyadic bases*; then each tensor $\mathbf{A}$ of the second rank can be written in the form

$$\mathbf{A} = A^{sk}\mathbf{R}_s\otimes\mathbf{R}_k = A_{sk}\mathbf{R}^s\otimes\mathbf{R}^k = A^s_k\mathbf{R}_s\otimes\mathbf{R}^k = \overset{\circ}{A}{}^{sk}\overset{\circ}{\mathbf{R}}_s\otimes\overset{\circ}{\mathbf{R}}_k,$$

where A^{sk}, A_{sk}, A^s_k are respectively the *contravariant*, *covariant* and *combined components* of the tensor $\mathbf{A}$, in the corresponding dyadic basis.

Scalar products of vectors of the local bases have the forms

$$\mathbf{R}_k\cdot\mathbf{R}_s = g_{ks}, \quad \mathbf{R}^k\cdot\mathbf{R}^s = g^{ks}, \quad \mathbf{R}_k\cdot\mathbf{R}^s = g^s_k = \delta^s_k,$$

$$\overset{\circ}{\mathbf{R}}_k\cdot\overset{\circ}{\mathbf{R}}_s = \overset{\circ}{g}_{ks}, \quad \overset{\circ}{\mathbf{R}}{}^k\cdot\overset{\circ}{\mathbf{R}}{}^s = \overset{\circ}{g}{}^{ks}, \quad \overset{\circ}{\mathbf{R}}_k\cdot\overset{\circ}{\mathbf{R}}{}^s = \overset{\circ}{g}{}^s_k = \delta^s_k,$$

where δ^s_k is the *Kronecker symbol*, g_{ks} and $\overset{\circ}{g}_{ks}$ are the components of the *metric (unit) tensor* $\mathbf{E}$ in the configurations V and $\overset{\circ}{V}$, respectively. The tensor E is determined as follows:

$$\begin{aligned}\mathbf{E} =& \mathbf{R}^k\otimes\mathbf{R}_k = g^{ks}\mathbf{R}_k\otimes\mathbf{R}_s = g_{ks}\mathbf{R}^k\otimes\mathbf{R}^s = \\ &= \overset{\circ}{\mathbf{R}}{}^k\otimes\overset{\circ}{\mathbf{R}}_k = \overset{\circ}{g}{}^{ks}\overset{\circ}{\mathbf{R}}_k\otimes\overset{\circ}{\mathbf{R}}_s = \overset{\circ}{g}_{ks}\overset{\circ}{\mathbf{R}}{}^k\otimes\overset{\circ}{\mathbf{R}}{}^s = \mathbf{e}_k\otimes\mathbf{e}_k.\end{aligned}$$

Passage from the general basis vectors to the coupling basis vectors is realized with the help of the components g^{ks} and $\overset{\circ}{g}{}^{ks}$:

$$\mathbf{R}^k = g^{ks}\mathbf{R}_s, \quad \mathbf{R}_k = g_{ks}\mathbf{R}^s,$$

$$\overset{\circ}{\mathbf{R}}{}^k = \overset{\circ}{g}{}^{ks}\overset{\circ}{\mathbf{R}}_s, \quad \overset{\circ}{\mathbf{R}}_k = \overset{\circ}{g}_{ks}\overset{\circ}{\mathbf{R}}{}^s.$$

Scalar products of vector $\mathbf{a}$ and tensor $\mathbf{A}$, and also tensors $\mathbf{A}$ and B with respect to one or two indices are determined as follows:

$$\mathbf{a}\cdot\mathbf{A} = a^k A_{ks}\mathbf{R}_s, \quad \mathbf{A}\cdot\mathbf{B} = A^{kr}B_{rs}\mathbf{R}_k\otimes\mathbf{R}^s, \quad \mathbf{A}\cdot\cdot\mathbf{B} = A^{ks}B_{sk},$$

where

$$\mathbf{B} = B^{sk}\mathbf{R}_s \otimes \mathbf{R}_k, \qquad \mathbf{a} = a^k\mathbf{R}_k.$$

Tensor and vector products of vector $\mathbf{a}$ and tensor $\mathbf{A}$ are defined by the formulae

$$\mathbf{a} \otimes \mathbf{A} = a^k A^{rs}\mathbf{R}_k \otimes \mathbf{R}_r \otimes \mathbf{R}_s, \quad \mathbf{a} \times \mathbf{A} = \sqrt{g}\epsilon_{ksp}a^k A^{sr}\mathbf{R}^p \otimes \mathbf{R}_r,$$

here $\mathbf{R}_k \otimes \mathbf{R}_r \otimes \mathbf{R}_s$ is the *triad* of the local basis vectors, $g = \det(g_{sk})$ is the determinant of the metric matrix and ϵ_{ksp} are the *Levi-Civita symbols.*

The *inverse tensor* $\mathbf{A}^{-1}$ and *transpose tensor* $\mathbf{A}^T$ are defined as follows:

$$\mathbf{A} \cdot \mathbf{A}^{-1} = \mathbf{E}, \qquad \mathbf{A}^T = A^{sk}\mathbf{R}_k \otimes \mathbf{R}_s.$$

Sequential application of the inversion and transposition to tensor $\mathbf{A}$ gives the *inverse-transpose tensor* $\mathbf{A}^{-T}$:

$$(\mathbf{A}^{-1})^T = (\mathbf{A}^T)^{-1} = \mathbf{A}^{-T}.$$

The differential *Hamilton nabla-operator* ∇ of the scalar θ, vector $\mathbf{a}$ and tensor $\mathbf{A}$ in the initial configuration are determined as follows:

$$\nabla\theta = \mathbf{R}^k\frac{\partial\theta}{\partial X^k} = \frac{\partial\theta}{\partial x^k}\mathbf{e}_k, \qquad \nabla \cdot \mathbf{a} = \mathbf{R}^k \cdot \frac{\partial\mathbf{a}}{\partial X^k} = \frac{\partial\mathbf{a}}{\partial x^k} \cdot \mathbf{e}_k,$$

$$\nabla \otimes \mathbf{a} = \mathbf{R}^k \otimes \frac{\partial\mathbf{a}}{\partial X^k} = \mathbf{e}_k \otimes \frac{\partial\mathbf{a}}{\partial x^k}, \qquad \nabla \cdot \mathbf{A} = \mathbf{R}^k \cdot \frac{\partial\mathbf{A}}{\partial X^k} = \mathbf{e}_k \cdot \frac{\partial\mathbf{A}}{\partial x^k},$$

$$\nabla \times \mathbf{A} = \mathbf{R}^k \times \frac{\partial\mathbf{A}}{\partial X^k} = \mathbf{e}_k \times \frac{\partial\mathbf{A}}{\partial x^k}.$$

We will apply these definitions to a continuum, the physical features of which at passage from one material point to another do not change or change continuously. If the functions describing physical properties of continuum are disconnected in the domain V, then the medium is called composite. This mathematical definition was given in [91].

Suppose that, at time $t > 0$ the domain $V(t)$ of the composite consists of n subdomains, each of which is a physically *homogeneous* medium and called a *phase* of the composite. Each ith phase of the composite occupies a domain $\overset{\circ}{V}_i$ in the initial configuration and $V_i(t)$ in the actual configuration. Let the interface between phases i and j be Σ_{ij}. Here and below indices i and j run over letter values $\{a, b, p, g, l\}$ and correspond to certain phases, and n=5. There is no summation over repeated indices i and j.

We assume the following:

- phase $i = a$ is an amorphic phase of the reinforcing filler;
- phase $i = b$ is a polymer phase of the matrix;
- phase $i = p$ is a pyrolytic phase of the matrix;
- phase $i = g$ is gaseous products of pyrolysis in pores of the composite;
- phase $i = l$ is a crystalline phase of the reinforcing filler.

The transformation of a local vicinity of a material point M from domain V_i into domain V_j at time $t = t_*$ is called a *phase transformation* of the type '$i \longrightarrow j$'.

For a multiphase medium, in addition to configurations $V(t)$ and $\overset{\circ}{V}$ there also exists an undeformed configuration $\overset{\circ}{V}(t)$ which differs from $\overset{\circ}{V}$ only in that boundaries of domains $\overset{\circ}{V}(t)$ of phases change according to the changing domains $V_i(t)$ therein, and $\overset{\circ}{V}(0) = \overset{\circ}{V} = V(0)$. When phase transformations are absent $\overset{\circ}{V}(t) = \overset{\circ}{V}$.

Within the domains $V_i(t)$, all functions $\overset{\circ}{\mathbf{x}}$, $\mathbf{x}$, $\mathbf{R}_k$, $\overset{\circ}{\mathbf{R}}_k$ and others describing a state of the composite have continuous partial derivatives with respect to x^k and t for all requested orders.

The *displacement vector* $\mathbf{u}_i(x^k, t)$ for material point M, at time t and point $x^k \in V_i(t)$, is determined as follows

$$\mathbf{u}_i(x^k, t) = \mathbf{x} - \overset{\circ}{\mathbf{x}}\left(X^s(x^k, t)\right), \quad x^k \in V_i(t).$$

Let us introduce the *strain gradient* $\mathbf{F}_i$, being an unsymmetric second-rank tensor

$$\mathbf{F}_i = \mathbf{R}_s(x^k, t) \otimes \overset{\circ}{\mathbf{R}}{}^s\left(X^s(x^k, t)\right), \quad x^k \in V_i(t),$$

which transforms local basis vectors and infinitesimal elements of radius-vectors $d\mathbf{x}$ and $d\overset{\circ}{\mathbf{x}}$ from the initial configuration to the actual one:

$$\mathbf{R}_s = \mathbf{F}_i \cdot \overset{\circ}{\mathbf{R}}_s, \qquad \mathbf{x} = \mathbf{F}_i \cdot d\overset{\circ}{\mathbf{x}} \quad x^k \in V_i(t).$$

The inverse strain gradient $\mathbf{F}_i^{-1}$ is expressed by the formula

$$\mathbf{F}_i^{-1} = \overset{\circ}{\mathbf{R}}_s \otimes \mathbf{R}^s = \frac{\partial \overset{\circ}{x}{}^s\left(X^p(x^k, t)\right)}{\partial x^r} \mathbf{e}_s \otimes \mathbf{e}_r, \quad x^k \in V_i(t).$$

2.1.2 SYSTEM OF CONSERVATION LAWS FOR PHASES

On the base of relationships given in subsection 2.1.1 we derive the connection between the inverse strain gradient $\mathbf{F}_i^{-1}$ and the displacement vector $\mathbf{u}_i$

$$\mathbf{F}_i^{-T} = \mathbf{E} - \nabla \otimes \mathbf{u}_i. \tag{2.1}$$

In the domain V_i occupied by the ith phase, there is a set of *conservation equations for continuum mechanics* which can be written as follows:

$$\frac{\partial \rho_i A_{i\zeta}}{\partial t} + \nabla_x \cdot (\rho_i \mathbf{v}_i A_{i\zeta} - B_{i\zeta}) = C_{i\zeta}, \tag{2.2}$$

$$\mathbf{x} \in V_i, \qquad i = a, b, p, g, l, \qquad \zeta = 1, \ldots 6.$$

Here $A_{i\zeta}$, $B_{i\zeta}$ and $C_{i\zeta}$ are generalized vectors with the following components:

$$A_{i\zeta} = \begin{pmatrix} 1 \\ \mathbf{v}_i \\ e_i + \frac{v_i^2}{2} \\ \eta_i \\ \mathbf{u}_i \\ \mathbf{F}_i \end{pmatrix}, \quad B_{i\zeta} = \begin{pmatrix} 0 \\ \boldsymbol{\sigma}_i \\ \boldsymbol{\sigma}_i \cdot \mathbf{v}_i - \mathbf{q}_i \\ -\mathbf{q}_i/\theta_i \\ 0 \\ \rho_i \mathbf{F}_i^T \otimes \mathbf{v}_i \end{pmatrix}, \quad C_{i\zeta} = \begin{pmatrix} 0 \\ 0 \\ 0 \\ \frac{W_i^*}{\theta_i} - \frac{\mathbf{q}_i \cdot \nabla \theta_i}{\theta_i^2} \\ \rho_i \mathbf{v}_i \\ 0 \end{pmatrix}, \quad (2.3)$$

$$i = a, b, p, g, l.$$

Equation system (2.2) consists of

- $\zeta = 1$ - scalar *equation of continuity* for the ith phase;
- $\zeta = 2$ - vector *equation of momentum* for the ith phase;
- $\zeta = 3$ - scalar *equation of energy* for the ith phase;
- $\zeta = 4$ - scalar *equation of entropy balance* for the ith phase;
- $\zeta = 5$ - vector *kinematic equation* for the ith phase;
- $\zeta = 6$ - tensor *equation of strain compatibility* for the ith phase.

In equations (2.2) there are the following notations: ρ_i is the density, $\mathbf{v}_i$ is the velocity, $\boldsymbol{\sigma}_i$ is the stress tensor (also called *Cauchy's stress tensor* or true stress tensor), e_i is the *internal energy*, $\mathbf{q}_i$ is the *heat flux vector*, η_i is the *entropy*, θ_i is the *temperature*, W_i^* is the *energy dissipation function* of the composite phases ($W_i^* \geqslant 0$), $\mathbf{F}_i$ is the strain gradient, $\mathbf{u}_i$ is the displacement vector and $v_i = \mathbf{v}_i \cdot \mathbf{v}_i$. It should be noted that terms of density of internal energy, vector of heat flux density etc. are often used for functions e_i, η_i, $\mathbf{q}_i$ and W_i^*, and terms of internal energy, heat flux vector etc. are applied for integral values averaged over volumes or surfaces of phases. However when one has no need to use the integral values, the word 'density' can be omitted for brevity.

The *conservation laws* set (2.2) has a *divergent form*. Nine scalar equations of strain compatibility in the system at $\zeta = 6$ are equivalent to nine relationships (2.1), and can be used instead of them. The compatibility equations in the divergent form were derived from the relationships for the first time in [68].

2.1.3. RELATIONSHIPS ON SURFACES OF STRONG BREAKS

The interface Σ_{ij} between the ith and jth phases at phase transformation is a *surface of a strong break*, i.e. when passing through the surface, some of the functions describing the motion of the composite have discontinuities. Values of an arbitrary function $\Omega_i(\mathbf{x}, t)$ at the break surface Σ_{ij} are determined by passing to the limit from the side of corresponding phase, V_i or V_j:

$$\Omega_i(\mathbf{x}, t)|_{\mathbf{x} \in \Sigma_{ij}} = \lim_{\substack{\mathbf{x}' \to \mathbf{x} \\ \mathbf{x}' \in V_i}} \Omega_i(\mathbf{x}', t).$$

The difference between the function values Ω_i and Ω_j at the phase interface Σ_{ij} is called the *jump of a function*:

$$[\Omega] = \Omega_i|_{\mathbf{x}\in\Sigma_{ij}} - \Omega_j|_{\mathbf{x}\in\Sigma_{ij}}. \tag{2.4}$$

For the generalized vectors of the set (2.2) at the phase interface Σ_{ij}, the following jump conditions of the functions hold:

$$[\rho(D - \mathbf{v}\cdot\mathbf{n})A_\zeta + \mathbf{n}B_\zeta] = H_\zeta, \quad \zeta = 1, \dots 6, \quad \mathbf{x} \in \Sigma_{ij}. \tag{2.5}$$

In the relations (2.5) there are some notations: $\mathbf{n}$ is the normal vector to the surface Σ_{ij}, being external with respect to the jth phase; D is the *speed of the phase interface*; and H_ζ is the generalized vector having the following components:

$$H_\zeta = \{0,\ \mathbf{p}_\Sigma,\ E_\Sigma,\ H_\Sigma,\ \mathbf{u}_\Sigma,\ \mathbf{F}_\Sigma\}, \tag{2.6}$$

where
$\mathbf{p}_\Sigma$ is the vector of surface stresses;
E_Σ is the energy of surface stresses;
H_Σ is the entropy release on the phase interface;
$\mathbf{u}_\Sigma$ is the incompatibility vector on the phase interface;
$\mathbf{F}_\Sigma$ is the incompatibility tensor on the phase interface.

The conditions (2.5) hold true on the phase interface Σ_{ij} for the case when phase transformations are absent. For this $D_0 = 0$, where D_0 is the *rate of phase transformation*, $D_0 = D - \mathbf{v}_i \cdot \mathbf{n}_i$.

2.2. Constitutive Relations for Phases of Ablative Composites

2.2.1. THE FOURIER LAW

The conservation equation set (2.2) consists of $18n$ scalar equations for $29n$ functions, where n is the number of phases. To close the system one should complement it with additional relations. One of them is the relationship connecting the heat flux vector $\mathbf{q}_i$ with the temperature gradient $\nabla\theta_i$ of phases:

$$\mathbf{q}_i = -\mathbf{k}_i \cdot \nabla\theta_i, \qquad i = a, b, p, g, l \tag{2.7}$$

and called *Fourier's law*.

The tensors $\mathbf{k}_i$ are the *heat conductivity tensors* of the phases.

2.2.2. GENERAL THERMODYNAMICAL IDENTITY

The derivation of the remaining additional relations is based on thermodynamic equations (equations (2.2) at $\zeta = 3$ and 4).

Let us introduce the *total derivative with respect to time* of an arbitrary function Ω_i defined within the domain V_i:

$$\frac{d_i\Omega_i}{dt} = \frac{\partial\Omega_i}{\partial t} + \mathbf{v}_i \cdot \nabla\Omega_i. \tag{2.8}$$

Let us also introduce *Almancy's finite strain tensor* $\mathbf{\Lambda}_i$ for solid phases $i = \{a, b, p, l\}$ connected to the strain gradient $\mathbf{F}_i$, the metric tensors components g^{ks}, $\overset{\circ}{g}{}^{ks}$ and the displacement vector $\mathbf{u}_i$ by the relationships

$$\mathbf{\Lambda}_i = \frac{1}{2}(g^{ks} - \overset{\circ}{g}{}^{ks})\mathbf{R}_k \otimes \mathbf{R}_s = \frac{1}{2}\left(\mathbf{E} - \mathbf{F}_i^{-1} \cdot \mathbf{F}_i^{-T}\right) =$$
$$= \frac{1}{2}\left(\nabla \otimes \mathbf{u}_i + (\nabla \otimes \mathbf{u}_i)^T - (\nabla_x \otimes \mathbf{u}_i)^T \cdot \nabla \otimes \mathbf{u}_i\right). \tag{2.9}$$

By taking account of equation (2.8), we can write the system (2.2) in the form

$$\rho_i \frac{d_i A_{i\zeta}}{dt} = \nabla B_{i\zeta} + C_{i\zeta}, \quad \zeta = 2, \ldots 5. \tag{2.10}$$

From the momentum equation (2.10) at $\zeta = 2$ we obtain the *equation of kinetic energy* for the phases

$$\rho_i \frac{d_i}{dt}\left(\frac{v_i^2}{2}\right) = \mathbf{v}_i \cdot \nabla \cdot \boldsymbol{\sigma}_i, \quad v_i^2 = \mathbf{v}_i \cdot \mathbf{v}_i. \tag{2.11}$$

The new function

$$\psi_i = e_i - \eta_i\theta_i \tag{2.12}$$

is called the *Helmholtz free energy* of the phase. From the energy equation (2.10) at $\zeta = 3$ and entropy balance at $\zeta = 4$, and equation (2.11), we derive the equation for ψ_i

$$\rho_i \frac{d_i\psi_i}{dt} + \rho_i\eta_i \frac{d_i\theta_i}{dt} - \boldsymbol{\sigma}_i \cdot\cdot \nabla \otimes \mathbf{v}_i + W_i^* = 0. \tag{2.13}$$

The stress tensor $\boldsymbol{\sigma}_i$ can be written in the form

$$\boldsymbol{\sigma}_i = \mathbf{T}_i + \mathbf{T}'_i, \tag{2.14}$$

where $\mathbf{T}_i$ is the tensor of *elastic stresses* and $\mathbf{T}'_i$ is the tensor of *viscous stresses*. The expression for power of elastic stresses of the ith phase $W_i = \mathbf{T}_i \cdot\cdot \nabla \otimes \mathbf{v}_i$ can be rewritten as follows by taking account of formulae (2.9) and (2.1) and the definition of total derivative (2.8):

$$W_i = \mathbf{T}_i \cdot\cdot \nabla \otimes \mathbf{v}_i = -\mathbf{T}_i \cdot\cdot \mathbf{F}_i \cdot \frac{d_i\mathbf{F}_i^{-1}}{dt} = \mathbf{F}_i^T \cdot \mathbf{T}_i \cdot \mathbf{F}_i \cdot\cdot \frac{d_i\mathbf{\Lambda}_i}{dt}. \tag{2.15}$$

On substituting equations (2.14) and (2.15) into the formula (2.13), we obtain the so-called *general thermodynamical identity* for the ith phase

$$\rho_i \frac{d_i\psi_i}{dt} + \rho_i\eta_i \frac{d_i\theta_i}{dt} - \mathbf{F}_i^T \cdot \mathbf{T}_i \cdot \mathbf{F}_i \cdot\cdot \frac{d\mathbf{\Lambda}_i}{dt} - \mathbf{T}'_i \cdot\cdot \nabla \otimes \mathbf{v}_i + W_i^* = 0. \tag{2.16}$$

2.2.3. NATURAL CONFIGURATIONS OF PHASES

Further analysis is based on an additional assumption concerning the form of the functions ψ_i and tensors $\mathbf{T}'_i$ of viscous stresses for the phases. First we establish the general structure of the dependence of ψ_i on its arguments.

We assume that for each ith phase there exists a certain configuration V_i^* in which both stresses and strains are absent and the free energy ψ_i has a minimum. This configuration V_i^* is called *natural.* If in the composite the phase transformation '$i \longrightarrow j$' takes place, then for each phase, the natural configuration is assumed to be coincident with the initial one $\overset{\circ}{V}_i$. For the jth phase formed as a result of the phase transformation the configurations V_j^* and $\overset{\circ}{V}_j$ do not usually coincide. Moreover, since after the phase transformation in the jth phase a stress-strain state immediately appears, the natural configuration V_j^* can be unrealizable.

A scheme of the phase transformation '$i \longrightarrow j$' can be presented in the following way (Figure 2.1). For the same point M with material coordinates X^k let the coordinates of the radius-vectors be $\overset{\circ}{\mathbf{x}}_i$, $\mathbf{x}_j$, $\mathbf{x}_j^*$, and infinitesimal elements of the radius-vectors $d\overset{\circ}{\mathbf{x}}_i$, $d\mathbf{x}_j$, $d\mathbf{x}_j^*$ in the configurations $\overset{\circ}{V}_i$, V_j and V_j^*, respectively. In each configuration we introduce the local basis vectors

$$\mathbf{R}_k = \partial \mathbf{x}_j/\partial X^k, \qquad \overset{\circ}{\mathbf{R}}_k = \partial \mathbf{x}_i/\partial X^k, \qquad \mathbf{R}_k^* = \partial \mathbf{x}_j^*/\partial X^k$$

and components of metric tensors

$$g_{sk} = \mathbf{R}_s \cdot \mathbf{R}_k, \qquad \overset{\circ}{g}_{sk} = \overset{\circ}{\mathbf{R}}_s \cdot \overset{\circ}{\mathbf{R}}_k, \qquad g_{sk}^* = \mathbf{R}_s^* \cdot \mathbf{R}_k^*, \qquad \mathbf{x}' \in V_j(t). \tag{2.17}$$

Transition of a local vicinity of the material point M of the ith phase from the initial configuration into the actual one is described by the strain gradient $\mathbf{F}_j$, and from V_j^* into V_j, by the strain gradient $\mathbf{F}_{j*}$:

$$d\mathbf{x}_j = \mathbf{F}_j \cdot d\overset{\circ}{\mathbf{x}}_i, \qquad d\mathbf{x}_j = \mathbf{F}_{j*} \cdot d\mathbf{x}_j^*. \tag{2.18}$$

We introduce a strain gradient $\mathbf{F}_j^*$ called the *phase transformation gradient* which transforms a local vicinity of the material point M from the natural configuration $\overset{\circ}{V}_i$ of the ith phase into the natural configuration V_j^* of the jth phase:

$$d\mathbf{x}_j^* = \mathbf{F}_j^* \cdot d\overset{\circ}{\mathbf{x}}_i. \tag{2.19}$$

As follows from formulae (2.18) and (2.19), the transformation of the initial configuration $\overset{\circ}{V}_i$ into the actual one V_j can be presented by the superposition

$$\mathbf{F}_j = \mathbf{F}_{j*} \cdot \mathbf{F}_j^*. \tag{2.20}$$

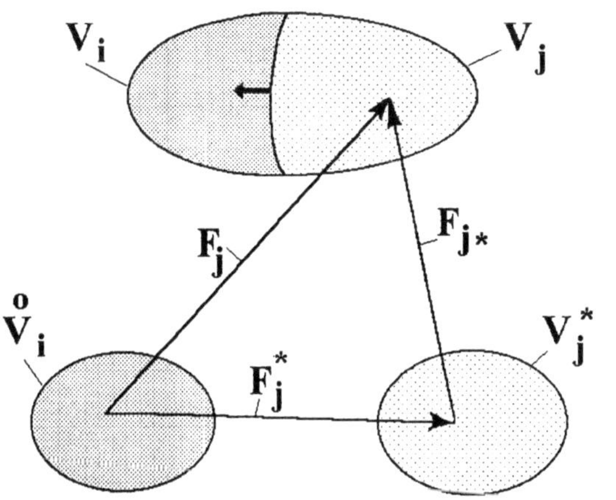

Figure 2.1. Schematic relation between natural and actual configurations of i-th and j-th phases

Introduce components of the inverse metric tensors and vectors of coupling bases in the configuration V_j^*

$$g^{*ks} g^*_{sr} = \delta^k_r, \qquad \mathbf{R}^{*k} = g^{*ks}\mathbf{R}^*_s. \tag{2.21}$$

Tensors $\mathbf{F}_{j*}$ and $\mathbf{F}^*_j$ are connected to the vectors of local bases by the relations

$$\mathbf{F}_{j*} = \mathbf{R}_k \otimes \mathbf{R}^{*k}, \qquad \mathbf{F}^*_j = \mathbf{R}^*_k \otimes \mathring{\mathbf{R}}^k, \qquad \mathbf{x}^k \in V_j. \tag{2.22}$$

Similarly to the formula (2.9), introduce Almancy's strain tensor $\mathbf{\Lambda}_{j*}$ describing a deformation of the jth phase at transition from the natural configuration V_j^* into the actual one V_j:

$$\mathbf{\Lambda}_{j*} = \frac{1}{2}(g^{*sk} - g^{sk})\mathbf{R}^*_s \cdot \mathbf{R}^*_k = \frac{1}{2}\left(\mathbf{E} - \mathbf{F}_{j*}{}^{-1} \cdot \mathbf{F}_{j*}{}^{-T}\right). \tag{2.23}$$

On expressing the tensor $\mathbf{F}_{j*}$ from (2.20) and substituting this into (2.23), we obtain the relationship between strain tensors $\mathbf{\Lambda}_j$ and $\mathbf{\Lambda}_{j*}$

$$\mathbf{\Lambda}_{j*} = \mathbf{F}^*_j \cdot \mathbf{\Lambda} \cdot \mathbf{F}^{*T}_j + \frac{1}{2}(\mathbf{E} - \mathbf{F}^*_j \cdot \mathbf{F}^{*T}_j). \tag{2.24}$$

The phase transformation gradient $\mathbf{F}^*_j$ involved in the equation (2.24) is a 'passport' characteristic of the phase transformation $i \longrightarrow j$ which should be given together with the model of the material. For the most actual problems the tensors $\mathbf{F}^*_j$ can be assumed to be constant and determined in independent specific experiments. Methods of definition of $\mathbf{F}_j^*$ will be given below. In fact, tensors $\mathbf{F}_j^*$ describe the appearance of shrinkage deformations at the phase transformations.

If the forming jth phase is gas ($j = g$), then one can also determine the natural configuration V_g^* of gas phase and the gradient of phase transformation $\mathbf{F}_g^*$ which has one independent component, being the density ρ_g^*:

$$\rho_g^* = \rho_g^0 \det(\mathbf{F}_g^{*-1}), \tag{2.25}$$

here det is the non-zero determinant of the tensor $\mathbf{F}_g^{*-1}$.

2.2.4. GENERAL FORM OF CONSTITUTIVE RELATIONS

Let us consider the form of the free energy ψ_i of phases. The free energy ψ_i is assumed to be a functional of Almancy's strain tensor $\mathbf{\Lambda}_{i*}$ and temperature θ_i with respect to time t:

$$\psi_i = \mathop{\psi_i}_{\tau=t_i^*}^{t} \left(\mathbf{\Lambda}_{i*}(\tau) - \overset{\circ}{\mathbf{\Lambda}}_i(\tau), \theta_i(\tau) \right), \qquad i = a, b, p, l, g, \tag{2.26}$$

where $\mathop{\psi_i}\limits_{\tau=t_i^*}^{t}$ means that the free energy ψ_i at time t depends on the prehistory of the tensor $\mathbf{\Lambda}_{i*}(\tau)$ and temperature $\theta_i(\tau)$ during the time period $t_i^* \leqslant \tau \leqslant t$; here $t_i^*(\mathbf{x})$ is the time of origin of a material particle of the ith phase with coordinate $\mathbf{x}$.

The choice of tensor $\mathbf{\Lambda}_{i*}$ as an argument in equation (2.25) under the condition $\psi_i(\mathbf{\Lambda}_{i*}, \theta_i) \geqslant \psi_i(0, \theta_0)$ ensures a minimum of the free energy in the natural configuration.

In the relationships (2.26) the following tensors are introduced

$$\overset{\circ}{\mathbf{\Lambda}}_i = \boldsymbol{\alpha}_i(\theta_i - \theta_i^*), \tag{2.27}$$

where $\boldsymbol{\alpha}_i$ are the *tensors of heat expansion*, θ_j^* is the temperature of the jth phase in the actual configuration at time $t_j^*(\mathbf{x})$ of its appearance: $\theta_j^* = \theta_j(t_j^*(\mathbf{x}), \mathbf{x})$.

The total derivative (2.8) of ψ_i with respect to time is determined in accordance with the rule of differentiation of functionals [22]:

$$\frac{d_i\psi}{dt} = \frac{\partial \psi_i}{\partial \mathbf{\Lambda}_i} \cdot\cdot \frac{d_i \mathbf{\Lambda}_i}{dt} + \frac{\partial \psi_i}{\partial \theta_i}\frac{d_i\theta_i}{dt} + \frac{\mathcal{D}\psi_i}{\mathcal{D}t}, \tag{2.28}$$

where $\mathcal{D}\psi_i/\mathcal{D}t$ is *Gateaux's derivative* [22], being a linear functional of prehistory of rates of change of the strain tensors $(d_i/d\tau)(\mathbf{\Lambda}_{i*} - \overset{\circ}{\mathbf{\Lambda}}_i)(\tau)$ and temperature $d_i\theta(\tau)/d\tau$.

On substituting the expression (2.28) into equation (2.16), from the condition that derivatives $(d_i \mathbf{\Lambda}_i/dt)$ and $(d_i\theta_i/dt)$ are independent, we derive the general thermodynamic identity (2.16) which is equivalent to the simultaneous relations

$$\mathbf{T}_i = \rho_i \mathbf{F}_i^{-T} \cdot \mathbf{F}_i^{*T} \cdot \frac{\partial \psi_i}{\partial \mathbf{\Lambda}_{i*}} \cdot \mathbf{F}_i^* \cdot \mathbf{F}_i^{-1}, \tag{2.29}$$

$$W_i^* = \mathbf{T}_i' : \nabla \otimes \mathbf{v}_i - \rho_i \frac{\mathcal{D}\psi_i}{\mathcal{D}t}, \tag{2.30}$$

$$\eta_i = -\frac{\partial \psi_i}{\partial \theta_i}. \tag{2.31}$$

called *constitutive relations for phases* of the composite.

The expression (2.12) for the internal energy of the ith phase e_i, on taking (2.31) into account, has the form

$$e_i = \psi_i - \theta_i \frac{\partial \psi_i}{\partial \theta_i}. \tag{2.32}$$

Formulae (2.23) and (2.24) connect tensors $\mathbf{\Lambda}_{i*}$ to $\mathbf{\Lambda}_i$ and $\mathbf{F}_i^*$ and hence to $\mathbf{F}_i^{-1}$ and $\mathbf{F}_i^*$; thus $\mathbf{T}_i$ in equation (2.29) depends on $\mathbf{F}_i^{-1}$ and $\mathbf{F}_i^*$.

In order to establish a specific form of the relations (2.29) – (2.31), one should define the following:

- the specific dependence (2.26) for the free energies ψ_i;
- the values of the phase transformation gradients $\mathbf{F}_i^*$;
- the specific expression for the viscous stress tensor depending on the velocity gradient: $\mathbf{T}_i'$ $(\nabla \otimes \mathbf{v}_i)$.

Below, for solid phases of the composite it is assumed that $\mathbf{T}_i' = 0$.

When there are no phase transformations,

$$\mathbf{F}_i^* = \mathbf{E}, \quad \mathbf{\Lambda}_i^* = \mathbf{\Lambda}_i, \quad \mathbf{\Lambda}_{i*} = 0, \quad t_i^* = 0, \quad \theta_i^* = \theta_0, \tag{2.33}$$

and constitutive relations (2.29), with account (2.26), are transformed as follows:

$$\mathbf{T}_i = \rho_i \mathbf{F}_i^{-T} \cdot \frac{\partial \psi}{\partial \mathbf{\Lambda}_i} \cdot \mathbf{F}_i^{-1}; \tag{2.34}$$

here $\mathbf{T}_i$ depend only on $\mathbf{\Lambda}_i$ and θ_i.

Gas phase $(i = g)$ of the composite is assumed to be *linear-viscous*, i.e. ψ_g and $\mathbf{T}_g'$ have the form

$$\psi_g = \psi_g(\rho_g, \rho_g^*, \theta_g), \tag{2.35}$$

$$\mathbf{T}_g' = \nu_g \mathbf{E} \nabla \cdot \mathbf{v}_g + \mu_g (\nabla \otimes \mathbf{v}_g + (\nabla \otimes \mathbf{v}_g)^T), \tag{2.36}$$

where ν_g, μ_g are the viscosity coefficients. The phase transformation gradient $\mathbf{F}_g^*$ for the gas phase has one independent component, the gas phase density ρ_g^* in its natural configuration introduced by formula (2.25).

Then constitutive relations (2.29) – (2.31) for the gas phase take the form

$$\mathbf{T}_g = -p\mathbf{E}, \quad p = \rho_g^2 \frac{\partial \psi_g}{\partial \rho_g}, \quad \eta_g = -\frac{\partial \psi_g}{\partial \theta_g}, \quad W_g^* = \mathbf{T}_g' \cdot\cdot \nabla \otimes \mathbf{v}_g. \tag{2.37}$$

Here it is taken into consideration that [74]

$$\partial \rho / \partial \mathbf{\Lambda} = -\rho \mathbf{F}^T \cdot \mathbf{F}, \tag{2.38}$$

and pressure of the gas phase is denoted by p.

Thus, constitutive relations for phases of the composite are defined completely by formulae (2.7), (2.29) – (2.32) and (2.35) – (2.37) connecting functions $\mathbf{q}_i$, W_i^*, η_i, e_i, $\boldsymbol{\sigma}_i$ to ρ_i, $\mathbf{v}_i$, $\mathbf{u}_i$, θ_i. Herein it is taken into account that strain gradients $\mathbf{F}_i^{-1}$ and strain tensors $\mathbf{\Lambda}_i$ are expressed in terms of displacement vectors $\mathbf{u}_i$ of phases by formulae (2.1), (2.9).

On substituting the constitutive relations (2.7), (2.29) – (2.32) and (2.35) – (2.37) into equations (2.2), we obtain the closed system of $8n$ scalar equations at $\zeta = 1, 2, 3, 5$ (equation at $\zeta = 4$ for entropy balance is excluded; the assumptions (2.26) and (2.35) for free energies of the phases make it equivalent to the constitutive relations) for $8n$ functions: ρ_i, $\mathbf{v}_i$, $\mathbf{u}_i$, θ_i, $i = a, b, p, g, l$, where $n = 5$ is the number of phases.

It should be noted that, for the gas phase, the kinematic equations (2.2) at $\zeta = 5$ for the displacement vector $\mathbf{u}_i$ can not be considered, thus, the number of equations and unknown functions in the set decreases by $3n_g$, where n_g is the number of gas phases in the multiphase medium (for the considered ablative composite $n_g = 1$).

2.3. Relations at the Phase Interface

2.3.1. MAIN EQUATIONS

Relations (2.5) at the phase interface Σ_{ij} between the ith and jth phases can be written in the explicit form

$$[\rho(D - \mathbf{v} \cdot \mathbf{n})] = 0, \tag{2.39}$$

$$\mathcal{M}[\mathbf{v}] + \mathbf{n} \cdot [\boldsymbol{\sigma}] = \mathbf{p}_\Sigma, \tag{2.40}$$

$$\mathcal{M}[e + \frac{v^2}{2}] + \mathbf{n} \cdot [\boldsymbol{\sigma} \cdot \mathbf{v} - \mathbf{q}] = E_\Sigma, \tag{2.41}$$

$$\mathcal{M}[\eta] - \mathbf{n} \cdot [\mathbf{q}/\theta] = H_\Sigma, \tag{2.42}$$

$$\mathcal{M}[\mathbf{u}] = \mathbf{u}_\Sigma, \tag{2.43}$$

$$\mathcal{M}[\mathbf{F}] + [\rho \mathbf{v} \otimes \mathbf{F}^T] \cdot \mathbf{n} = \mathbf{F}_\Sigma, \quad \mathbf{x} \in \Sigma_{ij}, \tag{2.44}$$

where $\mathcal{M}$ is the *mass rate of the phase transformation* connected to the rate D_0 by the relationship

$$\mathcal{M} = \rho_i(D - \mathbf{v}_i \cdot \mathbf{n}) = \rho_i D_0. \tag{2.45}$$

As the ith phase, for which the phase transformation rate D_0 is considered, a solid phase is usually chosen for phase transformation of the type 'solid phase $\leftrightarrows$ gas', and the 'old' solid phase is used for the phase transformation 'old solid phase $\leftrightarrows$ new solid phase'.

In addition, at the phase interface Σ_{ij} the following condition of continuity of the radius-vector is always satisfied:

$$[\mathbf{x}] = 0,$$

this means that there is no delamination between the phases.

The appearance of the functions $\mathbf{p}_\Sigma$, E_Σ in (2.40) and (2.41) determined on the phase interface is usually connected to the phenomenon of surface tension. In this case

$$\mathbf{p}_\Sigma = P_\Sigma \mathbf{n}, \qquad P_\Sigma = -\sigma_\Sigma \left(\frac{1}{R_1} + \frac{1}{R_2} \right), \tag{2.46}$$

where σ_Σ is the surface tension coefficient on the phase interface, and R_1, R_2 are the principal radii of curvature of the phase interface Σ_{ij} at the given point $\mathbf{x} \in \Sigma_{ij}$.

The surface tension energy E_Σ is determined by the formula

$$E_\Sigma = \mathbf{p}_\Sigma \cdot \{\mathbf{v}\}, \tag{2.47}$$

where $\{\Omega\}$ is the mean value of functions Ω_i and Ω_j at the phase interface:

$$\{\Omega\} = \frac{1}{2} \left(\Omega_i|_{\mathbf{x}\in\Sigma_{ij}} + \Omega_j|_{\mathbf{x}\in\Sigma_{ij}} \right). \tag{2.48}$$

Entropy production H_Σ at the phase transformation surface is always nonnegative $H_\Sigma \geqslant 0$.

We introduce a gradient of strain incompatibility $\mathbf{F}'_\Sigma$ and a vector of displacement break $\mathbf{u}'_\Sigma$ connected to $\mathbf{F}_\Sigma$ by the relationships

$$\mathbf{F}_\Sigma = (\mathcal{M}\mathbf{E} + \rho_j \mathbf{v}_j \otimes \mathbf{n}) \cdot \mathbf{F}'_\Sigma, \qquad \mathbf{u}_\Sigma = \mathcal{M}\mathbf{u}'_\Sigma. \tag{2.49}$$

2.3.2. CLASSIFICATION OF PHASE INTERFACES

The form of the surface functions H_Σ and $\mathbf{u}'_\Sigma$ does not generally follow from the general equations of thermomechanics, but is defined by the type of the phase interface.

A phase interface Σ_{ij} is called *nondissipative* if the following condition is satisfied:

$$H_\Sigma = 0, \qquad \mathbf{x} \in \Sigma_{ij}. \tag{2.50}$$

If the conditions

$$\mathbf{u}'_\Sigma = 0, \quad \mathbf{F}'_\Sigma = 0, \quad \mathbf{x} \in \Sigma_{ij}, \tag{2.51}$$

are satisfied, then Σ_{ij} is called the phase interface *with ideal contact.* At this surface two arbitrary neighboring material points on different sides of the surface remain neighboring for all deformations, and only the strain state in the vicinity of the points changes.

A phase interface *with slip* is a surface Σ_{ij} for which the following conditions are satisfied:

$$\mathbf{u}'_\Sigma = -[\overset{\circ}{\mathbf{x}}] \neq 0, \qquad \mathbf{F}'_\Sigma \neq 0, \quad \mathbf{x} \in \Sigma_{ij}. \tag{2.52}$$

For a phase interface with slip, neighboring points on different sides of the surface at time t_1 can displace by time t_2 along the surface Σ_{ij} without leaving the surface, i.e. delaminations between the phases are not permitted.

Surface Σ_{ij} is called the *surface of phase transformation* if

$$D_0 \neq 0.$$

Phase transformations with ideal contact at the surface are characteristic for transitions of the type: solid phase $\longrightarrow$ new solid phase, and also solid phase $\longrightarrow$ viscous gas (liquid).

Phase transformations with slip on their surfaces are, in particular, transitions: solid phase $\longrightarrow$ ideal gas (liquid). For surfaces with slip there appear additional unknown values, namely vector $\mathbf{u}'_\Sigma$ and tensor $\mathbf{F}'_\Sigma$. For case of the phase transformation: solid phase $\longrightarrow$ ideal gas values of $\mathbf{F}'_\Sigma$ and $\mathbf{u}'_\Sigma$ are determined from the relations (2.43) and (2.44), which are not involved in the problem statement for the determination of the main functions ρ_i, ρ_j, $\mathbf{v}_i$, $\mathbf{v}_j$, θ_i, θ_j and $\mathbf{u}_i$ for the set (2.2).

A phase interface Σ_{ij} is called *homothermal*, if there is no temperature jump on the surface:

$$[\theta] = 0, \qquad \mathbf{x} \in \Sigma_{ij}. \tag{2.53}$$

For most practical problems of phase transformations, including the problems for ablative composites, assumptions (2.53) and (2.50) on homothermicity and nondissipativity are justified, and allow us to describe adequately enough the actual processes of phase transformations.

2.3.3. CONSEQUENCES OF GENERAL EQUATIONS

Let us show several important consequences of relations (2.39) - (2.44) for a phase interface, which is assumed to be homothermal, nondissipative and with slip.

The scalar product of equation (2.40) and vector $\{\mathbf{v}\}$ is the analog of the equation of kinetic energy (2.11) for the phase interface:

$$\mathcal{M}[\frac{v^2}{2}] + \mathbf{n} \cdot [\boldsymbol{\sigma}] \cdot \{\mathbf{v}\} = \mathbf{p}_\Sigma \cdot \{\mathbf{v}\}. \tag{2.54}$$

From equations (2.41), (2.42), (2.50) and (2.53) we can derive the expression for the jump of the Helmholtz free energy ψ_i determined by formula (2.12)

$$\mathcal{M}[\psi] + \mathbf{n} \cdot \{\boldsymbol{\sigma}\} \cdot [\mathbf{v}] = E_\Sigma - \mathbf{p}_\Sigma \cdot \{\mathbf{v}\}. \tag{2.55}$$

We introduce new tensors $\mathbf{F}'_i$ and $\mathbf{F}'_j$ connected to $\mathbf{F}_i$ and $\mathbf{F}_j$ by the relations

$$\mathbf{F}'_i = \mathbf{F}_i, \qquad \mathbf{F}'_j = \mathbf{F}_j + \mathbf{F}'_\Sigma. \tag{2.56}$$

It is evident that $[\mathbf{F}'] = [\mathbf{F}] - \mathbf{F}'_\Sigma$.

Having multiplied scalarly equation (2.44) by normal vector $\mathbf{n}$ and substituted equations (2.45) and (2.39) into the product, we obtain the expression for the jump of the phase strain gradient

$$[\rho \mathbf{F}'^T] \cdot \mathbf{n} = 0. \tag{2.57}$$

On multiplying equation (2.44) on the right side by $\mathbf{F}_i^{-1} \cdot \mathbf{n}$ and using formula (2.57), we derive the expression for the jump of phase velocities through the phase interface

$$[\mathbf{v}] = -\frac{\mathcal{M}}{\rho_i}[\mathbf{F}'] \cdot \mathbf{F}_i^{-1} \cdot \mathbf{n}. \tag{2.58}$$

If relation (2.44) is multiplied on the left side by vector $\mathbf{F}_j'^{-1} \cdot \mathbf{n}$, then the formula (2.58) also follows, with index j replaced by i. On comparing this formula with equation (2.58), we find the relationship inverse to (2.57)

$$[\frac{\mathbf{F}'^{-1}}{\rho}] \cdot \mathbf{n} = 0. \tag{2.59}$$

On substituting equation (2.58) into formula (2.40), we obtain the expression for the jump of normal stress in the phases

$$\mathbf{n} \cdot [\boldsymbol{\sigma}] = \frac{\mathcal{M}^2}{\rho_i}[\mathbf{F}'] \cdot \mathbf{F}_i^{-1} \cdot \mathbf{n} + \mathbf{p}_\Sigma. \tag{2.60}$$

Then having substituted the expression for the velocity jump (2.58) and equation (2.47) for E_Σ into equation (2.55) and performed algebraic transformations of the type

$$\mathbf{n} \cdot \{\boldsymbol{\sigma}\} \cdot [\mathbf{F}'] \cdot \mathbf{F}_i^{-1} \cdot \mathbf{n} = \mathbf{n} \cdot [\boldsymbol{\sigma} \cdot \mathbf{F}'] \cdot \mathbf{F}_i^{-1} \cdot \mathbf{n} - \mathbf{n} \cdot [\boldsymbol{\sigma}] \cdot \{\mathbf{F}'\} \cdot \mathbf{F}_i^{-1} \cdot \mathbf{n},$$

we rewrite the relation (2.55) as follows:

$$[\psi] - \frac{1}{\rho_i}\mathbf{n} \cdot [\boldsymbol{\sigma} \cdot \mathbf{F}'] \cdot \mathbf{F}_i^{-1} \cdot \mathbf{n} + \frac{1}{\rho_i}\mathbf{n} \cdot [\boldsymbol{\sigma}] \cdot \{\mathbf{F}'\} \cdot \mathbf{F}_i^{-1} \cdot \mathbf{n} = 0. \tag{2.61}$$

On substituting equation (2.60) into relation (2.61) and taking the property (2.59) of tensor $\mathbf{F}'^{-1}_i/\rho$ into account, we find

$$[\psi] - \mathbf{n} \cdot \left[\frac{\boldsymbol{\sigma}}{\rho}\right] \cdot \mathbf{n} + \frac{\mathcal{M}^2}{2\rho_i}\mathbf{n} \cdot \mathbf{F}_i^{T-1} \cdot [\mathbf{F}^{T\prime} \cdot \mathbf{F}'] \cdot \mathbf{F}_i^{-1} \cdot \mathbf{n} + 2P_\Sigma\{\frac{1}{\rho}\} = 0. \tag{2.62}$$

2.3.4. TENSOR OF CHEMICAL POTENTIAL

We introduce two new tensors of the second rank for each αth phase $\alpha \in \{i,j\}$, where $i,j \in \{a,b,p,g,l\}$, namely the symmetric tensor $\boldsymbol{\chi}_\alpha$ called the *Bowen's tensor of chemical potential*, and the symmetric one $\mathbf{K}_\alpha$ called the *tensor of kinetic energy of phase transformation*:

$$\boldsymbol{\chi}_\alpha = \psi_\alpha \mathbf{E} - \frac{\boldsymbol{\sigma}_\alpha}{\rho_\alpha}, \quad \mathbf{K}_\alpha = \frac{\mathcal{M}^2}{2\rho_i}\mathbf{F}_i^{T-1} \cdot \mathbf{F}_\alpha^{T\prime} \cdot \mathbf{F}'_\alpha \cdot \mathbf{F}_i^{-1}, \tag{2.63}$$

$$\alpha \in \{i,j\}, \qquad i,j \in \{a,b,p,g,l\}.$$

Then the relation (2.62) takes the final form

$$\mathbf{n} \cdot [\boldsymbol{\chi} + \mathbf{K}] \cdot \mathbf{n} + 2P_\Sigma\{\frac{1}{\rho}\} = 0. \tag{2.64}$$

Tensors $\boldsymbol{\chi}_\alpha$ as well as internal energy, free energy etc. are thermodynamical characteristics of phases; they are independent of a phase geometry in explicit form and depend only on phase characteristics θ_i, $\mathbf{F}_i$, ρ_i etc.

The expression (2.63) for $\mathbf{K}_\alpha$ has been written for phase transformations with slip on their surface. For phase transformations without slip the relation (2.64) holds, but the expression for $\mathbf{K}_\alpha$ can be written in a simpler form

$$\mathbf{K}_\alpha = \frac{\mathcal{M}^2}{2\rho_i}\mathbf{F}_i^{T-1} \cdot \mathbf{F}_\alpha^{T} \cdot \mathbf{F}_\alpha \cdot \mathbf{F}_i^{-1}; \tag{2.63a}$$

as for $\boldsymbol{\chi}_\alpha$, this expression is independent of a phase geometry.

Relation (2.60) for surfaces without slip can be simplified as well

$$\mathbf{n} \cdot [\boldsymbol{\sigma}] = \frac{\mathcal{M}^2}{\rho_i}[\mathbf{F}] \cdot \mathbf{F}_i^{-1} \cdot \mathbf{n} + \mathbf{p}_\Sigma. \tag{2.60a}$$

2.4. Equation of Phase Transformation Rate

The phase transformation rate D_0 introduced in formula (2.4) as the rate of motion of the phase interface Σ_{ij} (in particular, of phases $i = b$ and $j = p$, and also $i = b$ and $j = g$) characterizes the rate of volumetric ablation of composite material according to the classification given in paragraph 1.2. This rate D_0 is not constant and depends on the temperature θ_i of the phases on their interface and, generally speaking, on stresses $\boldsymbol{\sigma}_i$ and $\boldsymbol{\sigma}_j$, densities ρ_i, ρ_j and possibly on other characteristics of the phases. This dependence

$$D_0 = D_0(\theta_i, \theta_j, \boldsymbol{\sigma}_i, \boldsymbol{\sigma}_j, \rho_i, \rho_j), \quad \mathbf{x} \in \Sigma_{ij} \tag{2.65}$$

as well as constitutive relations for phases (2.29) – (2.31), characterizes physical properties of the phase substance. The constitutive relations obey certain thermodynamical restrictions which are a consequence of the equation of entropy balance ($\zeta = 4$) from the set (2.2) and have a certain structure (2.29) – (2.31).

The phase transformation rate also satisfies some restrictions and cannot be given in an arbitrary way. These restrictions are a consequence of the equation (2.5) at $\zeta = 4$ for entropy balance on the phase interface.

As shown in paragraph 2.3, the relation (2.64) for chemical potential is a consequence of the equation of entropy balance (2.5) at $\zeta = 4$. We use the relation to derive an expression for the rate D_0.

Let us consider a phase interface without slip.

If the constitutive relations of phases (2.29) – (2.31) have been substituted into equations (2.64) for tensors $\boldsymbol{\chi}_\alpha$ and $\mathbf{K}_\alpha$, where functions ψ_i, ψ_j, $\mathbf{F}_i$, $\mathbf{F}_j$ have been expressed in terms of stresses $\boldsymbol{\sigma}_i$, $\boldsymbol{\sigma}_i$, temperature θ_i and density ρ_i, ρ_j of phases (one should invert the relations (2.29) and express $\mathbf{A}_i$ in terms of $\boldsymbol{\sigma}_i$ and then substitute them into (2.26)), and, in addition, $\mathcal{M}^2$ has been expressed from equation (2.60a) and substituted into (2.63), then the relation (2.64) takes the form of an equation connecting the functions $\boldsymbol{\sigma}_i$, $\boldsymbol{\sigma}_j$, ρ_i, ρ_j and θ_i of the phases

$$\Phi_n(\boldsymbol{\sigma}_i, \boldsymbol{\sigma}_j, \rho_i, \rho_j, \theta_i) = 0, \qquad \mathbf{x} \in \Sigma_{ij}. \tag{2.66}$$

Subscript n indicates that the equation depends on the geometry of the phase transformation surface.

The equation (2.66) can be used to express the normal stress of the jth phase on the phase interface $\sigma_{nj} = -\mathbf{n} \cdot \boldsymbol{\sigma}_j \cdot \mathbf{n}$ in terms of all the remaining arguments $\boldsymbol{\sigma}_i$, ρ_i, ρ_j, θ, and also of tangential stresses $\boldsymbol{\tau}_j$ on the phase interface determined as $\boldsymbol{\tau}_j = \boldsymbol{\sigma}_j - \sigma_{nj}\mathbf{n} \otimes \mathbf{n}$:

$$\sigma_{nj} = S_n(\boldsymbol{\sigma}_i, \boldsymbol{\tau}_j, \rho_i, \rho_j, \theta), \qquad \mathbf{x} \in \Sigma_{ij}. \tag{2.67}$$

In a similar way we determine σ_{ni} for the ith phase: $\sigma_{ni} = -\mathbf{n} \cdot \boldsymbol{\sigma}_i \cdot \mathbf{n}$. Having multiplied the equation (2.60) scalarly by vector $\mathbf{n}$ and solved the obtained equation for $\mathcal{M}^2$, and taken equation (2.67) into account, we get

$$\mathcal{M}^2 = \frac{\rho_i(S_n - \sigma_{ni} - P_\Sigma)}{\mathbf{n} \cdot [\mathbf{F}] \cdot \mathbf{F}_i^{-1} \cdot \mathbf{n}}. \tag{2.68}$$

This equation is the desired expression for the mass rate of the phase transformation. From this equation and equation (2.45) we can derive an equation of the type (2.65) for the linear rate of the phase transformation:

$$D_0 = \left(\frac{S_n - \sigma_{ni} - P_\Sigma}{\rho_i \mathbf{n} \cdot [\mathbf{F}] \cdot \mathbf{F}_i^{-1} \cdot \mathbf{n}}\right)^{1/2}. \tag{2.68a}$$

It should be noted, that the expression (2.68a) (or (2.68)) for phase transformation is not identical to the relation (2.40) (or (2.60)), as the normal stress σ_{nj} is replaced by the expression (2.67) obtained, in its turn, from the condition (2.42) for entropy balance on the phase interface. In fact, the equations (2.40) and (2.68a) are two different relations, the first of which is used for determination of a jump of normal stress of the phases, and the second is applied to define the phase transformation rate.

A specific expression for the function S_n and rate D_0 of phase transformations in ablative composite materials will be given below.

We now establish a connection between the location of the phase interface Σ_{ij} in the actual configuration, and the rate D_0. Let the shape of the smooth phase interface Σ_{ij} in the actual configuration be described by the equation

$$f_{ij}(\mathbf{x}, t) = 0. \tag{2.69}$$

On evaluating the total differential df_{ij} of the function, we obtain

$$df_{ij} = \frac{\partial f_{ij}}{\partial t} + D\mathbf{n} \cdot \nabla f_{ij} = 0.$$

If we use the definition of the normal vector $\mathbf{n}$ as a normalized gradient to the surface:

$$\mathbf{n} = \frac{\nabla f_{ij}}{|\nabla f_{ij}|}, \qquad |\nabla f_{ij}| = (\nabla f_{ij} \cdot \nabla f_{ij})^{1/2}, \tag{2.70}$$

and instead of the rate D its expression from (2.45), then we finally obtain

$$\frac{\partial f_{ij}}{\partial t} + \mathbf{v}_i \cdot \nabla f_{ij} + D_0 |\nabla f_{ij}| = 0, \qquad t > t_i^*. \tag{2.71}$$

This differential equation with the initial condition

$$t = t_j^*, \qquad f_{ij} = f_{ij}^0(\mathbf{x}) \tag{2.71a}$$

completely defines the location of the interface Σ_{ij} between the ith and jth phases in the actual configuration, if the phase transformation rate D_0 is known. Herein $f_{ij}^0(\mathbf{x})$ is the equation of the phase interface at the time of origin of the jth phase.

2.5. Infinitesimal Strains of Solid Phases

2.5.1. MAIN ASSUMPTIONS

Most composite materials are stiff, i.e. their maximum ultimate deformations do not exceed 1 - 3 %. For such materials an additional assumption on *infinitesimal strains* is admissible, which can be formulated as follows: typical values of the strain gradients $\mathbf{F}_i$ for all solid phases and also phase transformation gradients $\mathbf{F}_i^*$ are close to unity:

$$\Delta\mathbf{F}_i = \mathbf{E} - \mathbf{F}_i, \qquad \| \Delta\mathbf{F}_i \| \ll 1, \qquad i = a, b, p, l,$$

$$\boldsymbol{\beta}_i = \mathbf{E} - \mathbf{F}_i^*, \qquad \| \boldsymbol{\beta}_i \| \ll 1, \tag{2.72}$$

where $\| \Delta\mathbf{F}_i \|$ is the typical value of components of tensor $\Delta\mathbf{F}_i$.

From equations (2.72) we deduce the following:

a) in all equations of the general system (2.2) for the solid phases convective terms $\nabla \cdot \rho_i \mathbf{v}_i A_{ik}$ can be neglected as compared with the terms $\partial \rho_i A_{ik}/\partial t$ and $\nabla \cdot B_{ik}$; the form of the equations will be given below;

b) initial $\overset{\circ}{V}_i$ and actual V_i configurations for all solid phases are indistinguishable;

c) strain tensors $\mathbf{A}_i$ of phases depend linearly on the displacement gradients $\nabla \otimes \mathbf{u}_i$:

$$\mathbf{\Lambda}_i = \boldsymbol{\varepsilon}_i \equiv \frac{1}{2}\left(\nabla \otimes \mathbf{u}_i + (\nabla \otimes \mathbf{u}_i)^T\right), \tag{2.73}$$

d) strain tensors $\mathbf{\Lambda}_{i*}$ of phases depend linearly on $\nabla \otimes \mathbf{u}_i$ and $\boldsymbol{\beta}_i$:

$$\mathbf{\Lambda}_{i*} = \boldsymbol{\varepsilon}_i + \boldsymbol{\beta}_i. \tag{2.74}$$

The tensors $\boldsymbol{\beta}_j$ are called the *tensors of chemical shrinkage*, they characterize the deformation of the vicinity of the material point due to its phase transformation (when there is no mechanical load). For most ablative composites this deformation leads to a decrease in the volume of the material (shrinkage), which can be determined in experiments. The components of tensor $\boldsymbol{\beta}_j$ are evaluated by experimental values of composite shrinkage. This problem will be considered in detail below. When phase transformations are absent

$$\boldsymbol{\beta}_i = 0, \tag{2.75}$$

e) strain gradients $\mathbf{F}_i$ of solid phases can be assumed to be coincident with the unit tensor:

$$\mathbf{F}_i = \mathbf{E}. \tag{2.76}$$

2.5.2. CONSTITUTIVE RELATIONS

Solid Phases

Equations (2.26) and (2.29) show that for the case of infinitesimal strains the free energy ψ_i and stress tensor $\boldsymbol{\sigma}_i$ of solid phases depend on $\boldsymbol{\varepsilon}_i$ and θ_i.

The general thermodynamical identity (2.16) takes the form

$$\rho_i \frac{d\psi_i}{dt} - \boldsymbol{\sigma}_i \cdot \cdot \frac{d\boldsymbol{\varepsilon}_i}{dt} + \rho_i \eta_i \frac{d\theta_i}{dt} + W_i^* = 0, \quad \mathbf{x} \in V_i, \tag{2.77}$$

and the expression (2.26) for free energy ψ_i is rewritten as:

$$\psi_i = \psi_i^0 + \psi_i^\theta(\theta_i) + \underset{\tau=0}{\overset{t}{\psi}}{}_i^{\varepsilon}(\boldsymbol{\varepsilon}_i - \overset{\circ}{\boldsymbol{\varepsilon}}_i, \theta_i), \tag{2.78}$$

where ψ_i^θ depends only on current temperature $\theta_i(t)$ and $\overset{t}{\underset{\tau=0}{\psi}}{}_i^\varepsilon$ – only on the prehistory of changing temperature $\theta_i(\tau)$, $0 \leqslant \tau < t$ and strains $(\boldsymbol{\varepsilon}_i - \overset{\circ}{\boldsymbol{\varepsilon}}_i)(\tau)$, $0 \leqslant \tau \leqslant t$. The tensor $\overset{\circ}{\boldsymbol{\varepsilon}}_i = \overset{\circ}{\boldsymbol{\Lambda}}_i - \boldsymbol{\beta}_i$, introduced in equation (2.78), is called the *tensor of heat deformations* and in accordance with formula (2.27) can be written in the form

$$\overset{\circ}{\boldsymbol{\varepsilon}}_i = \boldsymbol{\alpha}_i(\theta_i - \theta_i^*) - \boldsymbol{\beta}_i, \tag{2.79}$$

The constitutive relations (2.29) – (2.31) for the case of infinitesimal strains take the form

$$\boldsymbol{\sigma}_i = \mathcal{F}_i(\boldsymbol{\varepsilon}_i, \theta_i) = \rho_i \frac{\partial \psi_i}{\partial \boldsymbol{\varepsilon}_i},$$

$$\eta_i = -\frac{\partial \psi_i}{\partial \theta_i}, \qquad W_i^* = -\rho_i \frac{\mathcal{D}\psi_i}{\mathcal{D}t}. \tag{2.80}$$

If solid phases are *linear-elastic* media, then the free energy ψ_i is a quadratic function of $\boldsymbol{\varepsilon}_i$ [107]:

$$\psi_i = \psi_i^0 + \psi_i^\theta + \overset{t}{\underset{\tau=0}{\psi}}{}_i^\varepsilon,$$

$$\psi_i^\theta = \int_{\theta_0}^{\theta_i} c_i d\theta_i - \theta_i \int_{\theta_0}^{\theta_i} \frac{c_i}{\theta_i} d\theta_i, \qquad \psi_i^0 = e_i^0 - \theta_i \eta_i^0, \tag{2.81}$$

$$\overset{t}{\underset{\tau=0}{\psi}}{}_i^\varepsilon = \frac{1}{2\rho_i}\left(\boldsymbol{\varepsilon}_i - \overset{\circ}{\boldsymbol{\varepsilon}}_i\right) \cdot\cdot \mathbf{C}_i \cdot\cdot \left(\boldsymbol{\varepsilon}_i - \overset{\circ}{\boldsymbol{\varepsilon}}_i\right),$$

where c_i are the *heat capacities of the phases.*

The expression (2.32) for the internal energy e_i of phases is now written as follows:

$$e_i = e_i^0 + \int_{\theta_0}^{\theta_i} c_i d\theta_i + \overset{t_\varepsilon}{\underset{\tau=0}{\psi}}{}_i + \frac{1}{\rho_i}\boldsymbol{\alpha}_i \cdot\cdot \boldsymbol{\sigma}_i \theta_i, \tag{2.82}$$

and the constitutive relations (2.80) take the simple form

$$\boldsymbol{\sigma}_i = \mathbf{C}_i \cdot\cdot (\boldsymbol{\varepsilon}_i - \overset{\circ}{\boldsymbol{\varepsilon}}_i), \qquad W_i^* = 0, \tag{2.83}$$

$$\eta_i = \eta_i^0 + \int_{\theta_0}^{\theta_i} \frac{c_i}{\theta_i} d\theta_i + \frac{1}{\rho_i}\boldsymbol{\alpha}_i \cdot\cdot \boldsymbol{\sigma}_i,$$

where e_i^0, η_i^0 = *const* are the initial values of internal energy and entropy of the ith phase.

The tensor $\mathbf{C}_i$ is called the *tensor of elastic moduli* of the solid phases.

Relations inverse to equations (2.83) have the form

$$\boldsymbol{\varepsilon}_i = \overset{\circ}{\boldsymbol{\varepsilon}}_i + \mathbf{\Pi}_i \cdot\cdot \boldsymbol{\sigma}_i, \tag{2.84}$$

where the tensor $\mathbf{\Pi}_i$, called the *tensor of elastic pliabilities*, is inverse to $\mathbf{C}_i$: $\mathbf{C}_i \cdot \cdot \mathbf{\Pi}_i = \mathbf{\Delta}$, here $\mathbf{\Delta}$ is the unit tensor of the fourth rank [91].

As a rule, phases of composite material can be considered as *isotropic* media. In this case, equations (2.83) and Fourier's law (2.7) can be simplified:

$$\boldsymbol{\sigma}_i = \left(\lambda_i \varepsilon_i - (3\lambda_i + 2\mu_i)\overset{\circ}{\varepsilon}_i\right) \mathbf{E} + 2\mu_i \boldsymbol{\varepsilon}_i, \tag{2.85}$$

$$\mathbf{q}_i = -k_i \nabla \theta_i, \qquad i = a, b, p, l, \tag{2.86}$$

where λ_i, μ_i are *Lamé's coefficients* of the ith phase, k_i is the heat conductivity coefficient of the ith phase and $\overset{\circ}{\varepsilon}_i$ is the heat deformation:

$$\overset{\circ}{\varepsilon}_i = \alpha_i(\theta - \theta_i^*) - \beta_i, \qquad i = a, b, p, l, \tag{2.87}$$

$$\beta_a = \beta_b = \beta_l = 0, \qquad \beta_p \neq 0.$$

Here α_i is the *heat expansion coefficient*, β_i is the *chemical shrinkage coefficient* of the ith phase. The equation (2.84) for an isotropic medium has the form

$$\boldsymbol{\varepsilon}_i = \overset{\circ}{\varepsilon}_i \mathbf{E} + \frac{1 + \nu_i}{E_i} \boldsymbol{\sigma}_i - \frac{\nu_i}{E_i} \sigma_i \mathbf{E}, \tag{2.84a}$$

where $\sigma_i = \boldsymbol{\sigma}_i \cdot \cdot \mathbf{E}$, $\varepsilon_i = \boldsymbol{\varepsilon}_i \cdot \cdot \mathbf{E}$ are the *first invariants* of tensors $\boldsymbol{\sigma}_i$ and $\boldsymbol{\varepsilon}_i$, ν_i and E_i are the *Poisson's ratios* and elastic moduli connected to λ_i and μ_i by the relations

$$\mu_i = \frac{E_i}{2(1 + \nu_i)}, \qquad \lambda_i = \frac{\nu_i E_i}{(1 + \nu_i)(1 - 2\nu_i)}.$$

Gas Phase

Gas phase is assumed below to be *perfect*, linear-viscous gas. Then the expression for free energy of the gas phase ψ_g (2.35) takes the form [107]

$$\psi_g = \psi_g^0 + \int_{\theta_0}^{\theta_g} c_g d\theta_g - \theta_g \int_{\theta_0}^{\theta_g} \frac{c_g}{\theta_g} d\theta_g - R\theta_g \ln \frac{\rho_g^*}{\rho_g}, \tag{2.88}$$

where $\psi_g^0 = e_g^0 - \eta_g^0 \theta_g$, and the relations (2.37) are written as follows:

$$\boldsymbol{\sigma}_g = -p\mathbf{E} + \mathbf{T}'_g,$$

$$p = R\rho_g \theta_g,$$

$$\mathbf{T}'_g = \nu_g \mathbf{E} \nabla \cdot \mathbf{v}_g + \mu_g (\nabla \otimes \mathbf{v}_g + (\nabla \otimes \mathbf{v}_g)^T),$$

$$e_g = e_g^0 + \int_{\theta_0}^{\theta_g} c_g d\theta_g, \tag{2.89}$$

$$\eta_g = \eta_g^0 + \int_{\theta_0}^{\theta_g} \frac{\bar{c}_g}{\theta_g} d\theta_g - R \ln \frac{p}{p^*}.$$

Here R is the *gas constant*, c_g is the heat capacity of the gas phase at constant volume, $\bar{c}_g = R + c_g$ is the heat capacity at constant pressure and ρ_g^*, $p^* = \rho_g^* R\theta_0$ are the density and pressure of the gas phase in the natural configuration under temperature θ_0.

2.5.3. QUASISTATIC PROCESSES

For current heat-loaded structures the most widespread case of high-temperature actions is the case of *quasistatic processes* of heating (see Chapter 1). The assumption on quasistaticity means that all processes in the composite propagate relatively slowly in time without dynamical (inertia) phenomena. Therefore:

- in the momentum equations (2.2) at $\zeta = 2$ for all n phases the inertial terms $\partial \rho_i \mathbf{v}_i / \partial t$ and $\nabla \rho_i \mathbf{v}_i \otimes \mathbf{v}_i$ can be neglected when compared with $\nabla \cdot \boldsymbol{\sigma}_i$;
- in the energy equations (2.2) at $\zeta = 3$ for all phases the kinetic energy of phase motion $\rho_i v_i^2/2$ can be neglected when compared with internal energy of phases $\rho_i e_i$;
- in the equations on the phase interface (2.5) at $\zeta = 2$ and $\zeta = 3$ analogous terms $\rho(D - \mathbf{v}_i \cdot \mathbf{n})\mathbf{v}_i$ and $\rho(D - \mathbf{v}_i \cdot \mathbf{n})v_i^2/2$ can be neglected when compared with $\boldsymbol{\sigma}_i \cdot \mathbf{n}$ and $\mathbf{n} \cdot \boldsymbol{\sigma}_i \cdot \mathbf{v}_i$, respectively.

2.5.4. CONSERVATION EQUATIONS

Solid Phases

The assumptions on quasistaticity of motion processes and smallness of strains of solid phases allow us to considerably simplify the equation system (2.2). For the solid phases

$$\frac{\partial \rho_i}{\partial t} = 0, \tag{2.90}$$

$$\nabla \cdot \boldsymbol{\sigma}_i = \mathbf{0}, \tag{2.91}$$

$$\rho_i \frac{\partial e_i}{\partial t} = -\nabla \cdot \mathbf{q}_i + \boldsymbol{\sigma}_i \cdot\cdot \frac{\partial \boldsymbol{\varepsilon}_i}{\partial t}, \tag{2.92}$$

$$\boldsymbol{\varepsilon}_i = \frac{1}{2}\left(\nabla \otimes \mathbf{u}_i + (\nabla \otimes \mathbf{u}_i)^T\right), \quad \mathbf{x} \in V_i, \quad i = a, b, p, l. \tag{2.93}$$

From equations (2.92), (2.82) and (2.7) one can derive the *equation of heat conductivity* for the solid phases

$$\rho_i c_i \frac{\partial \theta_i}{\partial t} = \nabla \cdot (\mathbf{k}_i \cdot \nabla \theta_i) - \boldsymbol{\alpha}_i \theta_i \cdot\cdot \frac{\partial \boldsymbol{\sigma}_i}{\partial t} + W_i^*, \quad \mathbf{x} \in V_i. \tag{2.92a}$$

Gas phase

The assumption on quasistaticity of motion processes allows to write the equation system (2.2) for gas phase in the form

$$\frac{\partial \rho_g}{\partial t} + \nabla \cdot \rho_g \mathbf{v}_g = 0, \tag{2.94}$$

$$\nabla \cdot \boldsymbol{\sigma}_g = \mathbf{0}, \tag{2.95}$$

$$\frac{\partial \rho_g e_g}{\partial t} + \nabla \cdot \rho_g \mathbf{v}_g e_g = -\nabla \cdot \mathbf{q}_g + \nabla \cdot (\boldsymbol{\sigma}_g \cdot \mathbf{v}_g), \quad \mathbf{x} \in V_g. \tag{2.96}$$

Heat conduction equation for the gas phase follows from equations (2.96), (2.89) and (2.94) at $c_g = \text{const}$:

$$\frac{\partial}{\partial t} \rho_g c_g \theta_g + \nabla \cdot \rho_g \mathbf{v}_g c_g \theta_g = \nabla \cdot (k_g \nabla \theta) + \nabla \cdot (\boldsymbol{\sigma}_g \cdot \mathbf{v}_g), \quad \mathbf{x} \in V_g. \tag{2.96a}$$

2.5.5. CONDITIONS ON THE PHASE INTERFACE

For of infinitesimal strains and quasistatic processes the conditions (2.39) – (2.41) on the nondissipative homothermal phase interface without slip are transformed in the following way.

On the surface of contacting gas and ith solid phase there are relations for the jumps of functions:

$$[\rho] D_0 + \rho_g \mathbf{v}_g \cdot \mathbf{n} = \mathbf{0}, \qquad [\boldsymbol{\sigma}] \cdot \mathbf{n} = \mathbf{0}, \tag{2.97}$$

$$\rho_i D_0 [e] - \mathbf{n} \cdot \boldsymbol{\sigma}_g \cdot \mathbf{v}_g + \mathbf{n} \cdot [\mathbf{k} \cdot \nabla \theta] = 0,$$

$$[\theta] = 0,$$

$$\mathbf{v}_g \cdot \mathbf{b}_I = 0, \quad \mathbf{x} \in \Sigma_{ig}.$$

For nonreacting phases (for example, gas and pyrolysis residue $i = p$ or gas and filler $i = a$) in the conditions (2.97) the phase transformation rate is $D_0 = 0$.

In addition, *adhesion conditions* are added to the first four conditions (2.97) for viscous gases, where $\mathbf{b}_I$ are the vectors in the surface tangent to the phase interface, which are orthogonal one to another and to the normal vector $\mathbf{n}$, i.e. $\mathbf{n} \cdot \mathbf{b}_I = 0$, $I = 1, 2$.

The conditions are the consequence of compatibility relations (2.44). Really, from (2.44) the equations (2.57) and (2.59) follow, which for the case of infinitesimal strains and the absence of slip can be written in the form

$$\mathbf{v}_g = -\frac{\mathcal{M}}{\rho_i}(\mathbf{F}_g \cdot \mathbf{n} - \mathbf{n}) = -D_0 \left(\frac{\rho_i}{\rho_g} - 1 \right) \mathbf{n}, \quad \mathbf{F}_g \cdot \mathbf{n} = \frac{\rho_i}{\rho_g} \mathbf{n}.$$

Multiplying the first equation scalarly by $\mathbf{b}_I$, we derive the condition of adhering (2.97).

At the interface of the ith and jth solid phases the conditions (2.39) – (2.43) under these assumptions for the interface take the form

$$[\mathbf{u}] = 0, \qquad [\theta] = 0, \qquad \mathbf{x} \in \Sigma_{ij}, \tag{2.98}$$

$$\mathbf{n} \cdot [\boldsymbol{\sigma}] = 0, \qquad \rho_i D_0 [e] + \mathbf{n} \cdot [\mathbf{k} \cdot \nabla \theta] = 0.$$

For most solid bodies

$$(e_i - e_i^0 - c_i \theta_i) \ll (e_i^0 + c_i \theta_i),$$

so that we may write $[e] = [e^0 + c\theta]$.

To describe the surface ablation one needs the conditions on the phase transformation surface Σ_{ge} for two ideal perfect gases:

$$\mathbf{n} \cdot [\rho \mathbf{v}] = 0, \qquad [p] = 0, \qquad \rho_g \mathbf{n} \cdot \mathbf{v}_g [I] - \mathbf{n} \cdot [\mathbf{k} \nabla \theta] = 0. \tag{2.99}$$

Here $[\Omega] = \Omega_g - \Omega_e$ and subscript e corresponds to the external gas medium contacting the gas phase of the composite; speed of the gas phase is much higher than the speed D_0 of the surface motion; $I_g = e_g + (p/\rho_g)$ and $I_e = e_e + (p_e/\rho_e)$ are the enthalpies of the gas media which in accordance with equations (2.88), (2.89) are expressed by formulae (1.6) and

$$I_g = I_g^0 + \int_0^{\theta_g} \bar{c}_g d\theta = I_g^0 + \bar{I}_g, \qquad I_e = I_g^0 + \bar{I}_e, \qquad I_g^0 = e_g^0 + \int_0^{\theta_0} c_g d\theta. \tag{2.100}$$

The conditions (2.99) follow from the general relations (2.39) – (2.41) of the phase interface in the framework of quasistatic approximation (i.e. when inertial terms are neglected).

2.5.6. A RATE OF THE PHASE TRANSFORMATION

Let us find an expression for the rate D_0 of phase transformation $i \longrightarrow j$ when the ith phase is solid with infinitesimal strains (for example, polymer resin $i = b$), and the jth phase is a perfect, linear-viscous gas (for example, gaseous pyrolysis products $j = g$).

Expressions (2.63) for chemical potential tensors $\boldsymbol{\chi}_i$ of solid and gas phases for this case have the form

$$\boldsymbol{\chi}_i = (\bar{\mu}_i^0 + \psi_i^\theta(\theta_i) + \underset{\tau=0}{\overset{t}{\psi}}{}_i^{\varepsilon})\mathbf{E} - \frac{\boldsymbol{\sigma}_i}{\rho_i}, \tag{2.101}$$

$$\boldsymbol{\chi}_g = (\bar{\mu}_g^0 + \bar{\psi}_g^\theta(\theta_g) + R\theta_g \ln \frac{p}{p^*})\mathbf{E} - \frac{\mathbf{T}_g'}{\rho_g},$$

where

$$\bar{\psi}_g^\theta(\theta_g) = \int_{\theta_0}^{\theta_g} \bar{c}_g d\theta_g - \theta_g \int_{\theta_0}^{\theta_g} \frac{\bar{c}_g}{\theta_g} d\theta_g, \tag{2.102}$$

$$\bar{\mu}_i^0 = \psi_i^0, \qquad \bar{\mu}_g^0 = \psi_g^0 + R\theta_0.$$

In accordance with (2.76), the strain gradient $\mathbf{F}_b$ of the solid phase coincides with the unit tensor $\mathbf{E}$; then from the relation (2.57) we can find the expression for the gradient $\mathbf{F}_g$ of the gas phase on the phase interface Σ_{bg}

$$\mathbf{n} \cdot \mathbf{F}_g \cdot \mathbf{n} = \frac{\rho_b}{\rho_g}. \tag{2.103}$$

For quasistatic processes, the tensors $\mathbf{K}_i$ and $\mathbf{K}_g$ can be neglected when compared with the chemical potential tensors. The surface tension P_Σ can also be neglected, as its influence is usually small for volumetric ablation. Under the assumptions from formulae (2.64) and (2.101) we derive the following equation of type (2.66)

$$\Delta\mu^0 + \Delta\psi^\theta + \Delta\psi_\sigma + R\theta_g \ln \frac{p}{p^*} = 0, \tag{2.104}$$

where the following notations are introduced: $\Delta\mu^0 = \bar{\mu}_g^0 - \bar{\mu}_i^0$, $\Delta\psi^\theta = \bar{\psi}_g^\theta - \psi_i^\theta$, $\Delta\psi_\sigma = \psi_i^\varepsilon + (\sigma_{ni}/\rho_i) + (T'_{gn}/\rho_g)$, $T'_{gn} = \mathbf{n} \cdot T'_g \cdot \mathbf{n}$; $\sigma_{ni} = -\mathbf{n} \cdot \boldsymbol{\sigma}_i \cdot \mathbf{n}$.

On resolving the equation (2.104) for p, we obtain a relation of type (2.67)

$$\frac{p}{p^*} = \bar{S}_n \equiv \exp\left(-\frac{\Delta\mu^0 + \Delta\psi^\theta + \Delta\psi_\sigma}{R\theta_g}\right). \tag{2.105}$$

On expressing density ρ_g in terms of gas pressure p, we can rewrite the formula (2.103) for the strain gradient $\mathbf{F}_g$ in the form

$$\mathbf{n} \cdot \mathbf{F}_g \cdot \mathbf{n} = \frac{\rho_b R\theta_g}{\bar{S}_n p^*}. \tag{2.106}$$

On substituting the expression $\mathbf{F}_i = \mathbf{E}$ and (2.106) into (2.68), one can find the expression for the rate D_0 of phase transformation '$i \longrightarrow j$'

$$D_0 = \left(\frac{p^*\bar{S}_n - \sigma_{ni}}{\rho_i\left(1 - \dfrac{\rho_i R\theta_g}{p^*\bar{S}_n}\right)}\right)^{1/2}. \tag{2.107}$$

For gas-phase transformations the following conditions are usually satisfied: $\Delta\psi^0 \gg \Delta\psi^\theta$, $\Delta\psi^0 \gg \Delta\psi_\sigma$, $p^*/(\rho_i R\theta_g) \ll 1$, and also $\bar{S}_n \leqslant 1$. Moreover, one can assume up to a sufficient accuracy that the hydrostatic stress in the solid phase σ_{ni} is equal to the external pressure p_e acting onto the whole composite material, and the pressure p^* is close to the current pressure of the gas phase, i.e.

$$\sigma_{ni} \approx p_e, \qquad p^* \approx p. \tag{2.108}$$

Under the assumptions, one can obtain the final expression for rate D_0 of phase transformation '$i \longrightarrow j$' from equation (2.107):

$$D_0 = D^0\left(\frac{p}{p_e}\right)^{1/2} \exp\left(-\frac{E_A}{R\theta_g}\right)\left(1 - \frac{p}{p_e}\exp\left(-\frac{2E_A}{R\theta_g}\right)\right)^{1/2}, \tag{2.109}$$

$$D^0 = \frac{p_e}{\rho_i (R\theta_g)^{1/2}}, \qquad E_A = \frac{\Delta\mu^0}{2},$$

where E_A is the so-called activation energy of the phase transformation.

The expression (2.109) gives the phase transformation rate D_0 as a function of temperature θ_g and pressure p of the gas phase.

The pressure p appearing in ablative composites usually satisfies the condition

$$\frac{p}{p_e} \exp\left(-\frac{2E_A}{R\theta_g}\right) \ll 1. \tag{2.110}$$

Then the expression (2.109) for the rate D_0 takes the simple form

$$D_0 = D^0 \exp\left(-\frac{E_A}{R\theta_g}\right). \tag{2.111}$$

The dependence of the phase transformation rate upon temperature θ_g in the form of $\exp\left(-E_A/(R\theta_g)\right)$ is called *Arrhenius dependence.* As follows from the above conclusion, dependence (2.111) of Arrhenius type takes place only if condition (2.110) is satisfied, i.e. at a relatively low level of temperature θ_g and pressure p. For higher θ_g and p formula (2.109) should be used.

Under infinitesimal strains the equation (2.71) describing a location of the phase interface Σ_{ij} takes the form

$$\frac{\partial f_{ij}}{\partial t} + D_0 |\nabla f_{ij}| = 0. \tag{2.112}$$

CHAPTER 3

MATHEMATICAL MODEL OF ABLATIVE COMPOSITES

The present chapter describes an averaging method for internal thermal and mechanical processes in ablative composites. The method is based on the assumption of a regularity in the internal structure of the ablative composite, and the concept of asymptotic expansion for partial differential equations with rapidly oscillating coefficients. This theory was formulated by N.S.Bakhvalov [8] and developed by B.E.Pobedria [91, 92], D.Lions [73] and others for composites without phase transformations, and by E.Sanchez-Palencia [104] for porous media without phase transformations. For heterogeneous media with phase transformations the averaging method has been developed in works [39, 41].

3.1. Main Assumptions

The equations in paragraph 2.5 were written for an arbitrary multiphase system. We now identify four possible phases: the amorphic phase of reinforcing filler, the polymer phase, the pyrolysis phase and the phase in which pores are filled with gas. We denote their volumes by V_i, where $i = a$ (amorphic), b (polymer), p (pyrolysis) and g (gas), respectively. Solving the equations for arbitrary configuration of the V_i (Figure 3.1) is complicated, but the procedure can be greatly simplified by assuming that the composite has a regular structure.

We make a number of assumptions.

a) Deformations of solid phases of the composite are small.

b) Ablative composite occupying a volume V in the Cartesian system $O\mathbf{x}$ at any time t has a *periodic structure*, i.e. there is an elementary recurring volume V_ξ, called the *periodicity cell*, and the whole volume V consists of $N \gg 1$ of these periodicity cells: $V = NV_\xi$.

c) The periodicity cell V_ξ at any time t consists of four phases $V_{\xi i}$, i.e. $V_\xi = \bigcup_\xi V_{\xi i}$, $i = a, b, p, g$, with non-zero volumes, $|V_{\xi i}| > 0$, and $V_{\xi i} = V_i \cap V_\xi$.

d) At each time $t \geqslant 0$ domains V_i and $V_{\xi i}$ are simply-connected (i.e. the porosity of the medium is 'open'); a collection of solid phases V_i $i = a, b, p$ forms a solid frame $V_s = V_a \cup V_b \cup V_p$ that occupies a volume $V_{\xi s} = V_s \cap V_\xi$ inside the periodicity cell.

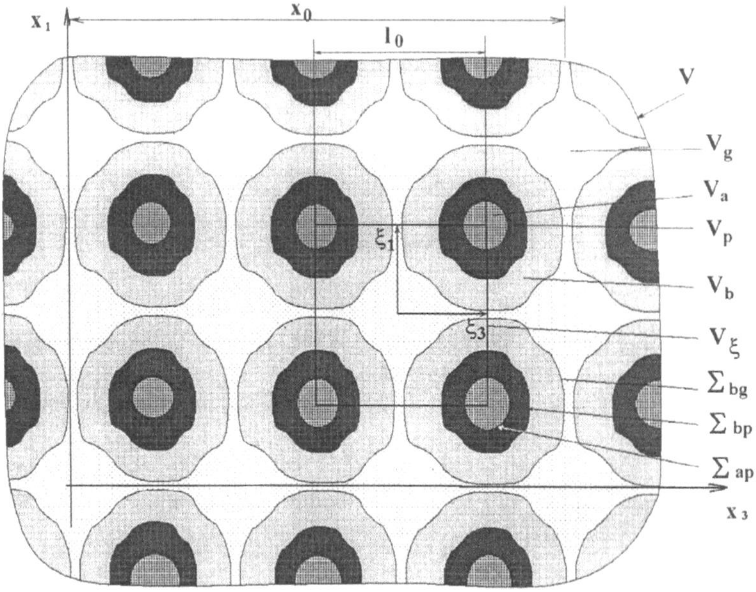

Figure 3.1. A scheme of a periodic structure of porous ablative material

e) Phases separation boundaries Σ_{ij} and $\Sigma_{\xi ij} = \Sigma_{ij} \cap V_\xi$ are smooth without corner points.

f) Only the process of volumetric ablation by thermodecomposition (pyrolysis) is considered, and the pyrolysis of composite material is simulated by parallel phase transformations of types $b \to p$ and $b \to g$.

g) Each of the phase transformations, $b \to p$ and $b \to g$, is considered as a process of continuously decreasing $V_{\xi b}(t)$ and increasing volumes of the other phases $V_{\xi p}(t)$ and $V_{\xi g}(t)$ so that the total volume V_ξ of the periodicity cell does not change.

h) Processes of motion of the phases are considered to be quasistatic, i.e. the assumptions of paragraph 2.5 still hold.

A geometric model of an ablative composite with a periodic structure is shown in Figure 3.1, and the periodicity cell V_ξ in Figure 3.2.

It should be noted that the assumption c) can be extended to composites for which volumetric ablation occurs not only in their matrix but also in their fibres (see paragraph 1.4), but then one should introduce an additional crystalline phase of the fibres. However, this addition does not contribute anything essentially new in principle, but only complicates calculations. Therefore, the main purpose of the present chapter is to study composites with non-ablative fibres, and only at the end of the chapter we will indicate the changes that occur when ablation of the fibres is taken into account.

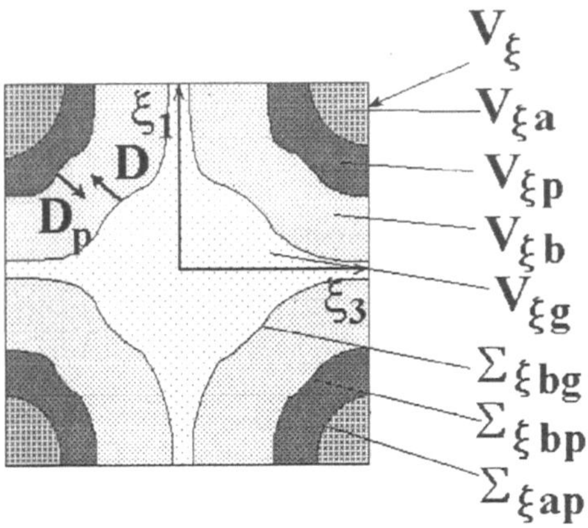

Figure 3.2. A periodicity cell V_ξ of porous ablative material

As shown in paragraph 2.5, phase interfaces are described by equations $f_{ij} = 0$. For the model given in Figure 3.1 there are three phase interfaces Σ_{bp}, Σ_{bg}, Σ_{ap}, two of which are mobile and described by equations: $f_{bg}(\mathbf{x}, t) = 0$, $f_{bp}(\mathbf{x}, t) = 0$, and one of them is fixed, i.e. $f_{ap}(\mathbf{x}) = 0$.

Let speeds of motion of the surfaces Σ_{bg} and Σ_{bp} be D_g and D_p, respectively. Assumption (f), that processes of forming the p and g phases are parallel, means that mass rates of the appearance of the phases are proportional:

$$\frac{D_p |\Sigma_{bp}|}{D_g |\Sigma_{bg}|} = \frac{1 - \Gamma}{\Gamma} = \text{const}, \tag{3.1}$$

where Γ is the *gasification coefficient* ($\Gamma = \text{const}$), $|\Sigma_{ap}|$ and $|\Sigma_{bp}|$ are the areas of phase interfaces. The assumption is quite justified, as will be shown below.

Functions f_{bg} and f_{bp} are solutions of the equations

$$\frac{\partial f_{bg}}{\partial t} + D_g \left(\nabla f_{bg} \cdot \nabla f_{bg}\right)^{1/2} = 0; \qquad f_{bg}(\mathbf{x}, 0) = f_{bg}^0(\mathbf{x}), \tag{3.2}$$

$$\frac{\partial f_{bp}}{\partial t} + D_p \left(\nabla f_{bp} \cdot \nabla f_{bp}\right)^{1/2} = 0; \qquad f_{bp}(\mathbf{x}, 0) = f_{bp}^0(0).$$

Assumption (b) on periodicity of the structure means that the functions f_{ij} are periodic in the coordinates $\mathbf{x}$ with period l_0, i.e.

$$f_{ij}(\mathbf{x}, t) \equiv f_{ij}(\mathbf{x} + l_0 \mathbf{a}_k, t), \tag{3.3}$$

$$k = 1, 2, 3, \qquad \mathbf{x} \in V,$$

where l_0 is the linear dimension of the periodicity cell V_ξ, and $\mathbf{a}_k$ are the constant vectors defining the periods of the structure. When a periodicity cell is a cube oriented along the coordinate axes Ox_k, vectors $\mathbf{a}_k$ are the basis vectors: $\mathbf{a}_k = \mathbf{e}_k$.

Let us introduce also a characteristic dimension x_0 of the whole domain V occupied by the composite.

Due to the assumption (a) we can introduce a small parameter $\varkappa$:

$$\varkappa = l_0/x_0, \qquad \varkappa \ll 1. \tag{3.4}$$

Let us also introduce a radius-vector $\boldsymbol{\xi}$ of *local coordinates* ('fast' coordinates), which changes within the limits of the periodicity cell V_ξ:

$$\boldsymbol{\xi} = \bar{\mathbf{x}}/\varkappa,$$

that can change within limits of periodicity cells V_ξ. Here $\bar{\mathbf{x}} = \mathbf{x}/x_0$ are the dimensionless coordinates changing within the whole region V occupied by the medium. The coordinates $\bar{\mathbf{x}}$ are called *global.*

Due to the periodicity condition (3.3), functions f_{ij} can be considered to depend on local coordinates $\boldsymbol{\xi}$:

$$f_{ij} = \tilde{f}_{ij}(\boldsymbol{\xi}, t), \qquad \boldsymbol{\xi} \in V_\xi. \tag{3.5}$$

Condition (3.5) means that geometric shapes of phases in all periodicity cells V_ξ are the same.

In a more general case, a geometric shape of phases can change (for example, under gradient heating) in going from one cell to another if these cells are at a distance of order x_0 one from another. Then the functions f_{ij} are *quasiperiodic*:

$$f_{ij} = f_{ij}\ (\mathbf{x}, \boldsymbol{\xi}, t), \quad \boldsymbol{\xi} \in V_\xi, \ \mathbf{x} \in V \tag{3.6}$$

i.e. depending not only on local coordinates $\boldsymbol{\xi}$ but also on global $\mathbf{x}$. The quasiperiodicity condition (3.6) means that the function f_{ij} is periodic with respect to argument $\boldsymbol{\xi}$ and changes 'slowly' with respect to argument $\mathbf{x}$, i.e. changes occur at distances of order x_0.

All functions Ω describing processes in such periodic structure are assumed to be quasiperiodic, differentiation of the functions is performed by the rule for differentiation of a composite function:

$$\nabla\Omega(\mathbf{x}, \boldsymbol{\xi}) \longrightarrow \nabla_x\Omega + (1/\varkappa)\nabla_\xi\Omega, \tag{3.7}$$

where ∇_x, ∇_ξ are gradients over coordinates $\mathbf{x}$ and $\boldsymbol{\xi}$, respectively.

Moreover, let us make one more assumption.

i) The viscosity of the gas phase is small: $\mathbf{T}'_g = \varkappa\boldsymbol{\tau}_g$, and, due to this, the constitutive relations (2.89) for gas phase can be written in the form

$$\boldsymbol{\sigma}_g = -p\mathbf{E} + \varkappa\boldsymbol{\tau}_g, \tag{3.8}$$

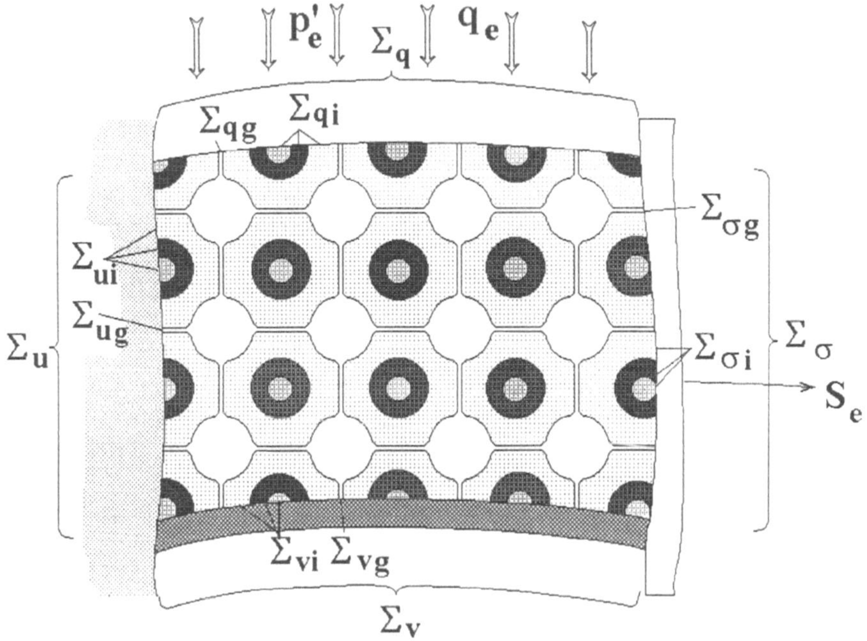

Figure 3.3 A scheme of four main types of boundary conditions on external surfaces Σ_u, Σ_v, Σ_q and Σ_σ of an ablative composite

where $\boldsymbol{\tau}_g$ is determined by formulae (2.89).

3.2. Statement of Thermomechanics Problem for Ablative Composites

3.2.1. BOUNDARY CONDITIONS

Boundary conditions at the external surface Σ of the whole composite should be joined to the equations of paragraph 2.5.4. In accordance with the assumptions (a) and (f), the surface Σ can be considered to be fixed (the case when this surface is mobile during surface ablation will be considered in paragraph 3.6).

Let us consider now the four common boundary conditions at the composite surface Σ contacting with the surroundings (Figure 3.3) –

Σ_u: the displacements are known at the surface;

Σ_v: the surface is covered by a tight coating hindering gas flowing out from material pores but transferring the pressure p_e of the external medium;

Σ_q: at the surface there is a direct contact of the porous composite with an external gas flow;

Σ_σ: the forces are known at the surface.

So the total external surface Σ of the composite consists of the following parts:

$$\Sigma = \Sigma_v \cup \Sigma_q \cup \Sigma_\sigma \cup \Sigma_u .$$

There is a mechanical contact of the composite with solid impenetrable bodies at the surfaces Σ_σ, Σ_u, the forces vector $\mathbf{S}_e$ being given at the Σ_σ and the displacements vector $\mathbf{u}_e$ being known at Σ_u. Each of the surfaces Σ_u, Σ_σ, Σ_q and Σ_v consists of the corresponding parts Σ_{ui}, $\Sigma_{\sigma i}$, Σ_{qi}, Σ_{vi} related to different phases $i = a, b, p, g$:

$$\Sigma_\sigma = \bigcup_i \Sigma_{\sigma i}, \qquad \Sigma_u = \bigcup_i \Sigma_{ui}, \qquad \Sigma_q = \bigcup_i \Sigma_{qi}, \qquad \Sigma_v = \bigcup_i \Sigma_{vi}.$$

Then with using the relations (2.97) – (2.99) at the phase interfaces, we can write the following boundary conditions:

$$\begin{gathered}
\mathbf{v}_g = \mathbf{0}; \qquad \mathbf{x} \in \Sigma_{vg}, \Sigma_{\sigma g}, \Sigma_{ug};\\
\mathbf{n} \cdot \boldsymbol{\sigma}_g = -p_e \mathbf{n}, \qquad \mathbf{x} \in \Sigma_{qg};\\
\mathbf{n} \cdot k_g \nabla \theta_g = q_e, \qquad \mathbf{x} \in \Sigma_{ug}, \Sigma_{vg}, \Sigma_{\sigma g};\\
\mathbf{n} \cdot k_g \nabla \theta_g = q_e - \rho_g \mathbf{n} \cdot \mathbf{v}_g (\bar{I}_e - I_g), \quad \mathbf{x} \in \Sigma_{qg};\\
\mathbf{n} \cdot \boldsymbol{\sigma}_i = \mathbf{S}_e, \qquad \mathbf{x} \in \Sigma_{\sigma i}, \Sigma_{qi}, \Sigma_{vi};\\
\mathbf{u} = \mathbf{u}_e, \qquad \mathbf{x} \in \Sigma_{ui};\\
\mathbf{n} \cdot k_i \cdot \nabla \theta_i = q_e, \qquad \mathbf{x} \in \Sigma_{vi}, \ \Sigma_{\sigma i}, \Sigma_{ui}, \Sigma_{qi}, \qquad i = a, b, p,
\end{gathered} \tag{3.9}$$

and $\mathbf{S}_e = -p_e \mathbf{n}$ on Σ_{qi}, Σ_{vi}.

In the system (3.9) the conditions on the surfaces Σ_{ui} and $\Sigma_{\sigma i}$ have been derived from equations (2.98) of two contacting solid phases when $D_0 = 0$; the conditions on the surfaces $\Sigma_{\sigma g}, \Sigma_{ug}, \Sigma_{vi}, \Sigma_{vg}$ have been obtained from the relations (2.97) for contacting solid and gas phases when $D_0 = 0$, and the condition on the surface Σ_{qg} is a consequence of the relations (2.99) for two interacting gas media when $D_0 = -\rho_g \mathbf{n} \cdot \mathbf{v}_g$. Values of the functions on one of the sides of the phase interface have been assumed to be given, in particular

$$q_e = \mathbf{n} \cdot k_e \nabla \theta_e = q_e^0 - q_w + q_R - q_{Rw}, \quad \mathbf{x} \in \Sigma_{qg}, \tag{3.10}$$

is the external flow to the composite, p_e is the external pressure on the surface Σ_{qg}, and $\bar{I}_e$ and I_g are the enthalpies of the external gas flow and pore gas on the surface Σ_{qg}. Constituents of the heat flux: q_e^0, q_w, q_R and q_{Rw} are assumed to be known and determined by formulae (1.11), (1.25) – (1.27), depending on the considered type of heating source.

3.2.2. INITIAL CONDITIONS

Initial conditions for the considered problem are

$$
\begin{aligned}
t = 0: \quad & \rho_g = \rho_{g0}, \quad \theta_g = \theta_0, \quad \theta_i = \theta_0, \quad \mathbf{x} \in V_i^0; \\
& f_{bg} = f_{bg}^0, \quad f_{bp} = f_{bp}^0,
\end{aligned} \tag{3.11}
$$

where V_i^0 are the domains occupied by the phases at $t = 0$.

3.2.3. STATEMENT OF THE PROBLEM

Equation system (2.91) - (2.96) and (3.2) with constitutive relations (2.84a), (2.86), (2.89), conditions (2.97)-(2.100) at the phase interface and (3.9) at the external boundary of the medium and initial conditions (3.11) constitutes the *thermomechanics problem for ablative composites* with a periodic structure for determination of the functions $\mathbf{u}_i$, $\mathbf{v}_g$, ρ_g, θ_i, θ_g $(i = a, b, p)$ and f_{bg}, f_{bp} depending on coordinates $\mathbf{x}$, $\boldsymbol{\xi}$ and time t.

3.3. Method of Asymptotic Averaging

3.3.1. ASYMPTOTIC EXPANSIONS

The problem formulated in paragraph 3.2.3 contains a small parameter $\varkappa$.

Solution of the problem is sought in the form of *asymptotic expansions*

$$
\begin{aligned}
\mathbf{u}_i &= \mathbf{u}^{(0)}(\mathbf{x}, t) + \varkappa \mathbf{u}_i^{(1)}(\mathbf{x}, \boldsymbol{\xi}, t) + \varkappa^2 \ldots ; \\
\mathbf{v}_g &= \mathbf{v}_g^{(0)}(\mathbf{x}, \boldsymbol{\xi}, t) + \varkappa \mathbf{v}_g^{(1)}(\mathbf{x}, \boldsymbol{\xi}, t) + \varkappa^2 \ldots ; \\
\rho_g &= \rho_g^{(0)}(\mathbf{x}, t) + \varkappa \rho_g^{(1)}(\mathbf{x}, \boldsymbol{\xi}, t) + \varkappa^2 \ldots ; \\
\theta_i &= \theta^{(0)}(\mathbf{x}, t) + \varkappa \theta_i^{(1)}(\mathbf{x}, \boldsymbol{\xi}, t) + \varkappa^2 \ldots , \quad i = a, b, p; \\
\theta_g &= \theta^{(0)}(\mathbf{x}, t) + \varkappa \theta_g^{(1)}(\mathbf{x}, \boldsymbol{\xi}, t) + \varkappa^2 \ldots ; \\
f_{bi} &= f_{bi}^{(0)}(\mathbf{x}, \boldsymbol{\xi}, t) + \varkappa f_{bi}^{(1)}(\mathbf{x}, \boldsymbol{\xi}, t) + \varkappa^2 \ldots , \quad i = g, p \;\; \boldsymbol{\xi} \in V_{\xi i}, \; \mathbf{x} \in V_i.
\end{aligned} \tag{3.12}
$$

On substituting functions (3.12) into equations (2.93), (2.85), (2.86) and (2.89), we can obtain asymptotic expansions for the remaining functions of the problem:

$$
\Omega = \Omega^{(0)}(\mathbf{x}, \boldsymbol{\xi}, t) + \varkappa \Omega^{(1)}(\mathbf{x}, \boldsymbol{\xi}, t) + \varkappa^2 \ldots , \tag{3.13}
$$

$$
\Omega = \{\boldsymbol{\sigma}_g, p, \boldsymbol{\tau}_g, \boldsymbol{\sigma}_i, \boldsymbol{\varepsilon}_i\}.
$$

3.3.2. ZERO-LEVEL LOCAL PROBLEMS OVER THE PERIODICITY CELL

Having substituted expansions (3.13) into equations (2.94), (2.95), and equated the zero order terms to zero, we obtain the *zero-level local gas dynamics problem over the periodicity cell*:

$$\begin{gathered}\nabla_\xi \cdot \mathbf{v}_g^{(0)} = 0;\\ -\nabla_\xi p^{(1)} + \nabla_\xi \cdot \boldsymbol{\tau}_g^{(0)} = \nabla_x p^{(0)}, \quad \boldsymbol{\xi} \in V_{\xi g};\\ \boldsymbol{\tau}_g^{(0)} = \nu_g \mathbf{E} \nabla_\xi \cdot \mathbf{v}_g^{(0)} + \mu_g (\nabla_\xi \otimes \mathbf{v}_g^{(0)} + (\nabla_\xi \otimes \mathbf{v}_g^{(0)})^T);\\ p^{(1)} = R(\rho_g^{(1)}\theta^{(0)} + \rho_g^{(0)}\theta_g^{(1)}); \quad \mathbf{v}_g^{(0)}|_{\Sigma_{\xi pg}} = 0, \quad \langle \rho_g^{(1)} \rangle_g = 0,\end{gathered} \tag{3.14}$$

the solution of which is the functions $\mathbf{v}_g^{(0)}$, $\theta^{(1)}$ and $\rho_g^{(1)}$. The remaining functions contained in (3.14), $p^{(0)}$ and $\theta^{(0)}$, are considered as known 'initial data' of the problem.

In a similar way, substituting expansions (3.13) into the equations (2.91), (2.93), (2.84a) and conditions on the phase interface (2.97), (2.98) and then collecting terms with the same powers of $\varkappa$ in these equations, and putting terms involving $(1/\varkappa)$ in equation (2.91) and terms with $\varkappa^0$ in the remaining equations equal to zero, we obtain the *zero-level local contact problem of solid deformable body mechanics over the periodicity cell*:

$$\begin{gathered}\nabla_\xi \cdot \boldsymbol{\sigma}_i^{(0)} = \mathbf{0};\\ \boldsymbol{\varepsilon}_i^{(0)} = \overset{\circ}{\varepsilon}{}_i^{(0)} \mathbf{E} + \frac{1+\nu_i}{E_i} \boldsymbol{\sigma}_i^{(0)} - \frac{\nu_i}{E_i} \sigma_i^{(0)} \mathbf{E},\\ 2\boldsymbol{\varepsilon}_i^{(0)} = 2\boldsymbol{\varepsilon} + \nabla_\xi \otimes \mathbf{u}_i^{(1)} + (\nabla_\xi \otimes \mathbf{u}_i^{(1)})^T, \quad \boldsymbol{\xi} \in V_{\xi i};\\ \boldsymbol{\sigma}_b^{(0)} \cdot \mathbf{n}^{(0)} = -p^{(0)} \mathbf{n}^{(0)}, \quad \boldsymbol{\xi} \in \Sigma_{\xi bg};\\ [\boldsymbol{\sigma}^{(0)}] \cdot \mathbf{n}^{(0)} = 0, \qquad [\mathbf{u}^{(1)}] = 0, \qquad \boldsymbol{\xi} \in \Sigma_{\xi ap}, \Sigma_{\xi bp};\\ \langle \mathbf{u}^{(1)} \rangle_s = 0; \qquad \overset{\circ}{\varepsilon}{}_i^{(0)} = \alpha_i(\theta^{(0)} - \theta_i^*) - \beta_i.\end{gathered} \tag{3.15}$$

Here $\boldsymbol{\varepsilon} = \frac{1}{2}\left(\nabla_x \otimes \mathbf{u}^{(0)} + (\nabla_x \otimes \mathbf{u}^{(0)})^T\right)$.

The functions $\theta^{(0)}$ and $\mathbf{u}^{(0)}$ in equations (3.15) are also assumed to be known data for the problem. The function $\mathbf{u}^{(1)}$, being a solution, depends on $\mathbf{u}^{(0)}$ and $\theta^{(0)}$ parametrically.

Finally, substitute the expansions (3.13) into equations (2.92a), (2.96a) and conditions (2.97), (2.98) and collect terms with the same powers of $\varkappa$. Then setting terms involving $(1/\varkappa)$ in equations (2.92a), (2.96a) and terms with $\varkappa^0$ in the boundary conditions equal to zero, we obtain the *zero-level local problem of heat conduction over the periodicity cell*:

$$\nabla_\xi \cdot (\nabla_\xi \theta_i^{(1)}) = 0, \quad \boldsymbol{\xi} \in V_{\xi i}, \quad i = a, b, p, g;$$

$$[\theta^{(1)}] = 0, \qquad \mathbf{n}^{(0)} \cdot [k\nabla_\xi \theta^{(1)}] = -\mathbf{n}^{(0)} \cdot [k]\nabla_x \theta^{(0)}, \quad \boldsymbol{\xi} \in \Sigma_{\xi ap}, \Sigma_{\xi bg}, \Sigma_{\xi bp};$$
$$\langle \theta^{(1)} \rangle = 0. \tag{3.16}$$

Here the function $\theta^{(0)}$ is also considered as known initial data, and the solution of the problem (3.16), functions $\theta_i^{(1)}$, $i = a, b, p, g$, depend on $\theta^{(0)}$ parametrically.

The following averaging operations have been introduced in the equations (3.14) - (3.16):

$$\langle \Omega \rangle_s = \varphi_a \langle \Omega \rangle_a + \varphi_b \langle \Omega \rangle_b + \varphi_p \langle \Omega \rangle_p; \tag{3.17}$$

$$\langle \Omega_i \rangle_i = \frac{1}{\varphi_i |V_\xi|} \int_{V_{\xi i}} \Omega_i dV_{\xi i}, \quad i = a, b, p, g; \qquad \langle \Omega \rangle = \langle \Omega \rangle_s + \varphi_g \langle \Omega \rangle_g;$$

$$\varphi_s = \varphi_a + \varphi_b + \varphi_p; \qquad \varphi_i = \frac{1}{|V_\xi|} \int_{V_{\xi i}} dV_{\xi i}, \quad i = a, b, p; \qquad \varphi_g = 1 - \varphi_s.$$

Solution of all the three problems (3.14) - (3.16) for functions $\rho_g^{(1)}$, $\mathbf{v}_g^{(0)}$, $\mathbf{u}_i^{(1)}$, $\theta_i^{(1)}$ and $\theta_g^{(1)}$ is sought in the class of functions which are periodic in the $\boldsymbol{\xi}$ coordinates.

Locations of the phase interfaces $\Sigma_{\xi bg}$ and $\Sigma_{\xi bp}$ within the periodicity cell are described by equations $f_{bi}^{(0)}(\mathbf{x}, \boldsymbol{\xi}, t) = 0$, $i = g, p$. Equations for functions $f_{bi}^{(0)}$ can be found by substituting the asymptotic expansions (3.13) into equations (3.2):

$$\frac{\partial f_{bi}^{(0)}}{\partial t} + D_i^{(0)} (\nabla_\xi f_{bi} \cdot \nabla_\xi f_{bi})^{1/2} = 0; \quad f_{bi}^{(0)}(\mathbf{x}, \boldsymbol{\xi}, 0) = f_{bi}^{0}(\mathbf{x}, \boldsymbol{\xi}), \qquad i = g, p, \tag{3.18}$$

where $D_i^{(0)}$ is the rate of the phase transformation in the zero approximation:

$$D_g = \varkappa D_g^{(0)};$$

$$D_g^{(0)} = D_0^{(0)} \exp\left(-\frac{E_A}{R\theta^{(0)}}\right)\left(1 - \frac{p^{(0)}}{p_0} \exp\left(-\frac{2E_A}{R\theta^{(0)}}\right)\right)^{1/2}; \tag{3.19}$$
$$D_0^{(0)} = \frac{p_e}{\rho_b \sqrt{R\theta^{(0)}}}.$$

The expression for the rate $D_p^{(0)}$ is defined in terms of $D_i^{(0)}$ by the formula (3.1).

Using the expression $\mathbf{n} = \nabla f_{bi} |\nabla f_{bi}|^{-1}$, we can present the normal vector $\mathbf{n}$ to the separation surface $\Sigma_{\xi bi}$ as follows:

$$\mathbf{n} = \mathbf{n}^{(0)} + \varkappa \mathbf{n}^{(1)} + \varkappa^2 \ldots; \tag{3.20}$$

$$\mathbf{n}^{(0)} = \frac{\nabla_\xi f_{bi}^{(0)}}{|\nabla_\xi f_{bi}^{(0)}|};$$

$$\mathbf{n}^{(1)} = \frac{(\nabla_x f_{bi}^{(0)} + \nabla_\xi f_{bi}^{(1)}) \cdot (\mathbf{E} - \mathbf{n}^{(0)} \otimes \mathbf{n}^{(0)})}{|\nabla_\xi f_{bi}^{(0)}|};$$

$$|\nabla_\xi f_{bi}^{(0)}| = (\nabla_\xi f_{bi} \cdot \nabla_\xi f_{bi})^{1/2}, \quad i = g, p.$$

3.3.3. FIRST-LEVEL LOCAL PROBLEMS OVER THE PERIODICITY CELL

Having collected terms in equations (2.94), (2.95) at the next powers of $\varkappa$ we obtain the *first-level local gas dynamics problem over the periodicity cell*:

$$\begin{gathered}
\nabla_\xi \cdot (\rho_g^{(0)} \mathbf{v}_g^{(1)} + \rho_g^{(1)} \mathbf{v}_g^{(0)}) = -\left(\frac{\partial \rho_g^{(0)}}{\partial t} + \nabla_x \cdot \rho_g^{(0)} \mathbf{v}_g^{(0)} \right); \\
-\nabla_\xi p^{(2)} + \nabla_\xi \cdot \boldsymbol{\tau}_g^{(1)} = \nabla_x p^{(1)} - \nabla_x \cdot \boldsymbol{\tau}_g^{(0)}; \qquad (3.21) \\
\boldsymbol{\tau}_g^{(1)} = \nu_g \mathbf{E} \nabla_\xi \cdot \mathbf{v}_g^{(1)} + \mu_g \left(\nabla_\xi \otimes \mathbf{v}_g^{(1)} + (\nabla_\xi \otimes \mathbf{v}_g^{(1)})^T \right); \\
p^{(2)} = R \rho_g^{(2)} \theta_g^{(1)} + \rho_g^{(1)} \theta_g^{(2)}; \\
(\rho_b - \rho_g^{(0)}) D_g^{(0)} + (\rho_g^{(0)} \mathbf{v}_g^{(1)} + \rho_g^{(1)} \mathbf{v}_g^{(0)}) \cdot \mathbf{n}^{(0)} + \rho_g^{(0)} \mathbf{v}_g^{(0)} \cdot \mathbf{n}^{(1)} = 0; \quad \boldsymbol{\xi} \in \Sigma_{\xi bg}; \\
\langle \rho_g^{(2)} \rangle_g = 0.
\end{gathered}$$

Functions $\rho_g^{(2)}$, $\theta_g^{(2)}$ and $\mathbf{v}_g^{(1)}$ in this problem are unknown, and the remainder are 'initial data'.

In a similar way, collecting terms at higher powers of $\varkappa$ in equations (2.91) and (2.97), (2.98), we obtain the *first-level local problem of solid deformable body mechanics over the periodicity cell*:

$$\begin{gathered}
\nabla_\xi \cdot \boldsymbol{\sigma}_i^{(1)} = -\nabla_x \cdot \boldsymbol{\sigma}_i^{(0)}, \quad \boldsymbol{\xi} \in V_{\xi i}; \\
[\boldsymbol{\sigma}^{(0)}] \cdot \mathbf{n}^{(1)} + [\boldsymbol{\sigma}^{(1)}] \cdot \mathbf{n}^{(0)} = 0; \quad \boldsymbol{\xi} \in \Sigma_{\xi ap}, \Sigma_{\xi bp}, \Sigma_{\xi bg}; \qquad (3.22) \\
\langle \mathbf{u}^{(2)} \rangle_s = 0
\end{gathered}$$

for determination of functions $\mathbf{u}_i^{(2)}$.

On collecting terms at powers of $\varkappa^0$ in equations (2.92a) and (2.96a) and conditions (2.97) at the phase interface, we obtain the *first-level local problem of heat conductivity over the periodicity cell* for determination of functions $\theta_i^{(2)}$:

$$\begin{aligned}
\nabla_\xi \cdot \left(k_i \cdot (\nabla_x \theta_i^{(1)} + \nabla_\xi \theta_i^{(2)}) \right) + \nabla_x \cdot (k_i \cdot \nabla_\xi \theta_i^{(1)}) = \rho_i c_i \frac{\partial \theta_i^{(0)}}{\partial t} - \\
- \nabla_x \cdot (k_i \cdot \nabla_x \theta^{(0)}) + \boldsymbol{\alpha} \theta^{(0)} : \frac{\partial \boldsymbol{\sigma}_i^{(0)}}{\partial t}; \quad \boldsymbol{\xi} \in V_{\xi i}, \quad i = a, b, p;
\end{aligned}$$

$$\begin{aligned}
\nabla_\xi \cdot \left(k_g (\nabla_x \theta_g^{(1)} + \nabla_\xi \theta_g^{(2)}) \right) + \nabla_x \cdot (k_g \nabla_\xi \theta_g^{(1)}) = \frac{\partial}{\partial t} \rho_g^{(0)} c_g \theta^{(0)} - \\
- \nabla_x \cdot (k_g \nabla_x \theta^{(0)}) + \nabla_x \cdot \rho_g^{(0)} \mathbf{v}_g^{(0)} c_g \theta^{(0)} + \\
+ \nabla_\xi c_g \cdot \left(\rho^{(1)} \mathbf{v}_g^{(0)} \theta_g^{(0)} + \rho^{(0)} (\mathbf{v}_g^{(1)} \theta^{(0)} + \mathbf{v}_g^{(0)} \theta^{(1)}) \right) - \\
- \nabla_x \cdot (\boldsymbol{\sigma}_g^{(0)} \cdot \mathbf{v}_g^{(0)}) + \nabla_\xi (\boldsymbol{\sigma}_g^{(1)} \cdot \mathbf{v}_g^{(0)} + \boldsymbol{\sigma}_g^{(0)} \cdot \mathbf{v}_g^{(1)}), \quad \boldsymbol{\xi} \in V_{\xi g}; \qquad (3.23)
\end{aligned}$$

$$\rho_p[e^0 + c\theta^{(0)}]D_g^{(0)} - \mathbf{n}^{(0)} \cdot (\boldsymbol{\sigma}_g^{(0)} \cdot \mathbf{v}_g^{(1)} + \boldsymbol{\sigma}_g^{(1)} \cdot \mathbf{v}_g^{(0)}) - \mathbf{n}^{(1)} \cdot \boldsymbol{\sigma}_g^{(0)} \cdot \mathbf{v}_g^{(0)} +$$
$$+ \mathbf{n}^{(0)} \cdot \left[k(\nabla_x \theta^{(1)} + \nabla_\xi \theta^{(2)})\right] + \mathbf{n}^{(1)} \cdot \left[k(\nabla_x \theta^{(0)} + \nabla_\xi \theta^{(1)})\right] = 0;$$

$$\langle \theta^{(2)} \rangle_s = 0.$$

3.4. Averaging of Processes in Ablative Composites

3.4.1. AVERAGED EQUATIONS

The averaging operation of quasiperiodic functions $\Omega(\mathbf{x}, \boldsymbol{\xi}, t)$ over one of the coordinates $\boldsymbol{\xi}$ was introduced by formula (3.17). To derive averaged equations, the following properties of averaging operation will be used:

$$\int_{V_{\xi b}(t)} \frac{\partial \Omega}{\partial t} dV_\xi = \frac{\partial}{\partial t} \varphi_b \langle \Omega \rangle_b + \int_{\Sigma_{\xi bg}} \Omega D_g^{(0)} d\Sigma_\xi + \int_{\Sigma_{\xi bp}} \Omega D_p^{(0)} d\Sigma_\xi; \tag{3.24}$$

$$\int_{V_{\xi b}(t)} \nabla_x \Omega dV_\xi = \nabla_x \varphi_b \langle \Omega \rangle_b - \int_{\Sigma_{\xi bg} \cup \Sigma_{\xi bp}} \Omega \mathbf{n}^{(1)} d\Sigma_\xi;$$

$$\int_{V_{\xi b}(t)} \nabla_\xi \Omega dV_\xi = - \int_{\Sigma_{\xi bg} \cup \Sigma_{\xi bp}} \Omega \mathbf{n}^{(0)} d\Sigma_\xi.$$

When averaging a function Ω over $V_{\xi g}(t)$ and $V_{\xi p}(t)$ we should change the sign before the integral over the phase interface $\Sigma_{\xi bg}$, for example:

$$\int_{V_{\xi g}(t)} \frac{\partial \Omega}{\partial t} dV_\xi = \frac{\partial}{\partial t} \varphi_g \langle \Omega \rangle_g - \int_{\Sigma_{\xi bg}} \Omega D_g^{(0)} d\Sigma_\xi.$$

With using relations (3.24), let us average the first equation from the set (3.21) over the domain $V_{\xi g}$, the momentum equation (3.22) over $V_{\xi i}$, and add that to the second equation from (3.14) averaged over $V_{\xi g}$. Two first equations from the system (3.23) are averaged over the domains $V_{\xi i}$ and $V_{\xi g}$, respectively, and then these equations are added. As a result, one can derive the following system of averaged equations:

$$\rho_b \frac{\partial \varphi_b}{\partial t} = -J; \tag{3.25}$$

$$\rho_p \frac{\partial \varphi_p}{\partial t} = J(1 - \Gamma); \tag{3.26}$$

$$\frac{\partial \varphi_g \rho_g^{(0)}}{\partial t} + \nabla_x \cdot \varphi_g \rho_g^{(0)} \langle \mathbf{v}_g^{(0)} \rangle_g = J\Gamma; \tag{3.27}$$

$$\nabla_x \cdot \boldsymbol{\sigma} - \nabla_x \varphi_g p^{(0)} = \mathbf{0}; \tag{3.28}$$

$$\rho c\frac{\partial\theta^{(0)}}{\partial t}+\varphi_g\rho_g^{(0)}c_g\langle\mathbf{v}_g^{(0)}\rangle_g\cdot\nabla_x\theta^{(0)}=-\nabla_x\cdot\mathbf{q}-\varphi_g\langle\mathbf{v}_g^{(0)}\rangle_g\cdot\nabla_x p^{(0)}-J\Delta e \quad (3.29)$$

for determination of functions φ_b, φ_p, $\rho_g^{(0)}$, $\mathbf{u}^{(0)}$ and $\theta^{(0)}$.

Equation (3.25) follows from equations (3.24) with $\Omega=\rho_b$. Equation (3.26) can be derived in the similar way.

3.4.2 AVERAGED CONSTITUTIVE RELATIONS

The following notations for parameters of phase interchange J and Δe have been introduced in equations (3.25) – (3.29):

$$\Delta e=[e^0]+[c]\theta^{(0)},\qquad J=\frac{\rho_b}{\Gamma}D_g^{(0)}\int_{\Sigma_{\xi bg}}d\Sigma_\xi. \quad (3.30)$$

In addition, the equations (3.27) – (3.29) contain averaged functions $\boldsymbol{\sigma}=\langle\boldsymbol{\sigma}^{(0)}\rangle_s$, $\mathbf{q}$ and ρc:

$$\mathbf{q}=-\langle\mathbf{k}\cdot(\nabla_\xi\theta_s^{(1)}+\nabla_x\theta_s^{(0)})\rangle_s=-\mathbf{k}\cdot\nabla_x\theta^{(0)};$$

$$\boldsymbol{\sigma}=\langle\frac{E}{1+\nu}\left(\frac{\nu}{1-2\nu}(\varepsilon^{(0)}-3\overset{\circ}{\varepsilon}{}^{(0)})\mathbf{E}+\boldsymbol{\varepsilon}^{(0)}-\overset{\circ}{\varepsilon}{}^{(0)}\mathbf{E}\right)\rangle_s=\boldsymbol{\sigma}(\boldsymbol{\varepsilon},\theta^{(0)},p^{(0)}); \quad (3.31)$$

$$\rho c=\sum_{i=a,b,p}\rho_i\varphi_i c_i+\rho_g^{(0)}\varphi_g c_g,\qquad \rho=\sum_{i=a,b,p}\varphi_i\rho_i,$$

here $\varepsilon^{(0)}=\boldsymbol{\varepsilon}_i^{(0)}:\mathbf{E}$ is the first invariant of the strain tensor $\boldsymbol{\varepsilon}_i^{(0)}$, and the averaged strain tensor $\boldsymbol{\varepsilon}$ is connected to the displacement vector by the relationship

$$\boldsymbol{\varepsilon}=\frac{1}{2}\left(\nabla_x\otimes\mathbf{u}^{(0)}+(\nabla_x\otimes\mathbf{u}^{(0)})^T\right). \quad (3.32)$$

To the equation system (3.25)-(3.29) one should add the relationship connecting the average velocity of gas phase flow $\langle\mathbf{v}^{(0)}\rangle_g$ to pressure $p^{(0)}$. This relationship can be found from a solution of the problem (3.14) over the periodicity cell, which can be presented in the form $\varphi_g\rho_g^{(0)}\mathbf{v}_g^{(0)}=\mathbf{W}_g(\boldsymbol{\xi})\cdot\nabla_x p^{(0)}$, where the tensor-function $W_g(\boldsymbol{\xi})$ is determined by a geometry of the pores.

After averaging the expression over the domain $V_{\xi g}$, one obtains the relationship

$$\varphi_g\rho_g^{(0)}\langle\mathbf{v}_g^{(0)}\rangle_g=-\mathbf{K}\cdot\nabla_x p^{(0)};\qquad \mathbf{K}=-\langle\mathbf{W}_g\rangle_g \quad (3.33)$$

known as *Darcy's Law.* Tensor $\mathbf{K}$ is called the *tensor of gas permeability.*

Equations (3.31), (3.33) are the *effective constitutive relations* for ablative composite materials.

3.4.3 STATEMENT OF THE AVERAGED PROBLEM

The composite boundary Σ is considered to consist of parts Σ_v, Σ_q, Σ_σ and Σ_u (see Figure 3.3). In order to obtain boundary conditions at these surfaces for the system (3.25) - (3.29), the asymptotic expansions (3.12) and (3.13) must be substituted into conditions (3.9), where $i = a, b, p$, and then these conditions must be averaged over the domain $V_{\xi i}$ in the corresponding way. As a result, we obtain

$$\mathbf{n} \cdot \nabla_x p^{(0)} = 0; \qquad \mathbf{x} \in \Sigma_v, \Sigma_\sigma, \Sigma_u; \tag{3.34}$$

$$p^{(0)} = p_e, \qquad \mathbf{x} \in \Sigma_q; \tag{3.35}$$

$$\boldsymbol{\sigma} \cdot \mathbf{n} - \varphi_g p^{(0)} = \mathbf{S}_e, \qquad \mathbf{x} \in \Sigma_\sigma, \Sigma_q, \Sigma_v; \tag{3.36}$$

$$\mathbf{u} = \mathbf{u}_e, \qquad \mathbf{x} \in \Sigma_u, \tag{3.37}$$

$$\mathbf{n} \cdot \mathbf{k} \cdot \nabla_x \theta^{(0)} = q_e - \gamma_{\rm bl} \varphi_g \rho_g^{(0)} \langle \mathbf{v}_g^{(0)} \rangle_g \cdot \mathbf{n}(\bar{I}_e - I_g), \qquad \mathbf{x} \in \Sigma_q; \tag{3.38}$$

$$\mathbf{n} \cdot \mathbf{k} \cdot \nabla_x \theta = q_e, \qquad \mathbf{x} \in \Sigma_v, \Sigma_\sigma, \Sigma_u. \tag{3.39}$$

In the equation (3.38), the coefficient $\gamma_{\rm bl}$ called the *injection coefficient* [95] has been introduced; this describes decreasing the heat flux $\varphi_g \rho_g^{(0)} \langle \mathbf{v}_g^{(0)} \rangle \cdot \mathbf{n}(\bar{I}_e - I_g)$ withdrawn from the surface Σ_q with the g-phase flowing out into the surroundings due to the difference of chemical composition of the g-phase and external gases, changing the form of pores at the surface Σ_q and other factors. Values of the coefficient $\gamma_{\rm bl}$ usually range between 0.1 and 1.

Surfaces Σ_u, Σ_σ, Σ_v are called *gas-tight*, if the condition (3.34) is satisfied on them; this means that there is no gas flowing out nor any gas penetrating material pores at the surface.

Surfaces Σ_u, Σ_σ, Σ_v are called *heat-insulated*, if the conditions (3.39) with $q_e = 0$ are satisfied on them.

Equations (3.25) - (3.29) with boundary conditions (3.34) - (3.39) and initial data

$$t = 0: \quad \varphi_b = \varphi_b^0, \quad \varphi_p = \varphi_p^0, \quad \rho_g^{(0)} = \rho_g^0, \quad \theta^{(0)} = \theta_0 \tag{3.40}$$

give the statement of the *averaged problem* of thermomechanics of ablative composites for determination of functions φ_b, φ_p, $\rho_g^{(0)}$, $\theta^{(0)}$ and $\mathbf{u}^{(0)}$ depending on global variables $\mathbf{x}$ and t.

3.5. Analysis of Asymptotic Averaging Method

The main idea of the asymptotic averaging method described above consists in the following. With the help of assumptions on a regularity of an internal structure of porous medium and asymptotic expansions of functions (3.12), the initial equation system of thermomechanics (2.91) - (2.96), accurately describing high-temperature processes in ablative composites, has been transformed to the averaged equations system (3.25) – (3.29) which describes the processes accurately for small values of $\varkappa$. Solving the system (3.25) – (3.29) is essentially simpler than solving the initial system (2.91) – (2.96), as the domain of definition of the system (3.25) – (3.29) is a 'homogenized' one, while the domain of definition of the system (2.91) – (2.96) is 'nonhomogeneous' one, V, with a very complicated internal geometry (see Figure 3.1).

As a result of the averaging procedure, the initial constitutive relations for phases (2.85) – (2.86), (2.89) have been transformed to averaged constitutive relations (3.31) – (3.33) connecting averaged function $\boldsymbol{\sigma}$ to $\boldsymbol{\varepsilon}$ and $\theta^{(0)}$, and also $\langle \mathbf{v}_g^{(0)} \rangle_g$ to $p^{(0)}$, $\mathbf{q}$ to $\nabla_x \theta^{(0)}$.

As a rule, averaged constitutive relations are determined empirically with use of additional experimental data. The method of asymptotic averaging allows us to establish a precise presentation of effective constitutive relations for a whole composite by properties of separate phases of the composite. For the purpose it is necessary to solve subsidiary local problems (3.14) – (3.16) over the periodicity cell, and then, having substituted solutions of the local problems $\rho_g^{(1)}$, $\mathbf{v}_g^{(0)}$, $\theta_g^{(1)}$, $\mathbf{u}^{(1)}$, $\theta_i^{(1)}$ into equations (3.31) – (3.33), to average them in the corresponding way over the periodicity cell.

Each the local problem is a set of stationary equations of conservation (i.e. in which all derivatives of the unknown functions with respect to time are equal to zero), solutions of these problems are sought within the domain V_ξ, i.e. within the periodicity cell. In addition, requirements on the periodicity of all the unknown functions $\rho_g^{(1)}$, $\mathbf{v}_g^{(0)}$, $\theta_g^{(1)}$, $\mathbf{u}^{(1)}$, $\theta_i^{(1)}$ are imposed. Moreover, conditions of an integral type (the last line in formulae (3.14) - (3.16)) are attached to the local system. These conditions set the averaged values of the functions to zero: $\langle \rho_g^{(1)} \rangle_g = 0$, $\langle \theta^{(1)} \rangle = 0$, $\langle \mathbf{u}^{(1)} \rangle_s = 0$. These equations are necessary for the existence of an unique solution of the problems over the periodicity cell [8, 104].

The method of asymptotic averaging allows us to set the order (i.e. zero, one, ...) of approximation. As follows from equations (3.12), the solution $\mathbf{u}^{(0)}$, $\mathbf{v}_g^{(0)}$, $\theta^{(0)}$, $p^{(0)}$ and $\rho_g^{(0)}$ obtained within the framework of the zero approximation differs from the exact solution $\mathbf{u}_i$, θ_i, $\mathbf{v}_g$, p, ρ_g by a small value of the order $\varkappa$, i.e. it converges to the exact one asymptotically at $\varkappa \to 0$.

In empirical theories on ablative composite materials, heat-mass-transport where only the equations derived from averages over phases are investigated, the question of how near the averaged discription is from the exact one remains open.

3.6. Statement of Problems for Composites with Ablative Matrix and Fibres

3.6.1. MAIN EQUATIONS

As in paragraph 3.1, we can conduct averaging processes for composites with ablative matrix and fibres. A fibre is assumed to consist of amorphic $i = a$ and crystalline $i = l$ phases. Let $\varphi_f = \varphi_a + \varphi_l$ be the total volumetric part of the whole fibre in the composite, where φ_l is the volumetric content of crystalline phase. Densities of the amorphic and crystalline phases are assumed to be the same $\rho_a = \rho_l = \rho_f$.

As a result of applying the methods of asymptotic averaging to the initial system of thermomechanics equations (2.91) – (2.99), as in paragraphs 3.2 - 3.4, one can derive the averaged system of thermomechanics equations for composites with ablative matrix and fibres:

$$\rho_f \frac{\partial \varphi_f}{\partial t} = -J_f \Gamma_f, \quad \mathbf{x} \in V \tag{3.41}$$

– the equation of changing mass of fibres;

$$\rho_b \frac{\partial \varphi_b}{\partial t} = -J, \quad \mathbf{x} \in V \tag{3.42}$$

– the equation of changing mass of polymer phase of the matrix;

$$\frac{\partial \rho_g \varphi_g}{\partial t} + \nabla_x \cdot \rho_g \varphi_g \langle \mathbf{v}_g \rangle_g = J\Gamma + J_f \Gamma_f, \quad \mathbf{x} \in V \tag{3.43}$$

– the equation of filtration of gas phase in pores of composite material;

$$\rho c \frac{\partial \theta}{\partial t} = -\nabla_x \cdot \mathbf{q} - c_g \nabla_x \theta \cdot \rho_g \varphi_g \langle \mathbf{v}_g^{(0)} \rangle_g - J_f \Delta e_f^0 - J \Delta e^0, \quad \mathbf{x} \in V \tag{3.44}$$

– the equation of heat transfer in ablative material;

$$\nabla_x \cdot \boldsymbol{\sigma} - \nabla_x \varphi_g p = 0, \quad \mathbf{x} \in V \tag{3.45}$$

– the momentum equation of ablative material;

$$\boldsymbol{\varepsilon} = \frac{1}{2} \left(\nabla_x \otimes \mathbf{u} + (\nabla_x \otimes \mathbf{u})^T \right), \quad \mathbf{x} \in V \tag{3.46}$$

– the Cauchy relation between the averaged strain tensor of ablative material and the averaged displacement vector.

Here and below, the superscript (0) is omitted for average values of functions over the phases: $p^{(0)}$, $\rho_g^{(0)}$, $\mathbf{u}^{(0)}$, $\theta^{(0)}$ are denoted by p, ρ_g, $\mathbf{u}$, θ, respectively.

Let us enumerate the functions appearing in the system (3.41) – (3.46):

ρ_f, ρ_b, ρ_p – densities of solid phases (they are constant, i.e. they do not change under heating);

ρ_g – averaged over $V_{\xi g}$ density of gas phase;

φ_f, φ_b, φ_p and φ_g – volumetric concentrations of the three solid phases and the gas phase, respectively;

φ_s – volumetric concentration of the solid frame ($\varphi_s = 1 - \varphi_g$);

p – averaged value of pore pressure of g-phase over $V_{\xi g}$;

$\mathbf{q}$ – averaged value of the heat flux vector;
θ – averaged over value of temperature of ablative material the periodicity cell V_ξ;
$\langle \mathbf{v}_g^{(0)} \rangle_g$ – averaged velocity of motion of gas phase;
$\mathbf{u}$ – averaged value of a displacement vector of ablative material over the periodicity cell V_ξ;
$\boldsymbol{\varepsilon}$ – averaged value of the strain tensor of ablative material;
$\boldsymbol{\sigma}$ – averaged value of the stress tensor of the solid frame;
c_i – heat capacities of phases;
Δe_f^0, Δe^0 – specific heats of volumetric ablation of fibres and matrix, respectively.

As follows from the equation (3.41), the volumetric concentration of fibres φ_f changes due to gasification of the fibres (for example, oxidizing). Concentrations of amorphic φ_a and crystalline φ_l phases of fibres are defined by function φ_f:

$$\varphi_a = \frac{\varphi_f - \varphi_f^0(1-\Gamma_f)}{\Gamma_f}; \qquad \varphi_l = \frac{1-\Gamma_f}{\Gamma_f}(\varphi_f^0 - \varphi_f). \tag{3.47}$$

Here Γ_f is the coefficient of gasification of the fibres, J_f is the rate of volumetric ablation of the fibres having the expression similar to (3.30). At the initial state under temperature $\theta_0 = 293$ K: $\varphi_f = \varphi_f^0$, $\varphi_a = \varphi_f^0$, $\varphi_l = 0$, where φ_f^0 is the initial concentration of the fibre in the composite.

As follows from equations (3.25), (3.26), volumetric concentration φ_p of pyrolysis phase of the matrix is expressed analytically in terms of φ_b:

$$\varphi_p = (\varphi_b^0 - \varphi_b)(1-\Gamma)\frac{\rho_b}{\rho_p}. \tag{3.48}$$

3.6.2. EQUATION OF MOTION OF THE ABLATION SURFACE Σ

In addition to volumetric ablation, surface ablation of a composite can also be taken into account. We suppose all equations (3.41) – (3.46) are written in the Cartesian coordinate system $O\mathbf{x}$ for the domain $V(t)$ occupied by ablative material at time t. Due to surface ablation this domain $V(t)$ and a part of its surface $\Sigma_q(t)$ are variables; the remainig parts Σ_u, Σ_v and Σ_σ do not change. Therefore an equation describing a location of the surface Σ should be added to the system (3.41) – (3.46). If we write the equation $f_\Sigma(\mathbf{x}, t) = 0$ describing a configuration of the surface Σ, then for the function f_Σ we have the equation of type (3.2):

$$\frac{\partial f_\Sigma}{\partial t} + D\left(\nabla_x f_\Sigma \cdot \nabla_x f_\Sigma\right)^{1/2} = 0, \tag{3.49}$$

where D is the rate of linear ablation of the composite, expression for which will be given in Chapter 10.

3.6.3. CONSTITUTIVE RELATIONS

Equations (3.41) - (3.49) should be supplemented by constitutive relations connecting functions $\mathbf{q}$, $\langle \mathbf{v}_g^{(0)} \rangle_g$, $\boldsymbol{\sigma}$ to $\nabla_x \theta$, $\nabla_x p$ and $\boldsymbol{\varepsilon}$ (or $\nabla_x \otimes \mathbf{u}$). As shown in paragraph 3.5, the relations can be written as follows:

$$p = R\rho_g \theta \tag{3.50}$$

– the constitutive relation for gas phase;

$$\mathbf{q} = -\mathbf{k} \cdot \nabla_x \theta \tag{3.51}$$

– the Fourier law,

$$\rho_g \varphi_g \langle \mathbf{v}_g^{(0)} \rangle_g = -\mathbf{K} \cdot \nabla_x p \tag{3.52}$$

– Darcy's law,

$$\boldsymbol{\sigma} = -\mathbf{f}p + \mathbf{C} \cdot\cdot (\boldsymbol{\varepsilon} - \overset{\circ}{\boldsymbol{\varepsilon}}) \tag{3.53}$$

– the constitutive relation for the solid frame.

The relation (3.53) follows readily from the second equation of the set (3.32), in which the tensor $\boldsymbol{\sigma}^{(0)}$ being a solution of the local problem (3.15) should be substituted. The tensor $\boldsymbol{\sigma}^{(0)}$ is a linear function of arguments $\boldsymbol{\varepsilon}$ and p. The linearity is a consequence of linearity of the constitutive relations of the phases. The tensor $\mathbf{f}$, called the *tensor of phase interaction*, describes the mechanical interaction of the gas phase and solid frame which is reduced to the action of the gas pressure p on the pore walls.

The constitutive relations (3.50) – (3.53) contain the following *effective characteristics* of the composite:

$\mathbf{k}$ – tensor of effective heat conductivity;
$\mathbf{K}$ – tensor of gas permeability;
$\mathbf{C}$ – tensor of effective elasticity moduli;
$\mathbf{f}$ – tensor of phase interactions;
$\overset{\circ}{\boldsymbol{\varepsilon}}$ – tensor of heat deformation;
J – rate of volumetric ablation of the matrix;
J_f – rate of volumetric ablation of the fibres;
D – rate of surface ablation of the composite.

The aim of Chapters 4 – 7 is to establish relationships between these effective characteristics of ablative composite and the characteristics of its separate phases.

3.6.4. BOUNDARY AND INITIAL CONDITIONS

In Chapter 1 different sources of heating a composite caused both volumetric and surface ablations were described, and also the main formulae for calculation of heat fluxes and local pressure of external gas medium were given. By taking account of the relations on the moving surface Σ_q, where linear ablation occurs, boundary conditions for equations (3.41) – (3.46) are written in the form

$$\boldsymbol{\sigma} \cdot \mathbf{n} - \varphi_g p \mathbf{n} = -p_e \mathbf{n}, \tag{3.54}$$

$$p = p_e, \tag{3.55}$$

$$\mathbf{k} \cdot \nabla_x \theta \cdot \mathbf{n} = q_e^0 + q_R - q_w - q_{\text{bl}} - q_l - q_{Rw}, \quad \mathbf{x} \in \Sigma_q. \tag{3.56}$$

Here
$\mathbf{n}$ is the outward drawn normal to the surface,
p_e is the local pressure of external gas flow on the surface Σ_q,
q_e^0 is the convective heat flux supplied to the surface Σ_q (determined by the formulae of paragraph 1.1),
q_{bl} is the heat flux withdrawn from the surface Σ_q due to the pyrolysis gaseous products flowing out from the composite,
q_l is the heat flux withdrawn from the surface Σ_q due to linear ablation (this will be described in detail in Chapter 10),
q_{Rw} is the heat flux withdrawn from the surface Σ_q by radiation from the heated ablative surface into the surroundings,
q_R is the radiant heat flux supplied to the surface Σ_q (determined by the formulae of paragraph 1.1),
q_w is the convective heat flux withdrawn from the surface Σ_q:

$$q_w = \left(\frac{\alpha}{c_p}\right)(I_w - I_w(\theta_0)), \qquad q_e = \left(\frac{\alpha}{c_p}\right)(\bar{I}_e - I_w(\theta_0)), \tag{3.57}$$

$$q_{\text{bl}} = \gamma_{\text{bl}} \varphi_g \rho_g \langle \mathbf{v}_g^{(0)} \rangle_g \cdot \mathbf{n}(\bar{I}_e - I_g), \qquad q_{Rw} = \varepsilon_w \sigma_{SB} \theta_w^4, \qquad q_l = \rho D \Delta e^*.$$

Here ρ is the density of the composite, Δe^* is the specific heat of surface ablation, ε_w is the coefficient of radiation of the composite surface, σ_{SB} is the Stefan-Boltzman constant, (α/c_p) is the coefficient of heat transfer between the surface Σ_q and the overrunning gas flow (see paragraph 1.1), $\bar{I}_e$ is the enthalpy of the overrunning gas flow, I_g and I_w are the enthalpies of gas and solid phases of the composite on the surface Σ_q, respectively, which are determined by the formulae (2.100), (1.4), (1.6), θ_w is the temperature of the surface Σ_q.

On the remaining parts Σ_u, Σ_σ and Σ_v of the surface Σ of the composite we have conditions (3.34), (3.36), (3.37) and (3.39).

Initial conditions for the equation system (3.41) – (3.49) are

$$t = 0: \quad \varphi_f = \varphi_f^0, \qquad \varphi_b = \varphi_b^0, \qquad \rho_g = \rho_g^0, \qquad \theta = \theta_0, \qquad f_\Sigma = f_\Sigma^0. \tag{3.58}$$

A statement of the averaged problem on thermomechanics of composites with volumetric and surface ablation (*problem 'A' in terms of displacements*) consists of the equations (3.41) – (3.45), (3.49), in which the constitutive relations (3.50) – (3.53) and kinematic relations (3.46) should be substituted, with boundary conditions (3.54) – (3.56) and initial data (3.58). The solution of problem 'A' consists of the functions φ_f, φ_b, ρ_g, θ, $\mathbf{u}$ defined within the moving domain $V(t)$, and function f_Σ defined on the surface Σ_q.

Problem 'A' consists of three independent problems: the *problem of mass transfer* of the gas phase in the porous ablative material (3.41) – (3.43), (3.50), (3.52), (3.55), (3.58) for functions φ_f, φ_b, ρ_g; the *problem of heat transfer* (3.44), (3.49), (3.51), (3.56) – (3.58) for θ and f_Σ; and the deformable mechanics problem (3.45), (3.46), (3.54) for the function $\mathbf{u}$, which will be called the *deformation problem*.

The problems of mass transfer and heat transfer cannot be solved separately: equation (3.44) has the term $c_g \nabla_x \theta \cdot \langle \mathbf{v}_g^{(0)} \rangle_g$, describing filtration of gas phase in pores, which essentially changes the temperature field. In the similar way, temperature affects the filtration regime of gas in pores, as pore pressure p, values of J_f and J and also the rate of linear ablation D and, hence, the shape of the ablative material depend on temperature. Thus, the *heat and mass transfer problem* is *coupled* and the equations must be solved simultaneously.

At the same time, the heat-mass-transfer problem can be solved separately from the deformation problem; within the framework developed above, mechanical stresses have no effect on parameters of internal heat and mass transfer. This situation, as a rule, is realized in practice. After determining the pore pressure p, concentrations φ_f, φ_b and temperature θ, we can solve the deformation problem.

3.6.5. STATEMENT OF THE PROBLEM IN TERMS OF STRESSES

Let us formulate the problem in terms of stresses. We need the *equation of strain compatibility* [92]

$$\text{Ink}\ \boldsymbol{\varepsilon} \equiv \nabla \times (\nabla \times \boldsymbol{\varepsilon})^T = \mathbf{0}, \tag{3.59}$$

be added, where Ink is the differential operator of incompatibility [74], which in Cartesian coordinates has the form

$$\epsilon_{pqs}\epsilon_{nkl}\frac{\partial \varepsilon_{sk}}{\partial x_q \partial x_l} = 0, \tag{3.60}$$

where ϵ_{pqm} are the Levi–Civita symbols (see 2.1.1.).

Constitutive relations (3.53) can be resolved with respect to strains:

$$\boldsymbol{\varepsilon} = \overset{\circ}{\boldsymbol{\varepsilon}} + \boldsymbol{\Pi} : (\boldsymbol{\sigma} + p\mathbf{f}), \tag{3.61}$$

where $\boldsymbol{\Pi}$ is the *tensor of elastic pliabilities* being inverse to $\mathbf{C}$: $\boldsymbol{\Pi} : \mathbf{C} = \boldsymbol{\Delta}$.

Tensor of stress functions $\mathbf{F}$ is introduced so that

$$\boldsymbol{\sigma} = \varphi_g p\mathbf{E} + \text{Ink}\ \mathbf{F}, \tag{3.62}$$

then the equilibrium equations (3.45) are satisfied identically, and a statement of the problem in terms of stresses consists of the set of equations (3.41) – (3.44) in which relations (3.50) – (3.52) and also the compatibility equation (3.59) with the constitutive relation (3.61) should be substituted, and to which the boundary conditions (3.54) – (3.56), the equation of ablation surface motion (3.49) and the initial data (3.58) should be added.

Solution of the problem (*problem 'B' in terms of stresses*) consists of the following functions:

$$\varphi_f, \quad \varphi_b, \quad \rho_g, \quad \theta, \quad \mathbf{F}, \quad f_\Sigma.$$

Stresses $\boldsymbol{\sigma}$ in the ablative composite are determined by formula (3.62).

CHAPTER 4

BEHAVIOUR OF MATRICES AT HIGH TEMPERATURES

The aim of the present chapter is to investigate the behaviour of unreinforced isotropic ablative matrices at high temperatures. We will consider the following questions:

- solving the local problems over a periodicity cell;
- calculating the effective characteristics of ablative isotropic materials with the help of these solutions;
- determining the relationship between microcharacteristics of phases (in the main, microstresses) and macrocharacteristics;
- formulating a failure criterion for ablative isotropic materials;
- investigating the phenomena caused by volumetric ablation inside matrices.

4.1. Varying Density of Matrices at High Temperatures

4.1.1. DETERMINATION OF A VOLUMETRIC ABLATION RATE OF MATRICES

As mentioned in paragraph 3.5, in order to determine effective characteristics of ablative materials, in accordance with the concept of the asymptotic averaging method it is necessary to solve preliminarily special local problems over a periodicity cell.

In accordance with the model developed in Chapter 3, isotropic materials under high temperatures consists of three phases: initial solid polymer phase ($i = b$), new solid pyrolytic phase ($i = p$) and gas phase (g-phase) inside pores, and

$$\varphi_f = 0, \qquad \varphi_b + \varphi_p + \varphi_g = 1. \tag{4.1}$$

Figure 4.1 shows a scheme of such a periodic structure for a three-phase matrix; Figure 4.2 presents its periodicity cell V_ξ.

Averaged stresses in a solid frame $\boldsymbol{\sigma}$, averaged strains $\boldsymbol{\varepsilon}$, density ρ and effective characteristics, Poisson ratio ν, elastic modulus E and heat deformation $\overset{0}{\varepsilon}$, for a nonreiforced matrix will be denoted as follows: $\boldsymbol{\sigma}_m$, $\boldsymbol{\varepsilon}_m$, ρ_m, ν_m, E_m and $\overset{0}{\varepsilon}_m$, respectively.

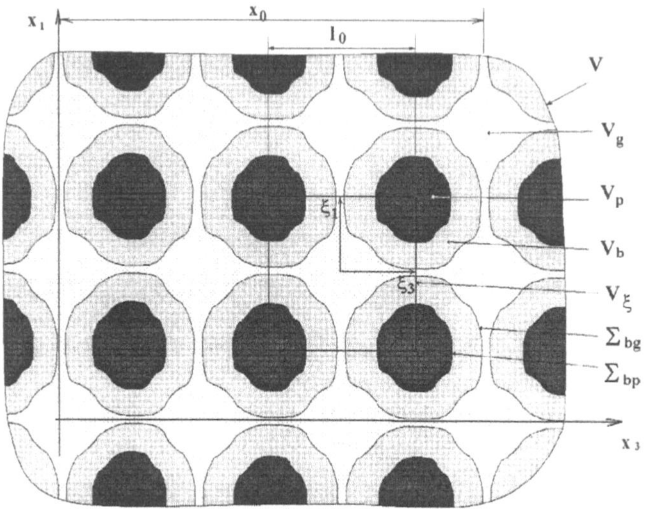

Figure 4.1. A scheme for the periodic structure of a three-phase matrix under high temperatures

Let us consider a model shape for the phases inside the periodicity cell. The pore is assumed to be a cube, and the polymer and pyrolytic phases are hollow cubes (Figure 4.3); we will call this the *cubic model*.

Domains occupied with phases $V_{\xi b}$, $V_{\xi p}$ and $V_{\xi g}$ for the cubic model of a periodicity cell (Figure 4.3) are determined as follows:

$$V_{\xi g}(t) = \left\{ 0 < |\xi_k| < \frac{S_b(t)}{2} \right\}, \qquad V_{\xi b}(t) = \left\{ \frac{S_b(t)}{2} < |\xi_k| < \frac{S_p(t)}{2} \right\},$$

$$V_{\xi p}(t) = \left\{ \frac{S_p(t)}{2} < |\xi_k| < \frac{1}{2} \right\}, \qquad k = 1, 2, 3.$$

These domains change due to motion of the phase boundaries $S_b(t)$ and $S_p(t)$ under heating.

As established in paragraph 3.4, the rate of volumetric ablation is described by the function J, which is connected to the linear rate of motion of the phase interface D_g by the formula (3.30). For cubic pores we can derive the expression for J from the formulae (2.109) for the rate D_g, and (3.30):

$$J = J_0 \varphi_g^\gamma \exp\left(-\frac{E_A}{R\theta}\right), \qquad J_0 = \frac{6}{l_0} \frac{p_e}{\sqrt{R\theta_0}}. \tag{4.2}$$

The cubic model gives a value of the coefficient $\gamma = 2/3$. Experimental data are described more accurately with the value $\gamma = 1$ which will be considered below.

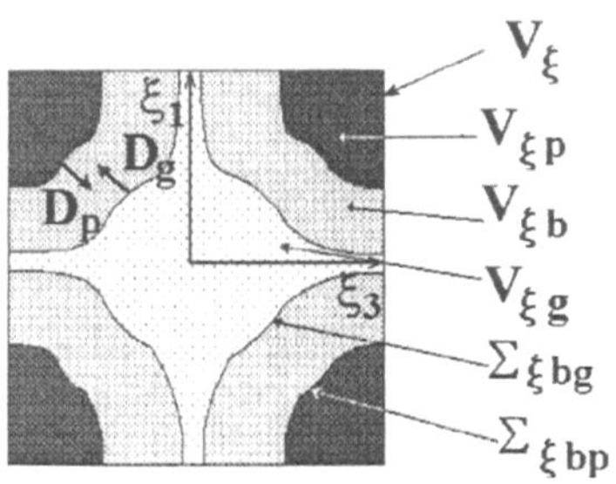

Figure 4.2. A periodicity cell V_ξ of a three-phase matrix under high temperatures

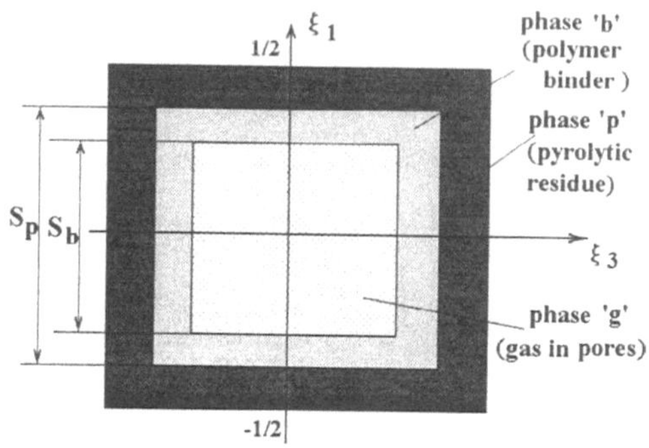

Figure 4.3. Model of periodicity cell V_ξ of a matrix under high temperatures with a cubic form of phases

The mass transfer equations (3.25), (3.26) for nonreinforced matrices under high temperatures have the form

$$\rho_b \frac{\partial \varphi_b}{\partial t} = -J, \qquad \varphi_p = (\varphi_b^0 - \varphi_b)\frac{\rho_b}{\rho_p}(1 - \Gamma). \tag{4.3}$$

4.1.2. EXPERIMENTAL DATA

Let us introduce the following notation: ρ_m – density of a nonreinforced ablative matrix; ρ_m^0 – density of the matrix at the initial state $t = 0$; ρ_m^∞ – density after completion of the volumetric ablation and

$$\rho_m = \rho_b\varphi_b + \rho_p\varphi_p, \qquad \rho_m^0 = \rho_b\varphi_b^0, \qquad \rho_m^\infty = (1 - \Gamma)\rho_m^0. \tag{4.4}$$

Then from equations (4.3), (4.4) one can derive the equation for the change in density ρ_m of the ablative matrix under heating:

$$\frac{\partial \rho_m}{\partial t} = -\frac{J_0}{\rho_b}\Gamma(\rho_m - \rho_m^\infty)\exp\left(-\frac{E_A}{R\theta}\right), \tag{4.5}$$

$$t = 0, \qquad \rho_m = \rho_m^0.$$

On solving equation (4.5), we get

$$\frac{\rho_m}{\rho_m^0} = 1 - \Gamma + \Gamma\exp\left(-\frac{J_0}{\rho_b}\Gamma\int_0^t \exp\left(-\frac{E_A}{R\theta}\right)d\tau\right). \tag{4.6}$$

Equation (4.6) shows that the density ρ_m of the ablative material depends not only on temperature θ but also on the time interval t of heating. The dependence

TABLE 4.1. Thermal characteristics of typical ablative matrices

Characteristics of matrices	Matrix					
	Polyethylene	Epoxy	Epoxy-phenolic	Phenolic	Polyimide	Siliconorganic
$\rho_b \cdot 10^{-3}$, $kg \cdot m^{-3}$	0.95	1.2	1.2	1.1	1.2	1.4
$\rho_p \cdot 10^{-3}$, $kg \cdot m^{-3}$	2	2	2	2	2	2
φ_g^0	0.01	0.06	0.15	0.1-0.25	0.01-0.2	0.2
$\mathbf{k}_b^0$, $W \cdot m^{-1} \cdot K^{-1}$	0.3	0.5	0.5	0.5	0.5	0.5
$\mathbf{k}_p^0$, $W \cdot m^{-1} \cdot K^{-1}$	5	5	5	5	5	5
c_b, $kJ \cdot kg^{-1} \cdot K^{-1}$	2.5	1.2	1.3	1.9	1.2	1.15
c_p, $kJ \cdot kg^{-1} \cdot K^{-1}$	0.6	0.6	0.6	0.6	0.6	0.6
$J_0 \cdot 10^{-5}$, $kg \cdot m^{-3} \cdot s^{-1}$	$2.8 \cdot 10^8$	3.6	3.6	3.3	3.6	5.0
For TD:						
$\frac{E_A}{R} \cdot 10^{-3}$, K	22	7.8	8.3	9	9.5	10.5
Γ	0.9	0.7	0.65	0.4	0.5	0.45
For TOD:						
$\frac{E_A}{R} \cdot 10^{-3}$, K	21.5	7.5	8.0	8.7	9.3	10.3
Γ	1.0	0.97	0.91	0.6	0.75	0.55

of ρ_m on heating time t appears, for example, in heating with different rates $\dot{\theta}$, when temperature grows by the linear law: $\theta = \theta_0 + \dot{\theta}_0 \tau$. On expressing the heating time t in terms of temperature θ: $t = \theta/\dot{\theta}_0$, and substituting this expression into equation (4.6), we derive the dependence of the relative density $(\rho_m/\rho_m^0)(\theta)$ on temperature at a fixed rate $\dot{\theta}_0$.

Equation (4.6) shows that the change in the relative density ρ_m/ρ_m^0 in heating is determined by three constants: E_A – activation energy (see definition in paragraph

2.5), Γ – gasification coefficient and J_0/ρ_b – pre-exponential multiplier.

Values of the parameters depend on the type of medium where heating of the material occurs: inert (vacuum, inert gas) or oxidizing (air, oxygen). In inert gas a process of thermodecomposition (TD) of the material occurs; in an oxidizing environment there is thermo-oxidative decomposition (TOD).

Table 4.1 presents numerical values of constants ρ_b, ρ_p, Γ, E and J_0 for different types of ablative resins for TD and TOD processes.

Figure 4.4,*a* shows dependences of relative density during heating for epoxy, epoxy-phenolic, phenolic, polyimide and silicon-organic resins. Such dependences are called curves of *thermogravimetric analysis* [69]. The experimental data have been obtained for TD in nitrogen (inert) atmosphere during heating with rate 0.1 K/s.

Of the materials listed, epoxy resin has the least thermostability, and polyimide and silicon-organic resins have the highest.

Figure 4.4,*b* shows graphs of relative density ρ_m/ρ_m^0 for the same types of ablative matrices under TOD. The figure shows that the relationships between thermostabilities for different types of matrices in an oxidizing environment are the same as in an inert one. However, in an oxidizing environment, $\rho_m/\rho_m^0(\theta)$ are displaced by approximately 50 - 150° C towards a lower temperature domain; this is caused by decreasing activation energy E_A in TOD (see Table 4.1). Besides, in TOD the proportion of solid pyrolysis residue sharply decreases, especially for epoxy resins; and the gasification coefficient of the matrices Γ increases (Table 4.1).

The change in the relative density ρ_m/ρ_m^0 during heating with constant rate $\dot{\theta}_0$ depends on the value $\dot{\theta}_0$.

Figure 4.5 shows curves of changing relative density $(\rho_m/\rho_m^0)(\theta)$ of epoxy matrix for heating with different rates. Dashed lines indicate experimental values of the density, and solid curves are computed results from equation (4.6). The figure shows that if the rate of heating changes by two orders, then the temperature of the beginning of intensive volumetric ablation (when $\rho_m/\rho_m^0 = 0.9$) changes by 200 °C.

An important characteristic of ablative matrices is the gasification coefficient Γ: the smaller the coefficient Γ, the less gas quantity is released during heating, and the greater the proportion of solid pyrolysis residue, which, as a rule, defines the strength of the material under high temperatures. Phenolic and silicon-organic resins have the smallest values of Γ (Figure 4.4,*a*), however, as it will be shown below, the strength of pyrolysis residue for silicon-organic resins is considerably lower than for phenolic ones.

4.1.3. PORE PRESSURE

The process of intensive release of the gas phase in pores of composite material during TD or TOD leads to the appearance of *pore gas pressure* p. The case of uniform distribution of pore pressure p within the material corresponds to conditions of complete tightness of the ablative material surface, so that there is

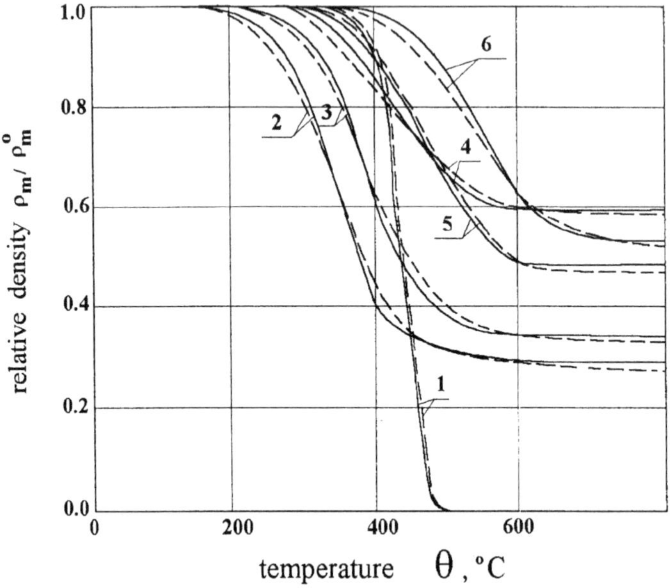

(a)

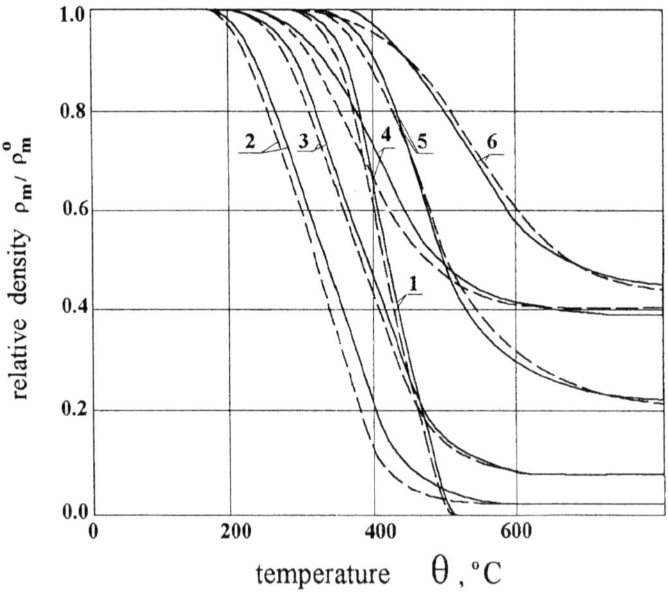

(b)

Figure 4.4. Change in the relative density ρ_m/ρ_m^0 in an inert environment (a) and in air (b) for different types of matrices (1 – polyethylene, 2 – epoxy, 3 – epoxy-phenolic, 4 – phenolic, 5 – polyimide, 6 – silicon-organic) in heating with constant rate $\dot{\theta}_0 = 0.1$ K/s. Solid curves - computations, dashed curves - experimental data [69]

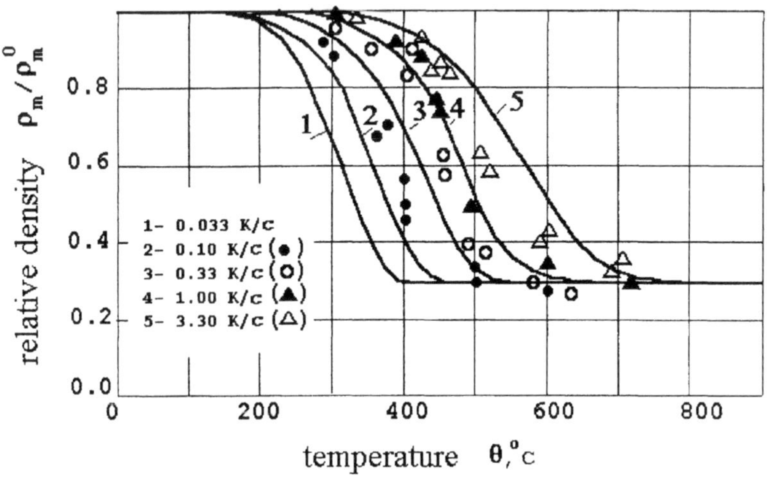

Figure 4.5. Change in the relative density ρ_m/ρ_m^0 of epoxy matrix during heating with different rates $\dot{\theta}_0$, in an inert environment; curves are computations by formula (4.6), points are experimental data

no filtrational motion of gases $\langle \mathbf{v}_g^{(0)} \rangle_g \equiv 0$. From the heat-mass-transfer equations (3.42), (3.43), (3.52) one can determine the pore pressure p dependence on temperature θ:

$$p = \frac{1}{1-\varphi_g} \left[R\theta\Gamma\rho_b(\varphi_b^0 - \varphi_b) + p_0(\theta/\theta_0)(1-\varphi_b^0) \right],$$

$$\varphi_g = 1 - \varphi_b - (\varphi_b^0 - \varphi_b)(1-\Gamma)\frac{\rho_b}{\rho_p}, \tag{4.7}$$

where p_0 is the initial pressure of gases in pores at time $t = 0$. Formula (4.7) defines the maximum value of the pore pressure within the ablative material. Actual values of the pressure p prove to be lower, due to filtration of gases into the surroundings (see paragraph 9.1).

4.2. Effective Elastic Properties of Ablative Matrices at High Temperatures

4.2.1. SOLVING THE MECHANICAL LOCAL PROBLEM OVER A PERIODICITY CELL

Effective mechanical relations for matrices under high temperatures are determined by formula (3.53), and contain three physical characteristics: the tensor of elastic moduli $\mathbf{C}$, the heat deformation tensor $\overset{0}{\boldsymbol{\varepsilon}}$ and the mechanical phase interaction tensor $\mathbf{f}$. In order to establish a relationship between the effective characteristics of the matrix and the corresponding characteristics of the phases, it is necessary to solve the local problem of mechanics (3.15) over the periodicity cell V_ξ

$$
\begin{aligned}
&\nabla_\xi \cdot \boldsymbol{\sigma}_i^{(0)} = \mathbf{0}, \\
&\boldsymbol{\varepsilon}_i^{(0)} = \overset{\circ}{\varepsilon}{}_i^{(0)} \mathbf{E} + \frac{1+\nu_i}{E_i} \boldsymbol{\sigma}_i^{(0)} - \frac{\nu_i}{E_i} \sigma_i^{(0)} \mathbf{E}, \\
&2\boldsymbol{\varepsilon}_i^{(0)} = 2\boldsymbol{\varepsilon}_m + \nabla_\xi \otimes \mathbf{u}_i^{(1)} + (\nabla_\xi \otimes \mathbf{u}_i^{(1)})^T, \quad \boldsymbol{\xi} \in V_{\xi i}, \\
&\overset{\circ}{\varepsilon}{}_i^{(0)} = \alpha_i(\theta - \theta_i^*) - \beta_i, \qquad i = b, p; \\
&\boldsymbol{\sigma}_b^{(0)} \cdot \mathbf{n}^{(0)} = -p\mathbf{n}^{(0)}, \quad \boldsymbol{\xi} \in \Sigma_{\xi bg}; \\
&[\boldsymbol{\sigma}^{(0)}] \cdot \mathbf{n}^{(0)} = 0, \qquad [\mathbf{u}^{(1)}] = 0, \qquad \boldsymbol{\xi} \in \Sigma_{\xi bp}; \\
&\langle \mathbf{u}^{(1)} \rangle_s = 0,
\end{aligned}
\tag{4.8}
$$

where E_i, ν_i, α_i are the elastic moduli, Poisson's ratios and coefficients of heat expansion of the phases, respectively;
$\boldsymbol{\varepsilon}_i^{(0)}$ are strain tensors of the phases;
$\boldsymbol{\varepsilon}_m$ is the tensor of averaged deformations within the matrix;
$\mathbf{u}_i^{(1)}$ are the local variations of phases displacements about the averaged value $\mathbf{u}^{(0)}$ of the whole matrix:

$$
\mathbf{u}_i - \mathbf{u}^{(0)} = \varkappa \mathbf{u}_i^{(1)},
$$

here $\mathbf{u}_i$ are the true values of the phase displacements;
β_i are the chemical shrinkage coefficients of the solid phases ($\beta_b = 0$, $\beta_p = \beta_{sh} > 0$);
$\sigma_i^{(0)}$ is the first invariant of the stress tensor $\boldsymbol{\sigma}_i^{(0)}$;
θ_i^* is the temperature of phase formation;
$\mathbf{E}$ is the unit tensor.

Note that in the present chapter, unlike Chapter 3, superscripts $^{(0)}$ at the averaged values $u^{(0)}$, $p^{(0)}$, $\theta^{(0)}$ have been omitted.

The problem (4.8) over the periodicity cell is, in essence, six separate problems corresponding to six components of the tensor $(\boldsymbol{\varepsilon}_m)_{kl}$ given as initial data.

The tensor $\boldsymbol{\varepsilon}_{kl}$ is connected to the vector of averaged displacements by the Cauchy relations (3.46)

$$
2\boldsymbol{\varepsilon}_m = \nabla_x \otimes \mathbf{u} + (\nabla_x \otimes \mathbf{u})^T.
$$

To solve the problem (4.8), let us consider a model shape of a periodicity cell V_ξ in the form of a hollow cube; Figure 4.6 shows a quarter of the periodicity cell. Axes of the local coordinate system $\boldsymbol{\xi}$ are oriented as shown in the figure.

Due to symmetry of the periodicity cell, it is sufficient to consider only two problems from (4.8): with the component ε_{m33} given as an initial value (and the remaining $\varepsilon_{mkl} = 0$) and with the given component ε_{m13}. The first problem can

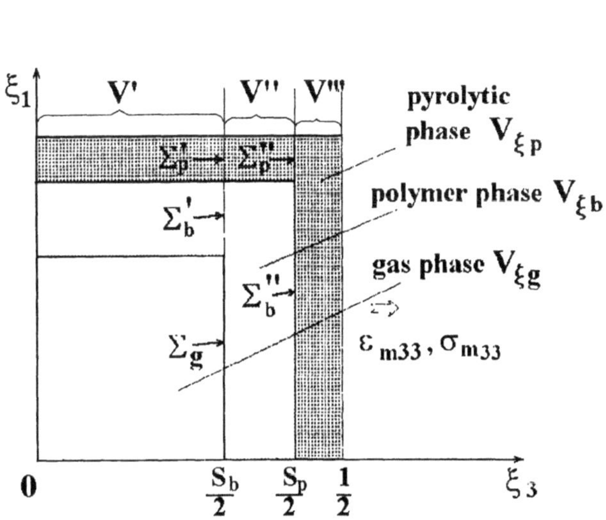

Figure 4.6. A scheme of a quarter of periodicity cell V_ξ for solving the local problem of mechanics (4.8)

be represented as the problem of tension of the cube (periodicity cell V_ξ) in the $O\xi_3$ axis direction by the stress $\tilde{\sigma}_{33}$; for this case the total deformation ε_{m33} of the cube is known. We consider a one-dimensional approximation of the problem, and divide the periodicity cell into three parts V', V'' and V''' (see Figure 4.6)

$$V' = \{|\xi_3| < \frac{S_b}{2}\}, \quad V'' = \{\frac{S_b}{2} < |\xi_3| < \frac{S_p}{2}\},$$

$$V''' = \{\frac{S_p}{2} < |\xi_3| < \frac{1}{2}\}. \tag{4.9}$$

In each of the domains V', V'', V''', within the scope of the one-dimensional approximation, under the action of a certain stress $\tilde{\sigma}_{33}$, there appears a homogeneous stress state in the solid phases which is characterized by stresses $\sigma_i^{','','''}$, strains $\varepsilon_i^{','','''}$ and displacements $u_i^{','','''}$. Let us introduce the notations

$$\sigma_i^{','',''''} = \sigma_{33i}^{(0)}(V^{','',''''}),\ \varepsilon_i^{','',''''} = \varepsilon_{33i}^{(0)}(V^{','',''''}),\ u_i^{','',''''} = u_{3i}^{(1)}(V^{','',''''}),\ i = b, p.$$

Then we can reduce the local problem (4.8) to solving the system of equations

$$\sigma_i^{','',''''} = \text{const}, \tag{4.10}$$

$$\varepsilon_i^{','',''''} = \frac{\sigma_i^{','',''''}}{E_i} + \overset{\circ}{\varepsilon}{}_i^{(0)}, \tag{4.11}$$

$$\varepsilon_i^{','',''''} = \varepsilon_{m33} + \frac{\partial u_i^{','',''''}}{\partial \xi_3}, \qquad i = b, p; \tag{4.12}$$

$$\int_{\Sigma_b'} \sigma_b' d\Sigma + \int_{\Sigma_p'} \sigma_p' d\Sigma - \int_{\Sigma_g} p d\Sigma = \widetilde{\sigma}_{33}, \tag{4.13}$$

$$\int_{\Sigma_b''} \sigma_b'' d\Sigma + \int_{\Sigma_p''} \sigma_p'' d\Sigma = \widetilde{\sigma}_{33}, \qquad \sigma_p''' = \widetilde{\sigma}_{33}, \tag{4.14}$$

$$u_i'(S_b/2) = u_i''(S_b/2), \qquad u_i''(S_p/2) = u_i'''(S_p/2), \qquad i = b, p; \tag{4.15}$$

$$u_b' = u_p', \quad u_b'' = u_p'', \quad \int_{V_{\xi s}} u_i dV = 0, \quad V_{\xi s} = V_{\xi b} + V_{\xi p},$$

where Σ_g is the side of the cube $V_{\xi g}$ simulating the pore;
Σ_b', Σ_p' are the 'b' and 'p' phase interfaces between domains V' and V'';
Σ_b'', Σ_p'' are the 'b' and 'p' phase interfaces between domains V'' and V''' (see Figure 4.6), and

$$|\Sigma_g| = S_b^2, \quad |\Sigma_b'| = S_p^2 - S_b^2, \quad |\Sigma_p'| = 1 - S_p^2, \quad |\Sigma_b''| = S_p^2, \quad |\Sigma_p''| = 1 - S_p^2,$$

$$|V'| = S_b, \qquad |V''| = S_p - S_b, \qquad |V'''| = 1 - S_p, \tag{4.16}$$

$$|V_{\xi g}| = S_b^3, \qquad |V_{\xi b}| = S_p^3 - S_b^3, \qquad |V_{\xi p}| = 1 - S_p^3,$$

$$S_b = \varphi_g^{1/3}, \quad S_p = (1 - \varphi_p)^{1/3}.$$

Exact conditions on continuity of stresses on the boundaries of the phase domains V', V'', V''' are replaced by the approximate integral relations (4.13), (4.14). Equations (4.15) are exact conditions on the continuity of displacements at the boundaries of the phase domains. From these conditions and equation (4.12), continuity of strains in the domains V' and V'' follows:

$$\varepsilon_b' = \varepsilon_p', \quad \varepsilon_b'' = \varepsilon_p''. \tag{4.17}$$

The elastic modulus of the pyrolysis phase is assumed to be independent of temperature.

The elastic modulus of the polymer phase E_b essentially depends on temperature and duration of heating. To take account of the phenomenon playing an important role in calculations of stresses in structures, we use the following dependence [27]:

$$E_b = E_b^0 a^{(0)}, \qquad a^{(0)} = \exp\left(-a(\Delta\widehat{\theta})^r\right). \tag{4.18}$$

$$\Delta\widehat{\theta} = \Delta\theta + P\int_0^t \exp\left(-Q(t-\tau)\right)\Delta\theta(\tau)d\tau, \quad \Delta\theta = \frac{\theta}{\theta_0} - 1.$$

TABLE 4.2. Mechanical characteristics of ablative matrices under high temperatures in an inert environment

Characteristics of matrices	Matrix					
	Polyether	Epoxy	Epoxy-phenolic	Phenolic	Polyimide	Silicon-organic
$\bar{a}$	0.8	0.7	0.4	0.3	0.15	0.1
$\alpha_b \cdot 10^6$, K^{-1}	100	60	70	150	40	87
$\alpha_p \cdot 10^6$, K^{-1}	6	6	6	6	6	6
E_b^0, GPa	2	3.2	3.2	4.1	3	3.8
ν_m	0.33	0.37	0.36	0.35	0.35	0.35
m_p	<0.01	0.3	0.3	1.0	0.6	0.01
β_{sh}	0	0.15	0.15	0.3	0.17	0.17
σ_T^0, MPa	16	74	62	35	85	11
σ_C^0, MPa	25	80	70	43	100	20
σ_S^0, MPa	15	48	42	27	65	10

Constants a, r and P, Q are evaluated in experiments, as described in [40, 45]. Values of the parameter a depend on the type of matrix; they are given in Table 4.2. Value of r is usually equal to $1 \ldots 3$.

We average the equation (4.12) over $V_{\xi s} = V_{\xi p} \cup V_{\xi b}$, then, on taking account of continuity of displacements $u_i^{\prime,\prime\prime}$, we obtain

$$(1 - S_b^2) \int_{-S_b/2}^{S_b/2} \varepsilon_b' d\xi_3 + \int_{V''} \varepsilon_b'' dV + \int_{V'''} \varepsilon_b''' dV = \\ = \varepsilon_{m33}(1 - S_b^3) - S_b^2 \left(u_b'(S_b/2) - u_b'(-S_b/2) \right). \tag{4.19}$$

On substituting the integrated expression (4.12) in place of the displacement difference $u_b'(S_b/2) - u_b'(-S_b/2)$ into the relation (4.19), finally we obtain

$$\int_{V'} \varepsilon_b' dV + \int_{V''} \varepsilon_b'' dV + \int_{V'''} \varepsilon_b''' dV = \varepsilon_{m33}. \tag{4.20}$$

The system of five equations (4.11), two equations (4.13), (4.14) and three equations (4.17), (4.20) allows us to express the eleven functions $\sigma_i^{\prime,\prime\prime}$, $\varepsilon_i^{\prime,\prime\prime,\prime\prime\prime}$, ε_{m33} in terms of the stress $\widetilde{\sigma}_{33}$.

In particular, we can derive the relationship between ε_{m33} and averaged stress σ_{m33} which is connected to $\tilde{\sigma}_{33}$ by the evident equation $\tilde{\sigma}_{33} = \sigma_{m33} - \varphi_g p$. This relation has the form

$$\varepsilon_{m33} = \frac{\sigma_{m33}}{E_m} + f_m p + \overset{\circ}{\varepsilon}_m. \tag{4.21}$$

Here σ_{m33} is the normal component of the tensor of averaged stresses $(\boldsymbol{\sigma}_m)_{kl}$ within the ablative isotropic matrix; E_m is the effective elastic modulus; f_m is the coefficient of phase interaction and $\overset{\circ}{\varepsilon}_m$ is the effective heat deformation. These values are expressed by the formulae

$$E_m = E_b^0 \tilde{a}_1, \tag{4.22}$$

$$\tilde{a}_1 = \left(\frac{S_b}{a^{(0)}(S_p^2 - S_b^2) + m_p(1 - S_p^2)} + \frac{S_p - S_b}{a^{(0)} S_p^2 + m_p(1 - S_p^2)} + \frac{1 - S_p}{m_p}\right)^{-1}, \tag{4.22a}$$

$$f_m = \frac{\varphi_g}{E_b^0}\left(\frac{1}{a^{(0)}(S_p^2 - S_b^2) + m_p(1 - S_p^2)} - \frac{1}{\tilde{a}_1}\right), \tag{4.23}$$

$$\overset{\circ}{\varepsilon}_m = \alpha_b \varphi_b(\theta - \theta_0) + \alpha_p \int_0^t (\theta(t) - \theta(\tau))\dot{\varphi}_p d\tau - \beta_p \varphi_p, \tag{4.24}$$

where $m_p = E_p/E_b$ is the ratio of elastic moduli of the phases.

4.2.2. EFFECTIVE CONSTITUTIVE RELATIONS

An effective constitutive relation connecting shear components ε_{m13} and σ_{m13} is derived in the same way, and has a form similar to (4.21), with the only difference that the elastic modulus E_m is replaced by the shear modulus G_m and there are no terms f_m and $\overset{\circ}{\varepsilon}_m$. Therefore the effective shear modulus G_m depends on concentrations φ_b and φ_p as does E_m (formula (4.22)). It follows that the one-dimensional approximation considered for the local problem (4.8) corresponds to the assumption that the effective Poisson's ratio ν_m is independent of temperature. This assumption usually proves to be quite acceptable for ablative porous materials.

Three-dimensional effective constitutive relations for an ablative matrix have the form

$$\varepsilon_{mkl} = \frac{1 + \nu_m}{E_m}\sigma_{mkl} - \frac{\nu_m}{E_m}\sigma_m \delta_{kl} + \overset{\circ}{\varepsilon}_m \delta_{kl} + f_m p \delta_{kl}, \tag{4.25}$$

where E_m, ν_m, $\overset{\circ}{\varepsilon}_m$ and f_m are the effective characteristics of the ablative matrix determined by formulae (4.22) – (4.24). Relation (4.21) is a consequence of the three-dimensional constitutive relations for uniaxial loading.

Thus, the averaged ablative material is isotropic as are the solid phases contained in it. Relations (4.25) are particular cases of the general effective constitutive relations for porous ablative material derived in paragraph 3.5 (formulae (3.53) and (3.61)).

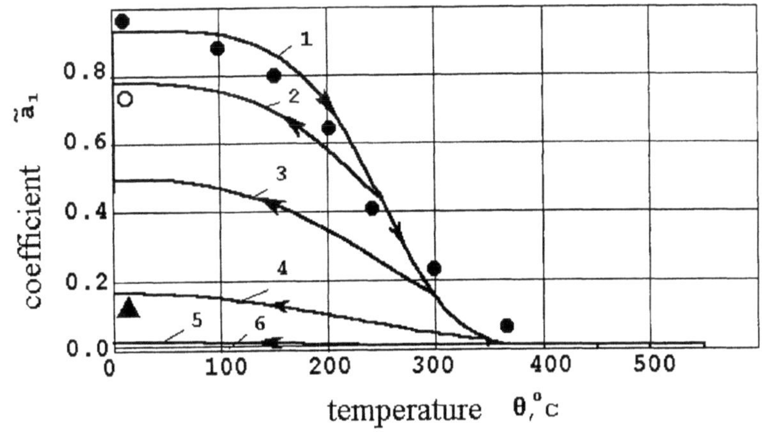

Figure 4.7. Change in the elastic property coefficient $\tilde{a}_1$ of an epoxy matrix *versus* temperature θ under heating and subsequent cooling with rate $\dot{\theta}_0$ =0.1 K/s: 1 - heating (•) and cooling from temperature: 2 - 250 °C (o), 3 - 300 °C, 4 - 350 °C (▲), 5 - 400 °C, 6 - 450 °C. Arrows indicate: heating - ($\longrightarrow$), cooling - ($\longleftarrow$). Lines are computations, points - experimental data.

4.2.3. EXPERIMENTAL INVESTIGATION OF CHANGING ELASTIC PROPERTIES OF MATRIX UNDER HEATING

The change in the elastic modulus of the ablative nonreinforced matrix is described by the function $\tilde{a}_1$ (see formula (4.22))

$$\tilde{a}_1 = E_m/E_b^0, \tag{4.26a}$$

where the function $\tilde{a}_1$ is determined by the formula (4.22a).

Let E_m^0 be a value of the elastic modulus of the matrix in its initial state at temperature $\theta_0 = 293$ K. In accordance with formula (4.22), we can express the modulus E_m^0 as follows:

$$E_m^0 = E_b^0 \tilde{a}_1^0, \tag{4.26b}$$

where

$$\tilde{a}_1^0 = \tilde{a}_1(\theta_0) = \frac{(\varphi_g^0)^{1/3}\left(1 - (\varphi_g^0)^{1/3}\right)}{(\varphi_g^0)^{2/3} + (1 - (\varphi_g^0)^{1/3})^2}.$$

Here φ_g^0 is the initial porosity of the matrix.

Taking formulae (4.26a) and (4.26b) into account, we obtain

$$\tilde{a}_1 = \tilde{a}_1^0 (E_m/E_m^0).$$

We can use this expression to compare experimental data on the change in the relative elastic modulus E_m/E_m^0 in heating with theoretical results of computations

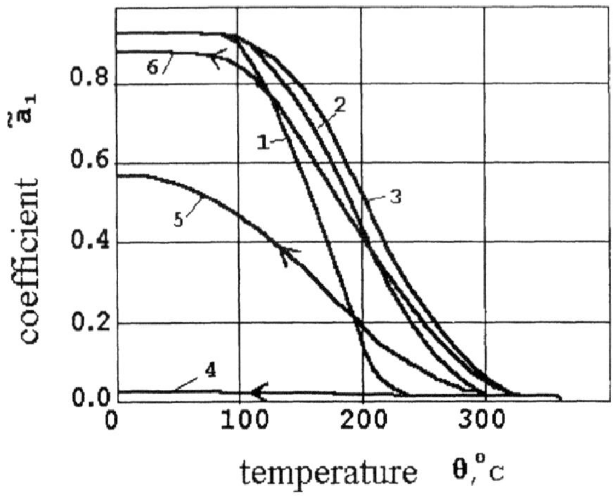

Figure 4.8. The change in the elastic property coefficient $\widetilde{a}_1$ of epoxy matrix *versus* temperature during heating and subsequent cooling with different rates: 1, 2, 3 – heating, 4, 5, 6 – cooling from temperature 350 °C; the rates of heating and cooling are: 1 and 4 - 0.033 K/s, 2 and 5 – 0.33 K/s, 3 and 6 – 3.3 K/s

of function $\widetilde{a}_1$ by formula (4.22a). The functions $\widetilde{a}_1$ and E_m/E_m^0 differ by the coefficient $\widetilde{a}_1^0$ which depends only on initial porosity φ_g^0 of the matrix; it is also determined in experiments (Table 4.2).

Figure 4.7 shows the general form of function $\widetilde{a}_1$ depending on temperature of heating θ for ablative matrices. We discuss some characteristics of this dependence.

For all ablative materials a considerable decrease of the function $\widetilde{a}_1$ occurs before the beginning of volumetric ablation (compare functions $\widetilde{a}_1$ and ρ_m/ρ_m^0 in Figure 1.21); the decrease is connected to peculiarities of features of the polymer phase.

At relatively low temperatures, when the ablation of a polymer matrix has not yet started, the function $\widetilde{a}_1$ is defined by the function $a^{(0)}$ which describes changing elastic-strength properties of the polymer phase; and $S_b = 1$, $S_p = \text{const}$.

The function $a^{(0)}$ is determined by the empirical formula (4.18), where the constants P and Q can be considered as universal, i.e. they are the same for all ablative materials ($P = 1.56 \cdot 10^{-4}$ s^{-1}, $Q = 2.5 \cdot 10^{-4}$ s^{-1}). Different materials are described by different constants a. Table 4.2 shows dimensionless values of the parameter $\bar{a} = a/\theta_0^4$ ($\theta_0 = 293$ K) for different materials.

Changing the coefficient $\widetilde{a}_1 = E_m/E_b^0$ during heating before the ablation beginning is reversible: under cooling, the coefficient $\widetilde{a}_1$ returns practically to its initial value (Figure 4.7).

For higher temperatures, when volumetric ablation of the matrix has started, the function $\widetilde{a}_1$ is determined, mainly, by the functions S_b and S_p. Decreasing the elastic modulus E_m during ablation proves to be irreversible: under cooling, the values of $\widetilde{a}_1$ are restored only partially (Figure 4.7). After volumetric ablation has finished and the material has been completely cooled, the coefficient $\tilde{a}_1$ does not return to its initial value. Figure 4.8 shows that the function $\tilde{a}_1$ describes the phenomena accurately.

Another peculiarity is the considerable dependence of the elastic modulus E_m during heating on the rate of heating $\dot{\theta}_0$; as the heating rate increases, the decrease in elastic modulus of an ablative matrix slows down.

Figure 4.8 shows that varying the elastic modulus of epoxy resin during heating with different rates is adequately described by the function $\tilde{a}_1$.

Figure 4.9,*a* presents the change in the relative elastic modulus $\tilde{a}_1(\theta)$ for different types of ablative matrices: polyether, epoxy, epoxy-phenolic, phenolic, polyimide and silicon-organic, during heating with constant rate $\dot{\theta}_0 = 0.1$ K/s in air; phenolic, polyimide and silicon-organic resins have the greatest thermostability.

We introduce the term of *temperature of thermostability* θ_{hE} of a matrix, as the temperature at which $\tilde{a}_1$ has reached the value 0.5 under heating at constant rate $\dot{\theta}_0$:

$$\tilde{a}_1(\theta_{hE}) = 0.5. \tag{4.26c}$$

Figure 4.9,*a* shows that for a polyether matrix, the thermostability temperature is $\theta_{hE} = 120°$ C; for epoxy matrix - $\theta_{hE} = 180°$ C; for epoxy-phenolic - $\theta_{hE} = 220°$ C; for phenolic - $\theta_{hE} = 280°$ C; for polyimide - $\theta_{hE} = 330°$ C; for silicon-organic - $\theta_{hE} = 380°$ C.

Another important parameter of thermostability is the coefficient m_p characterizing conservation of elastic properties after completion of volumetric ablation. The coefficient is the ratio of elastic moduli of polymer and pyrolysis phases.

Figure 4.9 shows that phenolic resins have the greatest value of m_p ($m_p \approx 1$ in an inert environment): they contain a strong carbon residue. Silicon-organic resins have the smallest value of $m_p \leqslant 0.01$; sometimes they do not form a mechanically strong pyrolysis residue at all (the residue is formed as dispersed particles).

For an inert environment (Figure 4.9,*b*), computed values of the coefficient $\tilde{a}_1$ (and, hence, of the elastic modulus E_m of the matrix) grow with increasing temperature from 500 to 600 °C. This phenomenon is observed also in experiments for phenolic and epoxy resins. It is connected to what happens in the interval of moderate temperatures from 350 to 500 °C; the stiffness of the polymer phase of the matrix is already small and the proportion of strong pyrolysis phase φ_p is not yet great, as the pyrolysis process only starts at these temperatures. At higher temperatures 500 - 600 °C, there occurs the most intensive part of thermodecomposition and the value φ_p reaches its maximum; this leads to increasing a stiffness of the whole matrix E_m.

Note that this phenomenon is observed only in an inert environment; under TOD, the proportion of the pyrolysis phase φ_p is considerably smaller, and porosity φ_g of the matrices is significantly higher; the matrix does not become stiffer (see Figure 4.9,*a*).

Figure 4.10 shows the dependence of the coefficient $\tilde{a}_1 = \tilde{a}_1^0(E_m/E_b^0)$ on temperature θ for a phenolic matrix, when the temperature ranges from 0 to 2700 °C, in an inert environment; the character of $\tilde{a}_1$ remains the same throughout the

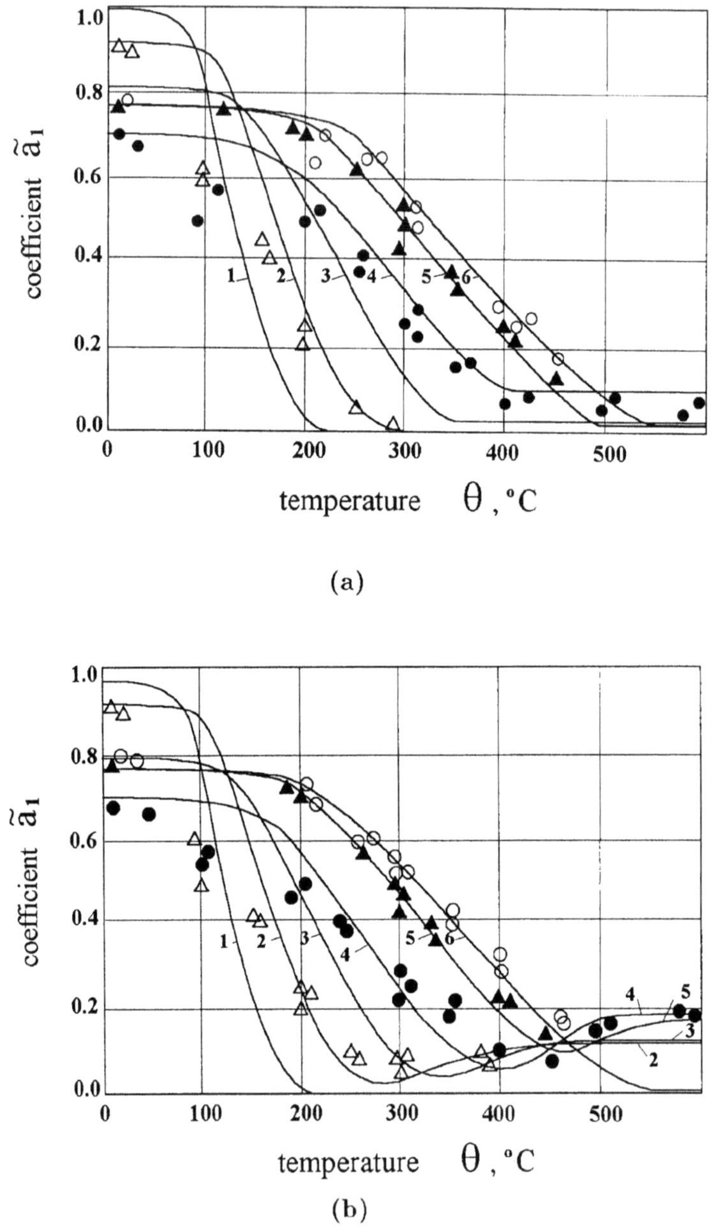

Figure 4.9. The change in the elastic property coefficient $\tilde{a}_1$ *versus* temperature θ during heating with rate $\dot{\theta}_0$ =0.1 K/s for different types of matrices (1 - polyether, 2 - epoxy ($\triangle$), 3 - epoxy, 4 - phenolic ($\bullet$), 5 - polyimide ($\blacktriangle$), 6 - silicon-organic ($\circ$)) (a) in air and (b) in an inert environment (solid curves are computations, points are experimental data)

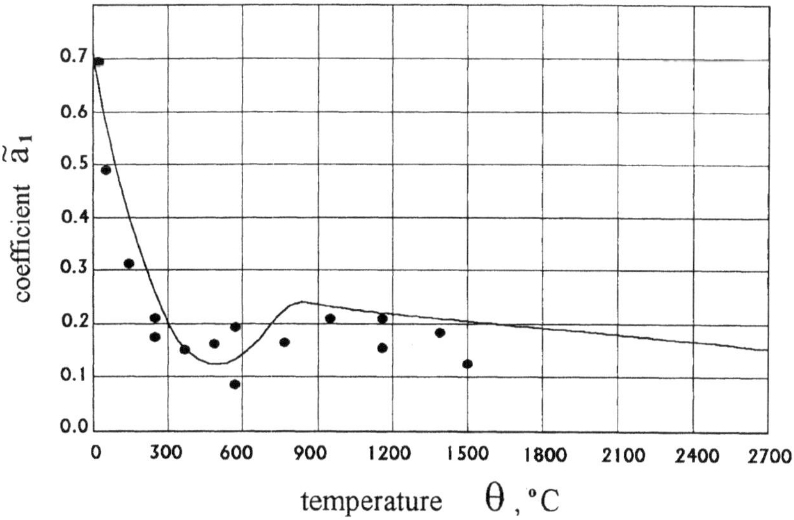

Figure 4.10. The change in the elastic property coefficient $\tilde{a}_1$ for a phenolic matrix *versus* temperature θ during heating with rate $\dot{\theta}_0$ =0.1 K/s in an inert environment (solid curves are computations, points are experimental data)

temperature interval.

4.3. Heat Expansion/Shrinkage of Matrices at High Temperatures

Heat deformation, which occurs only due to heat expansion or shrinkage during heating without mechanical loads applied, is described for ablative matrices by the function $\overset{o}{\varepsilon}_m$ (formula (4.24)). The first term in this formula describes heat expanding the matrix under a relatively low rate of heating, before the ablation begins, when

$$\varphi_b = \varphi_b^0, \qquad \varphi_p = 0$$

and

$$\overset{o}{\varepsilon}_m = \alpha_b(\theta - \theta_0).$$

The third term in (4.24) describes the phenomenon of so-called *chemical shrinkage* of the matrix during volumetric ablation; β_p is the shrinkage coefficient.

The shrinkage phenomenon is a general feature of all ablative polymer matrices. It is caused by decreasing the length of molecular chains due to their breaking into separate sections during heating. As will be shown below, chemical shrinkage is a cause of the appearance of shrinkage stresses in structures made of ablative composites under high temperatures.

The second term in equation (4.24) is a contribution of a pyrolysis residue (p-phase) to heat expansion of the matrix in ablation. The term is an integral because

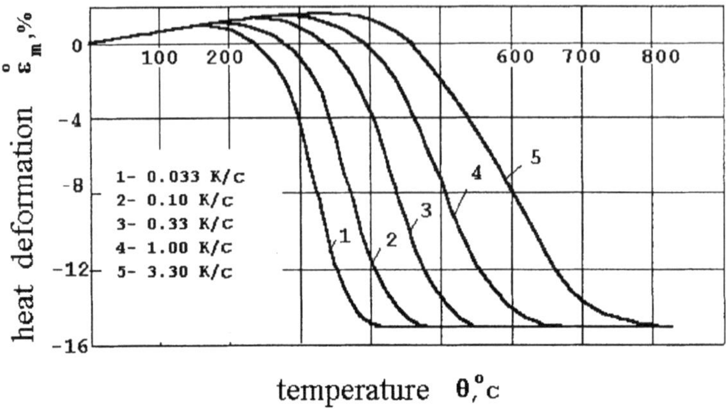

Figure 4.11. Dependences of heat deformation $\overset{\circ}{\varepsilon}_m$ of epoxy matrix on temperature θ during heating with different rates

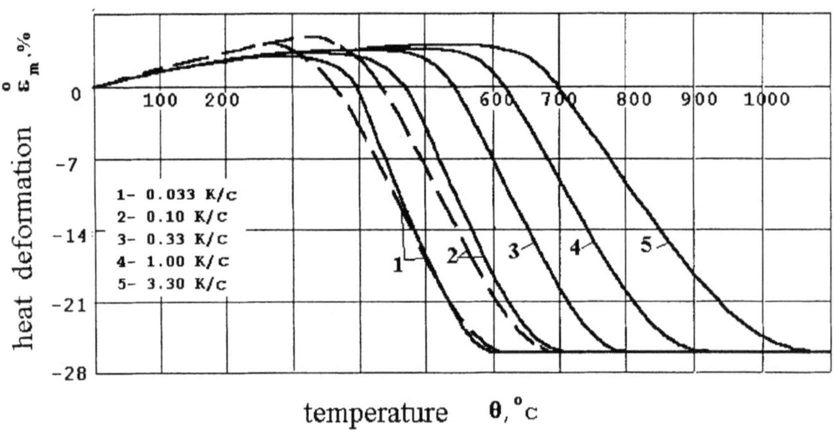

Figure 4.12. Dependences of heat deformation $\overset{\circ}{\varepsilon}_m$ of phenolic matrix on temperature θ during heating with different rates. Solid curves are computations by formula (4.24), dashed lines are experimental data [116]

different microscopic zones of the pyrolysis residue are formed at different values of temperature $\theta_p^* = \theta(\tau)$; this is in contrast to a polymer phase, for which there exists a fixed temperature $\theta_b^* = \theta_0$.

After completion of the volumetric ablation, the concentration of the pyrolysis phase reaches the value

$$\varphi_p^* = \varphi_b^0(1 - \Gamma)(\rho_b/\rho_p).$$

Now, heat deformation of the matrix is defined by heat expansion of the pyrolysis

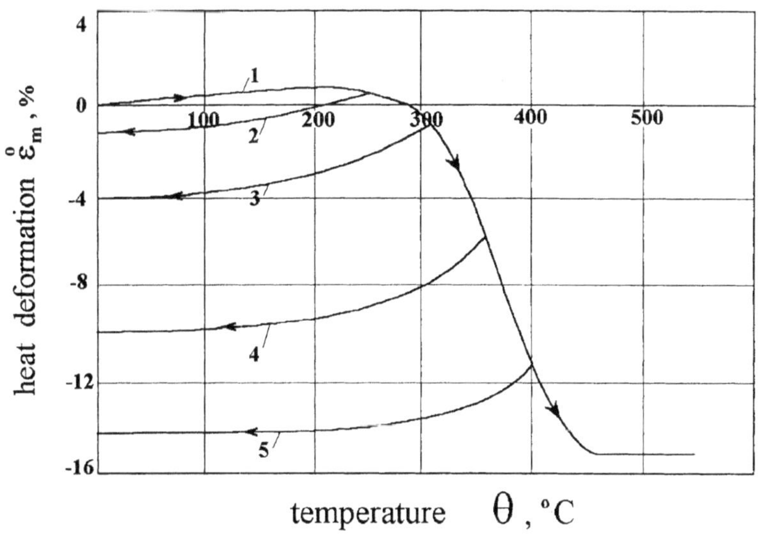

Figure 4.13. Dependences of heat deformation $\overset{o}{\varepsilon}_m$ of epoxy matrix on temperature θ during heating and subsequent cooling with rate $\dot{\theta}_0$ =0.1 K/s (1 – heating; cooling to temperature: 2 - 250 °C, 3 - 300 °C, 4 - 350 °C, 5 - 400 °C)

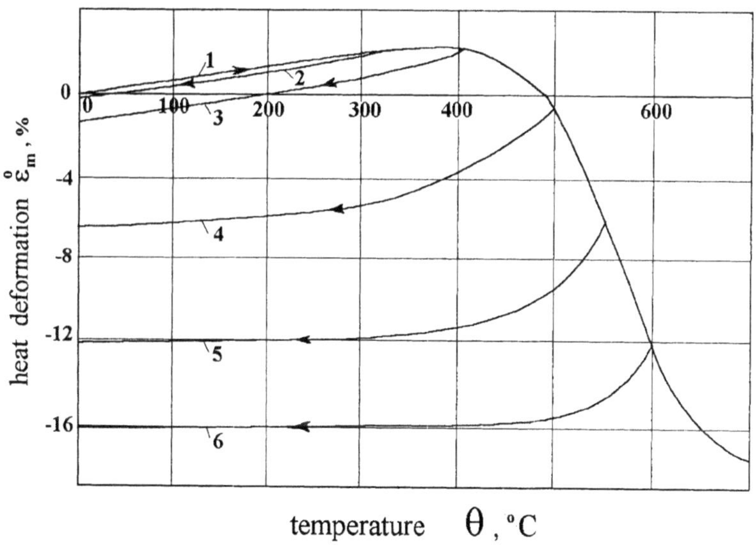

Figure 4.14. Dependences of heat deformation $\overset{o}{\varepsilon}_m$ of phenolic matrix on temperature θ during heating and subsequent cooling with rate $\dot{\theta}_0$ =0.1 K/s (1 – heating; cooling to temperature: 2 - 320 °C, 3 - 400 °C, 4 - 500 °C, 5 - 550 °C, 5 - 600 °C)

phase and by accumulated chemical shrinkage

$$\overset{\circ}{\varepsilon}_m = \alpha_p \varphi_p^*(\theta - \theta_*) - \beta_{sh}. \tag{4.27}$$

Here

$$\theta_* = \frac{1}{\varphi_p^*} \int_0^{t_*} \theta(\tau)\varphi_p d\tau, \qquad \beta_{sh} = \beta_p \varphi_p^*,$$

is the effective temperature of formation of pyrolysis residue, t_* is the time of completion of the volumetric ablation, and β_{sh} is the shrinkage coefficient of the matrix.

Figure 1.21 shows a general form of heat deformation $\overset{\circ}{\varepsilon}_m$ for an ablative matrix under uniform heating according to the linear law $\theta = \theta_0 + \dot{\theta}_0 t$. At first, $\overset{\circ}{\varepsilon}_m$ grows due to heat expansion of a polymer phase, then sharply falls due to the beginning of chemical shrinkage in ablation, then increases after completion of the volumetric ablation due to heat expansion of the pyrolysis residue.

Just as for $\widetilde{a}_1$, the heat deformation $\overset{\circ}{\varepsilon}$ of a matrix depends on heating rate. Figures 4.11, 4.12 show values of $\overset{\circ}{\varepsilon}_m$ for epoxy and phenolic resins during heating with different rates $\dot{\theta}_0$. With growing $\dot{\theta}_0$, the curve $\overset{\circ}{\varepsilon}_m(\theta)$ is displaced into the domain of higher temperatures; this is related to smaller warm-up times for the matrix at higher $\dot{\theta}_0$.

Like $\widetilde{a}_1$, the heat deformation $\overset{\circ}{\varepsilon}_m$ is reversible in heating only up to the temperature at the start of ablation (Figures 4.13, 4.14); for higher temperatures, the dependence $\overset{\circ}{\varepsilon}_m(\theta)$ proves to be irreversible; under cooling it does not return to its initial value, but is displaced into the domain of negative values.

Formula (4.24) shows that the heat deformation $\overset{\circ}{\varepsilon}_m$ of an ablative matrix is described by three constants α_b, α_p and β_{sh}; Table 4.2 gives their values for different types of ablative matrices.

4.4. Strength Properties of Matrices at High Temperatures

4.4.1. MICROSTRESSES IN PHASES

Let us return to the local problem (4.10) – (4.15) in the one-dimensional approximation.

The maximum microstresses in the solid phases of a matrix, called the *microstresses*, in tension along the $O\xi_3$ direction, appear within the domain V' of the periodicity cell (see Figure 4.6). In the unheated state this maximum is realized in a polymer phase; after completion of volumetric ablation of the material it is inside a pyrolysis phase. To take account of the transfer, we introduce average

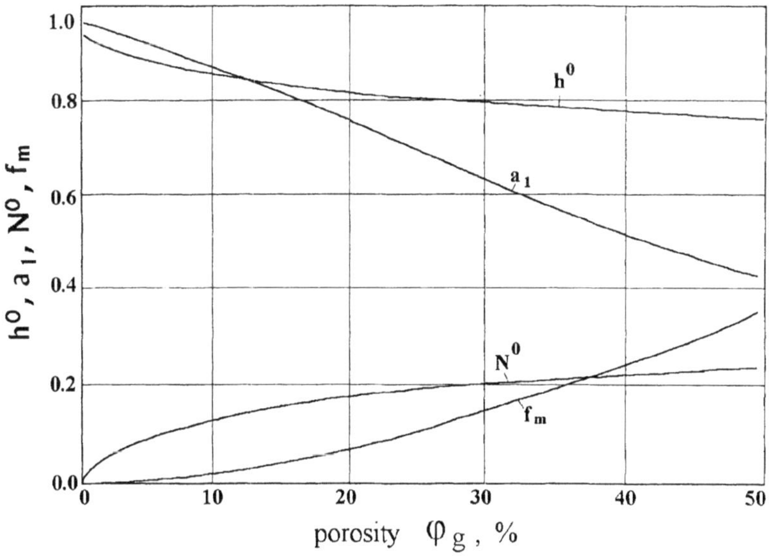

Figure 4.15. Functions a_1, h^0, f and N^0 *versus* porosity φ_g

stresses $\sigma_{kl}^{(0)}(V')$ in the domain V'

$$\sigma_{kl}^{(0)}(V') = \frac{1}{|\Sigma_b'| + |\Sigma_p'|}\left(\int_{\Sigma_b'} \sigma_{bkl}' d\Sigma + \int_{\Sigma_p'} \sigma_{pkl}' d\Sigma\right) =$$

$$= \frac{1}{h^0}\left(\sigma_{mkl} + pN^0\delta_{kl}\right). \tag{4.28}$$

These stresses $\sigma_{kl}^{(0)}(V')$ in the domain V' are the maxima of stresses in the subdomains V', V'' and V''' of the periodicity cells (see Figure 4.6). These stresses are connected to *macrostresses* within the matrix σ_{mkl} by two functions h^0 and N^0 depending on porosity φ_g:

$$h^0 = 1 - \varphi_g^{2/3}, \qquad N^0 = \varphi_g^{2/3} - \varphi_g. \tag{4.29}$$

Figure 4.15 shows that with growing porosity φ_g, N^0 increases while h^0 decreases.

4.4.2. FAILURE CRITERION FOR MATRICES AT HIGH TEMPERATURES

If microstresses $\sigma_{kl}^{(0)}(V')$ reach an ultimate strength in tension σ_T, in shear σ_S or in compression σ_C, then the whole isotropic material fails under high temperatures.

As $\sigma_{kl}^{(0)}(V')$ are averaged values of stresses in phases 'b' and 'p', we can consider ultimate strengths σ_γ ($\gamma = T, C, S$) as averaged values of the corresponding *ultimate strengths* $\sigma_{i\gamma}$ of the phases ($i = b, p$):

$$\sigma_\gamma = \frac{1}{|\Sigma_b'| + |\Sigma_p'|}\left(|\Sigma_b'|\sigma_{b\gamma} + |\Sigma_p'|\sigma_{p\gamma}\right), \quad \gamma = T, C, S. \tag{4.30}$$

To establish exact magnitudes of the ultimate strengths $\sigma_{i\gamma}$ of phases at different temperatures in experiments is very difficult. Therefore we make the following simplifying assumptions:

- ultimate strengths $\sigma_{b\gamma}$ of polymer phases change during heating in proportion to the change of the elastic modulus E_b

$$\sigma_{b\gamma} = \sigma_\gamma^0 a^{(0)}, \qquad \gamma = T, C, S; \tag{4.30a}$$

- ultimate strengths $\sigma_{p\gamma}$ of pyrolysis phases remain constant during heating and are proportional to the corresponding ultimate strengths of the polymer phases

$$\sigma_{p\gamma} = \sigma_\gamma^0 n_p, \qquad n_p = \text{const}, \tag{4.30b}$$

constants n_p and m_p can be different.

By using formulae (4.30a), (4.30b), we can rewrite expressions for the ultimate strengths σ_γ of the solid frame in the form

$$\sigma_\gamma = \sigma_\gamma^0 a_1, \qquad \gamma = T, C, S; \tag{4.31}$$

$$a_1 = \frac{1}{1 - S_b^2}\left(a^{(0)}(S_p^2 - S_b^2) + n_p(1 - S_p^2)\right).$$

For a multiaxial stress state (when several components of the tensor $\sigma_{kl}^{(0)}(V')$ are nonzero), we formulate a so-called *failure criterion*. We can choose the criterion in the invariant form [29]

$$\frac{1}{3}\left(\frac{Y_2}{\sigma_S}\right)^2 + Y_+^2\left(\frac{1}{(\sigma_T)^2} - \frac{1}{3(\sigma_S)^2}\right) + Y_-^2\left(\frac{1}{(\sigma_C)^2} - \frac{1}{3(\sigma_S)^2}\right) = 1, \tag{4.32}$$

where Y_1, Y_2 are the *first and second invariants* of the tensor $\sigma_{kl}^{(0)}(V')$, and Y_+, Y_- are *sign-constant invariants*

$$Y_2^2 = \frac{1}{2}\left(\sigma_{kl}^{(0)} - \frac{1}{3}Y_1\delta_{kl}\right)\left(\sigma_{kl}^{(0)} - \frac{1}{3}Y_1\delta_{kl}\right),$$

$$Y_1 = \sigma^{(0)}_{kk}, \quad Y_\pm = \frac{1}{2}\left(|Y_1| \pm Y_1\right). \tag{4.33}$$

On substituting (4.32) into the expression (4.28) for microstresses $\sigma^{(0)}_{kl}$, we can derive the failure criterion for the whole isotropic material under high temperatures:

$$\frac{1}{3}\left(\frac{Y_{m2}}{\sigma^0_S}\right)^2 + (Y_{m+})^2\left(\frac{1}{(\sigma^0_T)^2} - \frac{1}{3(\sigma^0_S)^2}\right) + \\ + (Y_{m-})^2\left(\frac{1}{(\sigma^0_C)^2} - \frac{1}{3(\sigma^0_S)^2}\right) = (h^0 a_1)^2, \tag{4.34}$$

where Y_{m2}, $Y_{m\pm}$ are the corresponding invariants of the tensor $(\sigma_{mkl} + pN^0\delta_{kl})$:

$$Y_{m2} = \frac{1}{\sqrt{2}}\left((\sigma_{mkl} - \frac{1}{3}Y_{m1}\delta_{kl})(\sigma_{mkl} - \frac{1}{3}Y_{m1}\delta_{kl})\right)^{1/2}, \quad Y_{m1} = \sigma_{mkk}, \tag{4.35}$$

$$2Y_{m\pm} = |Y_{m1} + 3pN^0| \pm (Y_{m1} + 3pN^0).$$

4.4.3. EXPERIMENTAL INVESTIGATION OF STRENGTH PROPERTIES OF MATRICES UNDER HIGH TEMPERATURES

Uniaxial Loading

We can write the second invariant Y_{m2} involved in the formula (4.34) in the explicit form

$$Y_{m2} = \frac{1}{\sqrt{2}}\Big((\sigma_{m11} - \sigma_{m22})^2 + (\sigma_{m22} - \sigma_{m33})^2 + \\ + (\sigma_{m11} - \sigma_{m33})^2 + 6({\sigma_{m12}}^2 + {\sigma_{m23}}^2 + {\sigma_{m13}}^2)\Big)^{1/2}. \tag{4.36}$$

For uniaxial loading of the material along the Ox_1 axis, i.e. when $\sigma_{m11} > 0$ and remaining $\sigma_{mkl} = 0$, we derive the failure criterion from the formula (4.34) in the form

$$\frac{1}{3}\left(\frac{\sigma_{m11}}{\sigma^0_S}\right)^2 + (\sigma_{m11} + 3N^0p)^2\left(\frac{1}{(\sigma^0_T)^2} - \frac{1}{3(\sigma^0_S)^2}\right) = (h^0 a_1)^2. \tag{4.37}$$

Formula (4.37) shows that in uniaxial tension, a condition for failure of the matrix is as follows:

$$\sigma_{m11} = \sigma_{mT} \quad \text{at} \quad \sigma_{m11} \geqslant 0, \tag{4.38}$$

where σ_{mT}, σ_{mC}, σ_{mS} are the strengths of the ablative matrix in tension, compression and shear, respectively:

$$\sigma_{mT} = \sigma^0_T h_{mT}, \quad \sigma_{mC} = \sigma^0_C h_{mC}, \quad \sigma_{mS} = \sigma^0_S h_{mS}. \tag{4.39}$$

Here and below, we use *coefficients of varying a strength* in tension, compression and shear

$$h_{mT} = h^0 a_1 \left((1 - \Delta_T \frac{(\sigma_T^0)^2}{3(\sigma_S^0)^2})^{1/2} - \Delta_T^{1/2} (1 - \frac{(\sigma_T^0)^2}{3(\sigma_S^0)^2})^{1/2} \right),$$

$$h_{mC} = h^0 a_1 \left(\left(1 - \Delta_C \frac{\sigma_C^{0\,2}}{3\sigma_S^{0\,2}} \right)^{1/2} - \Delta_C^{1/2} \left(1 - \frac{\sigma_C^{0\,2}}{3\sigma_S^{0\,2}} \right)^{1/2} \right), \quad (4.40)$$

$$h_{mS} = h^0 a_1 (1 - \Delta_T)^{1/2},$$

where

$$\Delta_T = \left(\frac{3N^0 p}{a_1 h^0} \right)^2 \left(\frac{1}{(\sigma_T^0)^2} - \frac{1}{3(\sigma_S^0)^2} \right),$$

$$\Delta_C = \left(\frac{3N^0 p}{a_1 h^0} \right)^2 \left(\frac{1}{(\sigma_C^0)^2} - \frac{1}{3(\sigma_S^0)^2} \right). \quad (4.41)$$

Thus, changing the strengths σ_{mT}, σ_{mC}, σ_{mS} of the matrix during heating is defined by three causes: varying strength features of the polymer phase described by function a_1, growing porosity of the matrix during thermodecomposition that is characterized by function h^0 and the presence of intrapore pressure p of gaseous products of pyrolysis described by functions Δ_T and Δ_C.

Figure 4.15 gives graphs of functions h^0 and a_1 for epoxy resin *versus* porosity φ_g.

At a low level of intrapore gas pressure, when $\Delta_{T,C} \ll 1$, formulae (4.40) give

$$h_{mT} = h_{mC} = h_{mS} = h^0 a_1. \quad (4.42)$$

The conditions for this case is realized will be considered in Chapter 11.

Figure 4.16 shows a typical form of the function h_{mT} (for example for an epoxy matrix) depending on temperature θ of heating for the case (4.42), when the pore pressure is assumed to be small; and for the general case (4.40) of the presence of pore gas pressure determined by the formula (4.7). When the pore pressure is absent, the function h_{mT} asymptotically tends to a very small but nonzero value with growing temperature. In the presence of pore pressure p, the function h_{mT} sharply falls as soon as volumetric ablation of the matrix starts. This means that, for the first case, if there is no external mechanical load on an ablative matrix (or, as is usual in practice, the mechanical loads are small), the matrix is not completely destroyed but does become significantly weaker; usually it is a weak porous charcoal. For the second case, an ablative matrix is destroyed mechanically at a certain temperature even without external mechanical loads, but only due to the presence of intrapore pressure of gaseous ablation products.

Comparison of theoretical and experimental data (Figure 4.16) shows that actual values of the coefficient of strength h_{mT} are between the two curves described above. This is because the level of pore pressure in actual materials is lower than

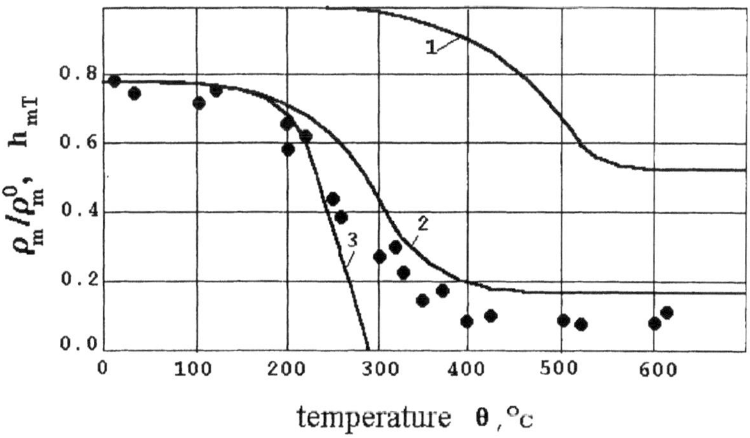

Figure 4.16. Dependences of the relative density ρ_m/ρ_m^0 (curve 1) and the coefficient h_{mT} (curves 2 and 3) on temperature θ during heating with rate $\dot{\theta}_0$ =0.1 K/s, for a phenolic matrix (curve 2 – calculation from formula (4.42), curve 3 – calculation from formula (4.40))

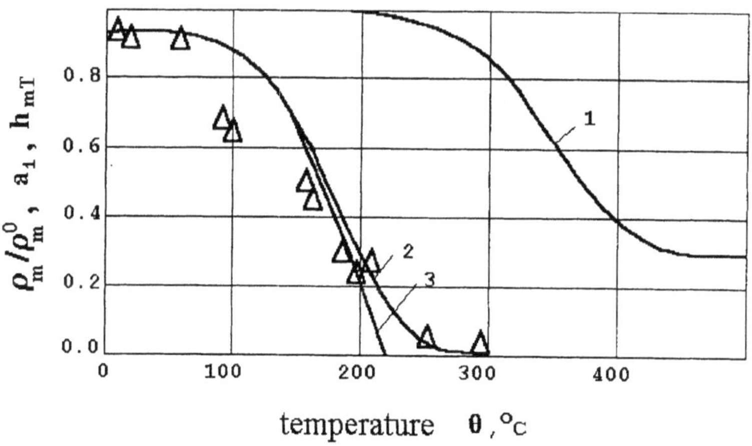

Figure 4.17. Dependences of relative density ρ_m/ρ_m^0 (1), relative strength a_1 (2) and function h_{mT} (3) of a phenolic matrix on temperature θ during heating with rate $\dot{\theta}_0$ =0.1 K/s. Curves are computations, points are experimental data

the maximum magnitude determined by the formula (4.7), but higher than the normal atmospheric pressure $p_0 = 0.1$ MPa. The exact level of pressure p can be found by solving the filtration equation (3.43) (see Chapters 9, 11).

The phenomenon of thermodecomposition by pore pressure is of the utmost danger for thermal protection of structures made of ablative materials [45]. Therefore, we will pay close attention to the conditions for the appearance of high intrapore pressure.

Figure 4.17 shows functions ρ_m/ρ_m^0, h_{mT} and a_1 depending on temperature θ of heating for a phenolic matrix. All the mentioned phenomena for this matrix occur in the field of lower temperatures.

The strength of the matrix under uniaxial compression, when $\sigma_{m11} \leqslant 0$, also has peculiarities at high temperatures. Formula (4.37) shows that the failure criterion for the ablative matrix under compression can have two different forms, depending on the value of pore gas pressure p:

$$|\sigma_{m11}| = \sigma_{mT} \quad \text{at} \quad -3N^0 p < \sigma_{m11} \leqslant 0, \tag{4.43}$$

and

$$|\sigma_{m11}| = \sigma_{mC} \quad \text{at} \quad \sigma_{m11} \leqslant -3N^0 p.$$

Hence:

- under certain external mechanical compressive loads, an ablative matrix can be destroyed by a tension mechanism (!) due to the presence of pore pressure;
- maximum stress of destruction under compression σ_{m11} grows in absolute magnitude with increasing the pore gas pressure p.

The strength of ablative matrices in shear ($|\sigma_{m12}| \neq 0$, the other $\sigma_{mkl} = 0$) also depends on the pore gas pressure p

$$\sigma_{m12} = \sigma_{mS}. \tag{4.44}$$

All these peculiarities for tension are valid for shear too.

Changing Strength Properties of Matrices under Multiaxial Loading and High Temperatures

Actual ablative matrices contained in composites are, as a rule, under conditions of multiaxial stress. The failure criterion (4.37) for the three-axial stress state, when $\sigma_{m11} \neq 0$, $\sigma_{m12} \neq 0$, $\sigma_{m22} \neq 0$ and the other $\sigma_{mkl} = 0$, has the form

$$\left(\frac{\sigma_{m12}}{\sigma_S^0}\right)^2 + \left(\frac{\sigma_{m11}+\sigma_{m22}}{\sigma_T^0}\right)^2 - \frac{\sigma_{m11}\sigma_{m22}}{\sigma_S^{0\,2}} + $$
$$+ (\sigma_{m11}+\sigma_{m22})6N^0 p\left(\frac{1}{\sigma_T^{0\,2}} - \frac{1}{3\sigma_S^{0\,2}}\right) = (h^0 a_1)^2(1-\Delta_T),$$

$$\text{if} \quad \sigma_{m11}+\sigma_{m22} \geqslant -3N^0 p, \tag{4.45}$$

$$\left(\frac{\sigma_{m12}}{\sigma_S^0}\right)^2 + \left(\frac{\sigma_{m11}+\sigma_{m22}}{\sigma_C^0}\right)^2 - \frac{\sigma_{m11}\sigma_{m22}}{\sigma_S^{0\,2}} +$$
$$+ (\sigma_{m11}+\sigma_{m22})6N^0p\Big(\frac{1}{\sigma_C^{0\,2}} - \frac{1}{3\sigma_S^{0\,2}}\Big) = (h^0a_1)^2(1-\Delta_C),$$
$$\text{if} \quad \sigma_{m11}+\sigma_{m22} \leqslant -3N^0p. \tag{4.46}$$

Failure conditions (4.38), (4.43) and (4.44) follow, in particular, from this criterion.

In the coordinate system σ_{m11}, σ_{m22}, σ_{m22} equations (4.45) and (4.46) describe a *strength surface*, being a combination of two surface parts of two ellipsoids joining along the plane

$$\sigma_{m11}+\sigma_{m22} = -3N^0p.$$

Crosssections of the strength surface by the plane by $O\sigma_{m11}\sigma_{m22}$ for epoxy resin are shown in Figure 4.18, and the crosssection by the plane $O\sigma_{m11}\sigma_{m12}$ is given in Figure 4.19 for several values of heating temperature θ. As before, heating is assumed to be linear $\theta^{(0)} = \theta_0 + \dot{\theta}_0 t$ with the rate $\dot{\theta}_0 = 0.1$ K/s.

Let us indicate some characteristics of the strength surface shape for the matrix under high temperatures.

- With increasing intrapore gas pressure p, the strength surface is displaced in the direction of the A axis determined by equations $\sigma_{m22} = \sigma_{m11}$, $\sigma_{m12} = 0$. (Figure 4.18).
- For a certain value of the intrapore pressure $p = p_{cr}$, the matrix under high temperatures and compressing stresses $\sigma_{m11} \leqslant 0$, $\sigma_{m22} \leqslant 0$ can be destroyed according to the tension mechanism.
- Strength under biaxial tension in the domain $(\sigma_{m11} \geqslant 0, \sigma_{m22} \geqslant 0)$ decreases during heating for all temperatures. Strength under compression in domains $(\sigma_{m11} \leqslant 0)$ and $(\sigma_{m22} \leqslant 0)$ decreases at relatively low temperatures before the beginning of ablation. When the ablation starts and pore gas pressure appears, destroying stresses in compression grow. We can explain this phenomenon by noting that compressive loads σ_{m11} and σ_{m22} applied to the specimen should first counterbalance tensile stresses which appeared due to the presence of gas in pores and then cause the material to fail according to the compression mechanism.
- Strength under combined shear and tension within the domain $(\sigma_{m11} \geqslant 0$, $\sigma_{m12} \geqslant 0)$ decreases during heating at any temperatures (Figure 4.19). But strength of ablative matrices under combined shear and tension within the zone $(\sigma_{m11} \leqslant 0, \sigma_{m12} \geqslant 0)$ decreases during heating up to the temperature corresponding to the beginning of ablation; at higher temperatures strength grows due to the appearance of pore gas pressure p. The physical essence of the phenomenon is the same as for biaxial compression.
- The strength surface in the plane $O\sigma_{m11}\sigma_{m12}$ moves during heating along the axis $O\sigma_{m11}$ in the direction of negative values of σ_{m11}. At certain values of

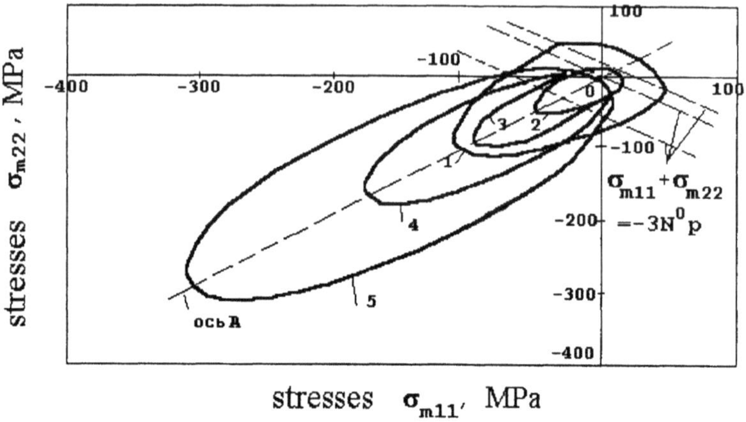

Figure 4.18. Crosssections of the strength surface of an epoxy matrix by the plane $O\sigma_{m11}\sigma_{m22}$ for different temperatures (1 - 100 °C, 2 - 200 °C, 3 - 240 °C, 4 - 260 °C, 5 - 280 °C) under heating with rate $\dot{\theta}_0$ =0.1 K/s

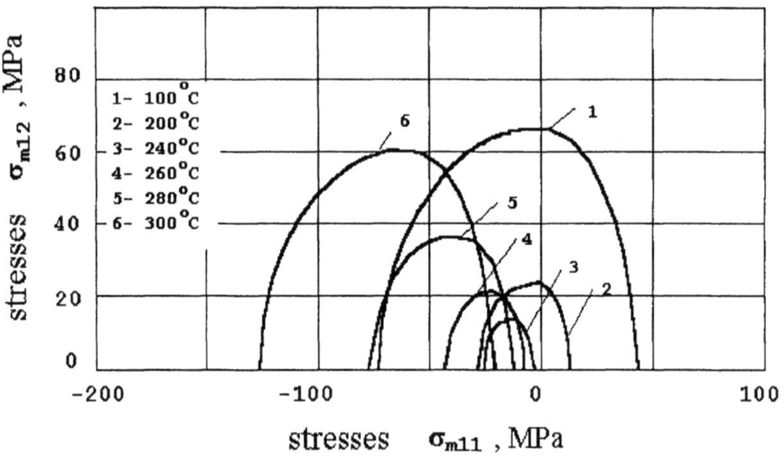

Figure 4.19. Crosssections of the strength surface of an epoxy matrix by the plane $O\sigma_{m11}\sigma_{m12}$ for different temperatures under heating with rate $\dot{\theta}_0$ =0.1 K/s

pore pressure $p \geqslant p_{cr}$, the matrix under high temperatures is destroyed under compression according to the tension mechanism.

- When $p = p_{cr}$, the strength surface in planes $O\sigma_{m11}\sigma_{m12}$ and $O\sigma_{m11}\sigma_{m22}$ passes through the origin of coordinates. The physical essence of this fact is that the ablative matrix in heating up to the state, when pore pressure p_{cr} appears, is destroyed without any external mechanical loads, where

$$p_{cr} = \frac{a_1}{3N^0}\left(\frac{1}{(\sigma_T^0)^2} - \frac{1}{3(\sigma_S^0)^2}\right)^{-1/2}.$$

Note that these effects take place for uniform distribution of fields of pore pressure p and stresses σ_{mkl} over the volume of the ablative matrix. Effects connected with non-uniformity of the fields will be considered in Chapter 10.

4.5. Heat Conductivity and Heat Capacity of Matrices at High Temperatures

4.5.1. SOLVING THE LOCAL PROBLEM OF HEAT CONDUCTION

Relations (3.51) show that the effective tensor of heat conductivity $\mathbf{k}_m$ of an ablative matrix connecting the averaged heat flux $\mathbf{q}$ and $\nabla_x\theta$

$$\mathbf{q} = -\mathbf{k}_m\nabla_x\theta \tag{4.47}$$

can be determined by solving the local problem of heat conductivity (3.16) over the periodicity cell. The governing equations are

$$\begin{gathered}
\nabla_\xi \cdot \mathbf{q}_i^{(0)} = 0, \\
-\nabla_\xi\theta^{(1)} = \nabla_x\theta - \frac{1}{k_i}\mathbf{q}_i^{(0)}, \qquad \boldsymbol{\xi} \in V_{\xi i}; \\
\mathbf{q}_b^{(0)} \cdot \mathbf{n}^{(0)} = 0, \qquad \boldsymbol{\xi} \in \Sigma_{\xi bg}; \\
[\mathbf{q}^{(0)}] \cdot \mathbf{n}^{(0)} = 0, \qquad [\theta^{(1)}] = 0, \qquad \boldsymbol{\xi} \in \Sigma_{\xi bp}, \quad i = b, p; \\
\langle\theta^{(1)}\rangle_s = 0.
\end{gathered} \tag{4.48}$$

Here, as for the problem (4.8), the gradient $\nabla_x\theta$ is assumed to be known. After determination of a solution of the local problem in the form of a function $\mathbf{q}_i^{(0)} = \mathbf{Q}_i(\boldsymbol{\xi}) \cdot \nabla_x\theta$, the effective heat conductivity tensor $\mathbf{k}_m$ is found by averaging:

$$\mathbf{k}_m = \langle\mathbf{Q}\rangle_s = k_m\mathbf{E}. \tag{4.49}$$

Since the composite phases are isotropic, the tensor $\mathbf{k}_m$ is isotropic too, i.e. it has one independent component k_m, that is shown in the formula (4.40).

We consider, just as for the local mechanics problem (4.8), a periodicity cell V_ξ in the form of a cube (see Figure 4.6).

As a rule, we can neglect the heat conductivity of the gas phase as compared with heat conductivity of the solid phase. Then the problem (4.48) for the determination of the local temperature field $\theta^{(1)}$ within the porous periodicity cell is analogous to the problem (4.8) for the determination of the displacements $\mathbf{u}^{(1)}$. Solving the problem is conducted similarly: $\mathbf{q}_i^{(0)}$ corresponds to $\boldsymbol{\sigma}_i^{(0)}$, $\nabla_x\theta$ to $\boldsymbol{\varepsilon}$, $\theta^{(1)}$ to $\mathbf{u}^{(1)}$, k_i to E_i, k_m to E_m, $\mathbf{q}$ to $\boldsymbol{\sigma}$, and the effective constitutive relations (4.47) correspond to (4.25).

Hence, we can solve the problem (4.48) like the problem (4.8); the effective coefficient of heat conductivity k_m can be determined like the elastic modulus E_m (formula (4.28)):

$$k_m = k_b^0 b_1 b^{(0)},$$

$$b_1 = \left(\frac{S_b}{(S_p^2 - S_b^2) + m_k(1 - S_p^2)} + \frac{S_p - S_b}{S_p^2 + m_k(1 - S_p^2)} + \frac{1 - S_p}{m_k} \right)^{-1}, \qquad (4.50)$$

where $m_k = k_p^0/k_b^0$ is the ratio of heat conductivities of the polymer and pyrolysis phases; $b^{(0)}$ is the function describing the change of the coefficients of phase heat conductivity k_i during heating. For most ablative matrices, we can express the function $b^{(0)}$ as a power with exponent (1/2) [27]:

$$k_i = k_i^0 b^{(0)}, \qquad b^{(0)} = (\theta/\theta_0)^{1/2}, \qquad i = p, b. \qquad (4.51)$$

4.5.2. EXPERIMENTAL DATA

Figure 4.20 shows the dependence of the heat conductivity coefficient k_m on temperature θ during heating by the linear law: $\theta = \theta_0 + \dot{\theta}_0 \tau$, for different types of ablative matrices. Points exhibit experimental data, and solid curves show approximation with the help of equation (4.50). As seen from Figure 4.20, an accuracy of the approximation is sufficiently high.

Increasing the heat conductivity k_m at the first stage of heating is connected to growing a heat conductivity of polymer phase k_b. During further heating the matrix up to temperatures of the beginning of volumetric ablation, the heat conductivity k_m diminishes; this is connected to the formation of secondary porosity φ_g. And, at last, after completion of the ablation, growth of k_m occurs again due to increasing the heat conductivity k_p of a solid residue. Figure 4.20 shows that the higher the gasification coefficient Γ of ablative material, the greater the heat conductivity k_m decreases within the ablation zone. For example, for an epoxy matrix, where $\Gamma = 0.7$, the coefficient k_m decreases by 1.5 times, and for phenolic resin, where $\Gamma = 0.4$, the heat conductivity k_m does not practically decrease at ablation temperatures.

Table 4.1 gives numerical values of constants k_b^0, k_p^0 for different types of ablative matrices.

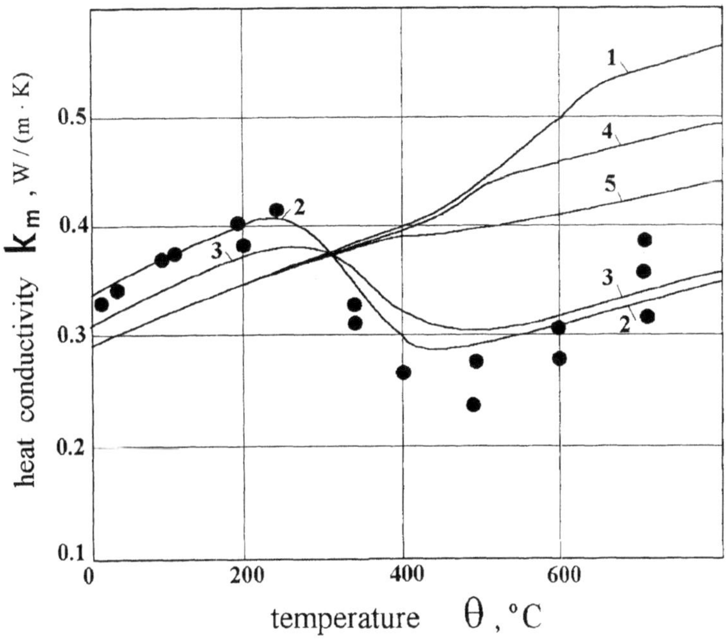

Figure 4.20. Changing heat conductivity coefficient k_m *versus* temperature during heating with rate $\dot{\theta}_0$ =0.1 K/s for different types of matrices (1 - silicon-organic, 2 - epoxy (•), 3 - epoxy-phenolic, 4 - phenolic, 5 - polyimide). Solid curves are computations, points are experimental data

4.5.3. HEAT CAPACITY

Variation of the heat capacity c for ablative matrices during heating is described by the simple relation

$$c(\theta) = \frac{\rho_b c_b \varphi_b(\theta) + \rho_p c_p \varphi_p(\theta)}{\rho_b \varphi_b(\theta) + \rho_p \varphi_p(\theta)}. \tag{4.52}$$

Figure 1.21 shows the graph of $c(\theta)$ for epoxy resin. Table 4.1 gives numerical values of constants c_b and c_p for different types of ablative matrices.

4.6. Gas Permeability of Matrices at High Temperatures

4.6.1. SOLVING THE LOCAL PROBLEM OF GAS DYNAMICS

As established in paragraph 3.5, the gas-permeability coefficient K of ablative porous material is determined by formula (3.33) with the help of the solution of

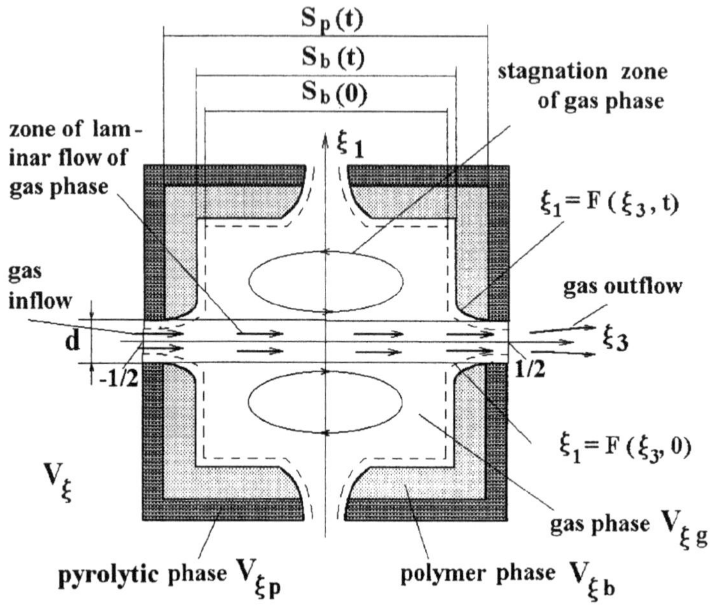

Figure 4.21. Periodicity cell V_ξ of a cubic form for solving the local problem of gas dynamics

the local problem of gas dynamics (3.14)

$$\nabla_\xi \cdot \mathbf{v}_g^{(0)} = 0,$$

$$-\nabla_\xi p^{(1)} + (\mu_g + \nu_g)\nabla_\xi \otimes \nabla_\xi \cdot \mathbf{v}_g^{(0)} + \mu_g \nabla_\xi \cdot \nabla_\xi \otimes \mathbf{v}_g^{(0)} = \nabla_x p; \qquad \boldsymbol{\xi} \in V_{\xi g}; \quad (4.53)$$

$$\mathbf{v}_g^{(0)} = 0, \qquad \boldsymbol{\xi} \in \Sigma_{\xi b g},$$

where $\mathbf{v}_g^{(0)}$ is the unknown velocity of the gas phase; p is the average gas pressure in the pores being an 'external' parameter of the problem (superscript $^{(0)}$ at p is omitted); μ_g and ν_g are the viscosity coefficients; ∇_ξ and ∇_x are the gradients with respect to local and global coordinates $\boldsymbol{\xi}$ and $\mathbf{x}$, respectively; $V_{\xi g}$ is the volume of gas phase in the periodicity cell V_ξ of porous ablative material; $p^{(1)}$ are the local pulsations of gas pressure within the pore about the average value p: $p_u - p = \varkappa p^{(1)}$, here p_u is the true pressure in the gas phase.

We assume that pores in the medium considered have a cubic form and are connected with each other by very thin channels oriented along the $O\xi_k$ coordinate axes $k = 1, 2, 3$ (Figure 4.21).

The shape of the channels at time t is assumed to be an axisymmetric surface of revolution about the corresponding coordinate axis. An equation for the surface element of the channels oriented along the $O\xi_3$ axis is chosen in the exponential form

$$r = F(\xi_3, t) \equiv A \exp\left(-n_k(\xi_3 - l_k(t)\right), \tag{4.54}$$

$$l_k(t) = \frac{S_b(t) - S_b(0)}{2}, \qquad \frac{S_b(t)}{2} \leqslant \xi_3 \leqslant 1/2,$$

$$r = \left(\xi_1^2 + \xi_2^2\right)^{1/2}.$$

For the other axes, there are analogical expressions. Here A and n_k are constants characterizing the geometric shape of the channels, $S_b(t)$ is the linear dimension of the pore (the side of the cube), r is the radial coordinate.

We now present a model of the change in an internal structure of pores in ablation.

At an arbitrary time $t > 0$ of the ablation, the pore size increases by the value $2l_k$. We assume that at the same time the function F giving the channel shape is displaced parallel to the axis of symmetry by the amount l_k. This fact is expressed in equation (4.54).

The minimal diameter of the open crossection of the channel $d = 2F(\frac{1}{2}, t)$ for this model at the arbitrary time is determined by the formula

$$d = 2A \exp\left(-\frac{n_k}{2}(1 + (\varphi_g^0)^{1/3})\right) \exp\left(\frac{n_k}{2}\varphi_g^{1/3}\right), \tag{4.55}$$

where $\varphi_g^0 = S_b^3(0)$ is the initial porosity of the material (see Figure 4.3).

One-dimensional gas flow moving, for example along the $O\xi_3$ axis, through such a porous structure is defined, in the main, by the minimal diameter d of the channels. According to this, a picture of the gas flow within the pore is assumed to consist of a zone of laminar flow and a stagnation zone. The domain of laminar flow is bounded by a cylindrical surface with diameter d, and for a steady-state flow the surface can be considered as impermeable.

Within the scope of these assumptions, the local problem of gas dynamics (4.53) for the case of one-dimensional gas flow along the $O\xi_3$ axis has the solution

$$p^{(1)} = 0, \qquad v_{gI}^{(0)} = 0, \quad I = 1, 2;$$

$$v_{g3}^{(0)} = v_{g3}^{(0)}(r), \tag{4.56}$$

where the function $v_{g3}^{(0)}(r)$ can be found from the equation

$$\mu_g \frac{1}{r}\frac{\partial}{\partial r}\left(r\frac{\partial v_{g3}^{(0)}}{\partial r}\right) = \frac{\partial p}{\partial x_3}, \qquad 0 < r < (d/2);$$

$$v_{g3}^{(0)} = 0, \qquad r = (d/2). \tag{4.57}$$

Averaging the equation (4.57) over the channel crosssection, we obtain

$$\langle v_{g3}^{(0)} \rangle_g = -\frac{d^2}{32\mu_g}\frac{\partial p}{\partial x_3}. \tag{4.58}$$

In a similar way, on considering two other directions $O\xi_2$ and $O\xi_3$, we get an expression of type (3.52)

$$\varphi_g \rho_g \langle \mathbf{v}_g^{(0)} \rangle_g = -K \nabla_x p \tag{4.59}$$

for the average velocity $\langle \mathbf{v}_g^{(0)} \rangle_g$ of the gas flow within the periodicity cell. Here K is the gas-permeability coefficient and

$$K = (\rho_g d^2)/(32\mu_g).$$

On substituting equations (4.54) into (4.59), we derive the expression for the gas permeability constant of the porous medium

$$K = K_0 \exp\,(n_k \varphi_g^{1/3}), \tag{4.60}$$

where

$$K_0 = \frac{A^2 \rho_g}{256\mu_g} \exp\left(-n_k(1 + (\varphi_g^0)^{1/3})\right). \tag{4.61}$$

Thus, the variation of the gas permeability K of a porous medium is determined by its porosity, and is described by two constants K_0 and n_k which can be calculated by taking account of the geometrical structure of real pores (that is a complicated problem) or with the help of an approximation to the experimentally determined dependence of K on φ_g.

Comparison of the theoretical dependence (4.61) with the experimental results for different matrices before and after heating is presented in Figure 4.22, where $\bar{K}$ is the dimensionless coefficient of gas permeability,

$$\bar{K} = \bar{K}_0 \exp\,(n_k \varphi_g^{1/3});$$

$$\bar{K}_0 = K_0 R \theta_0 t_0 / x_0^2. \tag{4.62}$$

Here t_0 is the characteristic time of the ablation, and x_0 is the characteristic dimension of the whole material. For further calculations we assume: $t_0 = 10^4$ s, $x_0 = 10^{-1}$ m, $\theta_0 = 293$ K.

Figure 4.22 shows the variation of the coefficient $\bar{K}$ for epoxy resin during heating. Comparison of theoretical and experimental results shown in the figure allows us to conclude that the dependence (4.62) adequately describes the change in the coefficient $\bar{K}$ in the ablation. Moreover, parameters $\bar{K}_0$ and n_k for a wide class of matrices under high temperatures can be considered as universal up to a certain accuracy:

$$\bar{K}_0 = 1.76 \cdot 10^{-12}, \qquad n_k = 50. \tag{4.63}$$

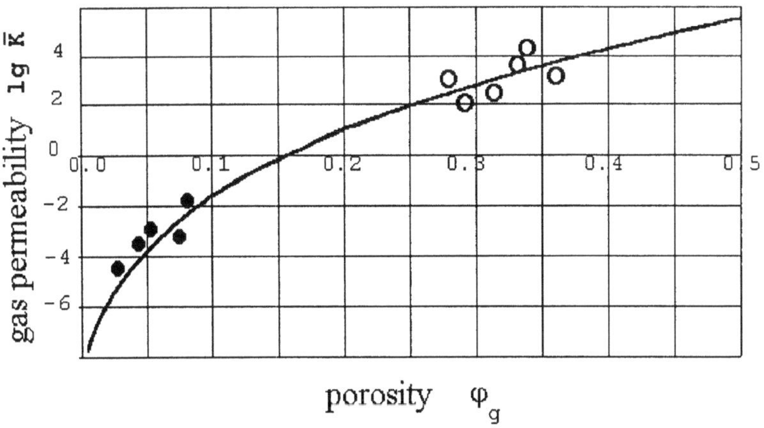

Figure 4.22. Dependence of dimensionless gas-permeability coefficient $\bar{K}$ on porosity φ_g for epoxy resin (● - at temperature $\theta_0 = 293$ K, ○ - after heating up to temperature 623 K). Curve corresponds to computations, points are experimental data

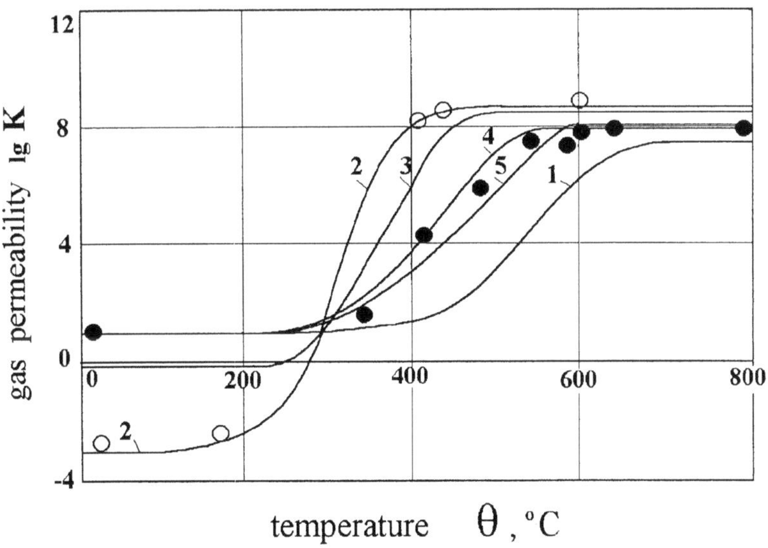

Figure 4.23. Dependence of dimensionless gas-permeability coefficient $\bar{K}$ on temperature under heating with rate $\dot{\theta}_0$ =0.1 K/s for different types of matrices (1 - silicon-organic, 2 (○) - epoxy, 3 - epoxy-phenolic, 4 (●) - phenolic, 5 - polyimide)

The gas permeability $\bar{K}$ at the initial state (before heating) in going from one material to another in accordance with (4.62) is described only by the different initial porosities φ_g^0 of the materials.

This conclusion will be widely used below. Typical magnitudes of initial porosity of ablative matrices are given in Table 4.1.

The change in the gas permeability $\bar{K}$ during heating with the constant rate $\dot{\theta}_0$ =0.1 K/s for different types of matrices is shown in Figure 4.23. It is seen that within the relatively restricted temperature interval from 200...300 °C up to 500...600 °C, when there is a volumetric ablation in the materials, the gas permeability changes by 8 – 12 orders; this is connected with formation of secondary porosity.

We give the formula connecting the gas-permeability coefficient K introduced in the model to the coefficient P usually used in scientific literature [7]:

$$K = \rho_g^0 P,$$

$$[K] = s, \qquad [\rho_g^0] = \mathrm{kg} \cdot \mathrm{m}^{-3};$$

$$[P] = \frac{\mathrm{cm}^3 \cdot \mathrm{mm}}{\mathrm{cm}^2 \cdot \mathrm{s} \cdot \mathrm{atm}} = 10^{-10}\ \mathrm{m}^3 \cdot \mathrm{s} \cdot \mathrm{kg}^{-1}.$$

CHAPTER 5

REINFORCING FIBRES UNDER HIGH TEMPERATURES

The following types of fillers are most widely used for reinforcing ablative matrices:

- continuous fibres (in the form of threads, bands, fabrics etc.);
- short fibres;
- dispersed particles (in the form of hollow microspheres or continuous balls etc.).

We consider the behaviour of the first type of fillers under high temperatures.

5.1. Changing Phase Composition of Fibres under Heating

Fibres used for reinforcing of ablative matrices are usually more thermostable than the matrices themselves. Therefore, at the lowest levels of temperatures of matrix ablation ($\approx 300-400$ °C) and for strength calculations fibres can be considered as thermostable, i.e. their density ρ_f does not change. Values of densities for typical ablative fibres are shown in Table 5.1.

However, for many problems of thermal protection design, for example for a vehicle re-entry from Earth's orbit, the level of ablation temperatures is considerably higher: from 600 °C to 2000 °C and even higher. At these temperatures, fibres of different types behave in various ways.

The density of glass fibres remains practically constant up to the melting point (~ 1800 °C), and for strength problems it can be considered as constant.

The densities of carbon and boron fibres in vacuum or inert environment may also be considered as constant. However, in air, these fibres are oxidized, with accompanying volumetric ablation (loss of mass).

The density of organic fibres of Kevlar type changes during heating in the same way as the density of all polymer materials, i.e. it decreases due to volumetric ablation.

In paragraphs 1.4, 3.6 a model of reinforcing fillers under high temperatures was constructed, the filler was considered as a three-phase system consisting of 'amorphic' solid phase, new 'crystalline' solid phase and pores filled with gas. We use parentheses, ' ', to indicate that this is the initial state of the phase; and then, under heating, the physical nature of the phase can differ from amorphic or crystalline.

TABLE 5.1. Thermal characteristics of fibres

Characteristics	graphitized carbon fibre	boron fibre	glass fibre (silica)	organic fibre (Kevlar-49)
ρ_f, $kg \cdot m^{-3}$	1800	2580	2200	1450
c_f, $kJ \cdot kg^{-1} \cdot K^{-1}$	0.6	0.6	0.65	1.2
k_f^0, $W \cdot m^{-1} \cdot K^{-1}$	42	32	1.3	0.4
k_l, $W \cdot m^{-1} \cdot K^{-1}$	42	32	1.3	5.0
Oxidizing environment:				
E_{Af}, K	$6.5 \cdot 10^3$	$9.5 \cdot 10^3$	$9.8 \cdot 10^3$	$8.6 \cdot 10^3$
J_f^0, $kg \cdot m^{-3} \cdot s^{-1}$	$5.4 \cdot 10^2$	$7.8 \cdot 10^3$	$6.6 \cdot 10^3$	$4.2 \cdot 10^4$
Γ_f	~ 0.7	~ 0.9	0	~ 0.8
Inert environment:				
E_{Af}, K	$7 \cdot 10^3$	$12 \cdot 10^3$	$9.8 \cdot 10^3$	$9 \cdot 10^3$
J_f^0, $kg \cdot m^{-3} \cdot s^{-1}$	$5.4 \cdot 10^2$	$7.8 \cdot 10^3$	$6.6 \cdot 10^3$	$4.2 \cdot 10^4$
Γ_f	0.05	0.07	0	0.35

The changing volumetric concentrations of the 'amorphic' and 'crystalline' phases of fibres are described by the kinetic equations (3.47)

$$\varphi_a = \frac{\varphi_f - \varphi_f^0(1-\Gamma_f)}{\Gamma_f},$$

$$\varphi_l = \frac{1-\Gamma_f}{\Gamma_f}(\varphi_f^0 - \varphi_f), \tag{5.1}$$

where φ_f is the volumetric comcentration of solid phases of the fibre ($\varphi_f = \varphi_a + \varphi_l$) determined from equation (3.41)

$$\rho_f \frac{\partial \varphi_f}{\partial t} = -J_f \Gamma_f. \tag{5.2}$$

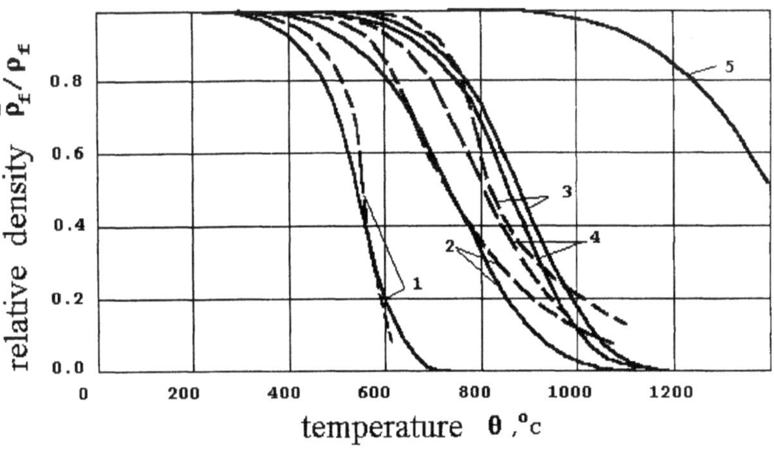

Figure 5.1. Dependence of relative density $\bar{\rho}_f/\rho_f$ of fibres on temperature during heating with rate 0.1 K/s in an oxidizing environment, for different types of fibres: 1 - organic; 2 - carbon; 3 - glass; 4 - boron; 5 - carbon with coating. Solid curves are theoretical results, dashed lines are experimental data

The rate of volumetric ablation J_f for all the types of fibres, as for matrices (see formula (4.21)), can be rewritten in the form

$$J_f = J_f^0 \varphi_a \exp\left(-\frac{E_{Af}}{R\theta}\right), \tag{5.3}$$

where J_f^0, Γ_f and E_{Af} are the constants determined in experiments by changing the densities of fibres under heating, and Γ_f is the coefficient of gasification of the fibres.

- For glass fibres $\Gamma_f = 0$.
- For organic, carbon and boron fibres $0 < \Gamma_f < 1$.

Equations (5.2) show that changing an average density $\bar{\rho}_f = \rho_f \varphi_f$ of fibres, as for matrices, depends on a heating rate (for linear heating).

Figure 5.1 shows graphs of the change in the average density of boron, organic, and carbon fibres during heating in air, the figure also shows the variation of the content of the amorphic phase of glass fibres described by the function $\bar{\rho}_f = \rho_f \varphi_a$.

Typical fibres are as follows: organic – Kevlar-49; carbon – graphitized carbon; boron – Avco (B/W), and glass – silica fibre.

This figure also shows a loss of density of carbon fibre with antioxidizing coating. All the curves are described well enough by equation (5.3). Constants E_{Af} and J_f^0 for different types of fibres are given in Table 5.1.

Figure 5.2 shows graphs of the variation of the content of the amorphic phase φ_a of glass fibres at different rates of heating.

In an inert gas environment (for example, very rarefied air, vacuum, helium, argon etc.) the volume-average density $\bar{\rho}_f$ of boron and carbon fibres hardly

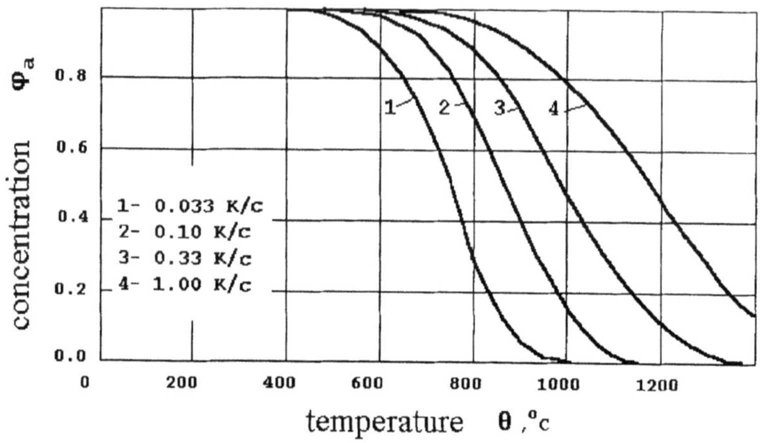

Figure 5.2. Dependence of content of amorphic phase φ_a of glass fibres on temperature for different rates of heating

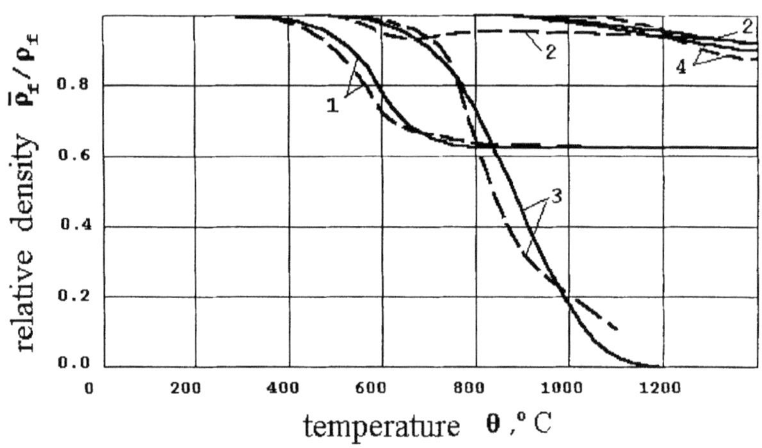

Figure 5.3. Dependence of relative density $\bar{\rho}_f/\rho_f$ of fibres on temperature under heating with rate 0.1 K/s in inert environment for different types of fibres: 1 - organic; 2 - carbon; 3 - glass; 4 - boron. Solid curves are theoretical computations, dashed lines are experimental data

changes (Figure 5.3) during heating, at least, up to 1400 °C, as there are no oxidizing processes. Therefore, for these fibres in an inert environment we can assume $J_f = 0$. Glass fibres at temperatures up to 1700 °C hardly react to hot gas surroundings, therefore, the variation of the content of amorphic phase φ_a for

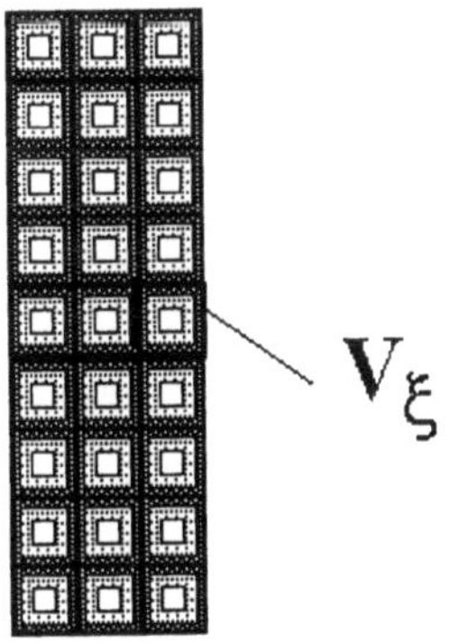

Figure 5.4. Model of a periodic structure of monofibre under high temperatures

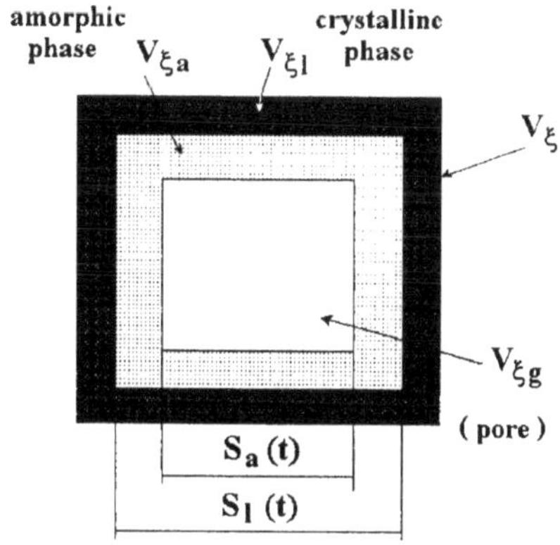

Figure 5.5. Model of a periodicity cell V_ξ of ablative fibres with cubic shaped phases

glass fibres occurs practically in the same way in oxidizing and inert environments.

Organic fibres ablate even if there is no oxygen present. However, for organic fibres $\Gamma_f \approx 0.8 - 0.9$ in an oxidizing environment (see Figure 5.1), while $\Gamma_f = 0.35$ in an inert environment.

5.2. Heat Conductivity and Heat Capacity of Ablative Fibres

The heat conductivity k_f of fibres varies under heating for two reasons: 1) variation of heat conductivity of amorphic and crystalline phases during heating and 2) volumetric ablation.

To calculate the heat conductivity k_f of fibres from the heat conductivity coefficients of their phases k_a and k_l, we apply the model of a periodicity cell with cubic shaped phases that was developed in detail in Chapter 4.

Fibres under high temperatures are assumed to consist of a large number of periodicity cells V_ξ (Figure 5.4); each cell contains three cubic shaped phases $V_{\xi a}$, $V_{\xi l}$ and $V_{\xi g}$ (Figure 5.5).

This model allows us to determine characteristics of fibres in the same way as for ablative matrices (see Chapter 4).

As for ablative matrices, heat conductivity k_f of fibres can be expressed by the equation

$$k_f = k_f^0(\theta/\theta_0)^{1/2}\widetilde{b}_f, \tag{5.4}$$

$$\widetilde{b}_f = \left((S_l - S_a) + (1 - S_l)\frac{k_l}{k_f^0}\right),$$

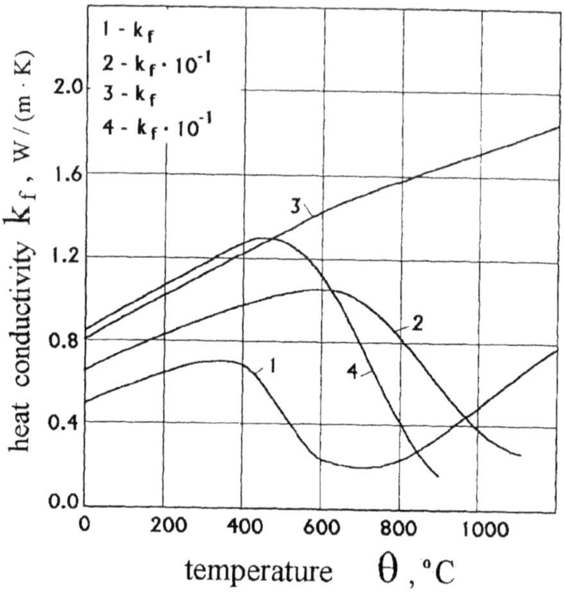

Figure 5.6. Dependence of heat conductivity k_f of fibres on temperature under heating with rate 0.1 K/s in air for different types of fibres: 1 - organic, 2 - carbon, 3 - glass, 4 - boron

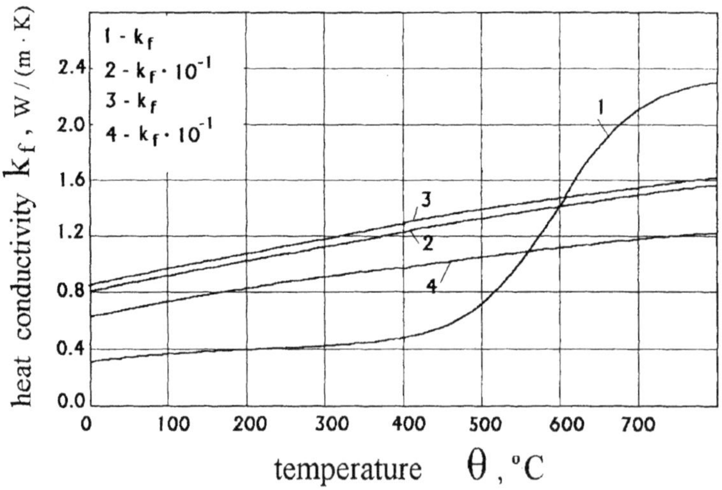

Figure 5.7. Dependence of heat conductivity k_f of fibres on temperature under heating with rate 0.1 K/s in inert environment for different types of fibres (1 - organic, 2 - carbon, 3 - glass, 4 - boron)

where

$$S_l = \left(1 - \frac{\varphi_l}{\varphi_f^0}\right)^{1/3}, \quad S_a = \left(1 - \frac{\varphi_f}{\varphi_f^0}\right)^{1/3}. \tag{5.5}$$

Functions S_l and S_a are similar to functions S_b and S_p for matrices.

Values of heat conductivity coefficients k_f^0 for fibres at temperature $\theta_0 = 293$ K are given in Table 5.1. Figure 5.6 shows dependences of heat conductivity coefficient k_f^0 of fibres on temperature θ during linear heating. The heating rate is $\dot{\theta} = 0.1$ K/s; the environment is air. Increasing the heat conductivity at the initial stage of heating is connected to increasing heat conductivity of the amorphic phase of fibres; oxidizing the fibres in an air environment under higher temperatures leads to decrease of the heat conductivity k_f for all fibres, except glass ones.

Figure 5.7 exhibits the variation of the heat conductivity k_f of fibres during heating in an inert environment. For this environment, as compared with an oxidizing environment, the intensity of ablation becomes smaller; this leads to another picture of the variation of k_f during heating for organic, carbon and boron fibres.

Values of heat capacity c_f of fibres are given in Table 5.1.

5.3. Varying Elastic Properties of Fibres under Heating

Heat produces changes in the elastic modulus of fibres. Just as for ablative matrices, there are two effects:

- changes in elastic properties of the fibre in the amorphic state under relatively low temperatures;
- physico-chemical processes of crystallization (for glass, carbon, boron fibres in an inert environment), oxidization (for carbon, boron, organic fibres in an oxidizing environment) or pyrolysis (for organic fibres in an inert environment) under higher temperatures.

The first factor causes a reversible change of the elastic modulus E_f of fibres; the second leads to changes which are irreversible under cooling.

Just as for an ablative matrix, we use a model of the periodicity cell V_ξ with cubic shaped phases. We can obtain expressions for E_f by taking account of both the effects listed above:

$$E_f = E_f^0 \widetilde{a}_f. \tag{5.6}$$

Here $\widetilde{a}_f$ is the coefficient specifying the change of elastic characteristics during heating:

$$\widetilde{a}_f = a_f^{(0)}(S_l - S_a) + (1 - S_l)n_l, \tag{5.7}$$

$$a_f^{(0)} = \exp\left(-a_f \Delta\widehat{\theta}\right),$$

$$\Delta\widehat{\theta} = \Delta\theta + P\int_0^t \exp\left(-Q(t-\tau)\right)\Delta\theta d\tau, \quad \Delta\theta = \theta - \theta_0,$$

TABLE 5.2. Physico-mechanical properties of fibres

Charac- teristics	graphitized carbon fibre	boron fibre (B/W)	glass fibre (silica)	organic fibre (Kevlar-49)
d_f, 10^{-3}mm	10	100	8	15
E_f^0, GPa	250	420	80	145
E_l, GPa	140	40	1	5
$\bar{a}_f$	0.001	0.001	0.005	0.1
γ	1	0.9	0.7	0.5
β	3.7	3.9	3.5	7.1
$\bar{\sigma}_f^0$, MPa	2200	2620	2100	3300
$\alpha_f \cdot 10^6$, K^{-1}	-0.5	8	4...6	-0.1...2
$\alpha_l \cdot 10^6$, K^{-1}	4.0	12	2...4	6
n_l	1.3	0.1	0.01	0.3

where $n_l = E_l/E_f^0$ is the ratio of elastic moduli of the crystalline E_l and amorphic E_f^0 phases.

The functions S_l and S_a are determined by formulae (5.5), and universal constants P and Q are those appearing in the expression (4.18).

Reversible changes of properties of fibres of different types during heating are determined by the coefficient a_f.

Values of the characteristics a_f, E_f^0 and E_l for carbon, boron, glass and organic fibres are given in Table 5.2, in which $\bar{a}_f = a_f/\theta_0$, $\theta_0 = 293$ K.

General three-dimensional constitutive relations for fibres, similar to expressions (4.25) for a matrix, are written in the form

$$\varepsilon_{fkl} = \frac{1+\nu_f}{E_f}\sigma_{fkl} - \frac{\nu_f}{\varepsilon_f}Y_f\delta_{kl} + \overset{\circ}{\varepsilon}_f\delta_{kl}. \tag{5.8}$$

Here ν_f is Poisson's ratio for the fibre, $\overset{\circ}{\varepsilon}_f$ is its heat deformation, ε_{fkl} and σ_{fkl} are the components of its strain and stress tensors, respectively, and $Y_f = \sigma_{fkk}$ is the first invariant of the tensor σ_{fkl}.

5.4. Heat Deformation of Fibres

Experimental data on heat expansion of reinforcing fibres under high temperaures are very limited, the existing results mainly fall into the interval of relatively small temperatures before the beginning of ablation.

In our model, the dependence of the heat deformation $\overset{\circ}{\varepsilon}_f$ of reinforcing fibres is described by an expression similar to equation (4.24):

$$\overset{\circ}{\varepsilon}_f = \alpha_f \varphi_a(\theta - \theta_0) + \alpha_l \int_0^t (\theta(t) - \theta(\tau)) \dot{\varphi}_l d\tau - \beta_l \varphi_l. \tag{5.9}$$

Here α_f and α_l are the coefficients of heat expansion of the amorphic and crystalline phases of the fibre, and β_l is the shrinkage coefficient. Under temperatures θ, for which there is no ablation of the fibres, equation (5.9) gives $\overset{\circ}{\varepsilon}_f = \alpha_f(\theta - \theta_0)$. Values of α_f are considerably different for various types of fibres. For most types of carbon fibres, negative values $\alpha_f \approx -0.5 \cdot 10^{-6}$ K^{-1} are typical ([117], see Table 5.2). Heat expansion across carbon fibres is positive and higher than α_f by 1 - 2 orders in absolute magnitudes. Some types of organic fibres also have negative values [121] $\alpha_f \approx -0.1 \cdot 10^{-6}$ K^{-1}. Glass and boron fibres have higher values of α_f: for glass $\alpha_f \approx 4 \cdot 10^{-6}$ K^{-1}, for boron $\alpha_f \approx 8 \cdot 10^{-6}$ K^{-1}. Under high temperatures in 'crystalline' states the coefficients of heat expansion α_l for all these fibres generally increase (see Table 5.2).

5.5. Strength Properties of Fibres under High Temperatures

5.5.1. STRENGTH OF IDEALIZIED FIBRE

Using the model of ablative fibre with a cubic shaped phases (see Figure 5.5) as for ablative matrices (formulae (4.39), (4.42)), one can derive the expression for fibre strength σ_f in tension:

$$\sigma_f = \sigma_f^0 \widetilde{a}_f. \tag{5.10}$$

Here σ_f^0 is the strength of the fibre in its initial 'amorphic' state at temperature θ_0, and $\widetilde{a}_f$, determined by formula (3.7), is an analog of the function $h_{mT} = a_1 h^0$ in equation (4.42).

5.5.2. MODEL OF A THREAD OF FIBRES WITH DEFECTS

Formula (5.10) describes a strength of an idealized fibre. In practice, there are cracks or other defects in the fibres which greatly reduce the strength.

Diameters d_f of individual fibres, so called *monofibres*, usually range between 1 and 100 mcm. In composites the fibres are bundled together to form *threads*.

Strength of monofibre and strength of the fibre bundle are considerably different: the strength of the bundle is considerably lower. This phenomenon is well

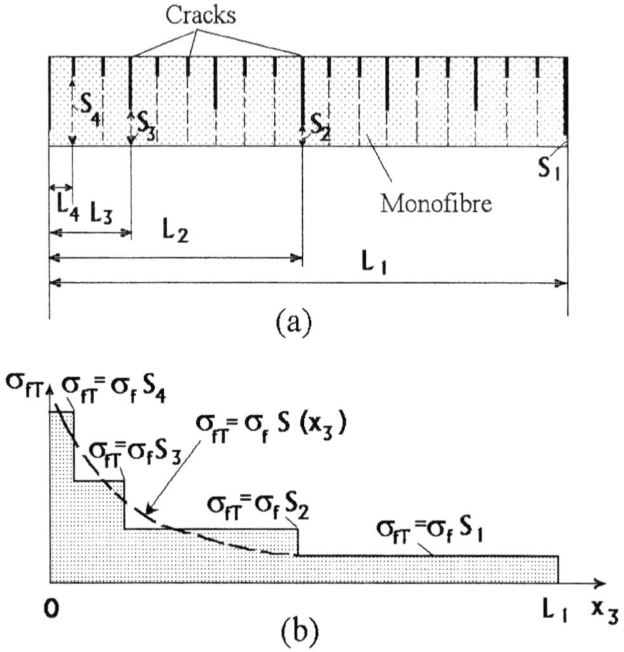

Figure 5.8. Model of monofibre with periodical system of transverse cracks (a), dependence of monofibre strength σ_{fT} on its length L (b)

known, and it plays an important role in the theory of composite strength (see Chapter 6). The cause of the difference in strength between monofibre and thread is the deficiency mentioned above of individual monofibres, mainly, in the form of surface cracks. This deficiency has a statistical character, therefore a statistically greater quantity of fibres has a large number of defects than the individual monofibre; this reduces an average strength of the thread.

Another important exhibition of the statistical character of fibres' deficiency is the phenomenon that the strength of both monofibres and threads depends on the length L of the fibres. The cause of this phenomenon is as before: increasing geometrical dimensions leads to increasing probability of the appearance of greater defects; this again lowers the strength of the fibres.

At present there exist many theoretical works devoted to description of these phenomena [47, 86, 100 etc.], but most of them are based on a statistical approach. However, the statistical approach does not allow us to formulate a mathematical model of strength of composites with a continuous structure, for example textile ones, in the form of compact analytical relations. At present researchers use statistical methods to simulate the strength; they can do this only for unidirectional composites, and the result of the simulation is a number, not an analytic formula.

We now develop a new deterministic model of a fibre bundle. Each monofibre in the thread is assumed to consist of N_1 segments (periodicity cells) with length

L_1 (Figure 5.8). Each segment with length L_1 consists in turn of N_2 segments with length L_2, etc., i.e.

$$L = L_1 N_1, \quad L_1 = N_2 L_2, \quad L_2 = N_3 L_3, \dots, \qquad \dots < L_3 < L_2 < L_1.$$

In each segment with length L_1 there is one defect in the form of a transverse crack with crosssection area $(1 - S_1)$; in each segment with length L_2 there is one crack with crosssection area $(1 - S_2)$ etc., where $\dots > S_3 > S_2 > S_1$. The crosssection area of the whole fibre is assumed to be equal to 1.

Let Ox_3 be the axis of the Cartesian coordinate system in the direction of the fibre axis, and let σ_{fT} be strength of the monofibre in tension. Due to the presence of defects within the monofibre, its strength depends on coordinate x_3.

If from the model of the fibre one separates any segment with length $x_3 = L_i$, then its strength will be equal to $\sigma_{fT}(x_3) = \sigma_f S_i$. Increasing the length L_i leads to the following consequences: the size of the periodicity cell increases, and hence the area $(1-S_i)$ of the maximum defect increases too. Passing to the limit, instead of the step function $\sigma_{fT} = \sigma_f S_i$, $L_{i-1} < x_3 < L_i$ we obtain the continuous one: $\sigma_f S_3(x_3)$ (see Figure 5.8). If the dependence $S(x_3)$ is known, then the dependence of fibre strength σ_{fT} on its length will be obtained too.

The function $S(x_3)$ is monotonically decreasing. Let it be chosen in the form

$$S(x_3) = S_0 \left(\frac{d_f}{x_3}\right)^{\gamma_0}, \qquad 0 < \gamma_0 < 1,$$

where γ_0 and S_0 are constants. The strength of the monofibre depending on the length L_3 is determined by the formula

$$\sigma_{fT} = \sigma_f S_0 \left(\frac{d_f}{x_3}\right)^{\gamma_0}. \tag{5.11}$$

For a thread we also apply a deterministic approach. Suppose that fibres in a thread are assembled by groups $V_{\xi f}$ with the same number of fibres. In each group, monofibres have the same locations of transverse cracks but different areas $(1 - S_0)$ of the cracks. In addition, assume that in each the group (which, in fact, is a periodicity cell of the thread) monofibres are ordered with respect to value $(1-S_0)$ of the crack areas: along the transverse coordinate x_2 the area $(1-S_0(x_2))$ of cracks of corresponding monofibres decreases.

Thus, in this model, strength of fibres in a thread σ_{fT} determined by formula (5.11) depends both on x_2 and x_3.

Keeping the dependence (5.11) for separate monofibres, in accordance with the assumptions made, we need the value of S_0 to be considered as depending on coordinate x_2. Choose function $S_0(x_2)$ in the form

$$S_0(x_2) = \left(\ln \frac{1}{1 - x_2}\right)^{1/\beta},$$

where $\beta > 0$. Here x_2 changes within the periodicity cell $V_{\xi f}$: $0 < x_2 < 1$. Substituting the function $S_0(x_2)$ into (5.11), we obtain the expression for strength of monofibre with coordinate $x_2 = W$ and length $x_3 = L$ in the thread:

$$\sigma_{fT} = \sigma_f \left(\frac{d_f}{L}\right)^{\gamma/\beta} \left(\lg \frac{1}{1-W}\right)^{1/\beta}, \qquad x_2 = W,\ x_3 = L,\ \gamma > 0,\ \beta > 0, \quad (5.12)$$

σ_f, γ, β are the experimental constants characterizing the given type of monofibres, and $\gamma = \gamma_0 \beta$.

On inverting (5.12) for W, we have

$$W = 1 - \exp\left(-\left(\frac{L}{d_f}\right)^{\gamma}\left(\frac{\sigma_{fT}}{\sigma_f}\right)^{\beta}\right). \quad (5.13)$$

When $\gamma = 1$, dependence (5.13) is the same as *the Weibull dependence* in the statistical theory of strength [13], where W is understood as a probability of failure of the fibre under stress equal to the value of σ_{fT}. In the deterministic model developed here, the parameter W is the volume ratio of broken monofibres at stress σ_{fT} in the thread. In fact, these two approaches are complementary; we can apply both. To retain a rigorous model, we will use the deterministic approach, and to determine constants γ, β and σ_f from experiments we will use the statistical approach.

Let us introduce the concept of average strength of fibres in a thread $\bar{\sigma}_f$ as follows:

$$\bar{\sigma}_f = \sigma_{fT}|_{L=L_0, W=0.5}, \quad (5.14)$$

where L_0 is the certain standard length of fibres which is assumed to be $L_0 = 10^{-2}$ m.

Then from (5.13) and (5.14), we derive

$$\sigma_{fT} = \bar{\sigma}_f^0 \tilde{a}_f \left(\frac{L_0}{L}\right)^{\gamma/\beta} \left(\frac{\ln 1/(1-W)}{\ln 2}\right)^{1/\beta}. \quad (5.15)$$

Values $\bar{\sigma}_f$ and σ_f^0 are connected by the relation

$$\bar{\sigma}_f^0 = \sigma_f^0 \left(\frac{d_f}{L_0}\right)^{\gamma/\beta} (\ln 2)^{1/\beta}, \quad (5.16)$$

$$\bar{\sigma}_f = \bar{\sigma}_f^0 \tilde{a}_f, \quad (5.17)$$

where $\bar{\sigma}_f^0$ is the average strength of monofibres at temperature θ_0 and $\bar{\sigma}_f$ is the one at temperature θ.

Values of constants $\bar{\sigma}_f$, γ, β characterizing strength of a separate monofibre as a component of carbon, boron, glass or organic thread are given in Table 5.2.

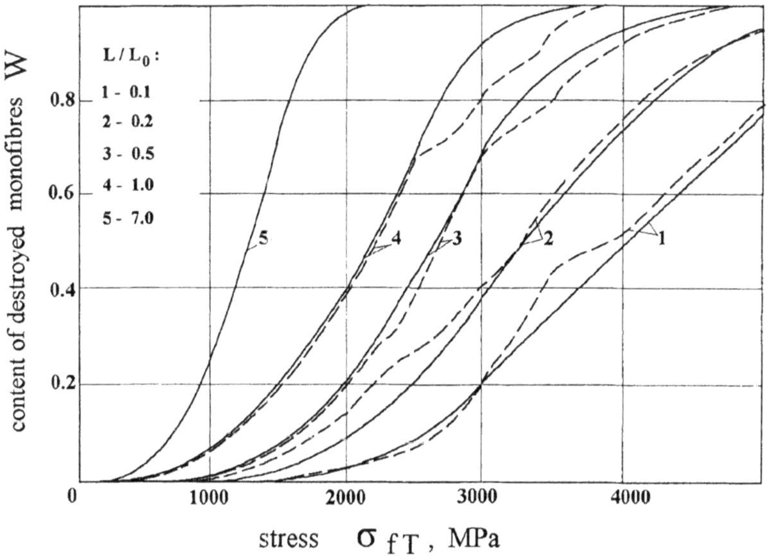

Figure 5.9. Dependence of content W of broken carbon monofibres in a thread on applied stress σ_{fT} for different ratios of length L to the standard length L_0. Solid curves are computations by (5.15), dashed lines are experimental data [113]

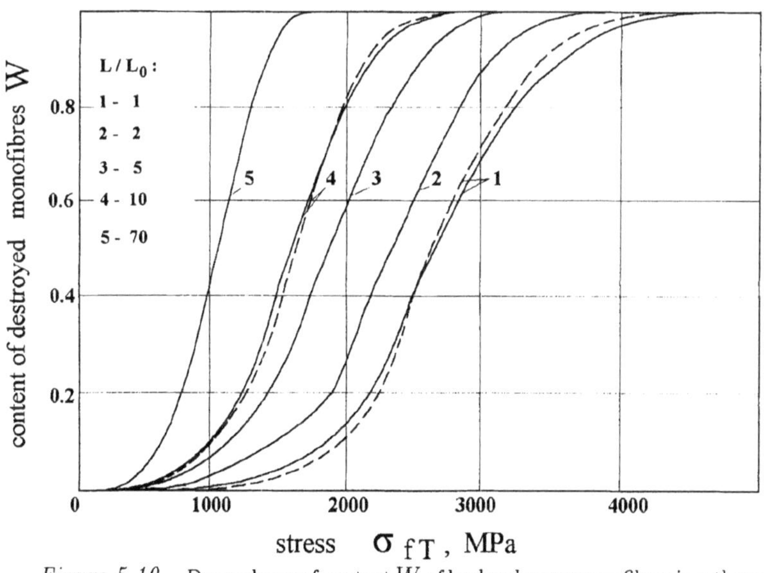

Figure 5.10. Dependence of content W of broken boron monofibres in a thread on applied stress σ_{fT} for different ratios L/L_0. Solid curves are computations, dashed lines are experimental data [113]

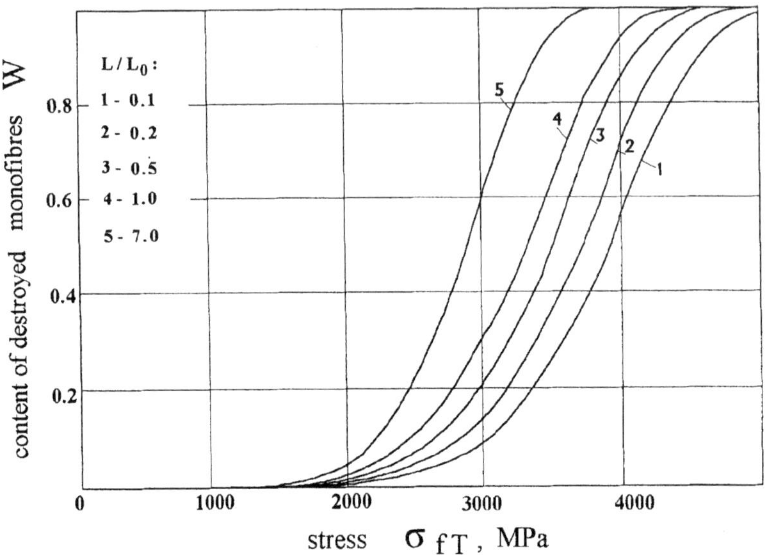

Figure 5.11. Dependence of content W of broken organic monofibres in a thread on applied stress σ_{fT} for different ratios L/L_0

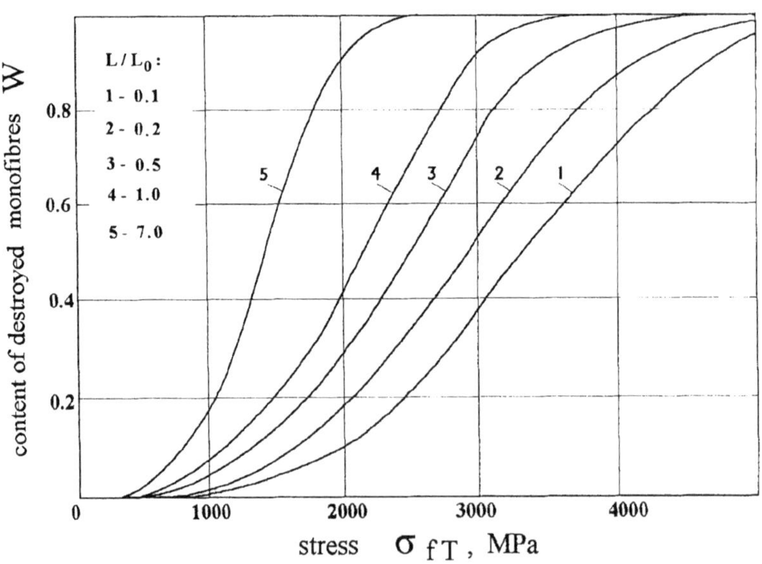

Figure 5.12. Dependence of content W of broken glass monofibres in a thread on applied stress σ_{fT} for different ratios L/L_0

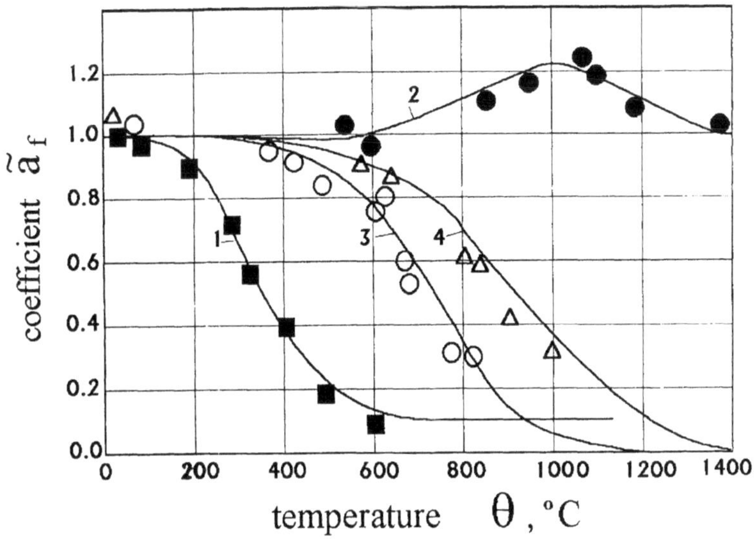

Figure 5.13. Dependence of elastostrength properties coefficient $\tilde{a}_f$ on temperature under heating with rate 0.1 K/s in inert environment for different types of fibres (1 (■) - organic, 2 (•) - carbon, 3 (○) - glass, 4 (△) - boron). Solid curves are computations by (5.7), points are experimental data [117]

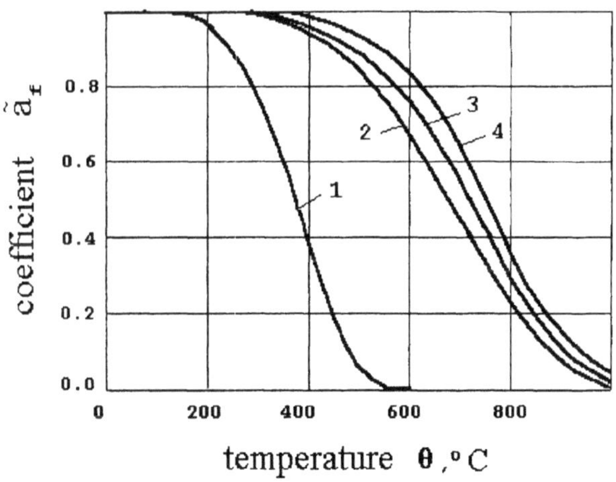

Figure 5.14. Dependence of elastostrength properties coefficient $\tilde{a}_f$ on temperature under heating with rate 0.1 K/s in air for different types of fibres (1 - organic, 2 - carbon, 3 - glass, 4 - boron)

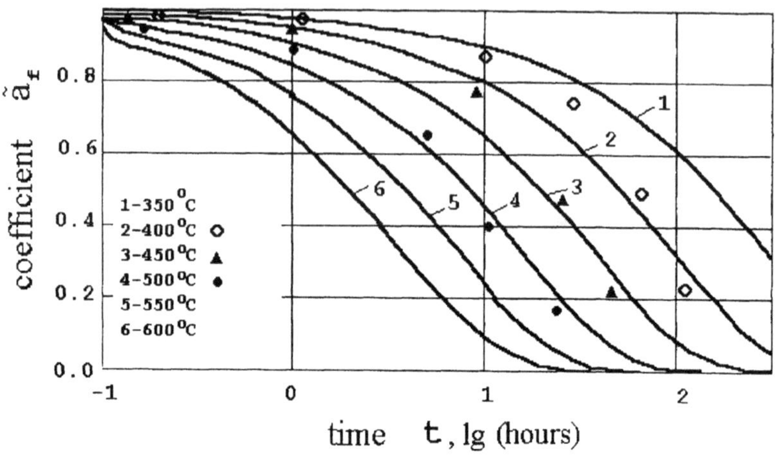

Figure 5.15. Dependence of elastostrength properties coefficient $\widetilde{a}_f$ of boron fibres on time of holding at different temperatures in air (solid curves are computations, points are experimental data [117])

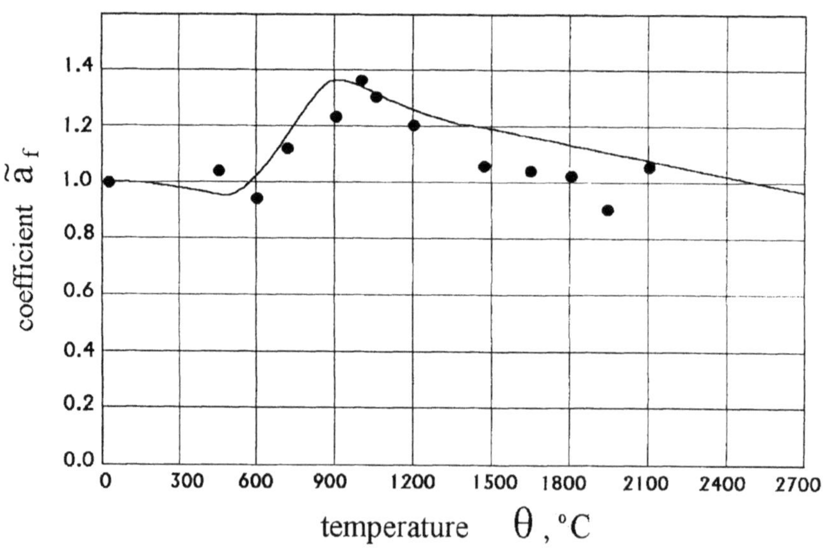

Figure 5.16. Dependence of coefficient $\widetilde{a}_f$ of carbon fibres on temperature under heating with rate 5 K/s in inert environment (solid curves are computations, points are experimental data [117])

Figures 5.9 - 5.12 show the dependence of content W of broken monofibres in the thread on the applied stress σ_{fT} (formula (5.15)) for different ratios L/L_0 of fibres. Fibres of the four aforementioned types were considered.

As seen from these figures, carbon fibres have the strongest dependence of monofibre strength on length L, organic fibres have the weakest dependence. This dependence is characterized quantitatively by the ratio of the parameters γ/β. For carbon fibre γ/β=0.27, for boron fibre γ/β =0.23, for glass fibre γ/β=0.2, for organic fibre γ/β=0.01. As it will be shown below, the higher the value of parameter γ/β, the longer strength properties are held for ablative composite under heating.

As follows from (5.6) and (5.16), changing elastic and strength properties of fibres under heating, within the frames of the suggested model, is described by the function $\widetilde{a}_f$. This assumption satisfactorily describes the actual situation.

Figure 5.13 shows the variation of the function $\widetilde{a}_f$ for the four types of fibres during heating with the constant rate $\dot{\theta}_0$ =0.1 K/s in inert environment. Experimental values of the function $\widetilde{a}_f$ were determined by changing the relative strength ($\widetilde{a}_f = \bar{\sigma}_f/\bar{\sigma}_f^0$) of fibres under heating. Theoretical values were determined by formulae (5.7), (5.1), (5.2) in which the constants E_{Af}, J_f^0 and Γ_f were found by the curves of ρ density loss (see Figure 5.1), and the constant a_f was determined from the best fit of theoretical and experimental values of the function $\widetilde{a}_f(\theta)$. As seen from Figure 5.13, function $\widetilde{a}_f$ in the form (5.7) satisfactorily approximates experimental values of the relative strength.

Figure 5.14 exhibits dependences of the coefficient $\widetilde{a}_f$ on the temperature of heating for the same types of fibres in air. Figures 5.13 and 5.14 show that the presence of the oxidizing environment leads to lowering of strength of carbon, organic and boron fibres. Strength of glass fibres is practically independent of the chemical composition of the surroundings. Figure 5.15 shows the dependence of the coefficient $\widetilde{a}_f$ on duration t of isothermic heating of boron fibres in air for different values of temperature θ. The figure shows that equation (5.7) accurately simulates the thermostrength of fibres.

Figure 5.16 shows the variation of the coefficient $\widetilde{a}_f$ of carbon fibres during heating to the temperature level of 2700 °C in an inert environment. A peculiarity of carbon fibres is the increase of high-temperature strength at temperature 1000 – 1200 °C [117] without oxidizing. For them, the value of the coefficient n_l characterizing the relation of strength properties of 'crystalline' and 'amorphic' phases in an inert environment is equal to 1.3.

5.6. Short Fibres and Dispersed Particles

Besides continuous fibres, other types of reinforcing fillers are used in composites, for example: short fibres oriented randomly in matrices, dispersed particles in the form of powder or in the form of hollow balls called microspheres.

Short fibres can be of the same types as continuous ones, and their length usually range between 1 and 10 mm.

Dispersed particles applied in high-temperature composites are usually glass, polymer, ceramic and carbon ones in the form of powder with grain dimensions between 10 and 1000 mcm; glass and polymer hollow microspheres are also used.

Elastic modulus E_f and heat conductivity coefficient k_f of the dispersed particles and short fibres under high temperatures are determined in the same way as for continuous fibres, i.e. by formulae (5.4), (5.6), (5.7). Strength of dispersed particles does not contribute much strength to the whole composite, as its failure occurs due to failure of its matrix, therefore for disperse-reinforced composites strength of their fillers is not considered.

Physical constants of glass, carbon and organic dispersed fillers for engineering problems can be considered to be the same as for continuous fibres (see Tables 4.1, 4.2, 5.1, 5.2).

CHAPTER 6

UNIDIRECTIONAL COMPOSITES UNDER HIGH TEMPERATURES

6.1. Structural Model of Unidirectional Composites

6.1.1. PECULIARITIES OF UNIDIRECTIONAL COMPOSITES UNDER HIGH TEMPERATURES

Unidirectional composites (UCs) are bundles of monofibres assembled as a thread and surrounded by a matrix. Properties of unidirectional composites differ considerably from features of a fibre bundle not impregnated by a matrix. The reason for this is the presence of fibre defects. As mentioned in Chapter 5, monofibres have a spread in strength properties due to the presence of microcracks in them. When the whole thread is loaded, its fibres break one at a time beginning with the weakest ones. In a non-impregnated bundle a destroyed fibre makes no contribution to the total strength of the thread; in a UC, due to tangential stresses in its matrix, the destroyed fibre does contribute; only some section L_{neff} of monofibre around the break does not contribute to the strength.

This section is called the *non-effective length*. The greater the non-effective length L_{neff}, the lower is the strength of the UC.

Under high temperatures, the elastic properties of a matrix responsible for involving a destroyed fibre into the work, decrease; therefore, non-effective length L_{neff} increases and, hence, composite strength decreases. This mechanism explains a well-known but non-evident phenomenon: strength properties of composite degrade under heating with degradation of features of its matrix, but not the fibres, i.e. heat-resistance of composites is more closely linked to the heat-resistance of its matrix than to its fibres.

Besides the breakage of fibres in UCs, there are other important types of defects, in particular delaminations within a matrix. Delaminations greatly lower the strength of the composite especially under high temperatures as shear strength of the matrix decreases. Theoretical estimates of composite strength, which neglect matrix delaminations, are too high [13].

As with strength, elastic properties of a UC under high temperatures are determined mainly by heat-resistance of its matrix rather than its fibre. The classical theory of mixtures does not describe this phenomenon, as according to the theory an elastic modulus of a composite in the reinforcing direction E'_{μ} is equal to the product of the elastic modulus of its fibre and the reinforcing coefficient: $\varphi_f^0 E_f^0$.

However, the most important factor for the elastic modulus of a UC under heating proves to be the *misalignment* of individual monofibres within a thread. According to experimental data, the misalignment angle Φ is equal to 1 - 5°.

This phenomenon was investigated in our work [29], where we developed a structural model for a UC. To simulate the effect of misalignment of monofibres, we introduced a special model of microcomposite, which is one nondestroyed monofibre curved with respect to the reinforcing direction and surrounded by a matrix.

Many works, for example [13, 113, 62], are devoted to theoretical description of properties of UCs under normal temperatures; in these works the mechanism of interaction of a matrix with destroyed fibres is simulated theoretically or numerically mainly with the help of statistical methods (beginning from the work of Rozen [100]).

A model of a composite under high temperatures may be found in our work [29]. Now we consider this structural model, which holds us within the scope of a deterministic approach and satisfies the main concept of modelling: it is possible to obtain final compact analytical relations while keeping a reasonable level of a modelling accuracy.

6.1.2. MULTILEVEL INTERNAL STRUCRURE OF UNIDIRECTIONAL COMPOSITE

In Chapter 3 the geometric form and orientation of the a-phase (filler) were taken to be arbitrary. Location of the filler in angles of the periodicity cell was taken to simplify intermediate mathematical formulae. Statements of local and averaged problems in Chapter 3 still hold for arbitrary orientation of the filler too.

To calculate effective characteristics of reinforced composites under high temperatures, we introduce *a hierarchy of internal structural levels.* We consider four structural levels for a UC (Figure 6.1).

At the 1-st structural level there are phases of a matrix (polymer phase - $i = b$, pyrolytic phase - $i = p$ and gas phase) and phases of a fibre (amorphic $i = a$, crystalline $i = l$ and gas phases).

The 2-nd level is formed by the matrix itself and a monofibre. The matrix and monofibre consist of a large number of periodicity cells PC1 and PC2, respectively; shapes of the cells and calculation methods for properties of matrix and fibre were discussed in Chapters 4 and 5.

The 3-rd structural level is formed by so-called microcomposite, which is represented by one monofibre curved periodically in one plane and surrounded by a matrix (see Figure 6.1). Let Φ be a maximum angle of misalignment of the monofibres. In the microcomposite, one can separate a characteristic cell of periodicity - PC3 consisting of one wave with length l_μ of the curved fibre inside the three-phase matrix.

Finally, the 4-th structural level corresponds to the UC itself. Periodicity cell (PC4) of this level is chosen in the form of a two-domain package (see Figure 6.1): the first domain is non-broken fibres surrounded by the matrix, the second domain is broken fibres with transverse cracks also surrounded by the matrix. The

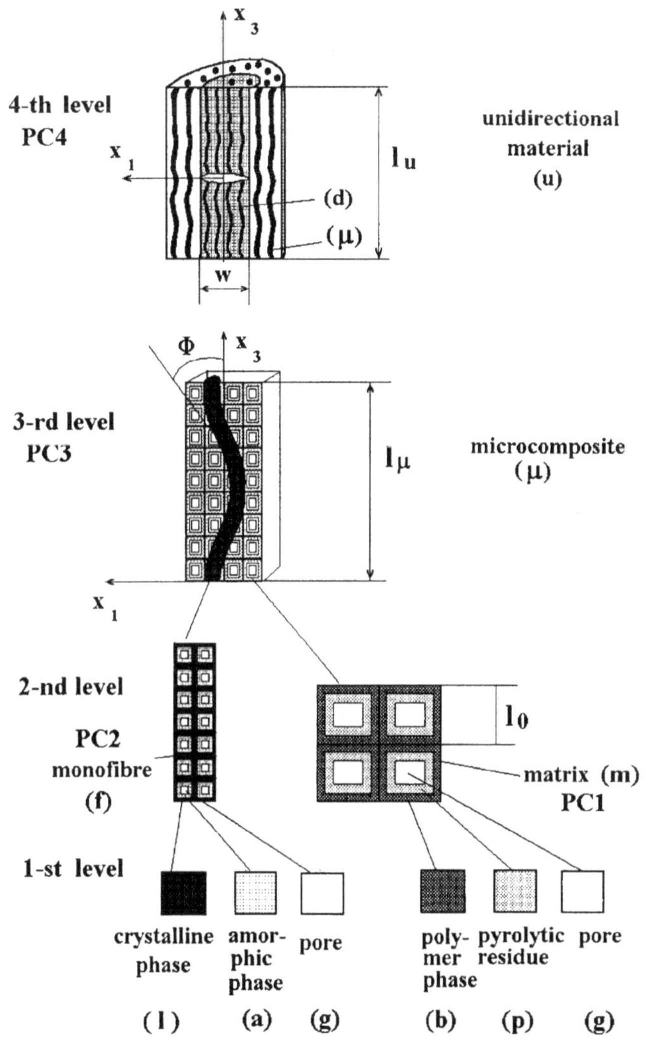

Figure 6.1. A scheme of 4-level internal structure of composite material under high temperatures

domains have indices (μ) and (d), respectively.

Justification of the adequacy of the four-level representation of a UC structure is the actual relationship between the geometric scales: x_0 - characteristic global dimension of the UC ($\sim 10^{-1}$ m), l_u - characteristic distance between transverse cracks in the composite ($\sim 10^{-3} \dots 10^{-2}$ m), l_μ - length of the curving wave in the monofibre ($\sim 10^{-4}$ m) and l_0 - characteristic dimension of phases of the matrix and fibre ($\sim 10^{-6}$ m):

$$l_0 \ll l_\mu \ll l_u \ll x_0. \tag{6.1}$$

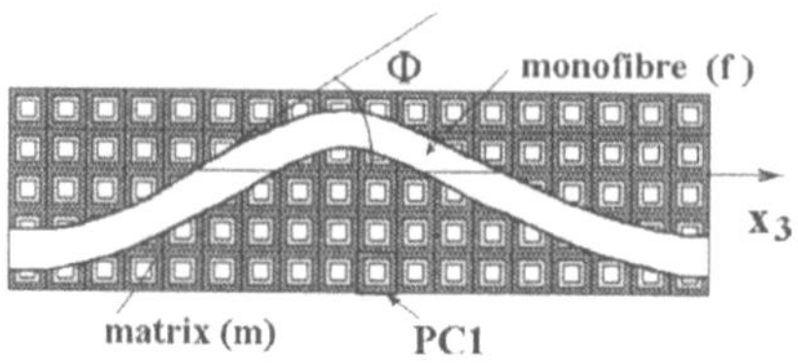

Figure 6.2. Periodicity cell PC3 of microcomposite consisting of misaligned monofibre and three-phase matrix

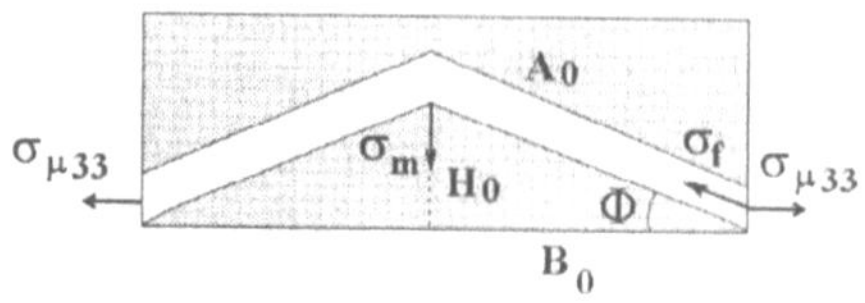

Figure 6.3. A scheme for calculation of the elastic modulus E'_μ in the longitudinal direction

6.2. Model of Microcomposite

6.2.1. ELASTIC PROPERTIES

The 3-rd structural level of unidirectional material, according to the scheme shown in Figure 6.2, is represented by a monofibre surrounded by a matrix. The fibre is assumed to be curved in the Ox_1x_3 plane. This model is called *microcomposite*, as it is, in essence, heterogeneous material; however, unlike a real composite, it does not contain structural defects of the monofibre breakage type.

Constitutive relations of the microcomposite determined by properties of the monofibre (5.8) and the matrix (4.25) are linear-elastic and correspond to a *transversally isotropic medium*, i.e. there is a plane of symmetry which is orthogonal to the axis [90].

The Ox_3 axis of transversal isotropy is coincident with the fibre axis and

$$\varepsilon_{\mu kl} = \Pi_{\mu klqr}\sigma_{\mu qr} + \overset{\circ}{\varepsilon}_{\mu kl} + pf_{\mu kl}, \tag{6.2}$$

where $\varepsilon_{\mu kl}$, $\sigma_{\mu qr}$ are the averaged strains and stresses of the microcomposite, μ is the indicator of properties of the microcomposite, k and l are indicators of tensor components $k, l = 1, 2, 3$ and $\Pi_{\mu klqr}$ are the components of the elastic flexibility tensor determined by the formulae

$$\Pi_{\mu kkll} = \begin{pmatrix} 1/E_\mu & -\nu_\mu/E_\mu & -\nu'_\mu/E'_\mu \\ -\nu_\mu/E_\mu & 1/E_\mu & -\nu'_\mu/E'_\mu \\ -\nu'_\mu/E'_\mu & -\nu'_\mu/E'_\mu & 1/E'_\mu \end{pmatrix} \tag{6.3}$$

$$\Pi_{\mu1212} = 1/(2G_\mu), \quad \Pi_{\mu2323} = \Pi_{\mu1313} = 1/(2G'_\mu).$$

Here there is no summation over repeated indices, and technical constants of the microcomposite are introduced as follows:

E'_μ and E_μ are the longitudinal and transverse moduli of elastic;
$\nu'_\mu = \nu'_{\mu 31}$, $\nu_\mu = \nu'_{\mu 12}$ are the Poisson's ratios;
G'_μ is the longitudinal shear modulus;
$G_\mu = E_\mu/2(1+\nu_\mu)$ is the transverse shear modulus.

Characteristics of microcomposite E'_μ, E_μ, ν'_μ, G'_μ, ν_μ are determined by the properties of its matrix and fibres:

$$E'_\mu = E_f^0 \widetilde{a}_2 \varphi_f, \qquad E_\mu = E_b^0 \widetilde{a}_5,$$

$$\nu_\mu = \nu_m, \qquad \nu'_\mu = \nu_m \widetilde{a}_4, \qquad G'_\mu = G_b^0 \widetilde{a}_3. \tag{6.4}$$

The terms of the order E_b^0/E_f^0 have been omitted as they are small as compared with 1.

Functions $\widetilde{a}_2$, $\widetilde{a}_3$, $\widetilde{a}_4$ and $\widetilde{a}_5$ depend on structural parameters φ_f, Φ; ratios of elastic characteritics of the phases E_f^0/E_b^0, ν_a/ν_m; and functions $\widetilde{a}_1$ and $\widetilde{a}_f$ (see (4.22a) and (5.7)) describing a change of elastic properties of the matrix and fibres:

$$\widetilde{a}_2 = \widetilde{a}_f \left(1 + \tan^2\Phi \frac{E_f^0}{E_b^0} \frac{\widetilde{a}_f \varphi_f}{\widetilde{a}_1}\right)^{-1},$$

$$\widetilde{a}_3 = \widetilde{a}_1 \left((1-\delta_f)\delta_f + \frac{1+\nu_a}{2} \delta_f^2 \frac{E_f^0 \widetilde{a}_f}{E_b^0 \widetilde{a}_1} \sin^2 2\Phi\right), \qquad \delta_f = 2\left(\frac{\varphi_f}{\pi}\right)^{1/2},$$

$$\widetilde{a}_4 = 1 - \varphi_f + \varphi_f \frac{\nu_a}{\nu_m}, \tag{6.5}$$

$$\widetilde{a}_5 = \widetilde{a}_1 \left((1-\delta_f)\delta_f + \frac{\nu_a}{2} \delta_f^2 \frac{E_f^0 \widetilde{a}_f}{E_b^0 \widetilde{a}_1} \sin^2 2\Phi\right).$$

Longitudinal Elastic Modulus

Evaluation of the elastic modulus E'_μ is performed by solving the problem of equlibrium of a flexible elastic bent thread in an elastic medium [29].

Figure 6.3 shows a calculation scheme for this problem; bending the monofibre is simulated by the triangle form, where A_0, $2B_0$ and H_0 are the lengths of two sides and the height of the triangle, respectively, and Φ is the maximum angle of bending. In accordance with this scheme, the elastic modulus E'_μ in tension of a specimen shown in Figure 6.3 by macrostress $\sigma_{\mu 33}$ is determined by the formula

$$E'_\mu = \frac{\sigma_{\mu 33}}{\varepsilon_{\mu 33}}, \tag{6.6}$$

and deformation $\varepsilon_{\mu 33}$ is combination of lengthening the monofibre $2\Delta A$ and transverse compression ΔH of the matrix that is responsible for straightening the monofibre:

$$\varepsilon_{\mu 33} = \left(\frac{A_0}{B_0}\right)^2 \frac{\Delta A}{A_0} - \left(\frac{H_0}{B_0}\right)^2 \frac{\Delta H}{H_0} \approx \frac{1}{\cos^2\Phi} \frac{\sigma_f}{E_f} - \tan\Phi \frac{\sigma_m}{E_m}, \tag{6.7}$$

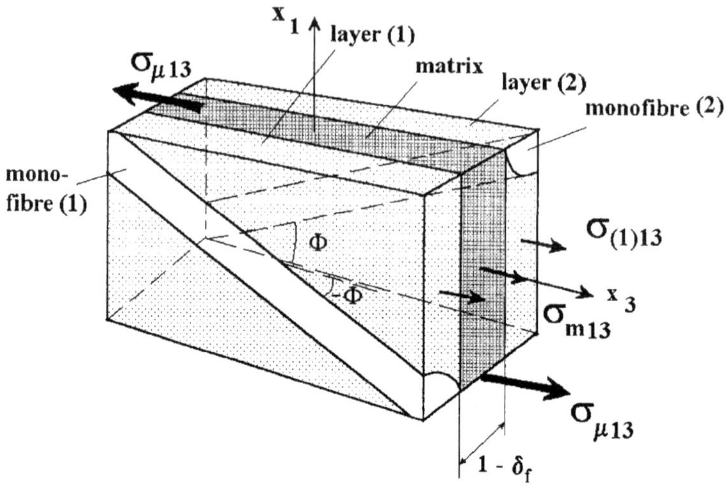

Figure 6.4. Model of periodicity cell PC3 of microcomposite for calculation of longitudinal modulus of shear G'_μ and stress concentration coefficient B_{m3} in longitudinal shear

where $\sigma_f = \sigma_{\mu 33}/(\varphi_f \cos^2 \Phi)$ is the stress in the monofibre, and $\sigma_m = -\sigma_{\mu 33} \tan \Phi$ is the transverse stress of compression in the matrix.

From (6.6) and (6.7) and expressions (4.22) and (5.6) for E_m and E_f one can obtain the expression (6.4) for E'_μ.

Longitudinal Shear Modulus and Transverse Elastic Modulus

The longitudinal shear modulus G'_μ of microcomposite (formula (6.4)) can be evaluated with the help of the following model (Figure 6.4).

The periodicity cell PC3 of a microcomposite can be presented in the form of a three-layered material, in which the first and third layers contain straight segments of monofibre deflected with angles Φ and $-\Phi$ from the reinforcing axis Ox_3, and the middle layer is the interlayer of the matrix. These layers are orthogonal to the Ox_2 direction.

The effective longitudinal shear modulus G'_μ of the three-layered model is calculated as the ratio of tangential stress $\sigma_{\mu 13}$ acting on the end surfaces and corresponding strain $\varepsilon_{\mu 13}$:

$$2G'_\mu = \frac{\sigma_{\mu 13}}{\varepsilon_{\mu 13}}. \tag{6.8}$$

The stress $\sigma_{\mu 13}$ is connected to stresses within each of the layers σ_{m13} and $\sigma_{(1)13}$ by the equilibrium relation (stress $\sigma_{(1)13}$ is the same for the first and third layers)

$$\sigma_{\mu 13} = \delta_f \sigma_{(1)13} + (1 - \delta_f)\sigma_{m13}, \tag{6.9}$$

where δ_f is the relative total thickness of the first and second layers with monofibres (Figure 6.4).

Expressions for stresses $\sigma_{(1)13}$ and σ_{m13} are determined with the help of the transformation formulae for tensors under rotation [3, 91] as follows:

$$\sigma_{(1)13} = 2\left(\delta_f \frac{(1+\nu_a)}{2} G_f \sin^2 2\Phi + (1-\delta_f)G_m\right)\varepsilon_{(1)13}, \qquad (6.10)$$

$$\sigma_{m13} = 2G_m\varepsilon_{13}.$$

In the expression for $\sigma_{(1)13}$ we have retained only principal terms. By taking account of $\varepsilon_{m13} = \varepsilon_{(1)13} = \varepsilon_{\mu 13}$, from equations (6.8), (6.9) and (6.10) we obtain the expression (6.4) for longitudinal shear modulus G'_μ.

The transverse elastic modulus E_μ of the microcomposite is determined similarly; in place of stresses $\sigma_{\mu 13}$ we consider stresses $\sigma_{\mu 11}$. By neglecting small terms of order (E_b^0/E_f^0), we obtain the expression (6.4) for E_μ that is similar to the expression for G'_μ.

Poisson's ratios ν_μ and ν'_μ are determined by the known mixture rules [20].

6.2.2. HEAT DEFORMATIONS AND PHASE INTERACTIONS

The heat deformation tensor $\overset{\circ}{\varepsilon}_{\mu kl}$ in (6.2) for a microcomposite has the following structure

$$\overset{\circ}{\varepsilon}_{\mu kl} = \begin{pmatrix} \overset{\circ}{\varepsilon}_\mu & 0 & 0 \\ 0 & \overset{\circ}{\varepsilon}_\mu & 0 \\ 0 & 0 & \overset{\circ}{\varepsilon}{}'_\mu \end{pmatrix} \qquad (6.11)$$

where $\overset{\circ}{\varepsilon}_\mu$ is the heat deformation of the microcomposite in the direction orthogonal to the fibre axis Ox_3 and $\overset{\circ}{\varepsilon}{}'_\mu$ is the heat deformation along the fibre axis. Deformations $\overset{\circ}{\varepsilon}_\mu$, $\overset{\circ}{\varepsilon}{}'_\mu$ are calculated by the mixture rule

$$\overset{\circ}{\varepsilon}_\mu = \varphi_f\overset{\circ}{\varepsilon}_f + (1-\varphi_f)\overset{\circ}{\varepsilon}_m,$$

$$\overset{\circ}{\varepsilon}{}'_\mu = \left(\varphi_f E_f\overset{\circ}{\varepsilon}_f + (1-\varphi_f)E_m\overset{\circ}{\varepsilon}_m\right)\frac{1}{E'_\mu}. \qquad (6.12)$$

Tensor $f_{\mu kl}$ is called the tensor of phase interaction in the microcomposite, it has a structure similar to $\overset{\circ}{\varepsilon}_{\mu kl}$:

$$f_{\mu kl} = \begin{pmatrix} f_\mu & 0 & 0 \\ 0 & f_\mu & 0 \\ 0 & 0 & f'_\mu \end{pmatrix}, \qquad (6.13)$$

where f_μ and f'_μ are calculated by the formulae

$$f_\mu = (1-\varphi_f)f_m, \quad f'_\mu = (1-\varphi_f)\frac{E_m}{E'_\mu}f_m. \qquad (6.14)$$

6.2.3. MICROSTRESSES

Microstresses in the monofibre and in the matrix can be determined as follows:

$$\sigma_{ikl} = \frac{1}{2} B_{i1} (\sigma_{\mu 11} + \sigma_{\mu 22}) (\delta_{k1}\delta_{l1} + \delta_{k2}\delta_{l2}) + \left(B_{i2}\sigma_{\mu 33} + \overset{\circ}{B}_i + N_i p \right) \delta_{k3}\delta_{l3} +$$
$$+ B_{i3} (\sigma_{\mu k3}\delta_{l3} + \sigma_{\mu l3}\delta_{k3} - 2\sigma_{\mu 33}\delta_{k3}\delta_{l3}) + B_{i1} \left(\sigma_{\mu kl} - \frac{1}{2}(\sigma_{\mu 11} + \sigma_{\mu 22}) \times \right.$$
$$\left. \times (\delta_{k1}\delta_{l1} + \delta_{k2}\delta_{l2}) + \sigma_{\mu 33}\delta_{k3}\delta_{l3} - \sigma_{\mu k3}\delta_{l3} - \sigma_{\mu l3}\delta_{k3} \right), \quad i = f, m. \tag{6.15}$$

This relationship between tensors σ_{mkl}, σ_{fkl} and $\sigma_{\mu kl}$ is *tensor-invariant* with respect to the group of transformations corresponding to transverse isotropy about the Ox_3 axis (i.e. it is independent of a change of the coordinate system in this group of transformations) [90].

Coefficients of stress concentration B_{ik}, $\overset{\circ}{B}_i$ and N_i in a matrix and monofibres are determined as follows:

$$B_{m1} = \tilde{a}_1/\tilde{a}_5, \quad B_{m2} = E_m/E'_\mu, \quad B_{m3} = \tilde{a}_1/\tilde{a}_3, \tag{6.16}$$

$$B_{f1} = 1, \quad B_{f2} = 1/\varphi_f, \quad B_{f3} = G_f/G'_\mu,$$

$$\overset{\circ}{B}_f = \frac{(1-\varphi_f)}{\varphi_f} E_m \left(\overset{\circ}{\varepsilon}_m - \overset{\circ}{\varepsilon}_f \right), \quad \overset{\circ}{B}_m = -E_m \left(\overset{\circ}{\varepsilon}_m - \overset{\circ}{\varepsilon}_f \right),$$

$$N_f = \frac{(1-\varphi_f)}{\varphi_f} E_m f_m, \quad N_m = -E_m f_m.$$

The coefficient of stress concentration within the matrix in longitudinal shear B_{m3} is determined with the help of the three-layered model of PC3 decribed in section 6.2.1 as follows:

$$B_{m3} = \frac{\sigma_{m13}}{\sigma_{\mu 13}} = \frac{G_m}{G'_\mu}. \tag{6.17}$$

On substituting in place of G'_μ its expression from (6.4) and also the expression $G_m = G_b^0 \tilde{a}_1$, we obtain the formula (6.16) for B_{m3}.

We can obtain expression for B_{m1} similarly. The remaining coefficients B_{ik} and N_i are calculated by the mixture rule.

6.2.4. HEAT CONDUCTIVITY

The heat conductivity of a microcomposite is described by the tensor of heat conductivity coefficients $k_{\mu kl}$; this is a transversally isotropic tensor with the following structure:

$$k_{\mu kl} = \begin{pmatrix} k_\mu & 0 & 0 \\ 0 & k_\mu & 0 \\ 0 & 0 & k'_\mu \end{pmatrix} \tag{6.18}$$

As a rule, the difference in the coefficients between a matrix and a fibre is not more than one order of magnitude (for elastic moduli this difference can reach three orders of magnitude). Due to this, the heat conductivity coefficients of a microcomposite in the reinforcing direction k'_μ and in the transverse direction k_μ depend weakly on small bendings of monofibres in the microcomposite.

The heat conductivity coefficients k'_μ and k_μ are determined mainly by the volumetric content of the fibre and the matrix, and are calculated by the mixture rule:

$$k'_\mu = k_f \varkappa_1, \qquad k_\mu = k_m \varkappa_2, \tag{6.19}$$

$$\varkappa_1 = \varphi_f + \frac{k_m}{k_f}\varkappa_0(1-\varphi_f), \qquad \varkappa_2 = \left(1 - \varphi_f + \frac{k_m}{k_f}\varkappa_0\varphi_f\right)^{-1}.$$

6.3. Thermo-Elastic Characteristics and Heat Conductivity of Unidirectional Composites

6.3.1. THEORETICAL RELATIONS

Microdefects of crack type in composites usually have a little effect on elastic moduli of composites. According to different estimations, this effect does not usually exceed 5 - 7 %. Only for stresses very close to the strength limit there can be a 15 - 20 % decrease of the instantaneous elastic modulus [13, 113].

Within the scope of the four-level structural scheme considered (Figure 6.1) a UC differs from a microcomposite only by the presence of defects in its matrix and fibres. Therefore, taking this into account, one can assume that constitutive relations of a UC are the same as the corresponding relations of the microcomposite (6.2), in which subscript μ is replaced by u for tensors σ_{kl} and ε_{kl}:

$$\varepsilon_{ukl} = \Pi_{\mu klqr}\sigma_{uqr} + \overset{\circ}{\varepsilon}_{\mu kl} + f_{\mu kl}p. \tag{6.20}$$

The tensor of elastic moduli $\Pi_{\mu klqr}$, the tensor of heat deformations $\overset{\circ}{\varepsilon}_{\mu kl}$ and the tensor of phase interactions $f_{\mu kl}$ are determined by formulae (6.3), (6.11) - (6.14).

The same can be said about the heat conductivity tensor k_{ukl}:

$$k_{ukl} = k_{\mu kl}. \tag{6.21}$$

6.3.2. EXPERIMENTAL DATA

Changing elastic properties of UCs in the reinforcing direction in heating, according to (6.4), is described by the function $\widetilde{a}_2$

$$\widetilde{a}_2 = E'_\mu/(E_f^0\varphi_f^0), \tag{6.22}$$

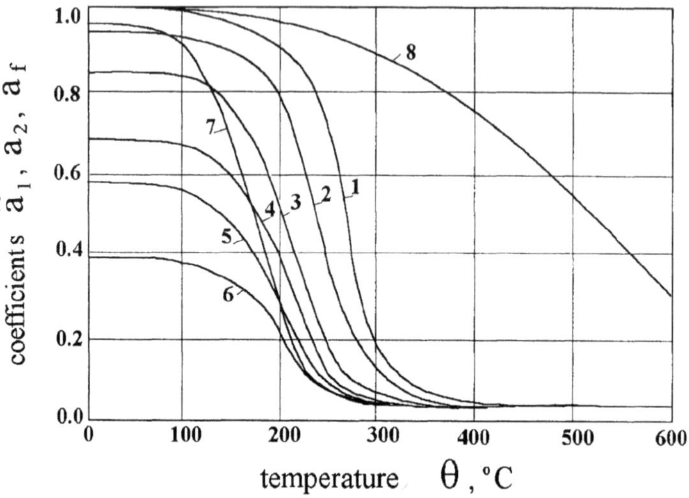

Figure 6.5. Varying elastic properties coefficient $\tilde{a}_2$ of unidirectional carbon/epoxy composite for different values of carbon monofibres misalignment angle Φ: 1 - $\Phi = 1^\circ$, 2 - $\Phi = 2^\circ$, 3 - $\Phi = 3^\circ$, 4 - $\Phi = 4^\circ$, 5 - $\Phi = 5^\circ$, 6 - $\Phi = 6^\circ$ and coefficients $\tilde{a}_1$ for epoxy matrix (curve 7) and $\tilde{a}_f$ for carbon fibres (curve 8) *versus* temperature of heating at 0.1 K/s rate in air

where E'_μ is the elastic modulus of the UC (being the same as the elastic modulus of the microcomposite) in the reinforcing direction at temperature $\theta \geqslant \theta_0$. The change with temperature of the elastic modulus E_μ in the transverse direction is described by the function $\tilde{a}_5$

$$\tilde{a}_5 = E_\mu / E_b^0. \tag{6.23}$$

Functions $\tilde{a}_2$ and $\tilde{a}_5$, on the one hand, can be determined by formulae (6.22), (6.23) if experimental values of E_f^0, E_b^0 at $\theta = \theta_0$ and E'_μ, E_μ at $\theta \geqslant \theta_0$ are known; on the other hand, they can be calculated by formulae (6.5).

Coefficient $\tilde{a}_2$ depends on the misalignment angle Φ of the monofibres. Special experiments have established that the angle Φ depends mainly on the quality of manufacturing technology of monofibres threads. Its values are within the interval: $1^\circ \leqslant \Phi \leqslant 5^\circ$ for actual UCs. The role of this angle Φ is of great importance. It is the presence of the fibre misalignment angle that explains why elastic properties of the composite in heating are defined by properties of its ablative matrix but not its fibres.

Figure 6.5 shows temperature dependences of coefficient $\tilde{a}_f$ for carbon monofibre, coefficient $\tilde{a}_1$ for an epoxy matrix and coefficient $\tilde{a}_2$ for unidirectional carbon-epoxy material with different values of the angle Φ: $\Phi = 1^\circ, 2^\circ, 3^\circ, 4^\circ, 5^\circ$ and 6°; the reinforcing coefficient is $\varphi_f^0 = 0.5$.

Figure 6.5 shows that even for minimum value $\Phi = 1^\circ$, the coefficient $\tilde{a}_2$ decreases in heating practically simultaneously with the coefficient $\tilde{a}_1$ of the matrix, while the coefficient $\tilde{a}_f$ has values greater than 0.9 at the same temperature.

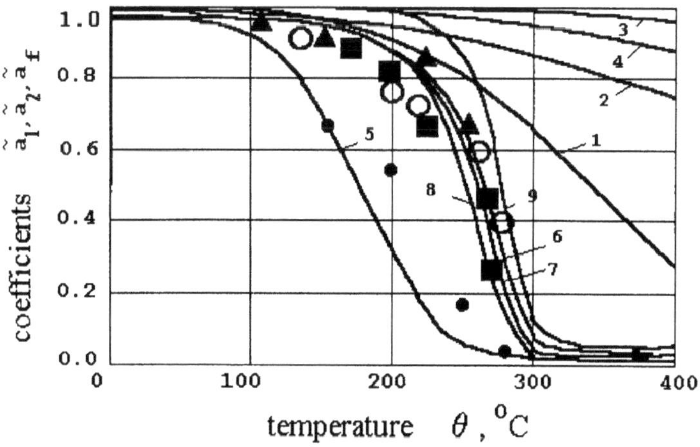

Figure 6.6. Elastic-strength properties coefficient $\widetilde{a}_f$ for different types of fibres (curves 1-4: 1 - organic, 2 - carbon, 3 - glass, 4 - boron), coefficient $\widetilde{a}_1$ for epoxy matrix (curve 5 (●)) and elastic properties coefficient $\widetilde{a}_2$ for a UC based on epoxy matrix with corresponding types of fibres (curves 6 - 9: 6 - organic, 7 - carbon (▲), 8 - boron (■), 9 - glass (○)) *versus* temperature at 0.1 K/s heating rate in air, $\Phi = 1°$. Lines are computations, points are experimental data [113, 117]

As a temperature characterizing a heat-resistance of a composite at a fixed rate of heating, we introduce the temperature θ_{hE} for which coefficient $\widetilde{a}_2$ decreases in heating by a factor of two:

$$\widetilde{a}_2(\theta_{hE}) = 0.5. \qquad (6.24)$$

At room temperature $\theta_0 = 20$ °C the coefficient $\widetilde{a}_2$ is close to 1: $\widetilde{a}_2(\theta_0) \approx 1$.

An analogous temperature of heat-resistance of a matrix is determined by formula (4.26c).

Heat-resistance of an epoxy matrix determined so that $\theta_{hE} = 180$ °C. At $\Phi = 1°$, the heat-resistance of the composite is $\theta_{hE} \approx 260$ °C. With increasing Φ, θ_{hE} decreases: $\theta_{hE} \approx 220$ °C at $\Phi = 3°$, and $\theta_{hE} \approx 200$ °C at $\Phi = 5°$.

Figure 6.6 shows temperature dependences of the function $\widetilde{a}_2$ for UCs based on epoxy matrix with different fibres: glass-, carbon, boron and organic; the misalignment angle for all the fibres is the same: $\Phi = 1°$.

Corresponding dependences $\widetilde{a}_f(\theta)$ for fibres are shown for comparison. This figure shows that the function $\widetilde{a}_2$ depends relatively weakly on the fibre type; they all have the same binder. The highest heat-resistance is for glass-epoxy material: $\theta_{hE} = 280$ °C, the smallest is for organo-epoxy material: $\theta_{hE} = 240$ °C.

The coefficient $\widetilde{a}_2$ of a UC is seen to be determined mainly by the type of its ablative matrix. Theoretical heat-resistance of carbon/polyether is $\theta_{hE} = 180$ °C, carbon/epoxy - $\theta_{hE} = 260$ °C, carbon/epoxy-phenol - $\theta_{hE} = 320$ °C, carbon/phenolic - $\theta_{hE} = 400$ °C, carbon/polyimide - $\theta_{hE} = 460$ °C, and carbon/silicon-organic -

$\theta_{hE} = 500$ °C.

These computed results are in a good agreement with experimental data on heat-resistance of composites with different binders. Thus, magnitudes of the coefficient $\widetilde{a}_2$ of a UC are determined rather by its matrix type than by its fibre type.

The coefficient $\widetilde{a}_2$ at temperature $\theta = \theta_0$ describes the so-called *realization of elastic properties of a monofibre* in a composite. According to the model, the realization of elastic properties is defined, in general, by the monofibre misalignment angle Φ. Figure 6.5 shows that values of $\widetilde{a}_2$ at $\theta = \theta_0$ decrease as Φ increases: $\widetilde{a}_2 = 0.9$ when $\Phi = 2^o$, $\widetilde{a}_2 = 0.8$ when $\Phi = 3^o$. These magnitudes correspond to experimental data on realization of fibre properties in unidirectional carbon and organic plastics.

One can make the following conclusion from these results: the higher the elastic properties of a fibre in a composite at normal temperature, the higher is the heat-resistance θ_{hE} of the composite in heating for the same matrix.

6.4. Strength Properties of Unidirectional Composite under High Temperatures

Introducing the third structural level in the multilevel structural model of a UC, as compared with the microcomposite model, has the purpose of taking account of the influence of structural defects (fibre breakages and matrix delaminations) on the strength of a material under high temperatures. These two types of defects affect elastic features of the material weakly, but strength properties greatly.

We will consider four main *types of destruction of a UC*:

(1) breakage of fibres and cracking of a matrix in longitudinal tension along the reinforcing direction Ox_3;
(2) loss of stability in longitudinal compression along the reinforcing direction Ox_3;
(3) delamination along fibres in transverse tensions in the Ox_1x_2 plane or in shear;
(4) failure due to shear of a matrix in transverse compressions of the composite in the Ox_1x_2 plane.

Schematically these four types of destruction are shown in Figures 6.7 - 6.10.

6.4.1. THERMAL STRENGTH OF UNIDIRECTIONAL COMPOSITE IN TENSION ALONG REINFORCING DIRECTION

The model of a UC is based on the model of a fibre bundle considered in Chapter 5. The UC consists of a large number of periodicity cells PC3 (Figure 6.11).

Each of the PC3 cells in the model is assumed to consist of domains of two types: 1) microcomposites with intact fibres (all parameters of this domain have the index (μ)) and 2) microcomposites with destroyed fibres (their parameters are indicated by index (d)).

As mentioned above, elastic properties of the both types of microcomposites (μ) and (d) are the same, therefore equation (6.3) hold for (d), in which (μ) is

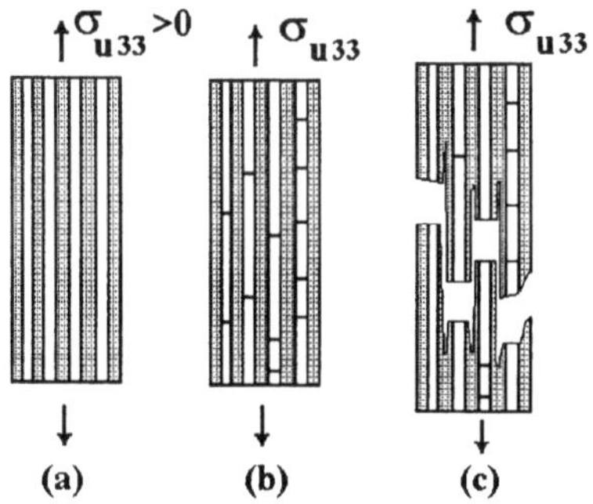

Figure 6.7. Three stages of failure of a UC in longitudinal tension: (a) - virgin composite; (b) - cracking of fibres; (c) - appearance of delaminations in a matrix and complete destruction

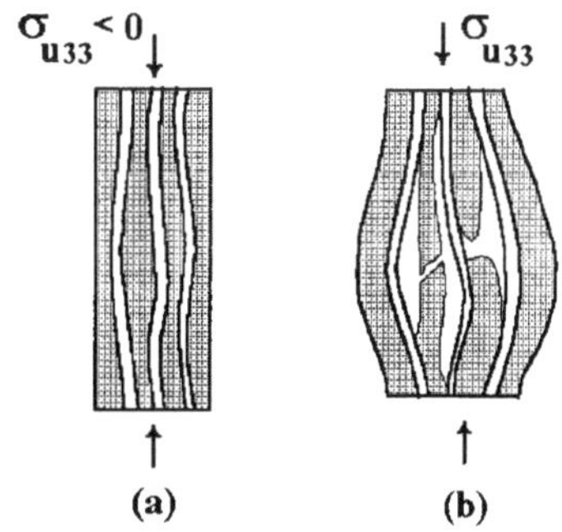

Figure 6.8. A scheme of failure of a UC in longitudinal compression: (a) - virgin composite with curved fibres; (b) - appearance of delaminations in a matrix and loss of stability of the composite

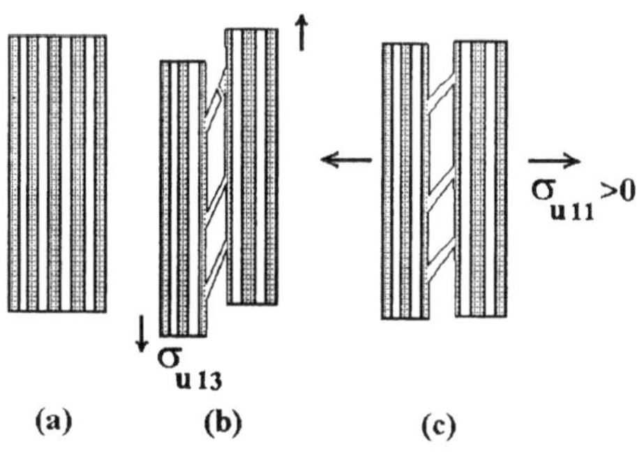

Figure 6.9. The third type of UC failure: (a) - virgin composite; (b) - delamination of a matrix in longitudinal shear; (c) - delamination of a matrix in transverse tension

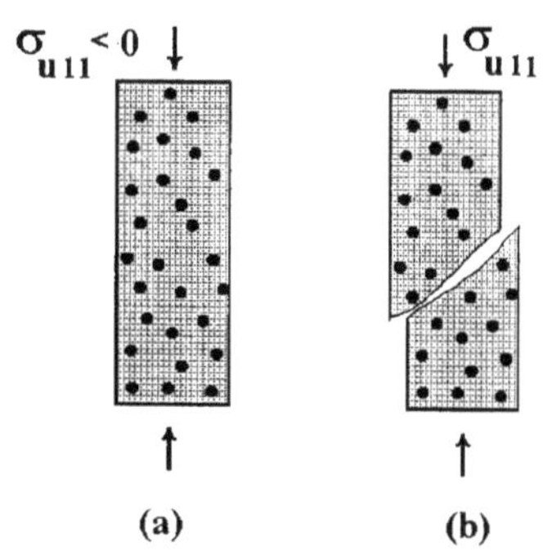

Figure 6.10. The fourth type of UC failure in transverse compression: (a) - virgin composite; (b) - shear in a matrix and destruction of the composite

replaced by (d).

In a real material reinforced by unidirectional threads consisting of a large number of monofibres, under loading along its reinforcing axis, there is a gradual process of individual monofibres breakage. This phenomenon is connected with the fact that separate monofibres have different strengths which, as was mentioned in Chapter 5, are determined by different contents of surface defects appearing after technological processing. At first the weakest fibres are broken, then the stronger

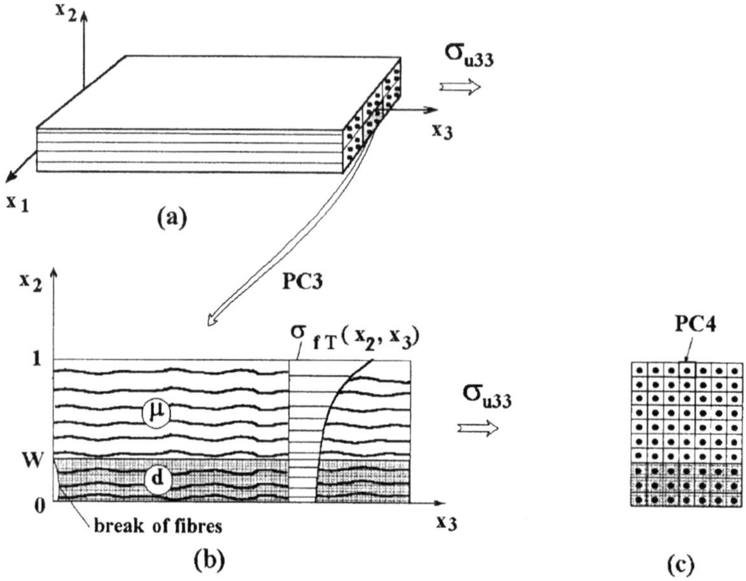

Figure 6.11. Structural model of a UC for calculation of strength in tension along fibres: (a) - unidirectional composite (UC); (b) - periodicity cell PC3 of UC consisting of microcomposite with intact fibres (μ) and destroyed fibres (d); (c) - crossection of PC3 with cells PC4 of the microcomposite

ones. Fibres in the destroyed zone take part only because the tangential stresses in the matrix transmit load to them; they do so only on part of their length except a section called the *non-effective length of the fibre* L_{neff} [13, 113].

This non-effective length depends mainly on properties of the matrix, and thus considerably increases under high temperatures.

If a delamination of a matrix occurs, then all destroyed fibres cease to bear a load; and there is a total destruction of the UC.

Real breakages of monofibres appear in a composite randomly, as the distribution of monofibres with different strengths in a material has a statistical nature.

There are many statistical models describing the phenomenon of fibre breakage in composites, beginning from the classical work of Rozen [100]. However, the Rozen model and its modifications do not take account of changing characteristics of matrix and fibres during heating, nor of the longitudinal delaminations of fibres from a matrix that is typical for high-temperature heating.

We will now present a deterministic (non-statistical) approach to modelling of monofibre breakage that allows us to obtain an analytical expression for strength of a UC. We thus have a deterministic approach for the whole multilevel scheme of composite material under high temperatures.

The essence of this method is as follows [29].

All fibres in microcomposites are assumed to have the same elastic characteristics, but different strengths (see Chapter 5). Their strength σ_{fT} depends on the

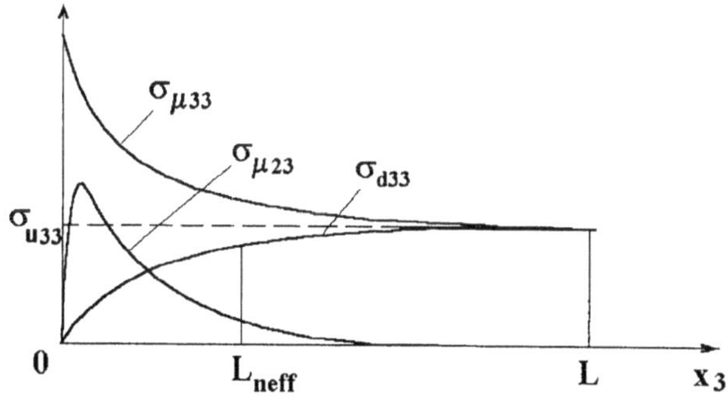

Figure 6.12. Schematic graphs of microstresses $\sigma_{\mu 33}$, $\sigma_{\mu 23}$ in microcomposite with intact fibres and σ_{d33} in microcomposite with destroyed fibres in tension of the UC along the fibres (axis x_3)

length L and coordinate x_2 within PC3 in accordance with the distribution (5.12):

$$\sigma_{fT}(x_2) = \sigma_f^0 \tilde{a}_f \left(\frac{d_f}{L}\right)^{\gamma/\beta} (\ln 1/(1-x_2))^{1/\beta}, \qquad 0 < x_2 < 1, \tag{6.25}$$

where d_f is the diameter, σ_f^0 is the strength of monofibres at temperature θ_0, and γ, β are the parameters characterizing the scatter of strengths of monofibres (see paragraph 5.5).

The microcomposites with fibres of different strengths σ_{fT} are considered to be distributed in the unidirectional material not randomly but ordered with respect to their length (in Figure 6.11 the longest of them, having by (6.25) the smallest strength, are in the middle of PC4 at $x_2 = 0$; and shorter fibres are in the periphery at $x_2 = 1$).

Thus, the strength is assumed to depend on the coordinates x_2 and x_3: $\sigma_{fT}(x_2, x_3)$ (or $\sigma_{fT}(x_1, x_3)$) (Figure 6.11) as was considered for the thread model (Chapter 5). (It is possible to consider a model with a radial distribution, however, in this case intermediate expressions prove to be more cumbersome and provide few improvements.)

When the unidirectional material is loaded along the Ox_3 axis by stress σ_{u33}, the stresses in the all fibres of microcomposites are the same before the fibres break; when the stress of the weakest fibre reaches the ultimate strength σ_{fT}, the fibre breaks. With further increase in load σ_{u33}, more and more fibres break. Thus for each value of σ_{u33}, a certain volume ratio W of the microcomposite has broken fibres. The ratio W is related to σ_{u33} by the equation

$$\sigma_{f33}(\sigma_{u33}, x_2, x_3) = \sigma_{fT}(x_2, x_3), \qquad x_2 = W, \qquad x_3 = L. \tag{6.26}$$

We assume that all breakages of fibres are on the same plane $x_3 = 0$.

Then the domain (μ) of the periodicity cell PC3 can be represented by the layer $W < x_2 < 1$ rigidly fastened in the plane $x_3 = 0$, and the domain (d) – $0 < x_2 < W$ with breaking plane $x_3 = 0$. Along the $x_2 = W$ plane the both layers are rigidly fastened.

Profiles of stresses $\sigma_{\mu 33}$, σ_{d33} and $\sigma_{\mu 23}$ in the layers (for the case when $W > 0$) are shown in Figure 6.12 and determined by the following formulae [24, 29]

$$\sigma_{\mu 33} = \sigma_{u33}\left(1 + \frac{W}{1-W}\exp\left(-\frac{k_L L}{d_f}\right)\right), \tag{6.27}$$

$$\sigma_{d33} = \sigma_{u33}\left(1 - \exp\left(-\frac{k_L L}{d_f}\right)\right), \tag{6.28}$$

$$\sigma_{\mu 23} = \sigma_{u33} W k_L \exp\left(-\frac{k_L L}{d_f}\right), \tag{6.29}$$

$$k_L = \left(\frac{2G'_\mu}{E'_\mu W(1-W)}\right)^{1/2}, \tag{6.30}$$

where L is the length of PC3 along the x_3 coordinate and d_f is the monofibre thickness.

Stress $\sigma_{\mu 33}$ in the intact layer and tangential stresses $\sigma_{\mu 23} = \sigma_{d23}$ on the boundary of the separation layers reach their maximum values at $x_3 = 0$:

$$\sigma_{\mu 33} = \frac{1}{1-W}\sigma_{u33}, \qquad x_3 = 0, \tag{6.31}$$

$$\sigma_{\mu 23} = \sigma_{u33} W k_L. \tag{6.32}$$

From equation (6.15) we can find the expressions for stresses in the monofibre σ_{f33} and matrix σ_{m23} in terms of stresses $\sigma_{\mu 33}$ and $\sigma_{\mu 23}$

$$\sigma_{f33} = \frac{1}{\varphi_f}\sigma_{\mu 33}, \tag{6.33}$$

$$\sigma_{m23} = (1-\varphi_f)\sigma_{\mu 23}, \tag{6.34}$$

here terms $\overset{\circ}{B}_f$ and N_f^0 were neglected as they are considerably smaller than stresses $(\sigma_{\mu 33}/\varphi_f)$.

Another phenomenon is characteristic for monofibres and cannot be neglected in strength calculations for UCs. It is the dependence of strength σ_{fT} on fibre length.

In Chapter 5 we described the deterministic model giving the dependence of strength σ_{fT} of monofibres with cracks on length L of the fibres. In accordance with the model, the strength of a monofibre situated within the zone of fibre breakage $x_2 = W$ in PC3 is determined by the dependence (6.25).

Equations (6.26), (6.31), (6.33) and (6.25) give the value of stress σ_{u33} which corresponds to the part W of the destroyed fibres in PC3:

$$\sigma_{u33} = \sigma_f \varphi_f (1-W)\left(\frac{d_f}{L_{neff}}\right)^{\gamma/\beta} 2^{1/\beta} W^{n/\beta}. \tag{6.35}$$

Here the following approximation was made: $-\lg(1-W) \approx 2W^n$, $n = 1.42$.

In equation (6.25) the whole length of monofibre L is used, as stresses in a separate monofibre are uniformly distributed along its axis Ox_3. For nonuniform distribution of stresses (6.27) - (6.29), in equation (6.25) instead of L we should use non-effective length L_{neff} for which stresses σ_{f33} and $\sigma_{\mu 33}$ have their maxima (see Figure 6.12); that was done in (6.35). Non-effective length of fibre is determined from the condition that, at the distance L_{neff} from the breakage, more than $(1-\chi)$ % of stresses in the destroyed fibre are recovered:

$$\frac{\sigma_{d33}(L_{neff})}{\sigma_{u33}} = 1 - \chi, \tag{6.36}$$

where $\chi \ll 1$ is a certain given constant. From (6.28) and (6.36) we obtain

$$\frac{L_{neff}}{d_f} = \left(\frac{E'_\mu}{2G'_\mu}\right)^{1/2} W^{1/2}(1-W)^{1/2}\lg\,(1/\chi). \tag{6.37}$$

On taking account of (6.37), we find that equation (6.35) takes the form

$$\begin{aligned}\sigma_{u33} =& \sigma_f \frac{\varphi_f^{1-\gamma/2\beta}}{(1-\varphi_f)^{\gamma/2\beta}} 2^{1/\beta} \left(\lg \frac{1}{\chi}\right)^{-\gamma/\beta} \times \\ &\times \left(\frac{2G_b^0\tilde{a}_1}{E_f^0\tilde{a}_2}\right)^{\gamma/\beta} (1-W)^{1-\gamma/2\beta} W^{(n/\beta)-(\gamma/2\beta)}.\end{aligned} \tag{6.38}$$

Equation (6.38) shows that σ_{u33} increases with W. However, the maximum value of σ_{u33} is not realized, as delamination of the matrix occurs at satisfaction of the condition of matrix strength disruption in shear (4.39)

$$\sigma_{\mu 23} = \sigma_S^0 h_{mS}(1-\varphi_f)^{-1}. \tag{6.39}$$

Here h_{mS} is determined by formula (4.40), and σ_S^0 is the strength of the polymer phase of the matrix in shear under temperature θ_0 =293 K.

Substituting expression (6.32) into (6.39), we have the limiting value σ_{u33}, for which delamination occurs:

$$\sigma_{u33} = \sigma_S^0 h_{mS} \left(\frac{E_f^0\tilde{a}_2}{2G_b^0\tilde{a}_1}\right)^{1/2} \left(\frac{1-W}{W}\right)^{1/2} \left(\frac{\varphi_f}{1-\varphi_f}\right)^{1/2}. \tag{6.40}$$

We will assume that complete destruction of the UC in tension occurs if fibre breakage and delamination occur simultaneously, i.e. both conditions (6.38) and (6.40) are satisfied. Equating the right-hand sides of the equations we find

$$W^{1/2+n/\beta-\gamma/2\beta}(1-W)^{1/2-\gamma/2\beta} = \Omega, \tag{6.40a}$$

where

$$\Omega \equiv \left(\frac{\sigma_S^0 h_{mS}}{\sigma_f^0 \tilde{a}_f}\right)\left(\frac{E_f^0 \tilde{a}_2}{2G_b^0 \tilde{a}_1}\right)^{1/2+\gamma/2\beta} 2^{1/\beta} \times$$
$$\times (\ln 1/\chi)^{-\gamma/\beta} \varphi_f^{-(1/2-\gamma/2\beta)} (1-\varphi_f)^{-(1/2+\gamma/2\beta)}.$$

The second multiplier on the left side of equation (6.40a) for actual composites is practically equal to 1, therefore an approximate solution of equation (6.40a) can be presented in the form

$$W = \Omega^{(1/2 + n/\beta - \gamma/2\beta)^{-1}}.$$

On substituting this expression into formula (6.39) and passing from σ_f^0 to average strength $\bar{\sigma}_f^0$ with the help of equation (5.16), one can derive the expression for the value of σ_{u33} for which the composite fails completely [29]:

$$\sigma_{u33} = \sigma'_{uT}, \quad \sigma'_{uT} = \bar{\sigma}_f^0 \tilde{a}_f H_0 \frac{\varphi_f^r}{(1-\varphi_f)^s} \left(\frac{E_f^0 \tilde{a}_2}{2G_b^0 \tilde{a}_1}\right)^s \left(\frac{\sigma_S^0 h_{mS}}{\bar{\sigma}_f^0 \tilde{a}_f}\right)^\omega, \tag{6.41}$$

where we introduced the constants

$$H_0 = \left(\frac{2}{\ln 2}\left(\frac{L_0}{d_f \ln 1/\chi}\right)^\gamma\right)^{(1-\omega)/\beta}, \quad s = \omega\left(\frac{1}{2} + \frac{\gamma}{2\beta}\right) - \frac{\gamma}{2\beta}. \tag{6.42}$$

$$\omega = \frac{(n/\beta) - (\gamma/2\beta)}{(1/2) + (n/\beta) - (\gamma/2\beta)}, \quad n = 1.42, \quad \chi = 0.01,$$

$$r = \left(\frac{1}{2} - \frac{\gamma}{2\beta}\right)(1-\omega).$$

Equation (6.41) gives the strength σ'_{uT} of a UC in tension.

If the condition $\sigma_{u33} = \sigma'_{uT}$ is satisfied, then there is a failure of the composite by destruction type (1) (see the classification mentioned above), i.e. due to breakage of fibres. We can write this condition in another form

$$\frac{|\sigma_{u33}| + \sigma_{u33}}{2\sigma'_{uT}} = 1. \tag{6.43}$$

This form is convenient, when we consider both tension ($\sigma_{u33} > 0$) and compression ($\sigma_{u33} < 0$) of a composite.

The exponent ω in equation (6.41) is usually substantially higher than s:

for carbon-plastics	$\omega = 0.33$,	$s = 0.07$,	$r = 0.25$,	$H_0 = 3$;
for boron-plastics	$\omega = 0.33$,	$s = 0.09$,	$r = 0.26$,	$H_0 = 2.5$;
for organoplastics	$\omega = 0.25$,	$s = 0.1$,	$r = 0.25$,	$H_0 = 1.5$;
for glass-plastics	$\omega = 0.38$,	$s = 0.13$,	$r = 0.35$,	$H_0 = 2.2$.

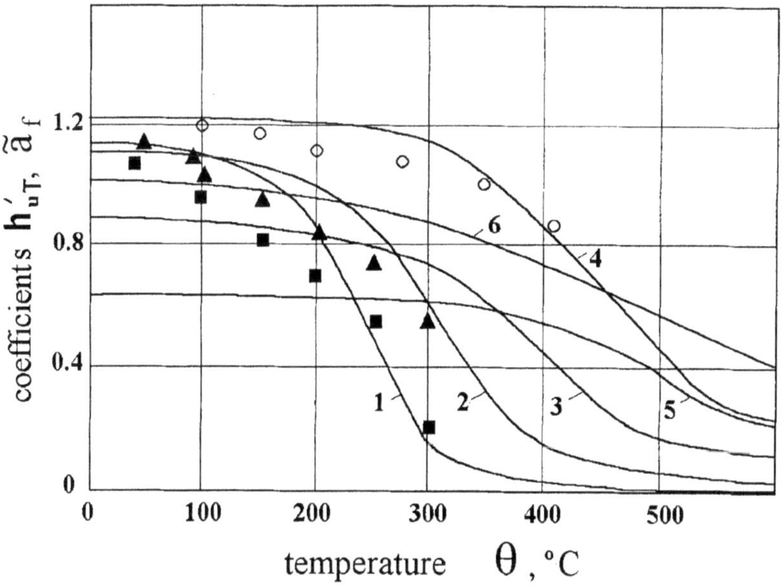

Figure 6.13. Varying strength properties coefficient h'_{uT} for unidirectional carbon composite with different types of matrix (curves 1-5: 1 (■) - epoxy, 2 (▲) - epoxy-phenolic, 3 – phenolic, 4 (o) - polyimide, 5 - silicon-organic) and elastic-strength coefficient $\widetilde{a}_f$ for carbon fibre (curve 6) *versus* temperature of heating at 0.1 K/s rate in air. Lines are computations, points are experimental data

Formula (6.41) shows that the strength in tension of a UC during high-temperature heating is determined mainly by the coefficient h_{mS} of decreasing the matrix strength in shear; the function $\widetilde{a}_f(\theta)$ is usually small at high temperatures.

6.4.2. EXPERIMENTAL DATA

Introduce a coefficient h'_{uT} relating the drop of strength σ'_{uT} of a UC under heating to the average strength of monofibres $\bar{\sigma}_f^0$:

$$h'_{uT} = \sigma'_{uT}/(\bar{\sigma}_f^0 \varphi_f^0), \tag{6.44}$$

where σ'_{uT} is calculated by formula (6.41), or determined in experiments.

Also introduce a *temperature* θ_h *of heat-resistance* of the composite for which the value of the coefficient h'_{uT} decreases in heating by a factor of two:

$$h'_{uT}(\theta_h) = 0.5. \tag{6.45}$$

As with the coefficient $\widetilde{a}_2$ of varying elastic properties, the dependence h'_{uT} on θ is defined, in the main, by the heat-resistance of the matrix. For epoxy matrix $\theta_h = 180$ °C and for carbon/epoxy composite $\theta_h = 240$ °C, but for the carbon fibres in air $\theta_h = 650$ °C.

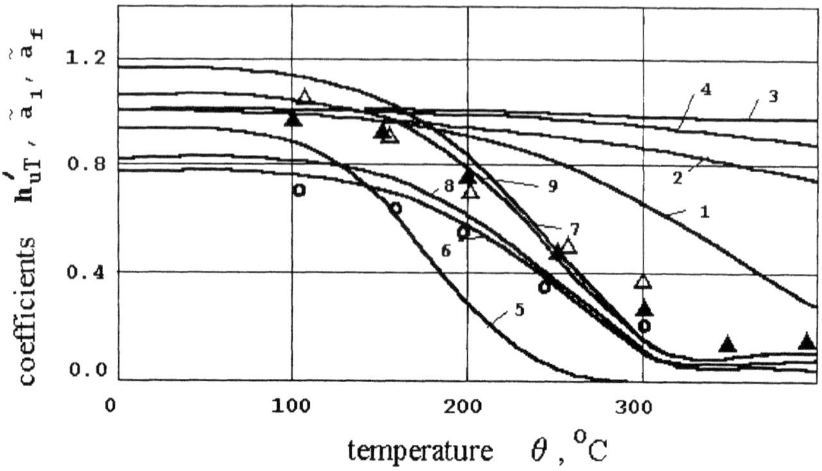

Figure 6.14. Varying elastic-strength properties coefficient $\widetilde{a}_f$ for different types of fibres (curves 1-4: 1 - organic, 2 - carbon, 3 - glass, 4 - boron), coefficient $\widetilde{a}_1$ for epoxy matrix (curve 5) and strength properties coefficient h'_{uT} for a UC based on epoxy matrix with corresponding types of fibres (curves 6 - 9: 6 - organic, 7 (▲) - carbon, 8 (○) - glass, 9 (△) - boron) *versus* temperature at 0.1 K/s heating rate in air. Lines are computations, points are experimental data

This conclusion is verified by experimental and computed results presented in Figure 6.13 in the form of the dependence of h'_{uT} on θ for unidirectional carbon plastics on different binders, and by values of $h'_{uT}(\theta)$ for composites on epoxy binder with different types of fibres, shown in Figure 6.14.

Figures 6.13, 6.14 show that if a type of a matrix considerably changes a temperature θ_h of heat-resistance of a composite (for example, for carbon/epoxy composite $\theta_h = 240$ °C, for carbon/epoxy-phenolic $\theta_h = 300$ °C, for carbon/phenolic $\theta_h = 350$ °C, for carbon/polyimide $\theta_h = 440$ °C, for carbon/silicon-organic $\theta_h = 430$ °C), then a type of fibre has a considerably smaller effect on θ_h: for organo/epoxy composite $\theta_h = 242$ °C, carbon/epoxy $\theta_h = 250$ °C, boron/epoxy $\theta_h = 255$ °C, glass/epoxy $\theta_h = 245$ °C.

As with the coefficient $\widetilde{a}_2$ at $\theta = \theta_0$ describing a realization of elastic properties of fibres in a composite, the function $h^0_{uT} = h'_{uT}(\theta_0)$ characterizes the strength of monofibres in the composite. These two important characteristics of a composite have an essential distinction.

The coefficient $\widetilde{a}_2$ is determined, in the main, by the magnitude of the angle Φ. Figure 6.6 shows that for the same angle $\Phi = 1°$, the value of the coefficient $\widetilde{a}_2$ at $\theta = \theta_0$ is practically unity for different types of fibres (see Figure 6.6) and matrices.

The coefficient h^0_{uT} depends on both the angle Φ and the properties of the matrix (strength σ^0_S, shear modulus G^0_b) and the fibre. Figure 6.13 shows that at the same angle $\Phi = 1°$ values h^0_{uT} are maximal for the strongest and stiffest polyimide matrix: h^0_{uT} =1.23, and they are minimal for silicon-organic matrix $h^0_{uT} = 0.62$ having low elastic-strength characteristics. For a composite with an epoxy matrix $h^0_{uT} = 1.08$; with an epoxy-phenolic matrix $h^0_{uT} = 1.1$; with a phenolic matrix $h^0_{uT} = 0.9$; with polyether matrix $h^0_{uT} = 0.82$ at $\theta = \theta_0$.

Unlike the matrix, the coefficient h^0_{uT} of fibre strength properties is influenced not only by the value of their average strength $\bar{\sigma}^0_f$ but by the width of the range of strengths of separate monofibres, characterized by the ratio of parameters γ/β. The higher the value γ/β, the greater is the spread of fibre strengths and the coefficient of fibre strength h^0_{uT} in the composite. For carbon fibres γ/β=0.27 and for carbon/epoxy composite $h^0_{uT} = 1.15$; for boron fibres γ/β =0.23 and for boron/epoxy composite $h^0_{uT} = 1.05$; for glass fibres γ/β =0.2 and for glass/epoxy composite $h^0_{uT} = 0.82$; for organic fibres γ/β =0.07 and for organo/epoxy composite $h^0_{uT} = 0.78$. These theoretical values agree well enough with experimental data [113].

The conclusion made for elastic properties is also valid for strength properties: among groups of the same composite material, the heat-resistance θ_h is higher if it has a higher coefficient h^0_{uT} of fibre strength.

6.4.3. THERMAL STRENGTH OF UNIDIRECTIONAL COMPOSITE IN COMPRESSION ALONG REINFORCING DIRECTION

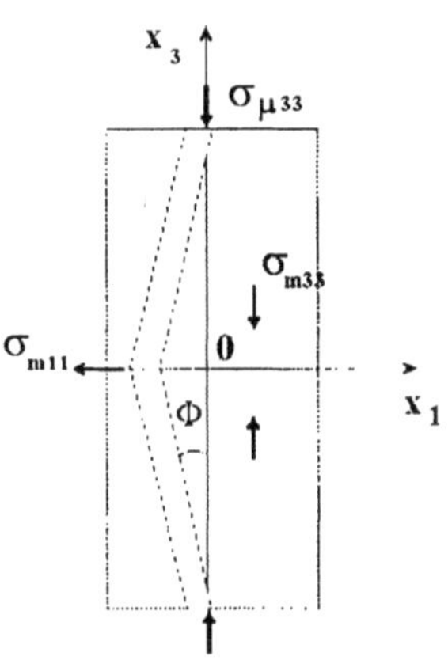

Figure 6.15. A scheme for calculation of strength of a UC in compression along the reinforcing direction

The strength σ_{uC} of unidirectional fibrous material in compression along the reinforcing direction Ox_3 (the 2nd type of destruction) is determined by the strength of the matrix and by the relation between the stiffnesses of the matrix and filler. A typical way in which a UC is destroyed in compression along its fibres is by a loss of stability of the fibres and delamination its matrix from fibres (see Figure 6.8).

Under heating, when stiffness of a matrix decreases, the strength of the composite sharply falls in compression.

Different criteria for strength in compression were suggested in [13]. Chamis in [13] showed that assuming that a composite is destroyed only due to stresses (or strains) of compression within its matrix overestimates the real compression strength by a factor of 2 or 3.

To consider the destruction of a UC in compression we apply the model of defects and misalignment of monofibres used before for tension.

Monofibres in a UC are assumed to be periodically curved in a certain plane, for example the Ox_1x_3 plane (Figure 6.15).

Fibre breaks are neglected, and stresses are assumed to be the same in both UC and microcomposite $\sigma_{u33} = \sigma_{\mu 33}$. Under the action of compressive stresses, i.e. $\sigma_{\mu 33} < 0$, on the microcomposite, its matrix will experience compressive stresses $\sigma_{m33} < 0$ and tensile transverse stresses $\sigma_{m11} > 0$ (Figure 6.15):

$$\sigma_{m11} = -B_{m4}\sigma_{\mu 33}, \qquad \sigma_{m33} = B_{m2}\sigma_{\mu 33}, \tag{6.46}$$

$$B_{m2} = E_m/E'_{\mu}, \qquad B_{m4} = \tan\,\Phi,$$

where Φ is the maximum angle of monofibre misalignment. Because the matrix is weaker in tension than in compression, the maximum stress value $|\sigma_{\mu 33}|$, when delamination of a matrix from fibres precedes destruction of the composite (Figure 6.8), is lower than for the case when only compressive stresses σ_{m33} are taken into account.

Substituting expressions for σ_{m11} and σ_{m33} (the remaining $\sigma_{mkl}0$ are zero) into the criterion of matrix strength (4.34), we obtain an expression for maximum stress $|\sigma_{\mu 33}|_{\max} = \sigma'_{uC}$; this is the strength of the composite in compression along its fibres under high temperatures:

$$\sigma'_{uC} = \sigma^0_S h'_{uC}, \tag{6.47}$$

where

$$h'_{uC} = \begin{cases} a_1 h^0 \left((\zeta_T(1-\Delta_T) + \Omega_T^2)^{1/2} - \Omega_T \right) / \zeta_T, \\ \qquad\qquad \text{if } \sigma^0_S h'_{uC}(B_{m4} - B_{m2}) < 3N^0 p, \\ a_1 h^0 \left((\zeta_C(1-\Delta_C) + \Omega_C^2)^{1/2} - \Omega_C \right) / \zeta_C, \\ \qquad\qquad \text{if } \sigma^0_S h'_{uC}(B_{m4} - B_{m2}) > 3N^0 p, \end{cases} \tag{6.48}$$

here $\Delta_{T,C}$ and $\Omega_{T,C}$ depend on pore pressure in the matrix p, and $\Delta_{T,C}$ are calculated by formulae (4.40) and $\Omega_{T,C}$ and $\zeta_{T,C}$ - by the formulae

$$\Omega^2_{T,C} = \left(\frac{(\sigma^0_S)^2}{(\sigma^0_{T,C})^2} - \frac{1}{3} \right) (B_{m2} - B_{m4}) \left(\frac{3N^0 p}{a_1 h^0} \right), \tag{6.49}$$

$$\zeta_{T,C} = (B_{m2} - B_{m4})^2 \left(\frac{\sigma^0_S}{\sigma^0_{T,C}} \right)^2 + B_{m2}B_{m4}.$$

If pore pressure in the composite is small: $p \ll \sigma^0_S$, then $\Omega_{T,C}$ and $\Delta_{T,C} \ll 1$ and expression (6.48) takes the form

$$h'_{uC} = \begin{cases} a_1 h^0 \zeta_T^{-1/2}, & B_{m2} > B_{m4}, \\ a_1 h^0 \zeta_C^{-1/2}, & B_{m2} < B_{m4}. \end{cases} \tag{6.50}$$

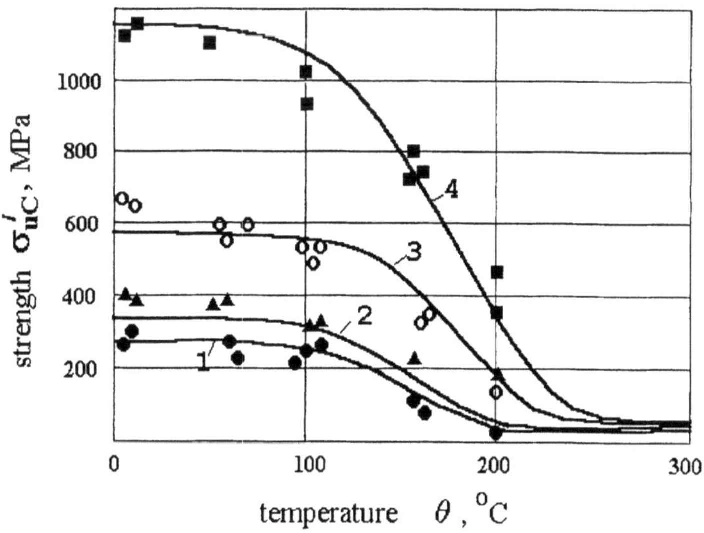

Figure 6.16. Varying strength σ'_{uC} of a UC based on epoxy matrix with different fibres (1 (●) - organic, 2 (▲) – carbon, 3 (○) – glass, 4 (■) – boron) in compression *versus* temperature under heating with rate 0.1 K/s in air

We can write the condition of failure of a UC under compression in the form:

$$\frac{|\sigma_{u33}| - \sigma_{u33}}{2\sigma'_{uC}} = 1, \tag{6.51}$$

where σ'_{uC} is determined by formulae (6.47) and (6.50).

Figure 6.16 shows experimental data for thermal strength σ'_{uC} for epoxy composites with different types of fibres under heating. Strength σ'_{uC} is determined mainly by the stiffness E_f of the fibres, by the angle of monofibre misalignment Φ and by the coefficient $\tilde{a}_1$ of changing elastic properties of its matrix.

Boron/epoxy has the highest strength σ'_{uC}, highest monofibre stiffness E_f^0 and smallest angle Φ; as boron monofibres are thicker (see Table 5.2) and have a smaller misalignment.

6.4.4. THERMAL STRENGTH OF UNIDIRECTIONAL COMPOSITE IN TRANSVERSE TENSION/COMPRESSION AND SHEAR

Failure criteria for unidirectional material according to the third and the fourth types are determined from the failure criterion for a matrix (4.34). This is used with the equation of stress concentration (6.15). We also should take account of that in transverse loadings and in shears of unidirectional material the stresses

$\sigma_{\mu kl}$ and σ_{ukl} are coincident. Then we have

$$z(Y_u) \equiv \frac{1}{12}\left(\frac{B_{m1}}{\sigma_S^0 a_1 h_0}\right)^2 \left(Y_{u2}^2 + 12\left(\frac{B_{m3}}{B_{m1}}\right)^2 Y_{u4}^2 + 3Y_{u3}^2\right) + \left(\frac{B_{m1}}{a_1 h^0}\right)^2 \times$$

$$\left(\frac{1}{(\sigma_C^0)^2} - \frac{1}{3(\sigma_S^0)^2}\right)Y_{u-}^2 + \left(\frac{B_{m1}}{a_1 h^0}\right)^2 \left(\frac{1}{(\sigma_T^0)^2} - \frac{1}{3(\sigma_S^0)^2}\right) Y_{u+}^2 = 1, \qquad (6.52)$$

where $Y_{u\alpha}$ are the invariants of the tensor σ_{ukl} with respect to the group of transformations of transverse isotropy and

$$2Y_{u\pm} = |Y_{u2} + \frac{3N^0 p}{B_{m1}}| \pm \left(Y_{u2} + \frac{3N^0}{B_{m1}}p\right),$$

$$Y_{u2} = \sigma_{u11} + \sigma_{u22}, \quad Y_{u3}^2 = (\sigma_{u11} - \sigma_{u22})^2 + 4\sigma_{u12}^2, \qquad (6.53)$$
$$Y_{u4}^2 = \sigma_{u13}^2 + \sigma_{u23}^2, \quad Y_{u1} = \sigma_{u33}.$$

Conditions of failure of the composite under just one non-zero stress follow from equation (6.52):

$$\text{pure tension:} \quad \sigma_{u11} = \sigma_{uT} \quad \text{if } \sigma_{u11} > 0,$$

$$\text{pure compression:} \quad |\sigma_{u11}| = \sigma_{uC} \quad \text{if } \sigma_{u11} < 0,$$

$$\text{pure shear:} \quad |\sigma_{u12}| = \sigma_{uS}. \qquad (6.54)$$

Here σ_{uT}, σ_{uC} and σ_{uS} are strengths in transverse tension, compression and shear:

$$\sigma_{uT} = \sigma_T^0 h_{uT}, \quad \sigma_{uC} = \sigma_C^0 h_{uC}, \quad \sigma_{uS} = \sigma_S^0 h_{uS}, \qquad (6.55)$$

$$h_{uT} = h_{mT}/B_{m1}, \quad h_{uC} = h_{mC}/B_{m1}, \quad h_{uS} = h_{mS}/B_{m1}.$$

Stress concentration coefficients B_{m1}, B_{m3} are expressed by formulae (6.16), (6.17), and functions $h_{mT,C,S}$ for the matrix by formulae (4.40).

6.4.5. THERMAL MICROSTRESSES AND MICROCRACKING

Due to the difference in heat deformations of matrix $\overset{\circ}{\varepsilon}_m$ and monofibre $\overset{\circ}{\varepsilon}_f$, there appear to be additional microstresses within the matrix and fibres called *thermal microstresses*.

At a lower level of heating before the beginning of thermodecomposition, heat deformations of the matrix are positive and substantially exceed the heat deformations of the fibres. Therefore, axial thermal microstress within the matrix is compressive, and the level of the stress $\overset{\circ}{\sigma}_m \approx E_m(\overset{\circ}{\varepsilon}_m - \overset{\circ}{\varepsilon}_f)$ is considerably smaller than the ultimate strength in compression (for carbon/epoxy: $\overset{\circ}{\sigma}_m \approx 1.5$ MPa and $\sigma_C^0 \approx 9.8$ MPa). This means that thermal stress does not cause a formation of microcracks and its effect on the composite strength can be neglected.

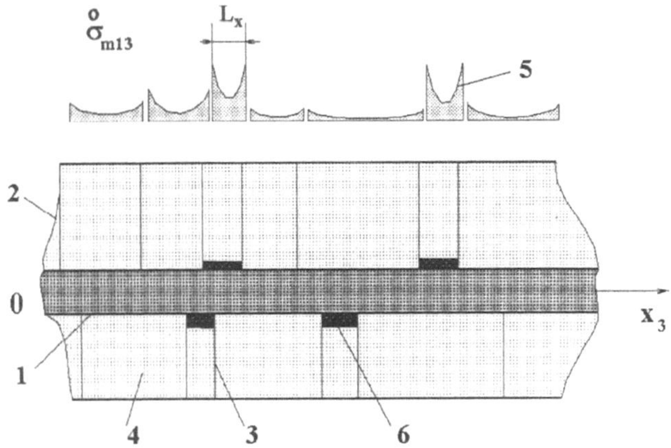

Figure 6.17. A model of cracking UCs due to the appearance of thermal microstress at high temperatures: 1 - monofibre, 2 - matrix, 3 - transverse microcrack, 4 - segment of the matrix between two cracks, 5 - distribution of the shearing of thermal microstress σ^0_{m13} *versus* the Ox_3 axis of the matrix segment, 6 - delamination of the short segments of the matrix from fibres

At high temperatures, when the charring process is the most intensive, the matrix can contract, and there will be substantial tensile axial thermal microstress within the matrix, which will exceed the matrix strength below these temperatures. Then there is cracking of the matrix, i.e. transverse cracks appear. These cracks have a most considerable effect on the strength in the transverse direction, σ_{uT}, and in the longitudinal shearing of the composite σ_{uS}.

The cause of the phenomenon is the appearance of tangential stresses $\overset{\circ}{\sigma}_{m13}$ and $\overset{\circ}{\sigma}_{m23}$ in segments of the matrix along fibres between the transverse cracks (Figure 6.17). They reach maximum values near the crack plane and their values depend on the segment length L_x:

$$\overset{\circ}{\sigma}_{m13} = \frac{E_m \Delta\overset{\circ}{\varepsilon}\Omega}{(1 - \exp(-K_L L_x/d_f))}, \qquad K_L = \frac{\Omega\varphi^0_f}{1 - \varphi^0_f}, \tag{6.56}$$

$$\Delta\overset{\circ}{\varepsilon} = \overset{\circ}{\varepsilon}_m - \overset{\circ}{\varepsilon}_f, \qquad \Omega = (2G_m/E_f)^{1/2}.$$

When the stress $\overset{\circ}{\sigma}_{m13}$ reaches the strength limit σ_S of the matrix at the considered temperature, there is delamination of all the segments of the matrix with a length smaller than L_x from the fibres. The delamination leads to a decrease in the effective contact surface of the matrix and fibres by the relative value $\Delta\Sigma$ and, as a result, the strengths σ_{uT} and σ_{uS} of the composite decrease.

The decrease of the contact surface is determined by $\Delta\Sigma = L_x/L_0$ where L_0 is the maximum magnitude of segment lengths of the cracking matrix. From the equation (6.56) we can find L_x/L_0 then the expression for the degradation coefficients of the mechanical properties in the transverse direction h_{uT}, in shearing

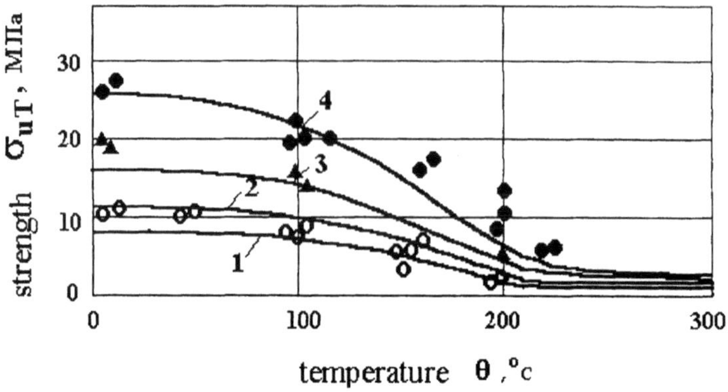

Figure 6.18. The varying strength σ_{uT} of UCs based on the epoxy matrix with different fibres (1 (○) - organic, 2 - glass, 3 (▲) - carbon, 4 (●) - boron) under transverse tension *versus* heaing temperature at a rate of 0.1 K/s in air. Curves are computations, points are experimental data

h_{uS} and in compression h_{uC} takes the form

$$h_{uT} = h_{uS} = h_{uC} = h^0 a_1(1 - \Delta\Sigma)/B_{m1}, \tag{6.57}$$

$$\Delta\Sigma \approx \frac{d_f}{L_0 K_L} \ln\left(1 + \frac{\Omega E_b^0 \Delta\overset{\circ}{\varepsilon}}{\sigma_S^0}\right). \tag{6.58}$$

If $\Delta\overset{\circ}{\varepsilon} > 0$, then $\Delta\Sigma > 0$, and if $\Delta\overset{\circ}{\varepsilon} \leqslant 0$ then $\Delta\Sigma = 0$.

The value d_f/L_0 determined by the experimental data is approximately 0.05. Thus, formulae (6.55) and (6.57) describe changing strengths σ_{uT} and σ_{uS} of UCs at high temperatures by taking account of the internal thermal microstresses and microcracking.

Figure 6.18 shows the computed dependences of the thermal strength σ_{uT} in transverse tension for the epoxy composite with fibres of different types. The reinforcing coefficient in computations was $\varphi_f = 0.6$, the misalignment angle of monofibres was $\Phi = 5°$. Figure 6.18 shows that the strength σ_{uT} depends strongly on the type of fibre: with increasing the ratio of stiffnesses E_f^0/E_b^0, the strength σ_{uT} grows. Temperature dependence $\sigma_{uT}(\theta)$ is determined by the heat-resistance of the matrix. Expressions (6.55) and (6.57) for σ_{uT} describe these phenomena adequately.

6.4.6. THERMAL STRENGTH OF UNIDIRECTIONAL COMPOSITE IN LONGITUDINAL SHEAR

From (6.52) we obtain the expression for strength σ'_{uS} of composite in longitudinal shear (when the only non-zero stress is σ_{u13})

$$\sigma_{u13} = \sigma'_{uS}, \qquad \sigma'_{uS} = \sigma_S^0 h'_{uS}, \qquad h'_{uS} = h_{mS}/B_{m3}. \tag{6.59}$$

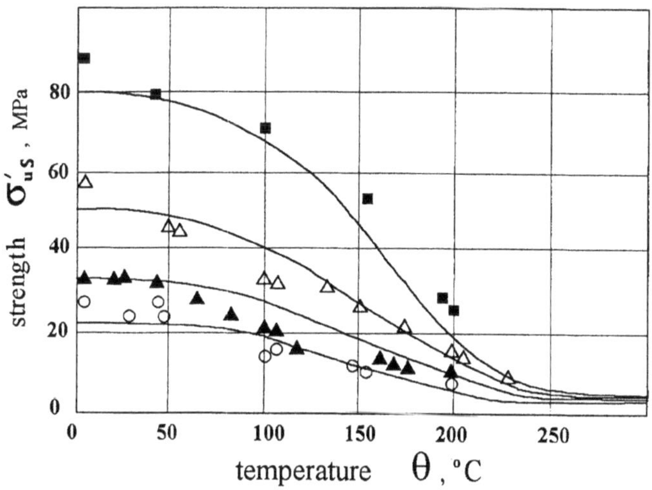

Figure 6.19. Varying strength σ'_{uS} of a UC based on epoxy matrix with different fibres (1 (■) - boron, 2 (△) - carbon, 3 (▲) - organic, 4 (○) - glass) in shear *versus* temperature under heating with rate 0.1 K/s in air

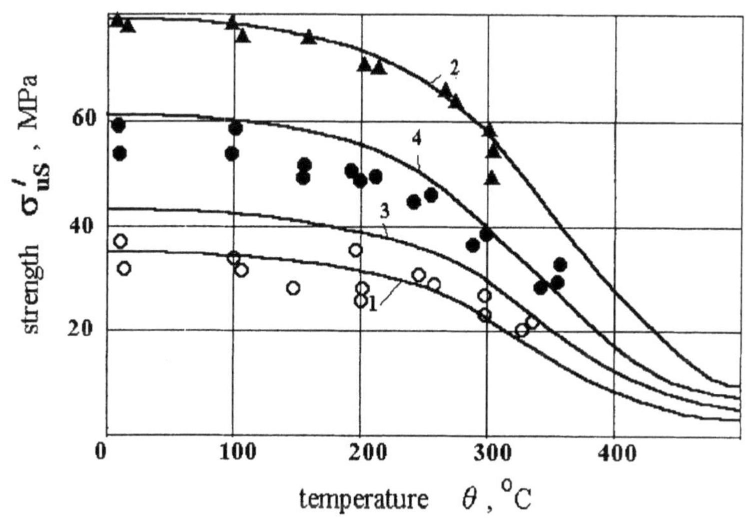

Figure 6.20. Varying strength σ'_{uS} of a UC based on polyimide matrix with different fibres (1 - organic, 2 (▲) - carbon, 3 (○) - glass, 4 (■) - boron) in shear *versus* temperature under heating with rate 0.1 K/s in air

Formulae (6.55) show that temperature dependences of strengths σ_{uT}, σ_{uC}, σ_{uS} and σ'_{uS}, as for σ'_{uT}, σ'_{uC}, are determined by the temperature dependence of the matrix strength described by the function $a_1(\theta)$.

Figures 6.19, 6.20 show theoretical strength σ'_{uS} values in longitudinal shear computed by formula (6.59) for epoxy UC and polyimide UC with fibres of different types under different temperatures. Fibre concentration was assumed to be $\varphi_f^0 = 0.6$, fibre misalignment angle was $\Phi = 5°$. For epoxy UC, initial porosity was $\varphi_g^0 = 0.03$, for carbon/polyimide $\varphi_g^0 = 0.01$, for boron/polyimide $\varphi_g^0 = 0.05$, for organic/polyimide and glass/polyimide $\varphi_g^0 = 0.1$. These values correspond to experimental data for UC; experimental values of thermal strength are also presented in Figures 6.19 and 6.20. These figures show that formula (6.59) allows us to accurately predict the strength of a composite in shear by the strength of its matrix and characteristics of its fibres. This is true not only for normal temperature but also for elevated temperatures; note that the formula describes the dependence of σ'_{uS} on fibre type. The effect of fibre type becomes apparent mainly for the dependence of σ'_{uS} on the ratio E_f^0/E_b^0 of stiffnesses of matrix and fibres. The strength σ'_{uS} also depends on matrix porosity φ_g^0. Figure 6.20 shows that since a matrix of boron/polyimide is more porous (i.e. weaker) than a matrix of carbon/polyimide, strength σ'_{uS} for carbon/polyimide is higher, although stiffness E_f^0 for boron fibres is greater than for carbon ones.

6.4.7. MULTIAXIAL LOADING OF UNIDIRECTIONAL COMPOSITES

Effect of Transverse Stresses on Thermal Strength in the Reinforcing Direction

The condition (6.43) was derived on the assumption of monoaxial loading of unidirectional fibrous material. If a thread of monofibres is in a composite with a complex non-unidirectional structure of reinforcing (for example, in a textile composite), then the stress state in the thread is multiaxial. To calculate the strength σ_{u33} of a unidirectional element as a part of the textile composite it is necessary to take account of the effect of tangential σ_{u13} and transverse σ_{u11} stresses. Stresses σ_{u13} and σ_{u11} reduce the 'reserve' of strength with respect to longitudinal delamination near desrtroyed fibres. Due to this, efficiency of involving the destroyed fibres into work decreases, and thus strength σ_{u33} decreases too.

An expression for strength of a composite in tension for multiaxial loading may be derived from (6.41). We must take account of the decreasing shear strength $\sigma_S^0 h_{mS}$ of a matrix because of the appearance of additional stresses under multiaxial loading $\sigma_S^0 h_{mS}(1 - z(Y_u))$. Formula (6.43) takes the form

$$z_u \equiv \left(\frac{|Y_{u1}| + Y_{u1}}{2\sigma'_{uT}}\right)^{1/\omega} + z(Y_u) = 1. \tag{6.60}$$

Transverse stresses σ_{u11}, σ_{u13} also have an effect on composite strength σ'_{uC} in compression. To take account of the influence, we combine (6.51), (6.60) and use the notation of (6.52); thus we obtain

$$(\frac{|Y_{u1}| - Y_{u1}}{2\sigma'_{uC}})^{1/\omega} + (\frac{|Y_{u1}| + Y_{u1}}{2\sigma'_{uT}})^{1/\omega} + \frac{1}{12}(\frac{B_{m1}}{\sigma_S^0 h^0 a_1})^2 (Y_{u2}^2 + \frac{12B_{m3}^2 Y_{u4}^{\prime 2}}{B_{m1}^2} +$$

$$+3Y_{u3}^2) + (\frac{B_{m1}}{a_1 h^0})^2 \Big((\frac{1}{(\sigma_T^0)^2} - \frac{1}{3(\sigma_S^0)^2}) Y_{u+}^2 + (\frac{1}{(\sigma_C^0)^2} - \frac{1}{3(\sigma_S^0)^2}) Y_{u-}^2 \Big) = 1. \quad (6.61)$$

When pore pressure is relatively small $p \ll \sigma_T^0$, the failure condition for the composite can be written in the form

$$\left(\frac{|Y_{u1}| + Y_{u1}}{2\sigma'_{uT}}\right)^{1/\omega} + \left(\frac{|Y_{u1}| - Y_{u1}}{2\sigma'_{uC}}\right)^{1/\omega} + \frac{1}{12\sigma_{uS}^2}(Y_{u2}^2 + 12k_Y Y_{u4}^2 + 3Y_{u3}^2) +$$
$$+ \left(\frac{1}{\sigma_{uC}^2} - \frac{1}{3\sigma_{uS}^2}\right) Y_{u-}^2 + \left(\frac{1}{\sigma_{uT}^2} - \frac{1}{3\sigma_{uS}^2}\right) Y_{u+}^2 = 1. \quad (6.62)$$

This equation contains only ultimate strengths σ_{uC}, σ_{uT}, σ_{uS} of the UC, determined by formulae (6.55), (6.57), σ'_{uC}, σ'_{uT}, and also coefficient k_Y:

$$k_Y = (B_{m3}/B_{m1})^2 = (\tilde{a}_5/\tilde{a}_3)^2. \quad (6.63)$$

This coefficient takes account of the effect of reinforcing of a composite on its thermoelastic properties in longitudinal shear and transverse tension.

Invariants $Y_{u\alpha}$ of tensor σ_{ukl} are determined by formulae (6.53), and for $p \ll \sigma_T^0$ we have $2Y_{u\pm} = |Y_{u2}| \pm Y_{u2}$.

The condition (6.62) is the *failure criterion of a UC* by destructions of types 1 and 2, i.e. due to breakage of fibres or loss of their stability under multiaxial loading.

Unfortunately, up to now there are no reliable data on thermal strength of a UC under multiaxial loading of the considered type. However, it is the type of destruction that plays a main role in determining a strength for composite with a more complicated structure of reinforcing than a UC, for example, for textile ones. This problem will be considered in detail in Chapter 7.

Destruction of Types 3 and 4 under Multiaxial Loading

The failure criterion for a UC at destructions of types 3 and 4 (delamination or cracking in compression) under multiaxial loading is the condition (6.52). When pore gas pressure is not high ($p \ll \sigma_T^0$), the criterion can be written in the form

$$\frac{1}{12\sigma_{uS}^2}(Y_{u2}^2 + 12k_Y Y_{u4}^2 + 3Y_{u3}^2) + (\frac{1}{\sigma_{uC}^2} - \frac{1}{3\sigma_{uS}^2}) Y_{u-}^2 + (\frac{1}{\sigma_{uT}^2} - \frac{1}{3\sigma_{uS}^2}) Y_{u+}^2 = 1. \quad (6.64)$$

6.5. Heat Expansion/Shrinkage

Formula (6.20) shows that the heat deformation of a UC coincides with heat deformation of a microcomposite $\overset{\circ}{\varepsilon}_{\mu kl}$, and is determined by relations (6.11).

The tensor $\overset{\circ}{\varepsilon}_{\mu kl}$ has two independent components: $\overset{\circ}{\varepsilon}{}'_{\mu}$ and $\overset{\circ}{\varepsilon}_{\mu}$, defining heat deformation of the UC in the reinforcing direction and in the transverse direction. Equation (6.12) implies that the deformation $\overset{\circ}{\varepsilon}_{\mu}$ is determined mainly by heat

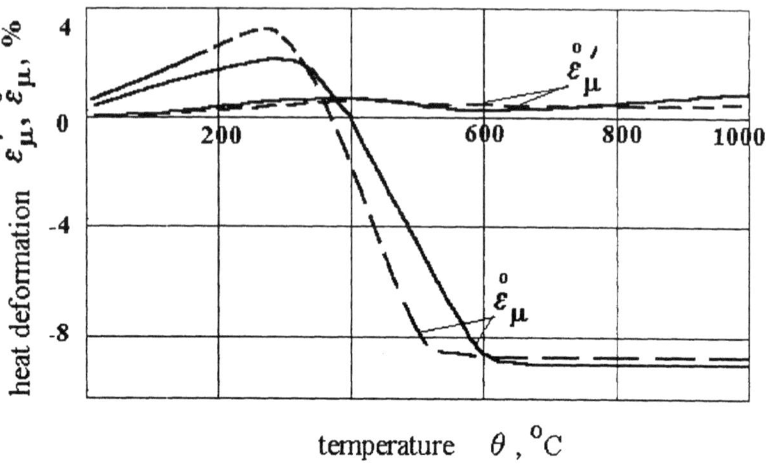

Figure 6.21. Dependence of heat deformations of unidirectional carbon/phenolic composite in the reinforcing direction $\overset{o}{\varepsilon}'_\mu$ and in the transverse direction $\overset{o}{\varepsilon}_\mu$ on temperature θ of heating. Solid curves are computations, dashed lines are experimental data

deformation $\overset{o}{\varepsilon}_m$ of the matrix, therefore the way in which $\overset{o}{\varepsilon}_\mu$ changes under heating is similar to the dependence $\overset{o}{\varepsilon}_m(\theta)$ (Figures 6.21 and 4.12): within the domain of high temperatures from 300 to 600 °C deformation $\overset{o}{\varepsilon}_\mu$ becomes negative, i.e. there appears an irreversible chemical shrinkage that does not vanish after cooling. The maximum absolute value of the shrinkage of the UC is $(1-\varphi_f)\beta_{sh}$, i.e. is smaller than the corresponding value β_{sh} for a non-reinforced matrix.

Deformation $\overset{o}{\varepsilon}'_\mu$ in the longitudinal direction depends mainly on the heat deformation of fibres (formula (6.12)), therefore shrinkage of a UC in the reinforcing direction under high temperatures is not usually observed (if heat expansion coefficient α_f of the fibre is positive).

TEXTILE ABLATIVE COMPOSITE MATERIALS

7.1. Model of a Structure of Ablative Textile Composite Material

Let us consider *textile composite materials* which are most widely used in structures working under high temperatures.

Typical ablative textile composite material consists of a fibrous filler (glass-, organo-, carbon-fabrics) and a polymer (epoxy, phenolic, silicon-organic) matrix. In accordance with the method of the hierarchy of internal levels (see Chapter 6), the textile material is considered to be a multilevel structure having six structural levels (Figure 7.1).

At the n-th structural level the composite is represented by a set of repeating elements (periodicity cells of the n-th type PCn). Each component in the PCn, in its turn, consists of periodicity cells of the $(n-1)$-th structural level (PC$(n-1)$) etc. $n = 2, 3, 4, 5, 6$.

At the 6-th structural level there is a whole ablative composite consisting of periodicity cells of the 6-th level, PC6 is reinforcing threads surrounded by a matrix and laid in two different directions $s = 1, 2$. The thread $s = 1$ is laid in a wavy shape in the Ox_1x_3 plane, and the thread $s = 2$ is laid in a wavy shape in the Ox_1x_2 plane. We make the following assumptions: each system of threads $s = 1$ and $s = 2$ is considered to be in the layer with number s, and both these layers are orthogonal to the Ox_1 axis. Each layer with curved fibres is an element of the 6-th structural level and consists of the collection of PC5s.

The periodicity cell PC5, in its turn, consists of a continuous collection of components characterized by the *angle of bending* ϑ_m in the Ox_1x_3 plane for $s = 1$ and in the Ox_1x_2 plane for $s = 2$ (Figure 7.1). Each component of PC5, being an element of the 4-th structural level, is represented by unidirectional material (u). The unidirectional material is a thread (bundle of fibres) impregnated by a binder.

As known, the impregnated bundle of fibres has different defects leading to a decrease in a strength of the fibre as compared with features of an elementary fibre.

The model of unidirectional material was considered in Chapter 6, it covers the first, second, third and fourth structural levels of the total scheme of textile composite shown in Figure 7.1.

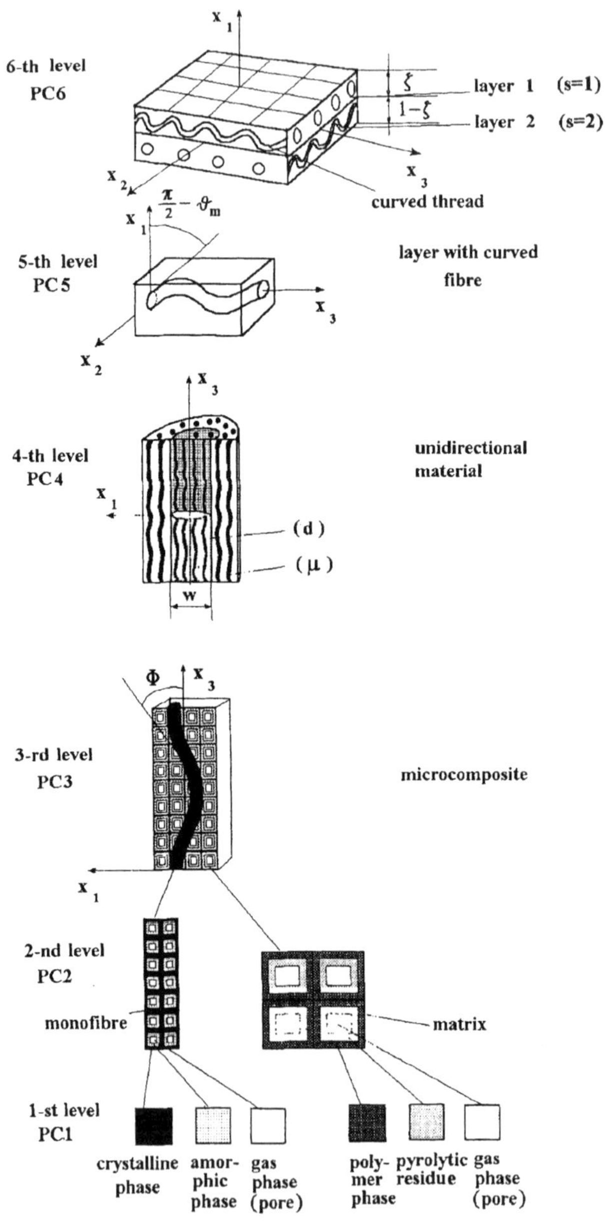

Figure 7.1. Schematic of a six-level structure for a textile composite material

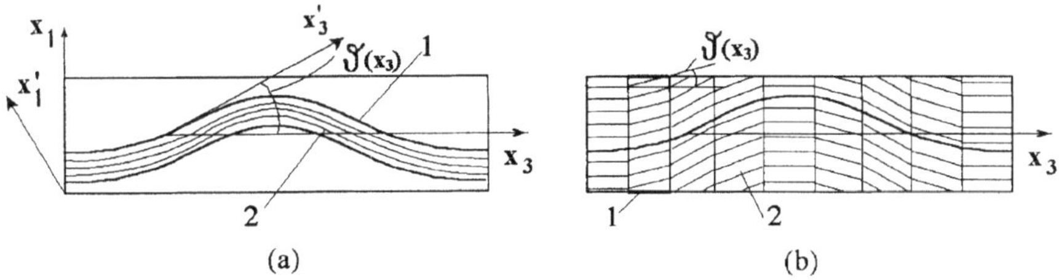

Figure 7.2. Model of PC5 of a layer with curved threads: (a) actual structure; (b) model: 1 - unidirectional component-layer curved with angle $\vartheta(x_3)$; 2 - monofibres in unidirectional component-layer

7.2. Model of a Layer with Curved Threads

7.2.1. ELASTIC PROPERTIES

At the fifth structural level of the multilevel model of an ablative textile material there is a layer oriented perpendicularly to the axis Ox_1 and represented by unidirectional material, the fibres of which, unlike the ones on the fourth structural level, are curved in the Ox_1x_3 plane.

To calculate elastic characteristics of the layer, the periodicity cell (PC5) is represented in the form of a continuous collection of *component-layers* orthogonal to the Ox_3 axis (Figure 7.2) and numbered by the x_3 coordinate. Each of the layers is represented by unidirectional material PC4 curved with a fixed angle $\vartheta(x_3)$ in the Ox_1x_3 plane. Strains and stresses in each component-layer ε^x_{kl} and $\sigma^x_{\alpha\beta}$ in the coordinate system $Ox_1x_2x_3$ and ε_{ukl} and σ_{uqr} in the rotated coordinate system $Ox'_1x'_2x'_3$ are connected with strains and stresses $\bar{\varepsilon}^{(1)}_{kl}$ and $\bar{\sigma}^{(1)}_{\alpha\beta}$ of the whole PC5 by the relationships

$$\varepsilon^x_{kl} = \varepsilon_{u\alpha\beta} Q^{(1)}_{k\alpha} Q^{(1)}_{l\beta}, \quad \sigma_{ukl} = \sigma^x_{\alpha\beta} Q^{(1)}_{k\alpha} Q^{(1)}_{l\beta},$$

$$\bar{\varepsilon}^{(1)}_{k3} = \int_0^1 \varepsilon^x_{k3} dx_3, \quad \sigma^x_{k3} = \bar{\sigma}^{(1)}_{k3}, \quad k = 1,2,3, \tag{7.1}$$

$$\varepsilon^x_{IJ} = \bar{\varepsilon}^{(1)}_{IJ}, \quad \bar{\sigma}^{(1)}_{IJ} = \int_0^1 \sigma^x_{IJ} dx_3, \quad I,J = 1,2, \quad k,l,\alpha,\beta = 1,2,3,$$

where $Q^{(s)}_{k\alpha}$ is the matrix of rotation by the angle $\vartheta(x_3)$ in the Ox_1x_3 plane (for $(s=1)$) and by the angle $\vartheta(x_2)$ in the Ox_1x_2 plane (for $(s=2)$):

$$Q^{(1)}_{k\alpha} = \begin{pmatrix} \cos\vartheta(x_3) & 0 & \sin\vartheta(x_3) \\ 0 & 1 & 0 \\ -\sin\vartheta(x_3) & 0 & \cos\vartheta(x_3) \end{pmatrix}, \quad 0 \leqslant x_3 \leqslant 1,$$

$$Q^{(2)}_{k\alpha} = \begin{pmatrix} \cos\vartheta(x_3) & \sin\vartheta(x_3) & 0 \\ -\sin\vartheta(x_3) & \cos\vartheta(x_3) & 0 \\ 0 & 0 & 1 \end{pmatrix}, \quad 0 \leqslant x_2 \leqslant 1, \tag{7.2}$$

$$\vartheta(0) = \vartheta(1).$$

Constitutive relations for PC5 can be obtained from formulae (7.1) and (6.20) for a unidirectional composite

$$\bar{\varepsilon}^{(s)}_{kl} = \sum_{\substack{\alpha=1 \\ \alpha\neq\beta\neq\gamma}}^{3} \left(\overset{o}{\bar{\varepsilon}}{}^{(s)}_{\alpha} + \bar{f}^{(s)}_{\alpha} p + \frac{1}{\bar{E}^{(s)}_{\alpha}} (\bar{\sigma}^{(s)}_{\alpha\alpha} - \bar{\nu}^{(s)}_{\alpha\beta}\bar{\sigma}^{(s)}_{\beta\beta} - \bar{\nu}^{(s)}_{\alpha\gamma}\bar{\sigma}^{(s)}_{\gamma\gamma}) \right) \delta_{k\alpha}\delta_{l\alpha} +$$

$$+ \frac{\bar{\sigma}^{(s)}_{12}}{2\bar{G}^{(s)}_{12}} (\delta_{k1}\delta_{l2} + \delta_{k2}\delta_{l1}) + \frac{\bar{\sigma}^{(s)}_{13}}{2\bar{G}^{(s)}_{13}} (\delta_{k1}\delta_{l3} +$$

$$+ \delta_{k3}\delta_{l1}) + \frac{\bar{\sigma}^{(s)}_{23}}{2\bar{G}^{(s)}_{23}} (\delta_{k2}\delta_{l3} + \delta_{k3}\delta_{l2}), \quad s = 1, \quad \beta, \gamma = 1, 2, 3, \tag{7.3}$$

where elastic characteristics $\bar{E}^{(1)}_{\alpha}$, $\bar{\nu}^{(1)}_{\alpha\beta}$ and $\bar{G}^{(1)}_{\alpha\beta}$ are determined with the help of the characteristics of the unidirectional material:

$$\bar{E}^{(1)}_{3} = \left(\frac{1}{E'_{\mu}} - \psi_1 \left(\frac{1}{E'_{\mu}} - \frac{1}{E_{\mu}} \right) - \frac{\psi_3}{4} \left(\frac{1+2\nu'_{\mu}}{E'_{\mu}} + \frac{1}{E_{\mu}} - \frac{1}{G'_{\mu}} \right) \right)^{-1},$$

$$\bar{E}^{(1)}_{2} = E_{\mu},$$

$$\bar{E}^{(1)}_{1} = E_{\mu} - \psi_1 (E_{\mu} - E'_{\mu}) - \frac{\psi_3}{4} (E_{\mu} + E'_{\mu} - 2\nu'_{\mu} E_{\mu} - 4G'_{\mu}),$$

$$\bar{\nu}^{(1)}_{32} = \bar{E}^{(1)}_{3} \left(\psi_1 \frac{\nu_{\mu}}{E_{\mu}} + \psi_2 \frac{\nu'_{\mu}}{E'_{\mu}} \right), \quad \bar{\nu}^{(1)}_{21} = \bar{E}^{(1)}_{2} \left(\psi_2 \frac{\nu_{\mu}}{E_{\mu}} + \psi_1 \frac{\nu'_{\mu}}{E'_{\mu}} \right),$$

$$\bar{\nu}^{(1)}_{31} = \bar{E}^{(1)}_{3} \left(\frac{\nu'_{\mu}}{E'_{\mu}} - \frac{\psi_3}{4} \left(\frac{1+2\nu'_{\mu}}{E'_{\mu}} + \frac{1}{E_{\mu}} - \frac{1}{G'_{\mu}} \right) \right),$$

$$\bar{G}^{(1)}_{12} = \psi_1 G'_{\mu} + \psi_2 G_{\mu}, \quad \bar{G}^{(1)}_{23} = \left(\frac{\psi_2}{G'_{\mu}} + \frac{\psi_1}{G_{\mu}} \right)^{-1}, \tag{7.4}$$

$$\bar{G}^{(1)}_{13} = \left(\frac{1}{G'_{\mu}} + \psi_3 \left(\frac{1+2\nu'_{\mu}}{E'_{\mu}} + \frac{1}{E_{\mu}} - \frac{1}{G'_{\mu}} \right) \right)^{-1}.$$

The coefficients ψ_{α} are introduced in the following way:

$$\psi_1 = \int_0^1 \sin^2\vartheta(x_3) dx_3, \quad \psi_2 = 1 - \psi_1, \quad \psi_3 = \int_0^1 \sin^2 2\vartheta(x_3) dx_3. \tag{7.5}$$

Here and below, the coordinates x_i are dimensionless.

For small curving angles $\vartheta(x_3)$ which are typical for textile composites [114]:

$$\vartheta(x_3) = \vartheta_{\max} \cos 2\pi x_3, \tag{7.6}$$

where $\vartheta_{\max} = A_{\max}/l$ is the maximum angle of curving of the thread.

Then

$$\psi_1 = \frac{\vartheta_m^2}{2}, \qquad \psi_2 = 1 - \frac{\vartheta_m^2}{2}, \qquad \psi_3 = 2\vartheta_m^2, \qquad \vartheta_m = \sin \vartheta_{\max} \tag{7.7}$$

and characteristics (7.4), after the small terms are neglected, can be expressed in the form

$$\begin{gathered} \bar{E}_1^{(1)} = E_\mu \bar{a}_2, \quad \bar{E}_2^{(1)} = E_\mu, \quad \bar{E}_3^{(1)} = E'_\mu \bar{a}_1, \\ \bar{G}_{13}^{(1)} = G'_\mu \bar{a}_3, \quad \bar{G}_{12}^{(1)} = G_\mu \bar{a}_2, \quad \bar{G}_{23}^{(1)} = G'_\mu \bar{a}_1, \\ \bar{\nu}_{32}^{(1)} = \nu'_\mu, \quad \bar{\nu}_{21}^{(1)} = \nu_\mu, \quad \bar{\nu}_{31}^{(1)} = \nu'_\mu, \end{gathered} \tag{7.8}$$

where the following notation is introduced:

$$\bar{a}_1 = \left(1 + \vartheta_m^2 \left(\frac{E'_\mu}{E_\mu} - 1\right)\right)^{-1}, \qquad \bar{a}_2 = 1 + \frac{\vartheta_m^2}{2}\left(\frac{G'_\mu}{G_\mu} - 1\right),$$

$$\bar{a}_3 = \left(1 - 2\vartheta_m^2\left(1 - \frac{G'_\mu}{E_\mu}\right)\right)^{-1}. \tag{7.9}$$

Constitutive relations for the second layer $s = 2$ in PC6, in which threads are curved in the Ox_1x_2 plane, differ from (7.3) only by substitution of index $s = 2$; the elastic characteristics in (7.3) are determined by the formulae

$$\begin{gathered} \bar{E}_1^{(2)} = E_\mu \bar{a}_2, \quad \bar{E}_2^{(2)} = E'_\mu \bar{a}_1, \quad \bar{E}_3^{(2)} = E_\mu, \\ \bar{G}_{12}^{(2)} = G'_\mu \bar{a}_3, \quad \bar{G}_{13}^{(2)} = G_\mu \bar{a}_2, \quad \bar{G}_{23}^{(2)} = G'_\mu \bar{a}_1, \\ \bar{\nu}_{23}^{(2)} = \nu'_\mu, \quad \bar{\nu}_{31}^{(2)} = \nu_\mu, \quad \bar{\nu}_{21}^{(2)} = \nu'_\mu. \end{gathered} \tag{7.10}$$

7.2.2. HEAT DEFORMATIONS AND COEFFICIENTS OF PHASE INTERACTION

In formulae (7.3), heat deformations $\overset{\circ}{\bar{\varepsilon}}{}_\alpha^{(s)}$ and coefficients of phase interactions $\bar{f}_\alpha^{(s)}$ in layers with curved threads can be determined as follows:

$$\overset{\circ}{\bar{\varepsilon}}{}_1^{(1)} = \overset{\circ}{\varepsilon}_\mu \psi_2 + \overset{\circ}{\varepsilon}{}'_\mu \psi_1, \qquad \overset{\circ}{\bar{\varepsilon}}{}_2^{(1)} = \overset{\circ}{\varepsilon}_\mu, \qquad \overset{\circ}{\bar{\varepsilon}}{}_3^{(1)} = \overset{\circ}{\varepsilon}_\mu \psi_1 + \overset{\circ}{\varepsilon}{}'_\mu \psi_2, \tag{7.11}$$

$$\bar{f}_1^{(1)} = f_\mu \psi_2 + f'_\mu \psi_1, \qquad \bar{f}_2^{(1)} = f_\mu \psi_2, \qquad \bar{f}_3^{(1)} = f_\mu \psi_1 + f'_\mu \psi_2. \tag{7.12}$$

Taking account of expressions (7.7), we can transform these formulae as follows:

$$\overset{\circ}{\bar{\varepsilon}}{}_3^{(1)} = \overset{\circ}{\varepsilon}_\mu \frac{\vartheta_m^2}{2} + \left(1 - \frac{\vartheta_m^2}{2}\right) \overset{\circ}{\varepsilon}{}'_\mu,$$

$$\overset{\circ}{\bar{\varepsilon}}{}_1^{(1)} = \overset{\circ}{\varepsilon}_\mu \left(1 - \frac{\vartheta_m^2}{2}\right) + \overset{\circ}{\varepsilon}{}'_\mu \frac{\vartheta_m^2}{2}, \qquad \overset{\circ}{\bar{\varepsilon}}{}_2^{(1)} = \overset{\circ}{\varepsilon}_\mu, \tag{7.13}$$

$$\bar{f}_1^{(1)} = f_\mu \frac{\vartheta_m^2}{2} + \left(1 - \frac{\vartheta_m^2}{2}\right) f'_\mu,$$

$$\bar{f}_2^{(1)} = f_\mu, \qquad \bar{f}_3^{(1)} = f_\mu \left(1 - \frac{\vartheta_m^2}{2}\right) + f'_\mu \frac{\vartheta_m^2}{2}. \tag{7.14}$$

For the second layer $s = 2$ of the textile composite we have similar formulae

$$\overset{\circ}{\bar{\varepsilon}}{}_1^{(2)} = \overset{\circ}{\varepsilon}_\mu \left(1 - \frac{\vartheta_m^2}{2}\right) + \overset{\circ}{\varepsilon}{}'_\mu \frac{\vartheta_m^2}{2},$$

$$\overset{\circ}{\bar{\varepsilon}}{}_2^{(2)} = \overset{\circ}{\varepsilon}_\mu \frac{\vartheta_m^2}{2} + \left(1 - \frac{\vartheta_m^2}{2}\right) \overset{\circ}{\varepsilon}{}'_\mu, \qquad \overset{\circ}{\bar{\varepsilon}}{}_3^{(2)} = \overset{\circ}{\varepsilon}_\mu, \tag{7.15}$$

$$\bar{f}_1^{(2)} = f_\mu \frac{\vartheta_m^2}{2} + \left(1 - \frac{\vartheta_m^2}{2}\right) f'_\mu,$$

$$\bar{f}_2^{(2)} = f_\mu \left(1 - \frac{\vartheta_m^2}{2}\right) + f'_\mu \frac{\vartheta_m^2}{2}, \qquad \bar{f}_3^{(2)} = f_\mu. \tag{7.16}$$

7.2.3. MICROSTRESSES

Microstresses $\sigma_{u\alpha\beta}$ in component-layers of PC5 (see Figure 7.2) depend on the coordinate x_3. Since each component-layer is unidirectional material, one can indicate four types of failure for the component-layer (see Chapter 6). Types (1) and (2) are connected with failure in the reinforcing direction (in the direction Ox'_3 for $s = 1$, and in the direction Ox'_2 for $s = 2$) and characterized by the parameter z_u introduced by (6.60). Types (3) and (4) are connected with failure of the component-layer in the $Ox'_1x'_3$ plane for $s = 1$ (or in $Ox'_1x'_2$ plane for $s = 2$) being orthogonal to the reinforcing direction; the types are characterized by the parameter z introduced by (6.52).

To determine the strength of the layer $s = 1$ of the textile composite it is necessary to establish the microstresses $\sigma_{ukl}(x_3)$ which would ensure maximum values of parameters $z_u(x_3)$, $z(x_3)$ in PC5 at a certain point $x_3 = x_{3(1)}$ where $0 \leqslant x_3 \leqslant 1$.

Let the microstresses be $\sigma_{ukl}^{(s)} = \sigma_{ukl}(x_{3(1)})$.

Microstresses $\sigma_{ukl}^{(1)}$ ensuring maximum values of parameter z for $s = 1$ appear at the point $x_{3(1)} = 0$ where $\vartheta(0) = 0$ and are calculated by (7.1) as follows:

$$\sigma_{ukl}^{(1)} = \bar{\sigma}_{kl}^{(1)}, \qquad \text{at} \ \ \vartheta(0) = 0. \tag{7.17}$$

Microstresses $\sigma_{ukl}^{(1)}$ ensuring maximum values of parameter z_u at $s = 1$ appear at the point x_3 of maximum curving of the thread, i.e. where $\vartheta(x_3) = \vartheta_{\max}$; therefore from (7.1)

$$\begin{aligned}
\sigma_{u33}^{(1)} &= \bar{\sigma}_{11}^{(1)}\vartheta_m^2 + (1-\vartheta_m^2)\bar{\sigma}_{33}^{(1)} + 2\vartheta_m\sqrt{1-\vartheta_m^2}\,\bar{\sigma}_{13}^{(1)},\\
\sigma_{u13}^{(1)} &= \bar{\sigma}_{13}^{(1)}(1-2\vartheta_m^2) + \vartheta_m\sqrt{1-\vartheta_m^2}\,(\bar{\sigma}_{33}^{(1)} - \bar{\sigma}_{11}^{(1)}),\\
\sigma_{u11}^{(1)} &= \bar{\sigma}_{11}^{(1)}(1-\vartheta_m^2) + \vartheta_m^2\bar{\sigma}_{33}^{(1)} + 2\vartheta_m\sqrt{1-\vartheta_m^2}\,\bar{\sigma}_{13}^{(1)}.
\end{aligned} \tag{7.18}$$

In (7.18) the effect of transverse $\bar{\sigma}_{11}^{(1)}$ and interlayer $\bar{\sigma}_{13}^{(1)}$ stresses can be neglected, as for actual textile composites the following relationships are satisfied:

$$\max\,\{|\bar{\sigma}_{11}^{(1)}|, |\bar{\sigma}_{13}^{(1)}|\} \leqslant \vartheta_m^2 \max\,|\bar{\sigma}_{33}^{(1)}|. \tag{7.19}$$

Then for small ϑ_m, stresses $\sigma_{ukl}^{(1)}$ are determined as follows:

$$\begin{gathered}
\sigma_{u33}^{(1)} = \bar{\sigma}_{33}^{(1)}, \qquad \sigma_{u22}^{(1)} = \bar{\sigma}_{22}^{(1)}, \qquad \sigma_{u23}^{(1)} = \bar{\sigma}_{23}^{(1)},\\
\sigma_{u12}^{(1)} = 0, \quad \sigma_{u11}^{(1)} = \vartheta_m^2\bar{\sigma}_{33}^{(1)}, \quad \sigma_{u13}^{(1)} = \vartheta_m\bar{\sigma}_{33}^{(1)}, \quad \text{at } \vartheta(x_{3(1)}) = \vartheta_{\max}.
\end{gathered} \tag{7.20}$$

Microstresses $\sigma_{u\alpha\beta}^{(2)}$ in the second layer ensuring maximum values of parameters z_u and z for $s = 2$ can be determined in a similar way

$$\sigma_{ukl}^{(2)} = \bar{\sigma}_{kl}^{(2)}, \quad \text{at } \vartheta(0) = 0, \tag{7.21}$$

$$\begin{gathered}
\sigma_{u22}^{(2)} = \bar{\sigma}_{22}^{(2)}, \quad \sigma_{u33}^{(2)} = \bar{\sigma}_{33}^{(2)}, \quad \sigma_{u23}^{(2)} = \bar{\sigma}_{23}^{(2)}, \quad \sigma_{u13}^{(2)} = 0,\\
\sigma_{u12}^{(2)} = \vartheta_m\bar{\sigma}_{22}^{(2)}, \quad \sigma_{u11}^{(2)} = \vartheta_m^2\bar{\sigma}_{22}^{(2)}, \quad \text{at } \vartheta(x_{2(2)}) = \vartheta_{\max}.
\end{gathered} \tag{7.22}$$

7.2.4. HEAT CONDUCTIVITY OF LAYERS WITH CURVED THREADS

Calculations of effective heat conductivity tensors $\bar{k}_{\alpha\beta}^{(s)}$ of layers $s = 1$ and $s = 2$ with curved threads are conducted in a similar way as for elastic moduli, in accordance with the model in Figure 7.2.

The heat conductivity tensor of each component-layer $k_{u\alpha\beta}^{x}$ in the coordinate system $Ox_1x_2x_3$ is determined by the formula

$$k_{u\alpha\beta}^{x} = k_{uqr}Q_{\alpha q}^{(1)}Q_{\beta r}^{(1)}. \tag{7.23}$$

Components of the heat conductivity tensor $\bar{k}_{\alpha\beta}^{(s)}$ of the layer with number $s = 1$ (in which fibres are bend in the Ox_1x_3 plane) can be determined as follows:

$$\begin{gathered}
\bar{k}_{11}^{(1)} = \left((k_{u11}\psi_2)^{-1} + (k_{u33}\psi_1)^{-1}\right)^{-1},\\
\bar{k}_{u22}^{(1)} = k_{u22},
\end{gathered} \tag{7.24}$$

$$\bar{k}_{u33}^{(1)} = k_{u11}\psi_1 + k_{u33}\psi_2,$$

the remaining of components are zero.

Quite similarly, components of the heat conductivity tensor are determined for the second layer $s = 2$ where fibres are bent in the Ox_1x_2 plane:

$$\bar{k}_{11}^{(2)} = \bar{k}_{11}^{(1)}, \quad \bar{k}_{22}^{(2)} = \bar{k}_{33}^{(1)}, \quad \bar{k}_{33}^{(2)} = \bar{k}_{22}^{(1)}. \tag{7.25}$$

The shape of the curved fibres was assumed to satisfy a sinusoidal law, i.e. formula (7.6); on integrating (7.24) and (7.25), we derive the following form of the heat conductivity tensors for layers

$$\bar{k}_{qr}^{(1)} = \begin{pmatrix} \bar{k}_1 & 0 & 0 \\ 0 & \bar{k}_2 & 0 \\ 0 & 0 & \bar{k}_3 \end{pmatrix}, \qquad \bar{k}_{qr}^{(2)} = \begin{pmatrix} \bar{k}_1 & 0 & 0 \\ 0 & \bar{k}_3 & 0 \\ 0 & 0 & \bar{k}_2 \end{pmatrix}, \tag{7.26}$$

where the principal coefficients of heat conductivity $\bar{k}_q$ of the layers are determined by the formulae

$$\bar{k}_1 = \left(k_u^{-1}(1 - \vartheta_m^2/2) + (k_u')^{-1}\vartheta_m^2/2\right)^{-1}, \quad \bar{k}_2 = k_u, \tag{7.27}$$
$$\bar{k}_3 = k_u'(1 - \vartheta_m^2/2) + k_u\vartheta_m^2/2.$$

7.3. Constitutive Relations for Ablative Textile Composites

To determine characteristics of the textile composite we make the following assumptions on the relationship between stresses $\bar{\sigma}_{\alpha\beta}^{(s)}$ and strains $\bar{\varepsilon}_{\alpha\beta}^{(s)}$ in layers $s = 1$ and $s = 2$ and stresses $\sigma_{\alpha\beta}$, $\varepsilon_{\alpha\beta}$ in the textile composite as the whole:

$$\begin{gathered} \bar{\sigma}_{1\alpha}^{(s)} = \sigma_{1\alpha}, \qquad \sigma_{IJ} = \zeta\bar{\sigma}_{IJ}^{(1)} + (1-\zeta)\bar{\sigma}_{IJ}^{(2)}, \\ \varepsilon_{IJ} = \bar{\varepsilon}_{IJ}^{(s)}, \qquad \varepsilon_{1\alpha} = \zeta\bar{\varepsilon}_{1\alpha}^{(1)} + (1-\zeta)\bar{\varepsilon}_{1\alpha}^{(2)}, \\ \alpha = 1,2,3, \qquad I, J = 2,3, \qquad s = 1,2, \end{gathered} \tag{7.28}$$

where ζ is the ratio of the number of threads laid in the Ox_3 axis direction to the total number of threads in PC6. The axis Ox_1, as usual, is orthogonal to the plane of the composite layers.

Then on substituting relations (7.3) for layers $s = 1$ and $s = 2$ into equations (7.28), we derive the following constitutive relations for the textile composite:

$$\begin{aligned} \varepsilon_{kl} = & \sum_{\substack{\alpha=1 \\ \alpha\neq\beta\neq\gamma}}^{3} \left(\overset{\circ}{\varepsilon}_\alpha + f_\alpha p + \frac{1}{E_\alpha}(\sigma_{\alpha\alpha} - \nu_{\alpha\beta}\sigma_{\beta\beta} - \nu_{\alpha\gamma}\sigma_{\gamma\gamma}) \right) \delta_{k\alpha}\delta_{l\alpha} + \\ & + \frac{\sigma_{12}}{2G_{12}}(\delta_{k1}\delta_{l2} + \delta_{k2}\delta_{l1}) + \frac{\sigma_{13}}{2G_{13}}(\delta_{k1}\delta_{l3} + \delta_{k3}\delta_{l1}) + \\ & + \frac{\sigma_{23}}{2G_{23}}(\delta_{k2}\delta_{l3} + \delta_{k3}\delta_{l2}), \end{aligned} \tag{7.29}$$

which correspond to *orthotropic* material with principal orthotropy axes Ox_α.

The relations (7.29) contain three groups of physical characteristics:

- elastic moduli E_α, G_α, $\nu_{\alpha\beta}$;
- heat deformations $\overset{\circ}{\varepsilon}_\alpha$;
- coefficients of mechanical interaction f_α of the solid and gas phases.

The heat conductivity tensor of textile composite k_{qr} is also orthotropic and has three independent components k_α

$$k_{qr} = \sum_{\alpha=1}^{3} k_\alpha \delta_{\alpha q} \delta_{\alpha r}. \tag{7.30}$$

7.4. Thermo-Elastic Moduli and Heat Conductivity Coefficients of Textile Composites

7.4.1. THEORETICAL RESULTS

Effective elastic constants E_α, $G_{\alpha\beta}$, $\nu_{\alpha\beta}$ in (7.29) can be calculated with the help of relations (7.3), (7.8), (7.10) and (7.28) and have the form

$$E_1 = E_\mu \bar{a}_2, \quad E_2 = (1-\zeta)E'_\mu \bar{a}_1, \quad E_3 = \zeta E'_\mu \bar{a}_1,$$

$$G_{13} = G_\mu \frac{\bar{a}_2}{1-\zeta}, \quad G_{12} = G_\mu \frac{\bar{a}_2}{\zeta}, \quad G_{23} = G'_\mu \bar{a}_1,$$

$$\nu_{31} = \nu_\mu \zeta, \quad \nu_{21} = \nu_\mu (1-\zeta), \quad \nu_{32} = \zeta(1-\zeta)\nu'_\mu.$$

Substituting expressions (6.4) into the relations in place of elastic characteristics of microcomposite E'_μ, E_μ, ν'_μ, G'_μ and G_μ, and expressions (7.9) in terms of functions of changing properties of matrix $\widetilde{a}_1$ and fibre $\widetilde{a}_2$ in place of functions $\bar{a}_1 \ldots \bar{a}_3$, we obtain

$$\begin{gathered} E_2 = E_f^0 (1-\zeta) a_{\theta 1}, \quad E_3 = E_f^0 \zeta a_{\theta 1}, \quad E_1 = E_b^0 a_{\theta 2}, \\ G_{13} = G_b^0 \frac{a_{\theta 2}}{1-\zeta}, \quad G_{23} = G_f^0 a_{\theta 1}, \quad G_{12} = G_b^0 \frac{a_{\theta 2}}{\zeta}, \\ \nu_{21} = \nu_m (1-\zeta) = \text{const}, \quad \nu_{31} = \nu_m \zeta = \text{const}, \\ \nu_{23} = \nu_m \widetilde{a}_4 (1-\zeta)/\zeta = \text{const}. \end{gathered} \tag{7.31}$$

The functions $a_{\theta 1}$ and $a_{\theta 2}$ are expressed by the formulae

$$a_{\theta 1} = \frac{\varphi_f^0 \widetilde{a}_2}{1 + \vartheta_m^2 \left((1-\varphi_f^0) \frac{\widetilde{a}_2}{m_E \widetilde{a}_1} - 1 \right)}, \tag{7.32}$$

$$a_{\theta 2} = \widetilde{a}_1 (1-\varphi_f^0)^{-1}.$$

Let us write now expressions for functions $\widetilde{a}_1$, $\widetilde{a}_2$ and $\widetilde{a}_f$ using formulae (6.5) and (5.7):

$$\widetilde{a}_f = a_f^{(0)}(S_l - S_a) + (1 - S_l)n_l,$$

$$\widetilde{a}_1 = \left(\frac{1-S_p}{m_p} + \frac{S_b}{a^{(0)}(S_p^2 - S_b^2) + m_p(1-S_p^2)} + \frac{S_p - S_b}{a^{(0)}S_p^2 + m_p(1-S_p^2)}\right)^{-1}, \quad (7.33)$$

$$\widetilde{a}_2 = \widetilde{a}_f \left(1 + \tan^2\Phi \, \varphi_f \frac{\widetilde{a}_f}{\widetilde{a}_1 m_E}\right)^{-1}.$$

Functions $a_f^{(0)}$ and $a^{(0)}$ describing reversible changes of elastic features of the polymer phase and fibre in heating are determined by the formulae (5.7), (4.18):

$$a_f^{(0)} = \exp\left(-a_f \Delta\hat{\theta}\right), \qquad a^{(0)} = \exp\left(-a(\Delta\hat{\theta})^r\right), \quad (7.34)$$

$$\Delta\hat{\theta} = \Delta\theta + P\int_0^t \exp\left(-Q(t-\tau)\right)\Delta\theta(\tau)d\tau,$$

and functions S_b, S_p and S_l, S_a describing irreversible changes of matrix and fibre features in heating are expressed by the formulae:

$$S_p = \left(\frac{1-\varphi_f^0 - \varphi_p}{1-\varphi_f^0}\right)^{1/3}, \qquad S_b = \left(\frac{\varphi_g}{1-\varphi_f^0}\right)^{1/3},$$

$$S_a = \left(1 - \frac{\varphi_l + \varphi_a}{\varphi_f^0}\right)^{1/3}, \qquad S_l = \left(1 - \frac{\varphi_l}{\varphi_f^0}\right)^{1/3}. \quad (7.35)$$

The formulae (7.31) – (7.35) contain the following physical constants:

$$E_f^0, \quad E_b^0, \quad m_p, \quad n_l, \quad m_E, \quad \nu_m, \quad \nu_a, \quad a, \quad a_f, \quad (7.36)$$

and also geometric constants

$$\varphi_f^0, \quad \zeta, \quad \vartheta_m, \quad \Phi. \quad (7.37)$$

Here E_f^0, ν_a are the elastic modulus and Poisson's ratio of monofibres in their initial state at temperature $\theta_0 = 293$ K; E_b^0, ν_m are the elastic modulus and Poisson's ratio of the polymer phase of the matrix. Coefficients a and a_f characterize the dependence of elastic properties of the polymer phase of the matrix and fibres on temperature. Coefficient $n_l = E_l/E_b^0$ is the ratio of elastic modulus of the monofibre in the crystalline state under high temperature to E_f^0. Coefficient $m_p = E_p/E_b^0$ is the ratio of elastic moduli of pyrolytical and polymer phases of the matrix, $m_E = E_b^0/E_f^0$ is the ratio of elastic moduli of the polymer phase of the matrix and the fibre in the amorphic state. Constants P and Q are universal, i.e. the same for matrices, fibres and composites, and their values are given in 4.2.3.

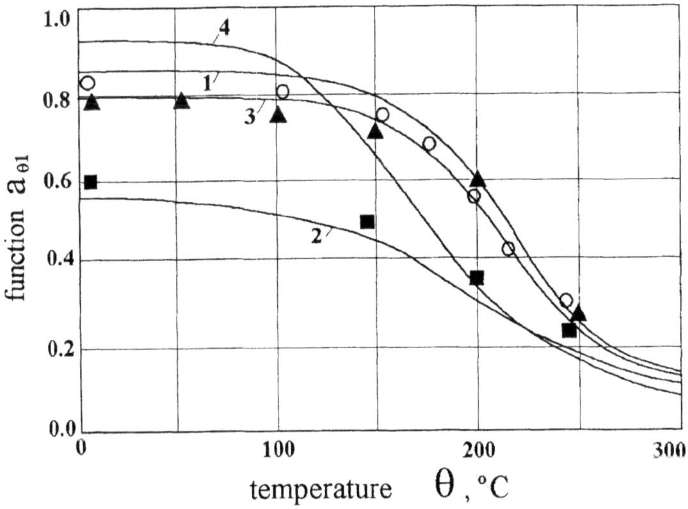

Figure 7.3. Dependence of function $a_{\theta 1}$ on temperature for epoxy textile composites with fibres of different types: 1 - organic fibre (○), 2 - carbon fibre (■), 3 - glass fibre (▲), 4 - epoxy matrix

7.4.2. EXPERIMENTAL DATA

The change of elastic properties of textile composites, in accordance with the model developed, is described by two functions $a_{\theta 1}$ and $a_{\theta 2}$, the expressions for which were given by formulae (7.32).

Function $a_{\theta 1}$ describes changes in the elastic properties of the composite in tension and shear in the fabric plane under heating, and $a_{\theta 2}$ – in tension in the transverse direction and interlayer shears.

At relatively low temperatures, when pyrolysis of the polymer matrix has not yet started ($\varphi_b^0 = 1$, $\varphi_p \equiv 0$), the change in the elastic modulus E_α during heating is reversible: on cooling, the elastic modulus returns practically to its initial value.

On heating up to higher temperatures, when pyrolysis of the matrix starts, the decrease of the elastic modulus E_α during heating is irreversible: on cooling, values of $a_{\theta 1}$ are only partially restored. After completion of the pyrolysis, when the material has been charred completely, the value of the coefficient $a_{\theta\alpha}$ cannot be restored entirely by cooling.

Figure 7.3 shows theoretical dependences of the function $a_{\theta 1}$ on temperature θ calculated by formula (7.32) for epoxy textile composites with fibres of different types. Points indicate experimental data for $a_{\theta 1}$ obtained by experimental values of elastic moduli E_2:

$$a_{\theta 1} = E_2/(E_f^0 \varphi_f^0 \zeta) \tag{7.38}$$

for $\zeta = 0.7$ and $\varphi_f^0 = 0.7$. The agreement between theoretical and experimental results is quite satisfactory.

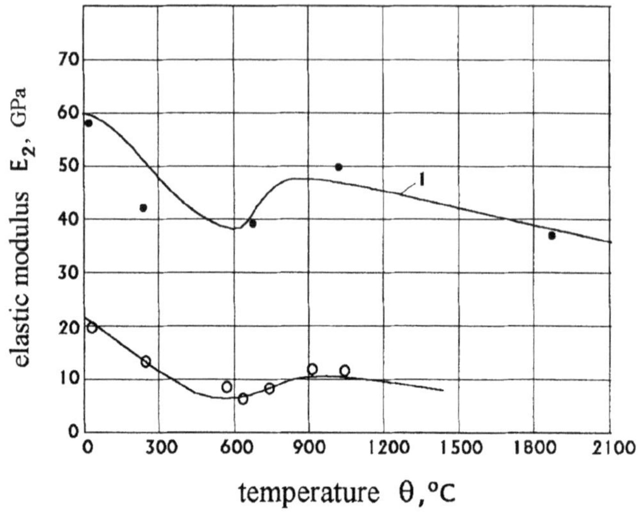

Figure 7.4. Dependence of elastic modulus E_2 of phenolic textile composites on temperature θ under heating with rate $\dot{\theta} = 5$ K/s in an inert environment (1 (•) - carbon/phenolic composite, 2 (○) - glass/phenolic composite [66]); points - experimental data, curves - computations

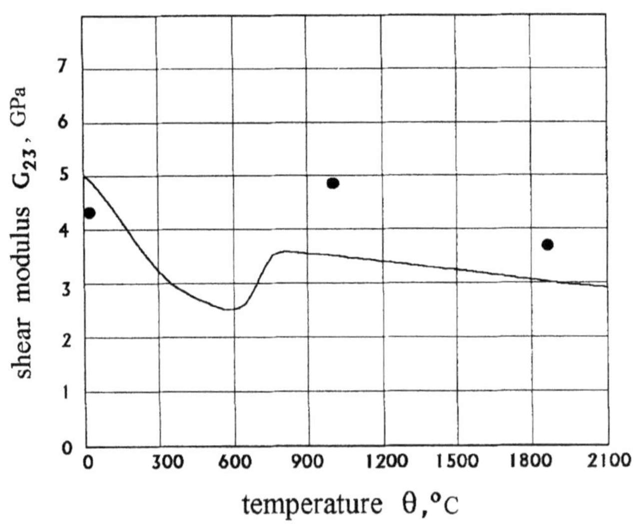

Figure 7.5. Dependence of shear modulus in fabric plane G_{23} for phenolic textile composites on temperature θ under heating with rate $\dot{\theta} = 5$ K/s in an inert environment; points - experimental data, curve - computations

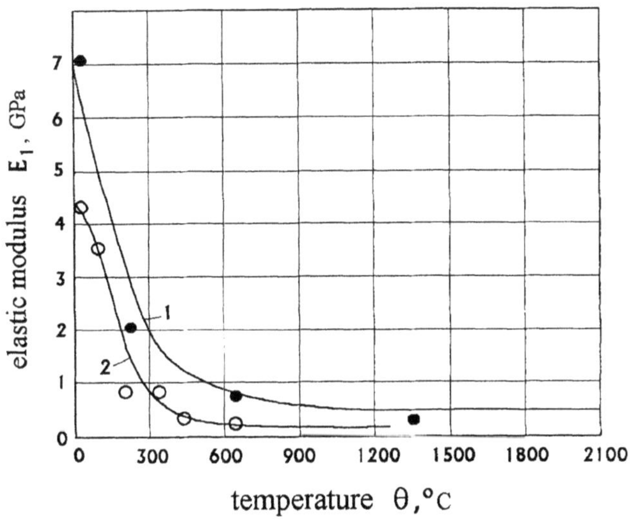

Figure 7.6. Dependence of elastic modulus E_1 of phenolic textile composites on temperature θ under heating in an inert environment (1 (o) - carbon/phenolic composite, 2 (o) - glass/phenolic composite); points – experimental data, curves – computations

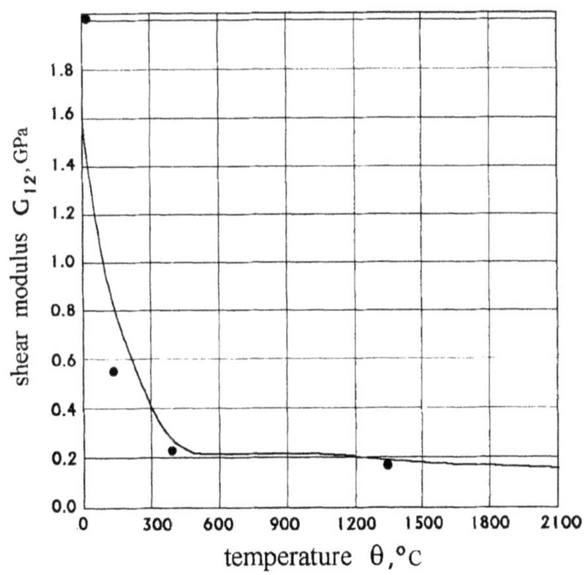

Figure 7.7. Dependence of interlayer shear modulus G_{12} of phenolic textile composites on temperature θ under heating in an inert environment

The function $a_{\theta 1}$ at $\theta = \theta_0$ also describes *realization of elastic properties of fibre* in textile composites. This means that, due to the influence of different causes: curving of threads, misalignment of monofibres, porosity etc., the actual elastic modulus E_1 of a textile composite proves to be lower than its expected theoretical value $(E_f \varphi_f \zeta)$.

Figure 7.3 shows that, for all types of composites, values of $a_{\theta 1}$ at $\theta = \theta_0$ are less than 1. The least value of the function of realization of elastic properties is for carbon fibres ($a_{\theta 1} \approx 0.6$ at $\theta_0 = 293$ K), the greatest value – for organic fibres ($a_{\theta 1} \approx 0.85$).

Figures 7.4 and 7.5 show the dependences of the elastic modulus E_2 and shear modulus G_{23} for textile carbon/ and glass/phenolic composites on temperature θ within the interval from 20 to 2100 °C in an inert environment. Points indicate experimental data, solid curves – computations by (7.32); here $\zeta = 0.5$, $\varphi_f^0 = 0.7$. A typical peculiarity of phenolic plastics is a certain increase of the elastic constants E_2 and G_{23} within the temperature interval from 600 to 1000 °C, that is connected with growth of the elastic modulus of carbon fibres, and also with the formation of a stiff coke residue of the phenolic matrix within the temperature interval.

Figures 7.6 and 7.7 show computed and experimental data for transverse elastic modulus E_1 and interlayer shear modulus G_{12} for the same composites. Unlike E_2 and G_{23}, values of moduli E_1 and G_{12} decrease sharply during heating up to temperatures higher than 300 °C, and $E_1/E_1(\theta_0) \sim 0.1$ at 1000 °C.

7.5. Heat Deformations

7.5.1. THEORETICAL RELATIONS

Heat deformations $\overset{\circ}{\varepsilon}_k$ of textile ablative composite material in equation (7.29) can be determined by the formulae

$$\overset{\circ}{\varepsilon}_k = (\alpha_f \varphi_f A_k + \alpha_b \varphi_b \Omega_k)(\theta - \theta_0) + \alpha_p \Omega_k \int_0^t (\theta(t) - \theta(\tau)) \dot{\varphi}_p \, d\tau - \beta_p \Omega_k \varphi_p, \quad (7.39)$$

$$k = 1, 2, 3,$$

where the following structural coefficients are denoted:

$$A_1 = 1 - \frac{\vartheta_m^2}{2} + \frac{\vartheta_m^2 \tilde{a}_f}{2\varphi_f \tilde{a}_2},$$

$$A_2 = \frac{\vartheta_m^2}{2} + \left(1 - \frac{\vartheta_m^2}{2}\right)\frac{\tilde{a}_f}{\varphi_f \tilde{a}_2} + \left(\frac{\zeta}{1-\zeta}\right)\frac{m_E \tilde{a}_5}{a_{\theta 1}},$$

$$A_3 = \frac{\vartheta_m^2}{2} + \left(1 - \frac{\vartheta_m^2}{2}\right)\frac{\tilde{a}_f}{\varphi_f \tilde{a}_2} + \left(\frac{1-\zeta}{\zeta}\right)\frac{m_E \tilde{a}_5}{a_{\theta 1}},$$

$$\Omega_1 = 1 - \frac{\vartheta_m^2}{2} + \frac{\vartheta_m^2}{2}\frac{m_E\tilde{a}_1}{\varphi_f\tilde{a}_2},$$

$$\Omega_2 = \frac{\vartheta_m^2}{2} + (1 - \frac{\vartheta_m^2}{2})\frac{m_E\tilde{a}_1}{\varphi_f\tilde{a}_2} + (\frac{\zeta}{1-\zeta})\frac{m_E\tilde{a}_5}{a_{\theta 1}}, \tag{7.40}$$

$$\Omega_3 = \frac{\vartheta_m^2}{2} + (1 - \frac{\vartheta_m^2}{2})\frac{m_E\tilde{a}_1}{\varphi_f\tilde{a}_2} + (\frac{1-\zeta}{\zeta})\frac{m_E\tilde{a}_5}{a_{\theta 1}}.$$

Here α_γ are the coefficients of heat expansion of the phases, and β_p is the coefficient of chemical shrinkage of the pyrolysis phase of the matrix.

7.5.2. EXPERIMENTAL DATA

Functions $\overset{\circ}{\varepsilon}_k$ in (7.29) describe heat deformations of composites without load. At low temperatures, when volumetric ablation of a matrix has not yet started, $\overset{\circ}{\varepsilon}_k$ depends linearly on the difference of temperatures:

$$\overset{\circ}{\varepsilon}_k = (\alpha_f\varphi_f A_k + \alpha_b\varphi_b\Omega_k)(\theta - \theta_0). \tag{7.41}$$

At high temperatures, the dependence of $\overset{\circ}{\varepsilon}_k$ on θ is nonlinear, as the pyrolysis process of polymer resin is accompanied by essential chemical shrinkage.

The shrinkage phenomenon is a general feature of all polymer matrices in pyrolysis. It is caused by decreasing the length of molecular chains due to break-off of their separate sections during heating. The chemical shrinkage is a cause of the appearance of shrinkage stresses in composite structures under high temperatures.

Figure 7.8 shows the heat deformation $\overset{\circ}{\varepsilon}_3$ for composites under uniform heating by the linear law $\theta = \theta_0 + \dot{\theta}_0 t$. At first, $\overset{\circ}{\varepsilon}_3$ grows due to heat expansion of the polymer phase, then sharply decreases due to the beginning of chemical shrinkage in pyrolysis, and finally increases again after completion of the pyrolysis due to heat expansion of a pyrolysis residue.

Just like the function $a_{\theta\alpha}$, the heat deformations $\overset{\circ}{\varepsilon}_k$ of a composite depend on the heating rate. Figure 7.8 shows values of $\overset{\circ}{\varepsilon}_3$ for glass/epoxy composite under heating with different values $\dot{\theta}_0$. With growing $\dot{\theta}_0$, the curve $\overset{\circ}{\varepsilon}_3(\theta)$ is displaced into a zone of higher temperatures that is connected with smaller warm-up times for the matrix at higher $\dot{\theta}_0$.

Figure 7.9 shows experimental [116] and calculated dependences of longitudinal heat deformation $\overset{\circ}{\varepsilon}_3$ for glass/phenolic textile composite on temperature θ for different rates of heating: $\dot{\theta} = 0.15$, 0.3 and 1.25 K/s. A peculiarity of glass/phenolic composites is the presence of two temperature domains in which shrinkage is sharply intensified: 300...600 °C and 900...1200 °C. In the first domain the shrinkage processes are caused by a shrinkage of the phenolic matrix; in the second they are caused by lowered elastic properties of the glass fibres; this lowers the reinforcing effect of the fibres and makes the heat deformation $\overset{\circ}{\varepsilon}_3$ of the whole composite closer to the heat deformation $\overset{\circ}{\varepsilon}_m$ of its matrix.

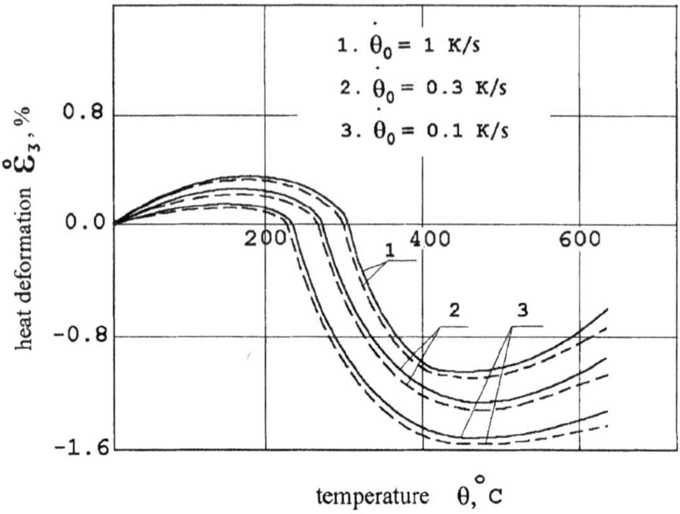

Figure 7.8. Dependence of heat deformation $\overset{\circ}{\varepsilon}_3$ of epoxy textile composite on temperature at different rates of heating; dashed lines – experimental data, solid curves – computations

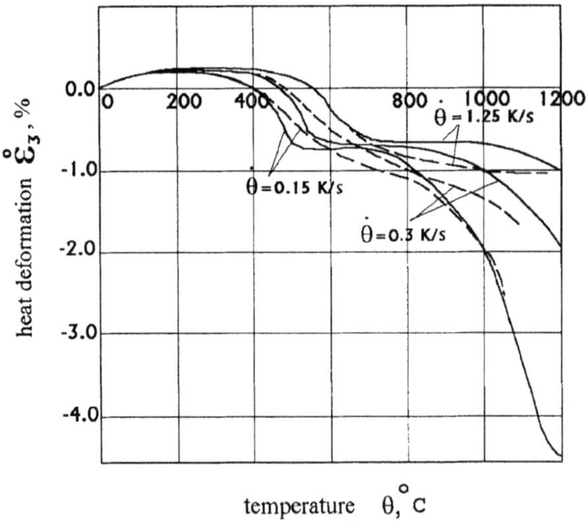

Figure 7.9. Dependence of heat deformation $\overset{\circ}{\varepsilon}_3$ of glass/phenolic textile composite on temperature θ at different rates of heating; solid curves – computations by formula (7.39), dashed lines – experimental data [116]

Heat deformations $\overset{\circ}{\varepsilon}_k$ of carbon/phenolic textile composites have another character (Figure 7.10). Elastic properties of carbon fibres remain practically the same up to temperatures 2000 °C in an inert environment, therefore shrinkage processes in the plastic are completed, in the main, at temperatures 600...800 °C; then there

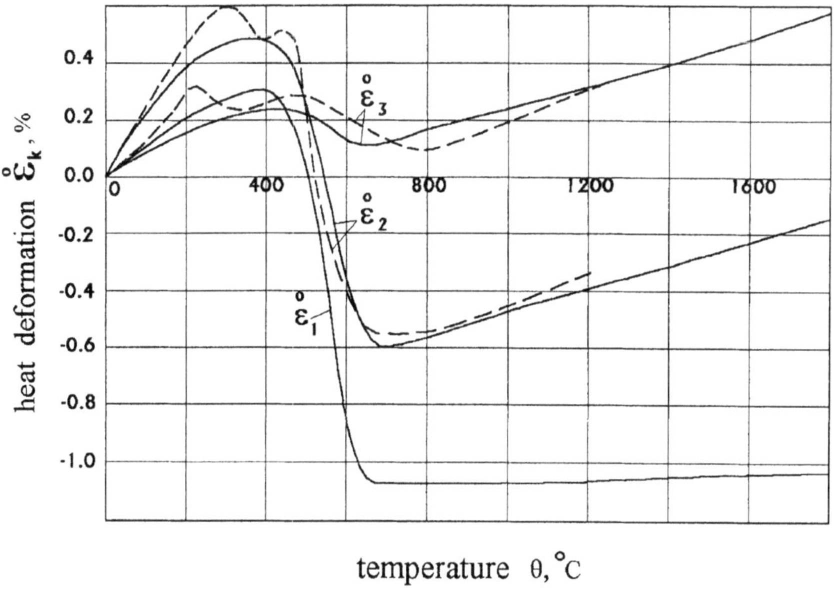

Figure 7.10. Dependence of heat deformations $\overset{o}{\varepsilon}_1$, $\overset{o}{\varepsilon}_2$, $\overset{o}{\varepsilon}_3$ of carbon/phenolic textile composite on temperature θ under heating with rate $\dot{\theta} = 1$ K/s; solid curves – computations, dashed lines – experimental data

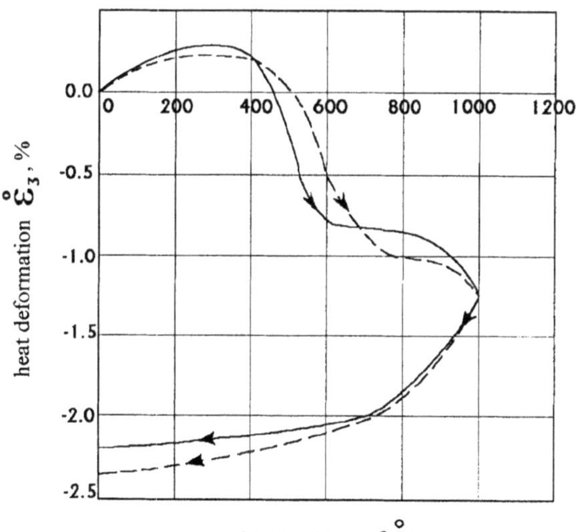

Figure 7.11. Dependence of heat deformation $\overset{o}{\varepsilon}_3$ of glasss/phenolic textile composite on temperature θ under heating with rate $\dot{\theta}_0 = 0.3$ K/s and subsequent cooling with rate $\dot{\theta}_0 = 0.3$ K/s; solid curves – computations, dashed lines – experimental data [116]

occurs a gradual growth of heat deformations $\overset{\circ}{\varepsilon}_2$ and $\overset{\circ}{\varepsilon}_3$ due to heat expansion of the carbon fibres and coke phase. The value of the shrinkage (negative) deformation $\overset{\circ}{\varepsilon}_2$ (or $\overset{\circ}{\varepsilon}_3$) at temperatures $500< \theta <1800$ °C is inversely proportional to the concentration of the carbon fibres in the corresponding direction Ox_2 (or Ox_3). Since there are more carbon fibres ($\zeta = 0.7$) in the *warp* direction (Ox_3) than in the *weft* direction Ox_2 ($1 - \zeta = 0.3$), the heat deformation $\overset{\circ}{\varepsilon}_3$ decreases at $500< \theta <1800$ °C but remains positive, while $\overset{\circ}{\varepsilon}_2$ becomes negative at these temperatures (i.e. the shrinkage occurs). At $\theta \approx 1900 \ldots 2000$ °C, deformation $\overset{\circ}{\varepsilon}_2$ again becomes positive.

Since there are no reinforcing fibres in the Ox_1 direction, the heat deformation $\overset{\circ}{\varepsilon}_1$ is defined by heat expansion and shrinkage of the matrix. At temperatures higher than 500 °C, there is a shrinkage of the composite in the Ox_1 direction (Figure 7.10).

Thus, the heat deformations $\overset{\circ}{\varepsilon}_k$ depend not only on heat deformations of the matrix and fibres $\overset{\circ}{\varepsilon}_m$ and $\overset{\circ}{\varepsilon}_f$ but also on changing elastic characteristics of fibres and matrix with temperature (i.e. on functions $\widetilde{a}_f$ and $\widetilde{a}_1$).

Figure 7.11 shows experimental [116] and calculated dependences of deformation $\overset{\circ}{\varepsilon}_3$ on temperature for glass/phenolic composite under heating and subsequent cooling. Cooling the composites, like matrices, after heating up to thermodecomposition temperatures leads to additional shrinkage of the material.

7.6. Coefficients of Phase Interaction

The coefficients of phase interaction f_k for textile composites in (7.29) are determined by formulae similar to (7.39)

$$f_k = f_m \Omega_k, \qquad k = 1, 2, 3, \tag{7.42}$$

where Ω_k are expressed by formulae (7.40).

Volumetric ablation of the polymer matrix leads to generation of a large quantity of gaseous products in the pores.

Under certain conditions (investigated in Chapter 9) the gas pressure in pores can prove to be considerably high (of order of 20 MPa). In this case gases will redistribute a stress state in phases of the composite. As seen from the formulae (7.29), the effect of pore pressure p on the composite is described by functions f_k for which we have expressions (7.42).

The function f_k depends strongly on the porosity φ_g of the composite. Dependences f_k on φ_g are shown in Figure 7.12.

7.7. Thermal Strength

Textile composite materials at high temperatures can be destroyed by different types of failure depending on the type of loading. In accordance with the structural

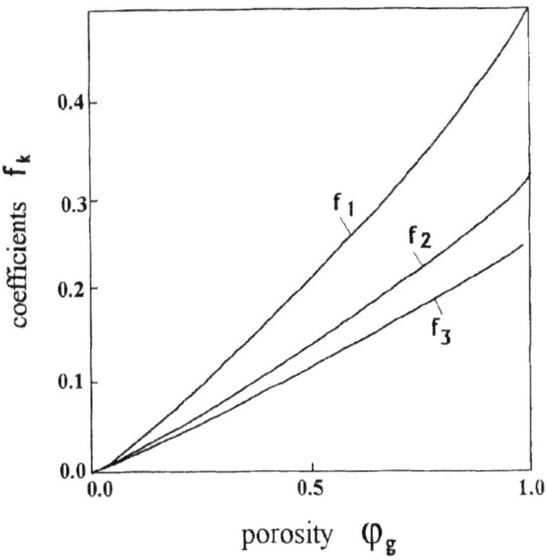

Figure 7.12. Dependence of phase interaction coefficients f_k for textile composite on porosity φ_g

model described in paragraph 7.1, there exist six different *types of destruction of textile composites* at normal and high temperatures:
(A_α) – breakage of fabric threads laid in one of the reinforcing directions Ox_α, $\alpha = 2, 3$;
(B_α) – loss of stability of threads laid in one of the directions Ox_α, $\alpha = 2, 3$;
(C) – delamination of the composite;
(D) – cracking under transverse compression in the direction Ox_1.

For brevity, both types (A_2) and (A_3) (or (B_2) and (B_3)) will be denoted by (A) (or (B), respectively).

Schematical models of destruction types (A), (B), (C) and (D) are given in Figures 7.13 – 7.16.

7.7.1. DESTRUCTION BY TYPES (A_α) AND (B_α)

Criterion of Failure

In accordance with the structural model (see Figures 7.1 and 7.13), the cause of failure of a textile composite under tension in its fabric plane is the appearance of shear cracks in its matrix between the monofibres of reinforcing threads. As mentioned in Chapter 6, the strength of a unidirectional composite (i.e. curved threads impregnated by a binder) depends on transverse σ_{u11} and shearing σ_{u13} stresses which lead to the appearance of longitudinal cracks and, hence, lower the strength σ'_{uT} in tension. Due to curving of the threads, the multiaxial state of the unidirectional component-layers is realized in the textile composite (see paragraph 7.2.1): within the domain of curving, besides longitudinal microstresses $\sigma^{(s)}_{u33}$ there appear transverse $\sigma^{(s)}_{u11}$ and shearing $\sigma^{(2)}_{u12}$ and $\sigma^{(1)}_{u13}$ stresses detemined

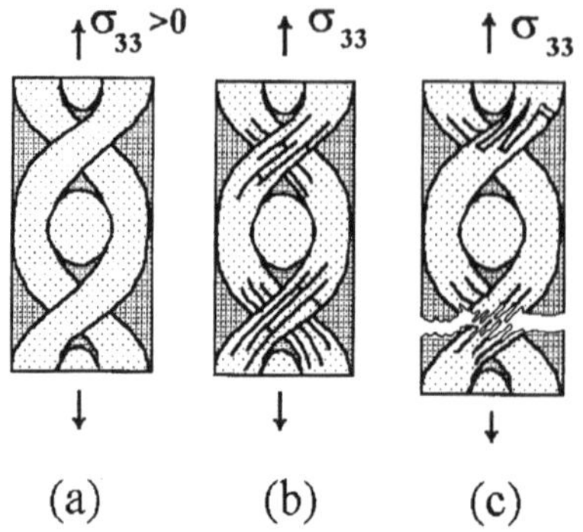

Figure 7.13. A model of destruction of textile composite by type (A) in longitudinal tension: (a) - virgin composite; (b) - appearance of shearing cracks within a matrix between fibres; (c) - faiilure of composite due to breakage of monofibres

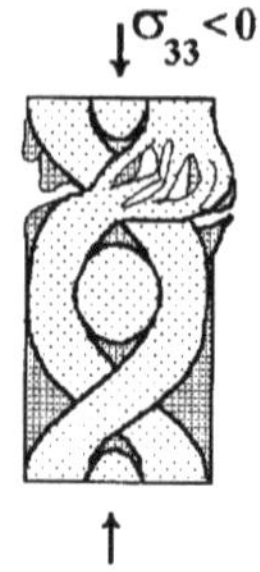

Figure 7.14. A model of destruction of textile composite by type (B) in longitudinal compression

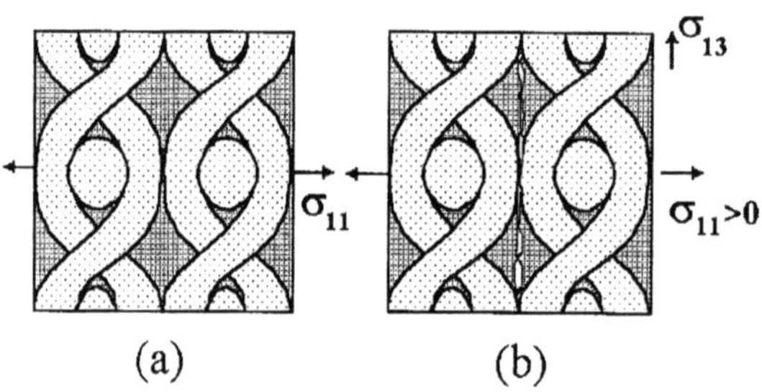

Figure 7.15. A model of destruction of textile composite by type (C) in transverse tension: (a) - virgin composite; (b) - appearance of interlayer cracks within a matrix and delamination

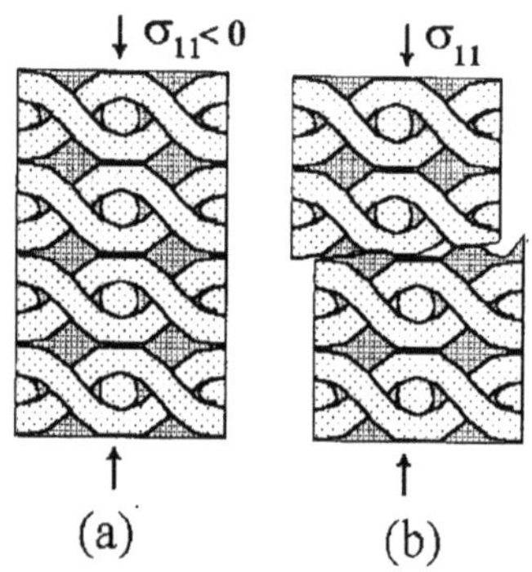

Figure 7.16. A model of destruction of textile composite by type (D) in transverse compression: (a) - virgin composite; (b) - shearing cracks within a matrix between layers of fabric and destruction

by the formulae (7.20), (7.22) and (7.28). These stresses reach their maximum within the domain of maximum curving of the threads at $\vartheta = \vartheta_{\max}$. Hence, the strength of the textile composite under tension in the Ox_α axis direction ($\alpha = 2, 3$) is determined by the strength of the corresponding unidirectional component-layer oriented by the angle $\vartheta = \vartheta_{\max}$ in the plane Ox_1x_α.

Then, writing the failure condition (6.60) separately in the form $z_u = 1$ for the unidirectional component-layer oriented in the direction Ox_3' (or Ox_2') and substituting expressions (7.20), (7.22), (7.28) connecting stresses $\sigma_{u\,kl}^{(s)}$ with stresses σ_{22}, σ_{33}, σ_{23} into the conditions

$$\sigma_{u\alpha\alpha}^{(s)} = \bar{\sigma}_{\alpha\alpha}^{(s)} = B_\alpha^{(s)}\sigma_{\alpha\alpha}, \quad \sigma_{u11}^{(1)} = \vartheta_m^2\bar{\sigma}_{33}^{(1)} = \vartheta_m^2 B_3^{(1)}\sigma_{33}, \quad \alpha = 2, 3,$$

$$\sigma^{(2)}_{u11} = \vartheta^2_m \bar{\sigma}^{(2)}_{22} = \vartheta^2_m B^{(2)}_2 \sigma_{22}, \qquad \sigma^{(1)}_{u12} = \sigma^{(2)}_{u13} = 0,$$

$$\sigma^{(1)}_{u13} = \vartheta_m \bar{\sigma}^{(1)}_{33} = \vartheta_m B^{(1)}_3 \sigma_{33}, \qquad \sigma^{(2)}_{u12} = \vartheta_m \bar{\sigma}^{(2)}_{22} = \vartheta_m B^{(2)}_2 \sigma_{22},$$

$$Y^{(1)}_{u1} = B^{(1)}_3 \sigma_{33}, \quad Y^{(2)}_{u1} = B^{(2)}_2 \sigma_{22}, \quad Y^{(1)}_{u2} = B^{(1)}_2 \sigma_{22} + \vartheta^2_m B^{(1)}_3 \sigma_{33},$$

$$Y^{(2)}_{u2} = B^{(2)}_2 \sigma_{33} + \vartheta^2_m B^{(2)}_3 \sigma_{22}, \quad Y^{(1)}_{u3} = |B^{(1)}_2 \sigma_{22} - \vartheta^2_m B^{(1)}_3 \sigma_{33}|,$$

$$Y^{(2)}_{u3} = |B^{(2)}_2 \sigma_{33} - \vartheta^2_m B^{(2)}_3 \sigma_{22}|,$$

$$Y^{(1)\,2}_{u4} = \sigma^2_{23} + \vartheta^2_m (B^{(1)}_3)^2 \sigma^2_{33}, \quad Y^{(2)}_{u4} = \sigma^2_{23} + \vartheta^2_m (B^{(2)}_2)^2 \sigma^2_{22},$$

we obtain the following two failure criteria for textile composite by the type (A) or (B):

$$z^2_\alpha \equiv \left(\frac{|\sigma_{\alpha\alpha}| + \sigma_{\alpha\alpha}}{2\sigma_{\alpha T}}\right)^2 + \left(\frac{|\sigma_{\alpha\alpha}| - \sigma_{\alpha\alpha}}{2\sigma_{\alpha C}}\right)^2 + \left(\frac{\sigma_{\beta\beta}}{S_\beta}\right)^2 -$$
$$- \frac{\sigma_{\alpha\alpha}\sigma_{\beta\beta}}{S^2} + \left(\frac{\sigma_{23}}{\sigma_{23S}}\right)^2 = 1, \quad \alpha, \beta = 2, 3, \quad \alpha \neq \beta. \tag{7.43}$$

Here $Y^{(s)}_{\alpha\alpha}$ are the invariants of tensors $\sigma^{(s)}_{ukl}$; they are determined by formulae (6.53) for $s = 1$ and $s = 2$.

The functions $\sigma_{\alpha T}$, $\sigma_{\alpha C}$, σ_{23S} in (7.43) are the ultimate strengths of the textile composite in tension–compression along the axis Ox_α and in shear in the plane Ox_2x_3:

$$\sigma_{3T,C} = \frac{1}{B^{(1)}_3} \left(\frac{1}{(\sigma'_{uT,C})^2} + \frac{\vartheta^2_m (B^2_{m1} + 3B^2_{m3})}{3(\sigma^0_{mS} \widetilde{a}_1 h^0)^2} \right)^{-1/2},$$

$$\sigma_{2T,C} = \frac{B^{(1)}_3}{B^{(2)}_2} \sigma_{3T,C}, \quad \sigma_{23S} = \frac{\sigma^0_{mS} \widetilde{a}_1 h^0}{B_{m3}}, \quad \beta = 2, 3. \tag{7.44}$$

The criterion (7.43) also contains three additional functions S_β and S:

$$S_\beta = \frac{3\sigma^0_{mS} \widetilde{a}_1 h^0}{B_{m1} B^{(\beta-1)}_\beta}, \quad S = \frac{3\sigma^0_{mS} \widetilde{a}_1 h^0}{B_{m1} (B^{(1)}_2 B^{(1)}_3)^{1/2}}, \quad \beta = 2, 3, \tag{7.45}$$

depending, mainly, on shearing properties of the matrix.

Coefficients of stress concentrations are determined as follows:

$$B_{m1} = \frac{\widetilde{a}_1}{\widetilde{a}_5}, \quad B_{m2} = \frac{m_E \widetilde{a}_1}{\varphi_f \widetilde{a}_2}, \quad B_{m3} = \frac{\widetilde{a}_1}{\widetilde{a}_3}, \tag{7.46}$$

$$B^{(1)}_2 = \frac{m_E a_{\theta 2}}{(1-\zeta) a_{\theta 1}}, \quad B^{(2)}_3 = \frac{m_E a_{\theta 2}}{\zeta a_{\theta 1}}, \quad B^{(1)}_3 = \frac{1}{\zeta}, \quad B^{(2)}_2 = \frac{1}{1-\zeta}.$$

Functions $\widetilde{a}_3$ and $\widetilde{a}_5$ are determined by formulae (6.5), functions $a_{\theta 1}$ and $a_{\theta 2}$ – by (7.32), and $\widetilde{a}_1$ and $\widetilde{a}_2$ – by (7.33).

As follows from (7.43), destruction occurs by type (A_α), if the following conditions are satisfied simultaneously:

$$z_\alpha = 1, \quad \sigma_{\alpha\alpha} > 0, \quad \alpha = 2, 3. \tag{7.47}$$

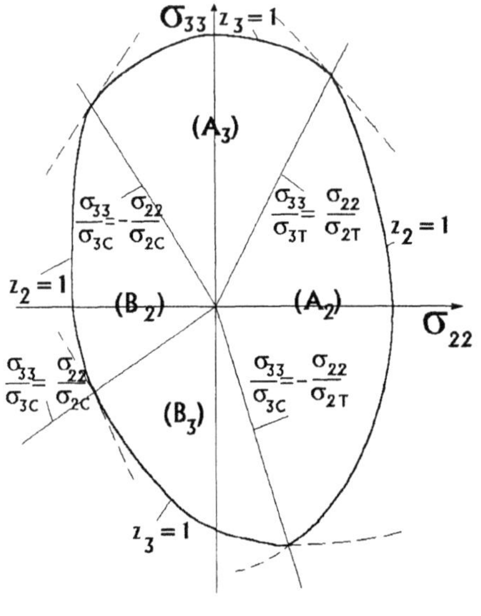

Figure 7.17. Crosssection of a strength surface of textile composite by plane $O\sigma_{22}\sigma_{33}$

Destruction of type (B_α) occurs if

$$z_\alpha = 1, \quad \sigma_{\alpha\alpha} < 0, \quad \alpha = 2, 3. \tag{7.48}$$

Functions z_α determined by expresiions (7.43) are called *damage parameters* of the textile composite. They characterize how near the stress state of the composite at point x and time t is to the ultimate state when failure of the composite occurs. Values of z_α are within the domain $[0, 1]$; as follows from (7.43), value $z_\alpha = 1$ corresponds to destruction by the type (A_α) or (B_α).

Thus, the three-dimensional space of stresses $(\sigma_{22}, \sigma_{33}, \sigma_{23})$ is divided into four domains, in each of which there occurs failure by one of the types (A_2), (A_3), (B_2) or (B_3) (Figure 7.17).

Experimental Results

Dependence of strength σ_{2T} on temperature in tension for both textile and unidirectional composites is determined, mainly, by heat-resistance of their matrices. Figure 7.18 shows experimental and calculated dependences of strength $\sigma_{2T}(\theta)$ for glass-plastics on different types of binders: epoxy, phenolic, polyimide and silicon-organic. Calculations of σ_{2T} were conducted by the formula (7.44); the parameters of the internal structure of the composite were as follows: $\varphi_f^0 \approx 0.55 \ldots 0.62$; $\zeta = 0.5$; $\Phi = 1 \ldots 3^\circ$; $\vartheta_{\max} = 5 \ldots 7^\circ$. Figure 7.18 shows that the *temperature of heat-resistance* of the composite θ_h, determined similarly to (6.45) as the temperature for which the parameter

$$\bar{h}(\theta) = \sigma_{2T}(\theta)/(\bar{\sigma}_f^0 \varphi_f^0 \zeta) \tag{7.49}$$

decreases by a factor of two $(\bar{h}(\theta_h)/\bar{h}(\theta_0) = 0.5)$ during heating from $\theta_0 = 20\ ^\circ$C to $\theta = \theta_h$, grows with increasing the heat-resistance of the matrix: for glass/epoxy composite – $\theta_h = 210\ ^\circ$C, for glass/phenolic – $\theta_h = 300\ ^\circ$C, for glass/polyimide – $\theta_h = 350\ ^\circ$C, for glass/silicon-organic – $\theta_h = 380\ ^\circ$C. However, for absolute values of the strength σ_{2T} there is another picture: within the temperature interval

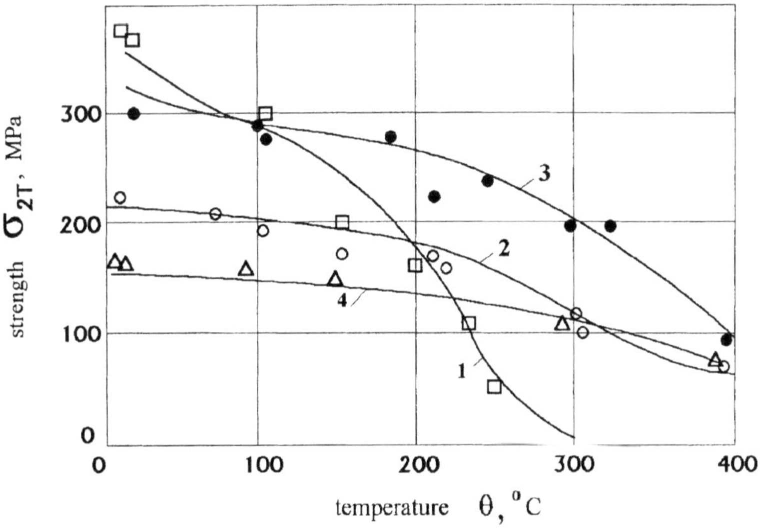

Figure 7.18. Strength in tension σ_{2T} for textile glass plastics based on binders of different types *versus* temperature θ during heating in an inert environment; points - experimantal data [54], curves - computations; matrices: 1 (▲) - epoxy, 2 (○) - phenolic, 3 (●) - polyimide, 4 (□) - silicon-organic

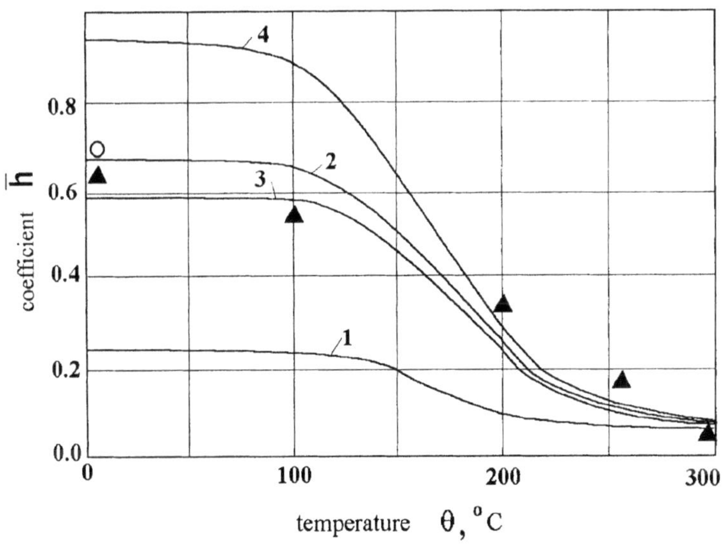

Figure 7.19. Coefficient of strength properties $\bar{h}$ for textile composites based on epoxy matrix and with fibres of different types (1 - organic, 2 (○) - carbon, 3 (▲) - glass) and coefficient $\tilde{a}_1$ for epoxy matrix (curve 4) *versus* temperature during heating in air. Curves - computations, points - experimental data

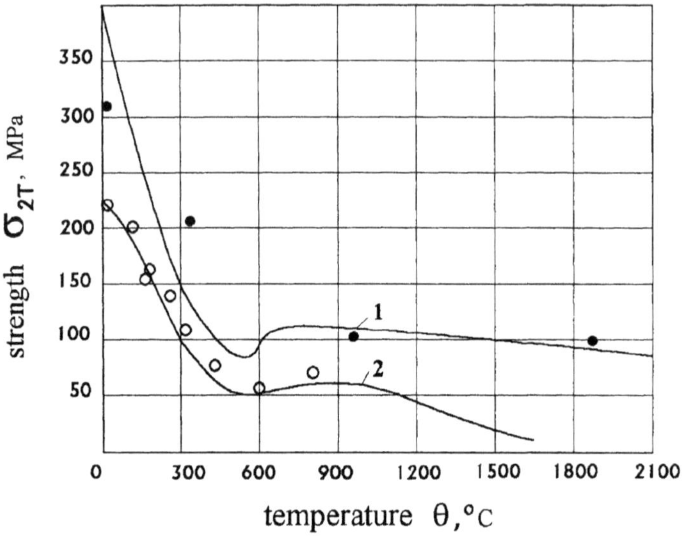

Figure 7.20. Strength in tension σ_{2T} for textile carbon/ (1 (•)) and glass/ (2 (○)) phenolic composites *versus* temperature θ during heating in an inert environment; points - experimental data, curves - computations

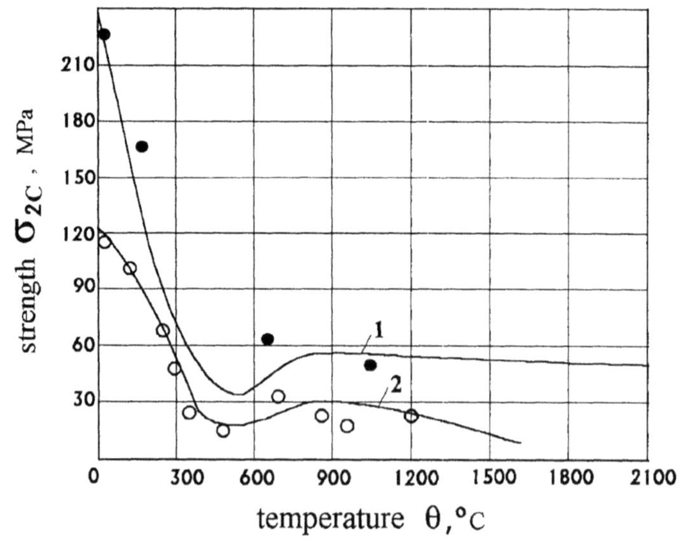

Figure 7.21. Strength in compression σ_{2C} for textile phenolic composites *versus* temperature θ during heating in an inert environment: 1 (•) - carbon fibre and 2 (○) - glass fibre. Points - experimental data [116], curves - computations

300 - 400 °C the maximum value of σ_{2T} was observed for polyimide and phenolic plastics, and silicon-organic composites, as was mentioned above, usually have non-high absolute values of the strength in tension. At the same time, the effect of the fibre type proves to be less considerable: Figure 7.19 shows experimental and calculated dependences of the coefficient of changing strength $\bar{h}$ on θ in air, for epoxy composites based on organo-, glass- and carbon-fabrics. The difference of values of θ_h for these types of composites is equal to 30 - 40 °C. These conclusions are of importance when one should make a choice of composites for airspace and other heat-loaded structures working within the temperature interval 300 - 400 °C.

For higher temperatures (up to 2000 °C and more, which are typical for re-entry vehicles (see Chapter 1)) the effect of the fibre type on the strength σ_{2T} is more considerable. Figure 7.20 shows experimental and calculated values of $\sigma_{2T}(\theta)$ for carbon/ and glass/phenolic textile composites in an inert environment at temperatures from 20 to 2100 °C. Within the interval of θ from 0 to 900 °C, the character of $\sigma_{2T}(\theta)$ for carbon- and glass-plastics is similar: sharp decrease of the strength within the interval 20 - 500 °C, and then a small growth of values of σ_{2T} in the interval from 500 to 900 °C caused by formation of a strong coke residue of the phenolic matrix. On further heating, the glass-plastic loses its strength properties again due to weakening of the glass fibres (because of formation of a weak crystalline phase), and at temperature 1600 - 1800 °C there appears a liquid phase in the composite. At the same time, phenolic carbon-plastic in an inert environment can keep its mechanical properties up to temperatures $\sim$ 2100 °C (see Figure 7.20) thanks to heat resistance of carbon fibres. These peculiarities play a considerable role in processes of linear ablation of composites (see Chapter 10).

The character of destruction of composites under compression is similar to that under tension; this is shown in Figure 7.21, for example, for the strength $\sigma_{2C}(\theta)$ of carbon/ and glass/phenolic composites.

7.7.2. DESTRUCTION BY THE TYPES (C) AND (D)

Destruction of textile composite by the type (C) or (D) occurs due to cracking of the matrix between the fabric layers (see Figures 7.10 and 7.11). The stress state of the matrix within the interlayer domain is determined, mainly, by transverse σ_{11} and interlayer shearing σ_{12}, σ_{13} stresses; the cracking criterion, in fact, coincides with the failure criterion (6.52) in the form $z = 1$ for unidirectional composite by the type 3 or 4. Substituting expressions (7.17) or (7.21) and (7.28) for stresses $\sigma_{ukl}^{(s)}$ into equation (6.52), taking account of

$$Y_{u2} = \sigma_{11}, \qquad Y_{u3}^2 = \sigma_{11}^2 + 4\sigma_{12}^2, \qquad Y_4 = \sigma_{13}^2 \tag{7.50}$$

and neglecting small terms, we obtain the failure criterion for the textile composite by the type (C) or (D)

$$z_1^2 \equiv \Big(\frac{|\sigma_{11} + \frac{3N^0}{B_{m1}}p| + \sigma_{11} + \frac{3N^0}{B_{m1}}p}{2\sigma_{1T}}\Big)^2 +$$
$$+ \Big(\frac{|\sigma_{11} + \frac{3N^0}{B_{m1}}p| - \sigma_{11} - \frac{3N^0}{B_{m1}}p}{2\sigma_{1C}}\Big)^2 + \Big(\frac{\sigma_{13}}{\sigma_{13S}}\Big)^2 + \Big(\frac{\sigma_{12}}{\sigma_{12S}}\Big)^2 = 1. \quad (7.51)$$

Here the functions σ_{1T}, σ_{1C}, σ_{13S} and σ_{12S} are the ultimate strengths of the textile composite in tension and compression across the fabric layers, and in interlayer shears, respectively:

$$\sigma_{1T} = \frac{\sigma^0_{mT}\tilde{a}_1 h^0(1-\Delta\Sigma)}{B_{m1}}, \qquad \sigma_{12S} = \frac{\sigma^0_{mS}\tilde{a}_1 h^0(1-\zeta)(1-\Delta\Sigma)}{B_{m3}},$$
$$\sigma_{13S} = \frac{\sigma^0_{mS}\tilde{a}_1 h^0\zeta(1-\Delta\Sigma)}{B_{m3}}, \qquad \sigma_{1C} = \frac{\sigma^0_{mC}\tilde{a}_1 h^0}{B_{m1}}, \quad (7.52)$$

where $\Delta\Sigma$ takes account of thermocracking of the matrix due to different heat deformations of the matrix and fibres, and is determined by the formula

$$\Delta\Sigma = \frac{d_f}{L_0 K_L}\ln\big(1 + \Omega\frac{E^0_b}{\sigma^0_S}|\overset{\circ}{\varepsilon}_f - \overset{\circ}{\varepsilon}_m|\big), \quad (7.53)$$

parameters K_L, d_f/L_0 and Ω are determined in (6.56). At $\overset{\circ}{\varepsilon}_f - \overset{\circ}{\varepsilon}_m < 0$, we have $\Delta\Sigma = 0$.

Destruction by the type (C) occurs, if the condition (7.51) is satisfied and $\sigma_{11} + 3N^0 p/B_{m1} \geqslant 0$; if $\sigma_{11} + 3N^0 p/B_{m1} \leqslant 0$, then conditions (7.52) define failure of the composite by the type (B).

7.7.3. EXPERIMENTAL DATA

Figure 7.22 shows experimental and calculated dependences of the strength in tension σ_{1T} for carbon/ and glass/phenolic textile composites in heating up to 2100 $°C$ in an inert environment; Figure 7.23 shows the strength in interlayer shearing $\sigma_{12S}(\theta)$. These data verify the conclusion that the strength of a composite at high temperatures is determined not only by heat-resistance of its matrix but also by interaction of the matrix with the fibres. In particular, the difference of heat deformations $\overset{\circ}{\varepsilon}_f$ and $\overset{\circ}{\varepsilon}_m$ of the fibre and matrix leads to cracking of the plastic, to delaminations of the matrix from the fibres and, hence, to lowering the transverse σ_{1T} and shearing σ_{12S} strengths; this is in spite of the fact that the phenolic matrix in an inert environment at temperatures 600 - 1200 °C is rather hardened and the strength σ_{2T} increases within the temperature interval (see Figure 7.20).

Figures 7.22 and 7.23 show that the calculated results are sufficiently accurate for forecasting the strengths σ_{1T} and σ_{12S}; the error is about 15 - 20 %.

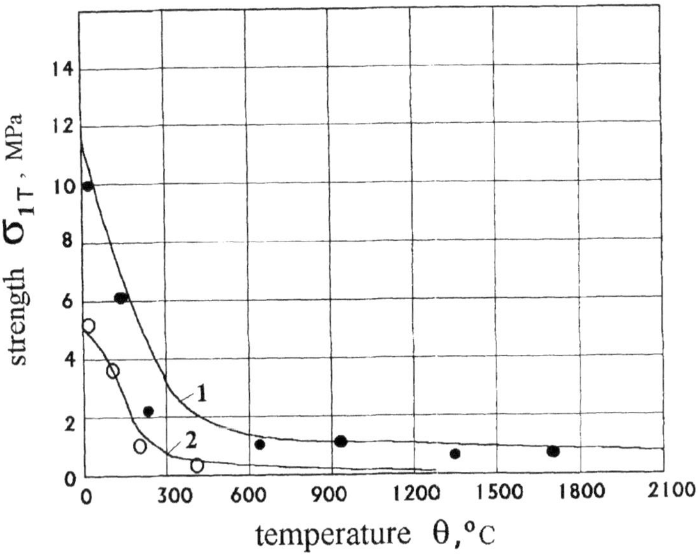

Figure 7.22. Strength in tension σ_{1T} for textile carbon/ (1 (•)) and glass/ (2 (○)) phenolic composites *versus* temperature θ during heating in an inert environment; points - experimental data, curves - computations

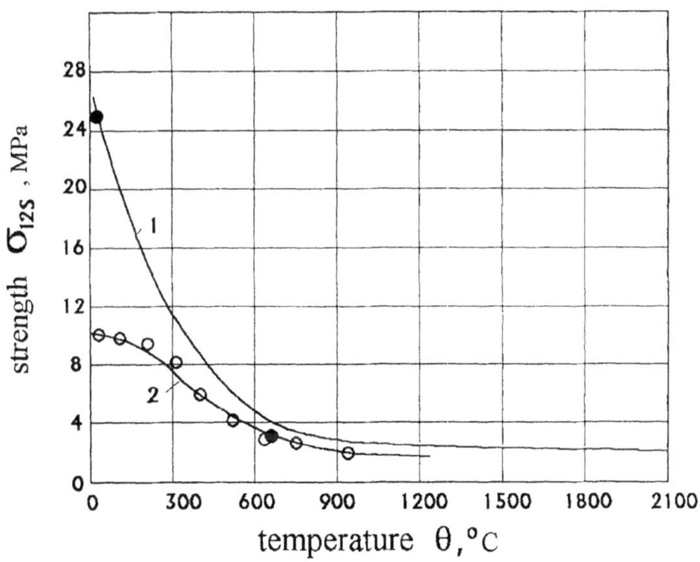

Figure 7.23. Strength under interlayer shear σ_{12S} for textile phenolic composites *versus* temperature θ during heating in an inert environment: 1 (•) - carbon fibre and 2 (○) - glass fibre. Points - experimental data, curves - computations

7.8. Thermal Properties of Textile Composites

7.8.1. HEAT CONDUCTIVITY

The principal coefficients of heat conductivity k_α in (7.30) are determined as for elastic moduli:

$$k_1 = k_b^0 \varkappa_{\theta 2}, \qquad k_2 = k_f^0 \zeta \varkappa_{\theta 1}, \qquad k_3 = k_f^0 (1-\zeta)\varkappa_{\theta 1},$$

where

$$\varkappa_{\theta 1} = \left(\varkappa_1\left(1 - \frac{\vartheta_m^2}{2}\right) + \varkappa_0 \varkappa_2 m_h \left(\frac{\zeta}{1-\zeta} + \frac{\vartheta_m^2}{2}\right)\right)\widetilde{b}_f b^{(0)}, \tag{7.54}$$

$$\varkappa_{\theta 2} = \left(\frac{1 - \frac{\vartheta_m}{2}}{\varkappa_2} + \frac{\vartheta_m^2 \varkappa_0 m_h}{2\varkappa_1}\right)^{-1} b^{(0)} b_1,$$

and functions $\varkappa_1$, $\varkappa_2$, $\varkappa_0$, b_1, $\widetilde{b}_f$ and $b^{(0)}$ are expressed by the formulae

$$b_1 = \left(\frac{S_b}{S_p^2 - S_b^2 + m_k(1 - S_p^2)} + \frac{S_p - S_b}{S_p^2 + m_k(1 - S_p^2)} + \frac{1 - S_p}{m_k}\right)^{-1},$$

$$b^{(0)} = \left(\frac{\theta}{\theta_0}\right)^{1/2}, \qquad \widetilde{b}_f = S_l - S_a + (1 - S_l)\frac{k_l}{k_f^0}, \tag{7.55}$$

$$\varkappa_0 = \frac{b_1}{\widetilde{b}_f}, \qquad \varkappa_1 = \varphi_f + \varkappa_0 m_h (1 - \varphi_f), \qquad \varkappa_2 = (1 - \varphi_f + \varkappa_0 m_h \varphi_f)^{-1}.$$

Thus, similarly to functions $a_{\theta 1}$, $a_{\theta 2}$, two functions $\varkappa_{\theta 1}$ and $\varkappa_{\theta 2}$ describe the change of heat conductivity coefficient k_α with temperature: function $\varkappa_{\theta 1}$ is for the reinforcing direction and $\varkappa_{\theta 2}$ for the transverse direction.

Formulae (7.54) contain four physical constants (heat conductivities of four solid phases, where $m_h = k_b^0/k_f^0$, $m_k = k_p/k_b^0$):

$$k_f^0, \qquad m_h, \qquad k_l, \qquad m_k \tag{7.56}$$

and geometrical constants (7.37).

Figure 7.24 shows computed results by formula (7.54) for the heat conductivity coefficient k_1 and experimental data for textile glass/epoxy composite. Deviation of the theoretical curve from the experimental one does not exceed 8 %.

Growth of k_1 within the interval $\theta = 20 \ldots 200$ °C is connected to the thermodynamical feature of many materials, that the heat conductivity increases with heating [108]; this is described within the scope of the model by the function $b^{(0)}$. Lowering heat conductivity k_1 within the interval $\theta = 200 \ldots 500$ °C is connected to the formation of secondary porosity within the matrix during thermodecomposition; increasing k_1 at $\theta \geqslant 500$ °C is caused by thermodynamical growth of the

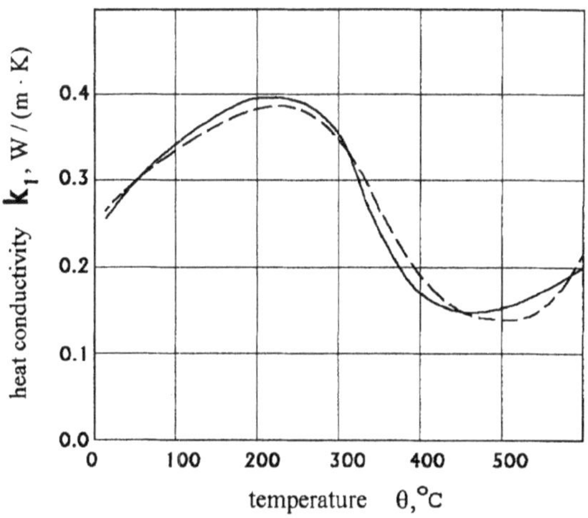

Figure 7.24. Heat conductivity coefficient k_1 for textile glass/epoxy composite *versus* temperature θ during heating with rate $\dot{\theta}_0 = 0.7$ K/s; solid curves - experimental data, dashed lines - computations by formula (7.54)

heat conductivity of the coke phase after the thermodecomposition process of the composite has practically finished.

Dependence $k_1(\theta)$ for carbon/phenolic composite has another character (Figure 7.25).

Because the heat conductivity coefficient of carbon fibres increases during heating (from the value $k_f^0 = 9$ W/(m·K) to $k_f = 27$ W/(m·K)), the heat conductivity coefficient k_1 does not decrease during thermodecomposition. This conclusion is verified by experimental data obtained for an inert medium at $\dot{\theta}_0 = 5$ K/s (see Figure 7.25).

The longitudinal heat conductivity coefficient k_2 of a carbon/phenolic composite changes in a similar way (Figure 7.26).

7.8.2. DENSITY

The change in the relative density ρ/ρ_0 of a composite under heating is determined by the formula

$$\frac{\rho}{\rho_0} = \frac{\rho_f\varphi_f + \rho_b\varphi_b + \rho_p\varphi_p}{\rho_f\varphi_f^0 + \rho_b\varphi_b^0}, \tag{7.57}$$

where changing phase concentrations φ_f, φ_b and φ_p are described by formulae (3.41), (3.42) and (3.48).

The presence of reinforcing fibres leads to an increase in the solid residue of thermodecomposition of the composite as compared with the non-reinforced matrix.

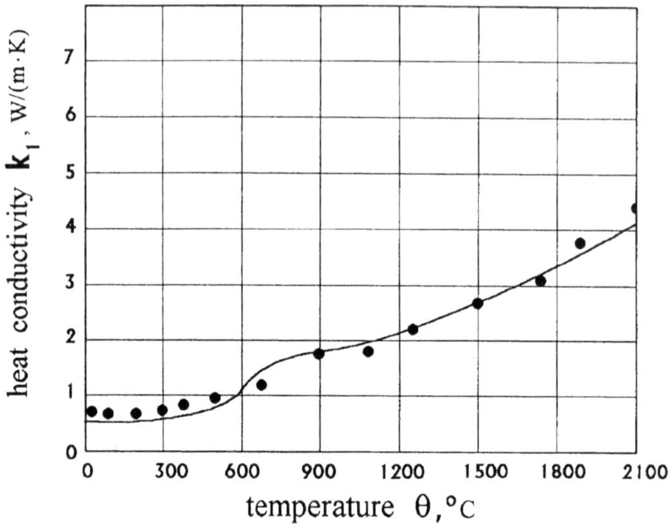

Figure 7.25. Heat conductivity coefficient k_1 of carbon/phenolic composite *versus* temperature during heating in an inert environment; curves – computations, points – experimental data

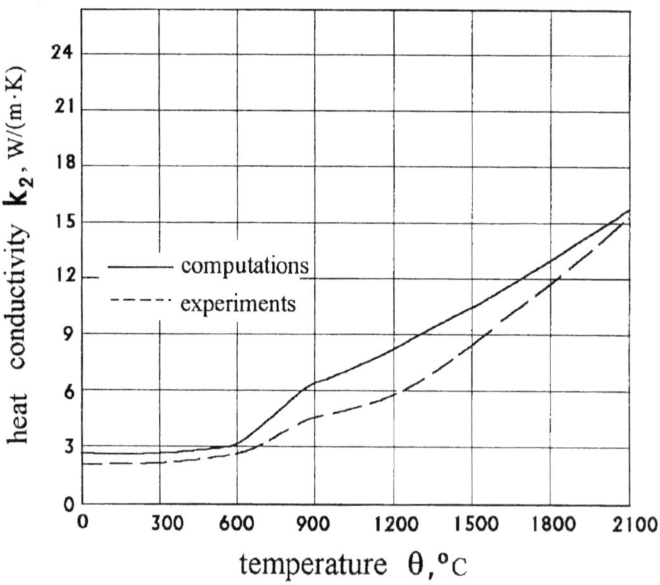

Figure 7.26. Heat conductivity coefficient k_2 of carbon/phenolic composite *versus* temperature during heating in an inert environment

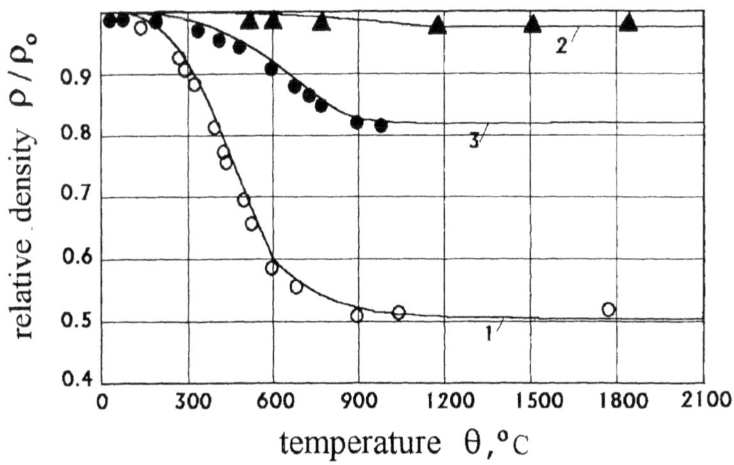

Figure 7.27. Relative density ρ/ρ_0 of non-reinforced phenolic matrix (1, o), carbon fibres (2, ▲) and carbon/phenolic textile composite (3, •) *versus* temperature during heating with rate $\dot{\theta}_0 = 5$ K/s, in an inert environment; curves - computations, points - experimental data

For the matrix it is $\rho_m^\infty/\rho_m^0 = 1 - \Gamma$ (by formula (4.4)), while for the composite

$$\frac{\rho^\infty}{\rho_0} = 1 - \frac{\xi_\Gamma \Gamma_f + \Gamma}{1 + \xi_\Gamma}, \qquad \xi_\Gamma \equiv \frac{\rho_f \varphi_f^0}{\rho_b \varphi_g^0}. \tag{7.58}$$

Usually $\Gamma_f \leqslant \Gamma$, so that $\rho^\infty/\rho_0 \geqslant \rho_m^\infty/\rho_m^0$.

Increasing the content of thermodecomposition solid residue, as a rule, is a positive factor; for this case, internal gas generation decreases and thermostrength of the composite grows.

Figure 7.27 shows the dependence of the relative density ρ/ρ_0 for carbon/phenolic composite, phenolic matrix and carbon fibres on temperature θ during heating in an inert medium. Equation (7.57) adequately simulates the change in density: the error does not exceed 5 - 7 %.

Figure 7.28 shows theoretical and experimental dependences of the relative density ρ/ρ_0 for glass/epoxy composite on temperature during heating with different rates. As for non-reinforced ablative matrices, increasing the heating rate leads to displacement of curves $(\rho(\theta)/\rho_0)$ towards higher values of temperature.

Figure 7.29 shows the effect of the type of ablative matrix on changing density ρ/ρ_0 for textile composite with glass-reinforcement under heating. Increasing thermostability of the matrix leads to growing thermostability of the whole composite. Due to reinforcing by fibres, the curves $\rho(\theta)/\rho_0$ for the composite are placed higher than the corresponding curves $\rho_m(\theta)/\rho_m^0$ for the matrix (see Figures 4.4,*a* and 7.29), therefore *temperature of thermostability* $\theta_{h\theta}$, for which the density has changed by a given magnitude, e.g. by 10 %:

$$\rho(\theta_{h\theta})/\rho_0 = 0.9 \tag{7.59}$$

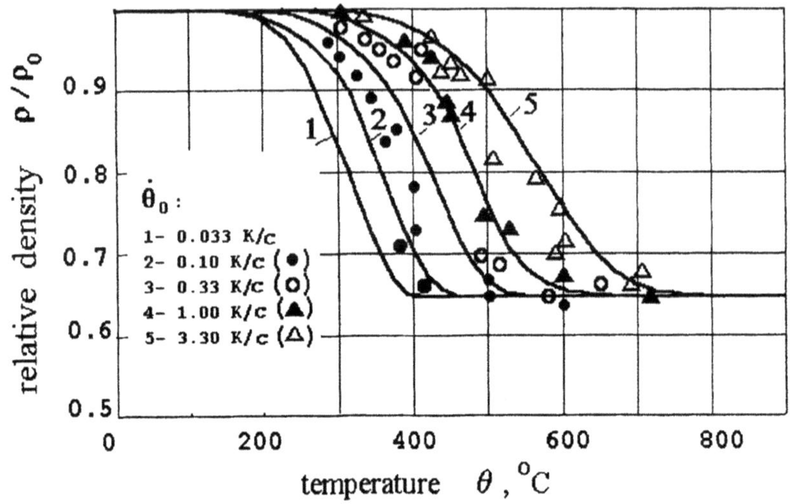

Figure 7.28. Relative density ρ/ρ_0 of glass/epoxy composite *versus* temperature during heating with different rates; curves - computations, points - experimental data

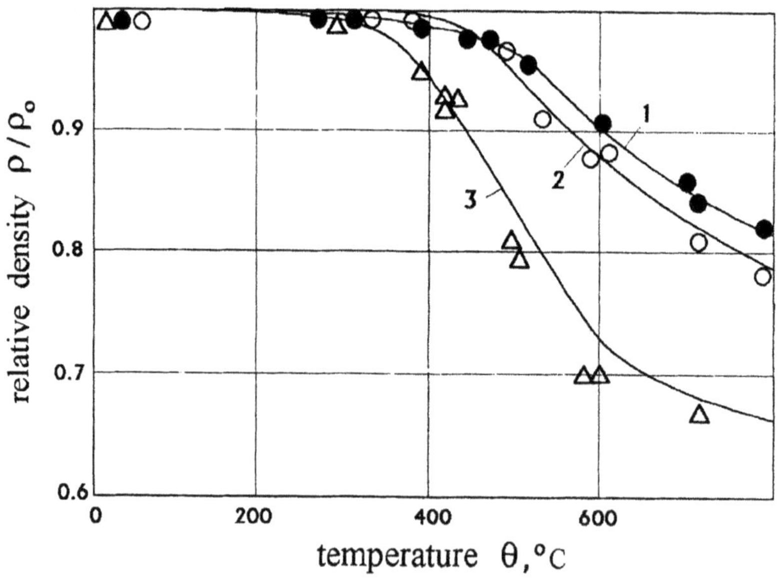

Figure 7.29. Relative density of glass plastics based on different matrices *versus* temperature during heating with rate $\dot{\theta} = 3$ K/s, in air. Curves - computations by (7.58), points - experimental data; matrices: curve 1 and points (•) - phenolic, curve 2 and points (○) - polyimide, curve 3 and points (△) - epoxy

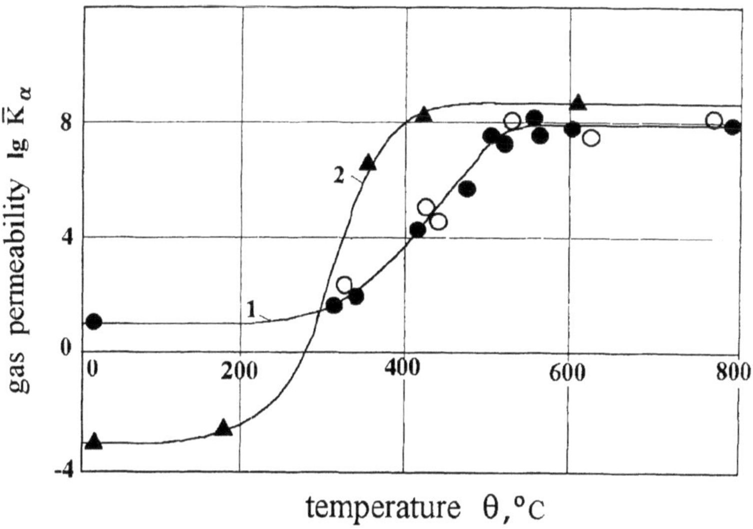

Figure 7.30. Gas-permeability coefficient for carbon/phenolic (1, (○), (●)) and glass/epoxy (2, (▲)) composites *versus* temperature θ. Curves - computations by (7.60), points - experimental data; points (○) correspond to $\bar{K}_1$, (●) and (▲) - to $\bar{K}_2$

for the composite proves to be higher than for the corresponding non-reinforced matrix.

7.9. Gas Permeability

The gas-permeability tensor **K**, introduced in relations (3.52), for textile orthotropic composites has a structure similar to the heat conductivity tensor **k** (7.30):

$$K_{qr} = \sum_{\alpha=1}^{3} K_\alpha \delta_{q\alpha} \delta_{r\alpha}. \tag{7.60}$$

Like the gas-permeability coefficient of matrix K, the principal gas-permeability coefficients K_α depend mainly on porosity φ_g of the composite and are determined similarly to (4.60)

$$K_\alpha = K_0 \exp\,(n_k \varphi_g^{1/3}), \qquad \alpha = 1, 2, 3. \tag{7.61}$$

Constants K_0 and n_k are determined by formula (4.63).

As follows from (7.61), gas permeability of the composite is independent of the direction of gas filtration in the pores with respect to the orthotropy axes. This conclusion is verified by experimental data. Figure 7.30 shows dependences of dimensionless gas permeabilities $\bar{K}_1$ and $\bar{K}_2$ for carbon/phenolic and glass/epoxy textile composites on temperature θ, where $\bar{K}_\alpha = K_\alpha(R\theta_0 t_0/x_0^2)$ (see formula (4.62)).

These data show that the gas permeability of composites considerably changes (by several orders) mainly within the domain of thermodecomposition temperatures and is practically independent of the structure of reinforcing.

CHAPTER 8

COMPOSITES REINFORCED BY DISPERSED PARTICLES

8.1. Model of the Composite

We consider a composite material consisting of two phases in the initial state: polymer binder (b-phase) and filler in the form of dispersed particles (a-phase).

We apply the dispersed filler to optimize the composite material properties such as heat conductivity, ablation rate, elastic modulus etc.

Under heating, the polymer binder undergoes pyrolysis, leaving a solid residue (p-phase) and gaseous products (g-phase).

The dispersed particles can also be subjected to physico-chemical transformations under high temperatures. According to the model described in Chapter 5, they consist of three phases: a, l and g.

Composite material reinforced by particles is simulated by a three-level structure (Figure 8.1). At the third structural level the composite material consists of periodicity cells of the third type (PC3). Each PC3 consists of two components: matrix and filler, like the isotropic matrix in Chapter 4. The filler is represented by a hollow cube, and the matrix is also modelled as a cube surrounding the filler (Figure 8.1). This model allows us to consider both solid dispersed particles (glass microballs, carbon powder etc.) and hollow particles (glass microspheres, phenolic microspheres etc.).

The polymer matrix itself is a collection of periodicity cells of the second level. PC1 is a hollow cube of solid residue (p-phase) surrounded by initial polymer substance (b-phase). The cavity of this combined cube is represented as a pore with gaseous products (g-phase). The filler (dispersed particles) consists of periodicity cells PC2. Each PC2 is a set of l-, a- and g-phases (see Figure 8.1).

This suggested model of a composite material structure allows us to evaluate all the main characteristics of the material: elastic modulus, strength in tension, heat conductivity, density, ablation rate.

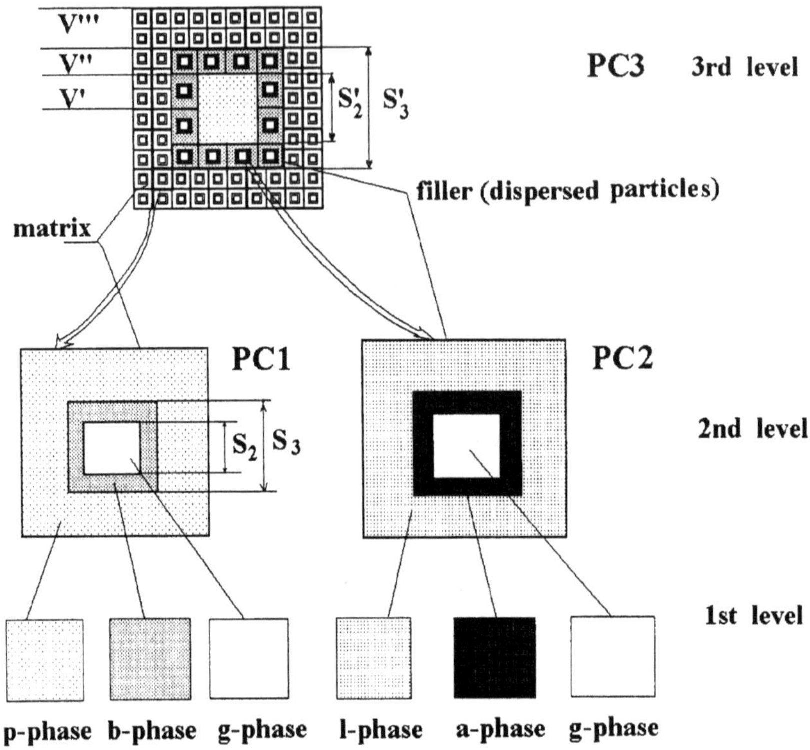

Figure 8.1. A model of the three-level structure of a composite material reinforced by dispersed particles

8.2. Thermo-Elastic Characteristics

8.2.1. ONE-DIMENSIONAL MODEL

All the characteristics of elements at the second structural level (see Figure 8.1) are determined as those for the matrix (Chapter 4) and the fibre (Chapter 5).

To evaluate the elastic modulus E of the composite at the third structural level, one can apply the same calculation method as for a non-reinforced matrix, namely by dividing the PC3 into domains V', V'' and V''' (Figure 8.2) which are layers parallel to the direction of the acting stress σ.

Let S_3' be a dimensionless length (a cube edge) of the dispersed filler related to the linear size of PC3, h_f – a dimensionless wall thickness of the hollow filler, and

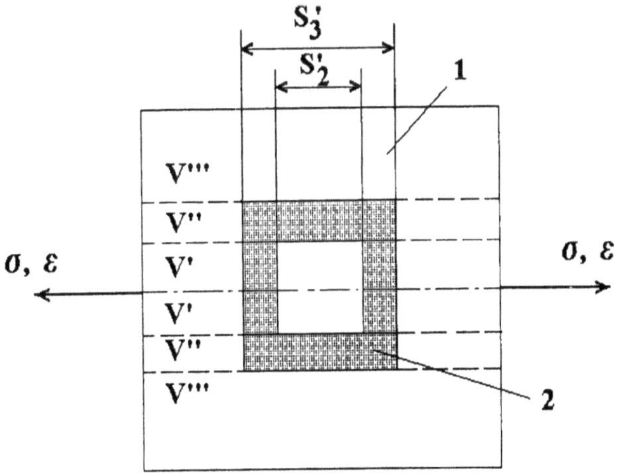

Figure 8.2. A model of PRC for calculation of the elastic modulus E and strength σ_T in tension: 1 - matrix, 2 - filler

$S_2' = S_3' - 2h_f$. Let stresses and strains in the domain V''' occupied by the matrix be σ_m''' and ε_m'''; stresses and strains in the domain V'' be σ_m'' and ε_m'' for the matrix and σ_f'' and ε_f'' for the fibre; stresses and strains in the domain V' be σ_m', ε_m' and σ_f', ε_f', respectively.

For the simplest model, which was formulated in Chapter 4, all the stresses $\sigma_{m,f}^{','',''' }$ and strains $\varepsilon_{m,f}^{','',''' }$ are constant but different, and they are connected by the structural relationships of the mixture equation type:

$$
\begin{aligned}
\sigma_m''' &= \sigma''', & \varepsilon_m''' &= \varepsilon''', \\
\sigma_m'' &= \sigma_f'' = \sigma'', & \varepsilon_m''(1 - S_3') + \varepsilon_f'' S_3' &= \varepsilon'', \qquad (8.1)\\
\sigma_m' &= \sigma_f' = \sigma', & \varepsilon_m'(1 - S_3') + \varepsilon_f'(S_3' - S_2') &= \varepsilon',
\end{aligned}
$$

where σ', σ'', σ''' and ε', ε'', ε''' are the average (effective) stresses and strains in the domains V', V'' and V''', respectively, which are connected with stress σ and strain ε of the whole PC3 by the relations

$$
\varepsilon' = \varepsilon'' = \varepsilon''' = \varepsilon, \qquad (8.2)
$$

$$
\sigma'''(1 - (S_3')^2) + \sigma''((S_3')^2 - (S_2')^2) = \sigma.
$$

Stresses and strains in the domains relating to the matrix are connected by the relations (see Chapter 4)

$$
\sigma_m^{','',''' } = E_m(\varepsilon_m^{','',''' } - \overset{\circ}{\varepsilon}_m - f_{mp}), \qquad (8.3)
$$

where E_m, $\overset{\circ}{\varepsilon}_m$ and f_m are expressed by formulae (4.22) - (4.24), and stresses and strains relating to the domains of PC3 occupied by the filler are connected by the similar relations

$$\sigma_f^{\prime,\prime\prime,\prime\prime\prime} = E_f(\varepsilon_f^{\prime,\prime\prime,\prime\prime\prime} - \overset{\circ}{\varepsilon}_f), \tag{8.4}$$

where E_f and $\overset{\circ}{\varepsilon}_f$ are expressed by formulae (5.6, (5.9).

On substituting the relations (8.3) and (8.4) into (8.1) and then into (8.2), we obtain effective constitutive relations for PRC

$$\sigma = E(\varepsilon - \overset{\circ}{\varepsilon}) - fp, \tag{8.5}$$

where E is the elastic modulus of the composite, $\overset{\circ}{\varepsilon}$ is the heat deformation, f is the phase interaction coefficient, and

$$E = E_b^0 a',$$

$$\overset{\circ}{\varepsilon} = \overset{\circ}{\varepsilon}_m \left(1 - \frac{a'_p}{a'_Q}\right) + \frac{a'_p}{a'_\varphi}\overset{\circ}{\varepsilon}_f, \tag{8.6}$$

$$f = f_m \left(1 - \frac{a'_p}{a'_Q}\right).$$

Functions a', a'_p and a'_Q in the formulae are introduced as follows:

$$a' = \tilde{a}_1 a'_Q, \tag{8.7}$$

$$a'_Q = 1 - S_3'^2 + \frac{a'_p}{S'_3}, \qquad a'_p = \frac{S'_3(S_3'^2 - S_2'^2)}{1 - S'_3(1 - m_E\tilde{a}_1/\tilde{a}_f)}.$$

Here E_b^0 is the elastic modulus of the polymer phase at temperature θ_0; a' is the coefficient describing the effect of filler on the elastic modulus of the material; $\tilde{a}_1$ is the coefficient describing the variation of the elastic modulus of the matrix on heating up to pyrolysis temperatures (it is determined by formula (4.22a)); $\tilde{a}_f$ is the coefficient describing the variation of the elastic modulus of filler (it is determined by the formula (5.7)); h_f is the thickness of the hollow particle wall (if particles have no cavities, then $h_f = S'_3/2$); m_E is the ratio of the elastic moduli of a-phase to the polymer phase: $m_E = E_f^0/E_b^0$. Functions S_2, S'_2, S_3, S'_3 are connected to the volumetric concentrations of phases by the relations

$$S_2 = \left(\frac{\varphi_g}{1 - \varphi_f^0}\right)^{1/3}, \qquad S_3 = \left(1 - \frac{\varphi_b}{1 - \varphi_f^0}\right)^{1/3},$$

$$S'_2 = S'_3 - 2h_f, \qquad S'_3 = \varphi_f^{0\,1/3}. \tag{8.8}$$

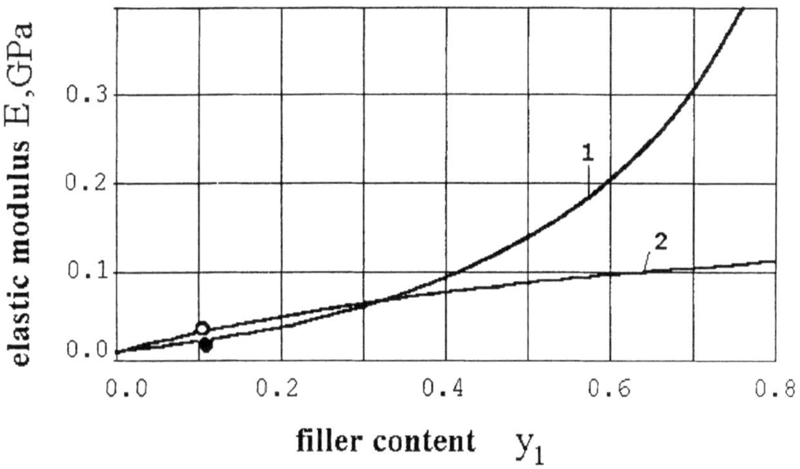

Figure 8.3. Relationships between elastic modulus E of a composite material and content y_1 of glass (1) and phenolic (2) microspheres

We introduced mass contents of phases y_i, $i = 1, 2$, which are connected to the volumetric phase concentrations by the relations

$$\varphi_b^0 = \frac{1 - \varphi_g^0}{1 + \frac{\rho_b}{\rho_f}\frac{y_1}{y_2}(1 - \varphi_g^0)},$$

$$\varphi_f = 1 - \varphi_b^0 - \varphi_g^0; \qquad y_1 + y_2 = 1.$$

Below we will consider composite materials based on a silicon-organic matrix and reinforced with polymer (phenolic) and glass microspheres.

Figure 8.3 shows the relationship between the elastic modulus of the composite material and the content of glass and phenolic microspheres. With an increase of the filler content, the elastic modulus of the material grows sharply, especially for glass filler, which has a greater stiffness than the phenolic one.

In computations we chose the following values of geometrical parameters:

$$h_f = 0.03, \qquad \varphi_g = 0.1.$$

The elastic moduli E_f^0 of glass and phenolic microspheres were taken to be these of glass fibres and phenolic matrices, respectively: 70 GPa and 2 GPa.

8.2.2. THREE-DIMENSIONAL RELATIONS

Three-dimensional constitutive relations for PRC are written as follows:

$$\varepsilon_{kl} = \frac{1+\nu}{E}\sigma_{kl} + \left(-\frac{\nu}{E}\sigma + \overset{\circ}{\varepsilon} + fp\right)\delta_{kl}, \qquad \sigma = \sigma_{kk}, \tag{8.9}$$

where E, f and $\overset{\circ}{\varepsilon}$ are determined by formulae (8.6), and the Poisson's ratio of the composite, like that for ablative matrices, is supposed to be constant and independent of temperature; the ratio is calculated by the mixture rule

$$\nu = \nu_m(1-\varphi_f) + \nu_f\varphi_f. \tag{8.10}$$

Thus, PRCs are isotropic media and have two elastic constants E and ν and one component $\overset{\circ}{\varepsilon}$ and f for each of the tensors of heat deformation $\overset{\circ}{\varepsilon}_{kl}$ and phase interaction f_{kl}, respectively.

8.3. Strength

8.3.1. STRENGTH IN TENSION

Maximum stresses in the matrix in tension are located in domains V'' (see Figure 8.2) and equal to

$$\sigma''_m = \sigma\frac{E_1}{E_m}\left(1 - S'_3(1 - E_m/E_f)\right). \tag{8.11}$$

Failure of the composite occurs when σ'' reaches the ultimate strength of the matrix $\sigma_{mT} = \sigma^0_T a_1$ (see Chapter 4). The value of σ, for which there occurs a failure of the composite, is called the *ultimate strength* σ_{1T} and calculated by (8.11) where instead of E we should substitute its expression (8.6), and take $E_m = E^0_b\tilde{a}_1$:

$$\sigma_{1T} = \sigma^0_T h_T, \tag{8.12}$$

$$h_T = a_1 a'_Q\left(1 - S'_3(1 - m\tilde{a}_1/\tilde{a}_f)\right);$$

the expression for a_1 is determined by formula (4.31)

Figure 8.4 shows the relationship between the relative strength $h_T = \sigma_T/\sigma^0_b$ of the composite material and the content of glass and phenolic microspheres in the material. With increasing filler content, the relative strength decreases sharply.

Figure 8.5 shows dependences of strength σ_T on temperature of heating θ (for linear heating in an inert environment) for two types of PRC with glass microspheres: one based on a silicon-organic matrix (2) and one on a phenolic matrix (1).

Characteristics for silicon-organic and phenolic matrices were given in Tables 4.1, 4.2, and for glass microspheres in Table 5.1. Also

$$\varphi_f = 0.4, \quad \varphi^0_g = 0.1, \quad h_f = 0.03, \quad \sigma^0_T = 5MPa \quad \text{for PRC-2},$$

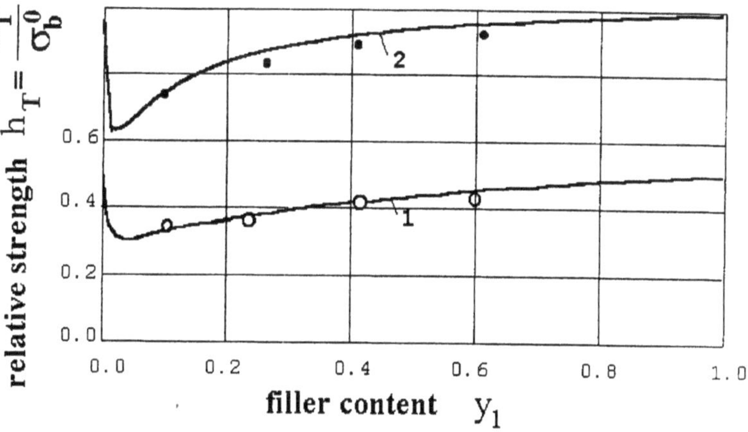

Figure 8.4. Relationships between relative strength $h_T = \sigma_T/\sigma_b^0$ of PRC and content y_1 of glass (curve 1 and experimental values - •) and phenolic (curve 2 and experimental values - ○) microspheres

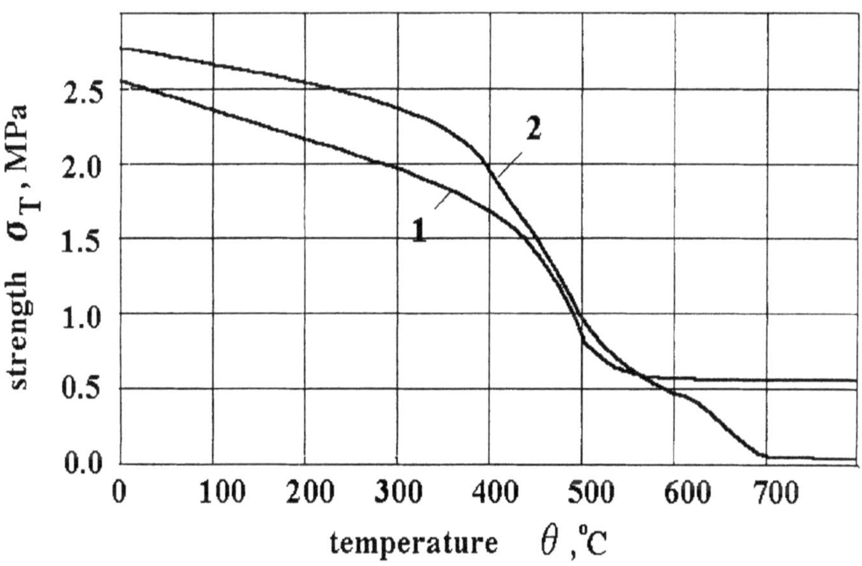

Figure 8.5. Dependences of relative strength σ_T of PRC with phenolic (1) and silicon-organic (2) matrices on temperature during heating with rate 0.1 K/s in an inert environment

$$\varphi_f = 0.2, \quad \varphi_g^0 = 0.3, \quad h_f = 0.02, \quad \sigma_T^0 = 5MPa \quad \text{for PRC-1.} \tag{8.13}$$

Composite PRC-2 has a higher strength at temperatures less than 500 °C due to lower porosity φ_g^0; however at temperatures higher than 600 °C, the strength of silicon-organic composite PRC-2 falls practically to zero due to low strength of the pyrolytical residue. At the same time the strength of phenolic composite PRC-1 remains relatively high because of the formation of strong coke residue.

8.3.2. STRENGTH IN COMPRESSION

Strength σ_{1C} of PRC in compression can be determined as for σ_{1T}:

$$\sigma_{1C} = \sigma_C^0 h_T, \tag{8.14}$$

where σ_C^0 is the strength of the polymer matrix under compression at $\theta = \theta_0$.

8.4. Thermal Properties

8.4.1. HEAT-CONDUCTIVITY

The heat conductivity of a PRC with a multilevel structure is determined like the elastic modulus E, by the formula

$$k = k_b^0 b', \qquad b' = b^{(0)} b_1 b'_Q, \tag{8.15}$$

where

$$b'_Q = 1 - S_3'^2 + \frac{b'_p}{S'_3},$$

$$b'_p = \frac{S'_3(S_3'^2 - S_2'^2)}{1 - S'_3(1 - m_h)}. \tag{8.16}$$

Functions $b^{(0)}$ and b_1 are determined by formulae (4.50), (4.51); $m_h = k_b^0/k_f^0$ is the ratio of the heat conductivity coefficients of the matrix and filler.

Figure 8.6 shows the dependence of heat conductivity k of the PRC on filler content y_1 of types 1 and 2. Since the filler is more heat-conductive, the coefficient k of the composite with filler 1 is twice as high as the coefficient of the composite with filler 2. With increasing y_1, heat conductivity decreases, as the heat conductivity of the filler is lower than that of the polymer.

Figure 8.7 presents the dependence of the heat conductivity k on the temperature θ (during linear heating) for PRC-1 and PRC-2 with characteristics (8.13). Due to a higher content of microspheres with thinner walls, PRC-2 has a considerably lower heat conductivity. At temperatures near the beginning of thermodecomposition, the coefficient k decreases because of growing porosity and then starts to increase again due to the growth of heat conductivity of the pyrolytical residue.

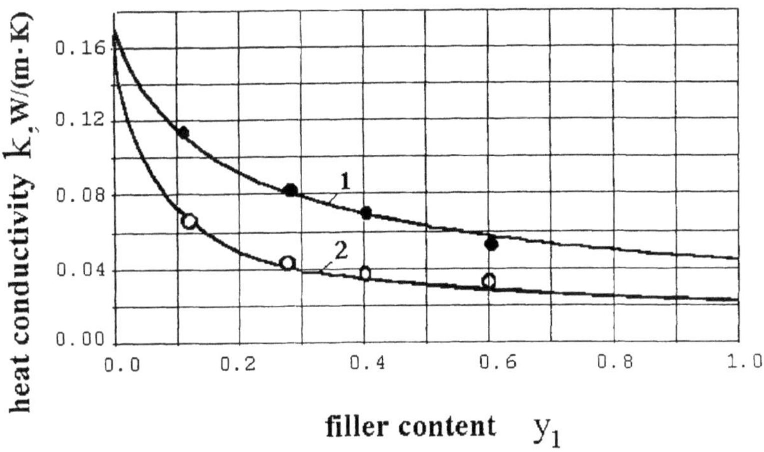

Figure 8.6. Relationships between heat conductivity k of PRC and content y_1 of fillers 1 (curve 1 and experimental values - ●) and 2 (curve 2 and experimental values - ○)

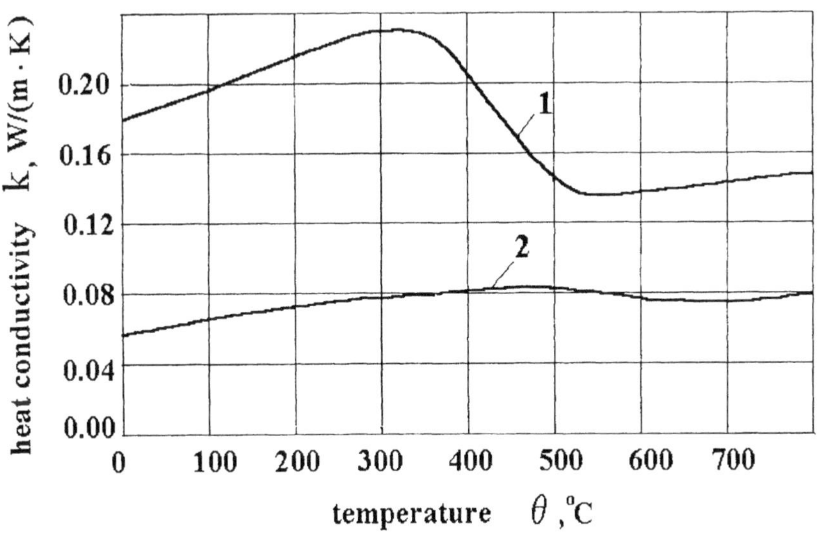

Figure 8.7. Dependence of heat conductivity k of PRC with phenolic (1) and silicon-organic (2) matrices on temperature in heating with rate 0.1 K/s in an inert environment

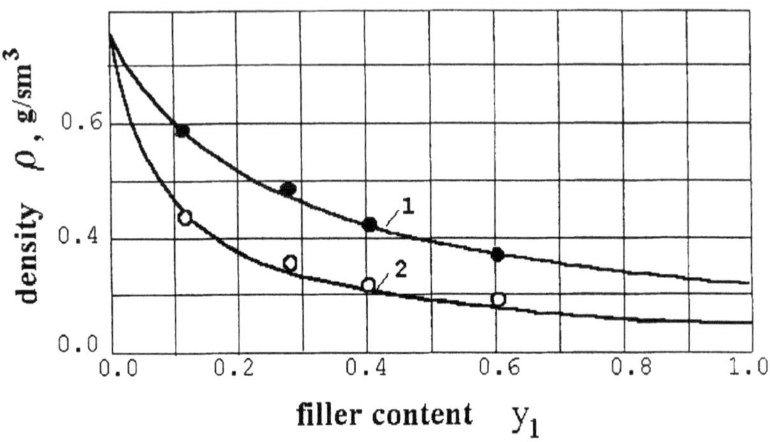

Figure 8.8. Relationships between density ρ of PRC and content y_1 of fillers 1 (curve 1 and experimental values - ●) and 2 (curve 2 and experimental values - ○)

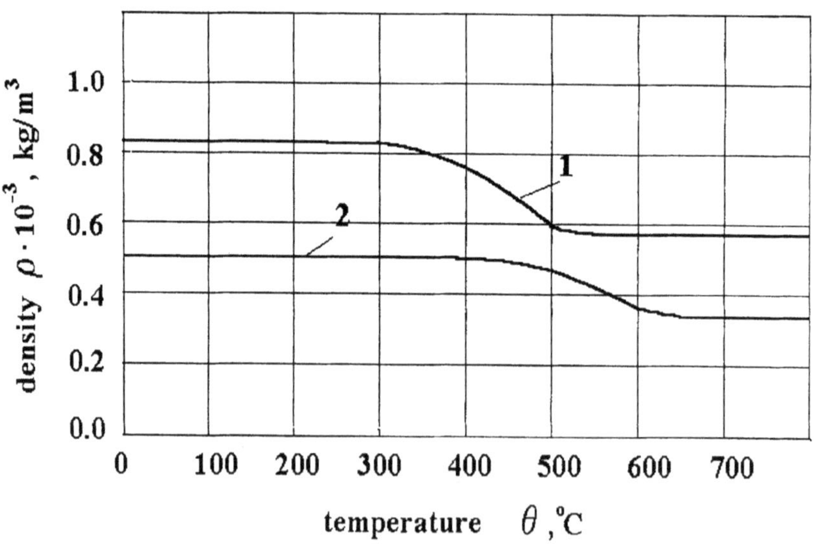

Figure 8.9. Dependence of density ρ of PRC with phenolic (1) and silicon-organic (2) matrices on temperature in heating with rate 0.1 K/s in an inert environment

8.4.2. DENSITY AND HEAT CAPACITY

The density and heat capacity of PRC are determined by the formulae

$$\rho = \rho_a \varphi_f \left(1 - (1 - h_f)^3\right) + \rho_b \varphi_b + \rho_p \varphi_p, \tag{8.17}$$

$$c = \frac{1}{\rho}\left(c_a \varphi_f \rho_a \left(1 - (1 - h_f)^3\right) + c_b \varphi_b \rho_b + c_p \varphi_p \rho_p\right),$$

where c_i are the heat capacities of phases.

Figure 8.8 shows the relationship between the density ρ of the material and filler content in the material at temperature θ_0. With increasing filler content, the material density decreases.

Figure 8.9 shows the variation of the density ρ of PRC-1 and PRC-2 in heating up to high temperatures. Composite PRC-2 has a higher thermal stability (the temperatures of loss of 5 % mass for the composites are 500 and 350 °C, respectively) due to a higher thermal stability of a silicon-organic matrix. Because of the presence of the filler, a solid residue content in the composite increases as compared with a nonreinforced matrix (in detail, see Chapter 10).

8.4.3. GAS PERMEABILITY

The gas permeability coefficient K of PRC is determined by the porosity coefficient φ_g (4.60):

$$K = K_0 \exp\left(n_k \varphi_g^{1/3}\right), \tag{8.18}$$

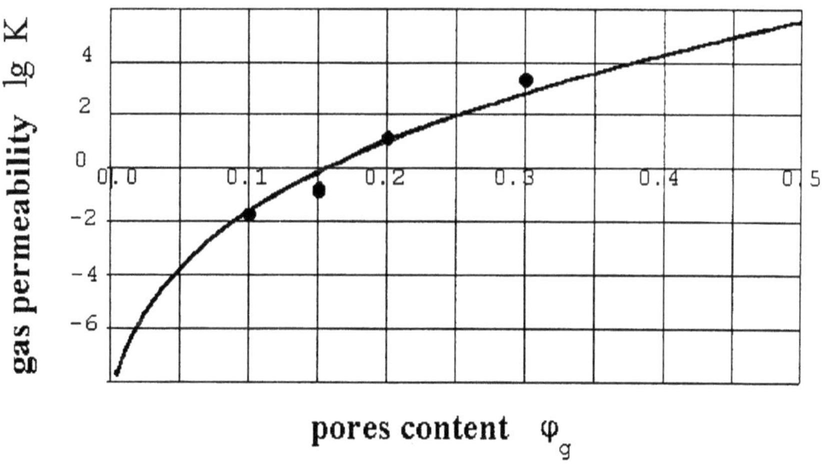

Figure 8.10. Relationship between material gas-permeability $\bar{K}$ and pores content φ_g. Points correspond to experimental values

where K_0 and n_k are universal constants (4.63), independent of the material type and the presence of fillers in the material.

Figure 8.10 shows the dependence of the dimensionless gas permeability coefficient $\bar{K} = KR\theta_0 t_0/x_0^2$ on the porosity coefficient φ_g; increasing porosity is accompanied by a sharp growth of gas permeability of the material, especially within the interval $0 \leqslant \varphi_g \leqslant 0.15$.

CHAPTER 9

PHENOMENA IN COMPOSITE MATERIALS CAUSED BY GRADIENT HEATING

In earlier chapters we assumed that the temperature θ of a composite was uniformly distributed over the whole domain V occupied by the material (or by a specimen, a structure element). This assumption was necessary in order to establish a dependence of effective properties of composites on temperature $\theta(\mathbf{x}, t)$ at a given material point $\mathbf{x}$. The expressions for thermomechanical characteristics of composites derived above are applicable for the case of *gradient heating*, i.e. when temperature $\theta(\mathbf{x}, t)$ changes in passing from one spatial point $\mathbf{x}$ to another within the domain V occupied by a composite structure at a fixed time t. Therefore all the aforementioned phenomena of changing (degradation) properties of composites under high temperatures will also take place under gradient heating. However under gradient heating there appear new specific phenomena in the behaviour of composites, which will be investigated below. As an example, the behaviour of composites is considered for textile materials, for which orthotropy axes are directed so that fabric layers are orthogonal to the Ox_1 axis.

9.1. Internal Heat-Mass-Transfer and Stresses in Ablative Composites under Gradient Heating

9.1.1. PROBLEM STATEMENT AND SOLUTION

Consider a plate made of textile ablative material with thickness h (Figure 9.1). As before, the textile composite is assumed to be orthotropic with orthotropy axes being axes of Cartesian coordinate system Ox_i. The axis Ox_1 is taken to be orthogonal to the plane of layers of the textile composite.

The plate undergoes a heat flux q_e acting along the Ox_1 axis, perpendicular to the plate surface $x_1 = h$; this heat flux causes a volumetric ablation of the composite (see Figure 9.1). Here the problem 'A' of internal heat-mass-transfer within the composite material is one-dimensional and has the form (see formulae (3.41) - (3.44))

$$\rho_b \frac{\partial \varphi_b}{\partial t} = -J, \qquad \frac{\partial \rho_g \varphi_g}{\partial t} = \frac{\partial}{\partial x_1}\left(K_{11} \frac{\partial p}{\partial x_1}\right) + J\Gamma, \tag{9.1}$$

$$\rho c \frac{\partial \theta}{\partial t} = \frac{\partial}{\partial x_1}\left(k_1 \frac{\partial \theta}{\partial x_1}\right) + c_g K_{11} \frac{\partial \theta}{\partial x_1} \frac{\partial p}{\partial x_1} - \Delta e^0 J, \qquad 0 < x_1 < h.$$

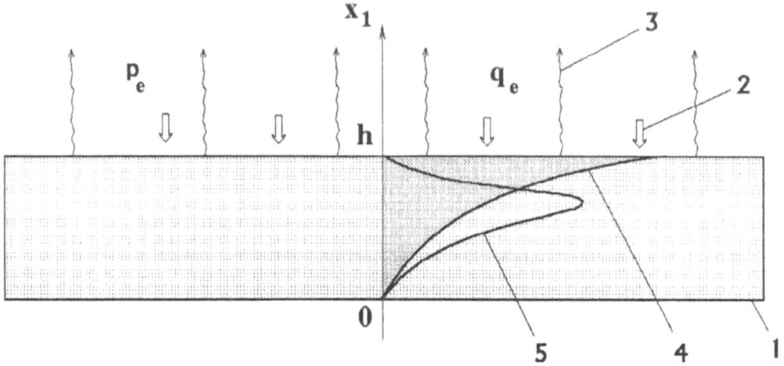

Figure 9.1. Schematic of the action of heat flux q_e on a plate of ablative textile composite: 1 – composite, 2 – heat flux, 3 – gas outflow from composite, 4 – temperature distribution, 5 – pore gas pressure distribution *versus* plate thickness

Boundary and initial conditions (3.54) – (3.55) have the form

$$x_1 = h: \quad p = p_e, \qquad k_1 \frac{\partial \theta}{\partial x_1} = \left(\left(\frac{\alpha}{c_p} \right) - K_{11} \frac{\partial p}{\partial x_1} \right) (\bar{I}_e - c\theta) - \varepsilon_w \sigma_{SB} \theta^4,$$

$$x_1 = 0: \quad \frac{\partial p}{\partial x_1} = 0, \qquad \frac{\partial \theta}{\partial x_1} = 0,$$

$$t = 0: \quad \varphi_b = \varphi_b^0, \qquad \rho_g = \frac{p_0}{R\theta_0}, \qquad \theta = \theta_0,$$

where ε_w is the coefficient of surface radiation; σ_{SB} is the Stefan-Boltzman constant; θ_e, p_e are the temperature and pressure of the overrunning gas flow; (α/c_p) is the heat transport coefficient; $p = \rho_g R_a \theta$ is the pore gas pressure; $\bar{I}_e = \bar{c}_g \theta_e$ is the flow enthalpy; K_{11} is the gas permeability coefficient, and k_1 is the heat conductivity coefficient in the x_1–direction.

The temperature θ in the composite is assumed to be lower than the level of ablation temperatures for the reinforcing fibres, therefore in this chapter the conditions $\varphi_f = \varphi_f^0$, $\varphi_a = \varphi_f$, $\varphi_l = 0$ are satisfied.

One can readily resolve the equations of mechanics (3.45) in the one-dimensional approximation and, as a result, determine macrostresses in the composite as follows:

$$\sigma_{11} = \varphi_g p - p_e, \tag{9.2}$$

the remaining stresses σ_{kl} are equal to 0.

Since only transverse stresses σ_{11} are nonzero, failure of the composite can occur only by type (C) or (D) (see Chapter 7), i.e. due to delamination or destruction of the matrix in compression.

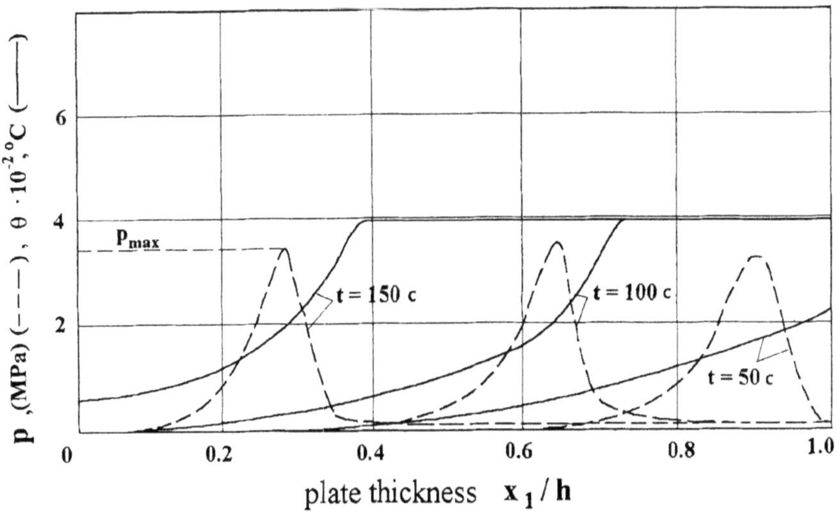

Figure 9.2 Distributions of pore pressure p and temperature θ *versus* thickness of the plate of composite for different times t of heating

The failure condition for textile composite (7.51) for a certain value $x_1 = x_1^*$ at time $t = t^*$ has the form

$$z_1(x_1^*, t^*) = 1, \tag{9.3}$$

where $z_1(x_1, t)$ is the damage parameter which for the considered problem has the form

$$z_1^2(x_1, t) = \left(\frac{|\sigma_{11} + 3\bar{N}^0 p| + \sigma_{11} + 3\bar{N}^0 p}{2\sigma_{1T}}\right)^2 + \\ + \left(\frac{|\sigma_{11} + 3\bar{N}^0 p| - \sigma_{11} - 3\bar{N}^0 p}{2\sigma_{1C}}\right)^2. \tag{9.4}$$

Expressions for ultimate strengths of the textile composite in tension σ_{1T} and compression σ_{1C} along the Ox_1 axis are determined by formulae (7.52), and $\bar{N}^0 = 3N^0/B_{m1}$, where N^0 are expressed by (4.29), and B_{m1} – by formula (7.46).

9.1.2. COMPUTED RESULTS

Solving the equation system of internal heat-mass-transfer (9.1) was conducted numerically, and values of the constants corresponding to glass/epoxy textile composite were chosen. Values of all material constants of heat-mass-transfer were taken from Tables 4.1, 4.2, 5.1, 5.2, and we assumed that $\varphi_f = 0.6$, $\zeta = 0.5$, $\Phi = 1°$, $\vartheta_{\max} = 10°$.

Pressure p_e was equal to 0.1 MPa, heat transport coefficient $(\alpha/c_p) = 0.72$ kg · m^{-2} · s^{-1}, $\varepsilon_w = 0.8$. Temperature θ_e of the flow passing over the plate ranges between 300 °C and 1200 °C.

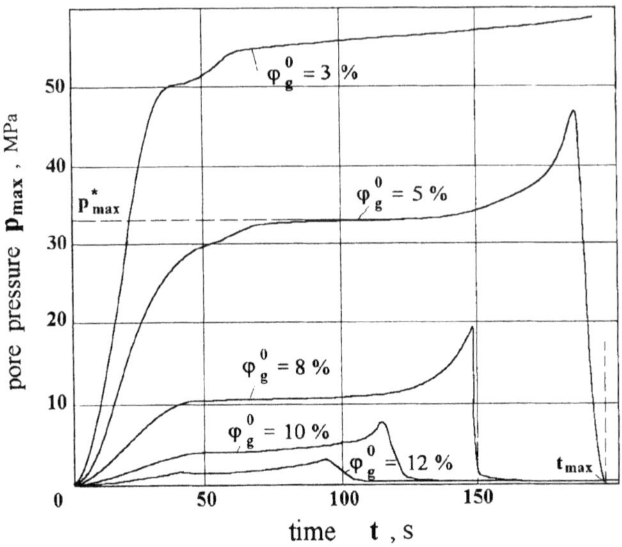

Figure 9.3 Dependence of maximum pore pressure $p_{\max}$ in the plate on time t of heating for different values of initial porosity φ_g^0 of composite material

Besides, initial porosity of the material φ_g^0 ranges between 3 and 12 %. The aim of calculations was to investigate the effect of initial porosity φ_g^0 and flow temperature θ_e on internal heat-mass-transfer processes in the plate.

Figures 9.2, 9.6 and 9.7 show distributions of temperature θ, pore pressure p, stress σ_{11} and damage parameter z_1 *versus* the plate thickness for different times. The action of high temperature leads to heat propagation into the porous plate.

A peculiarity of the behaviour of composite material under high temperatures is the presence of a specific profile of intrapore pressure p (Figure 9.2). Intensive pyrolysis of the composite material leads to growth of pore gas pressure p. Only gases near the external suface $x_1 = h$ of the plate, where porosity and gas permeability of the composite are maximum, have time to filter into the surroundings. In the depth of the plate, porosity and gas permeability are small, gases accumulate and form a local peak of pore pressure (Figure 9.2). This peak moves towards the internal cold surface $x = 0$ of the plate with time.

Maximum pore pressure $p_{\max}$ in the plate changes with time t under heating. Figure 9.3 shows dependences of maximum pore pressure $p_{\max}$ on time t for different values of initial porosity φ_g^0. As seen from this figure, the dependence $p_{\max}(t)$ has four typical stages. At the first stage there occurs a fast growth of pore pressure $p_{\max}$, and for all values of φ_g^0 this stage runs approximately for 40 - 50 s.

At the second stage the pore pressure $p_{\max}$ is stabilized at a certain level $p^*_{\max}$. Duration of this stationary stage decreases with increasing the porosity φ_g^0: e.g. for $\varphi_g^0 = 5$ % the stationary stage is approximately 100 sec, while for $\varphi_g^0 = 12$ % it is only 40 sec. At the third stage there occurs a sharp growth of pore pressure

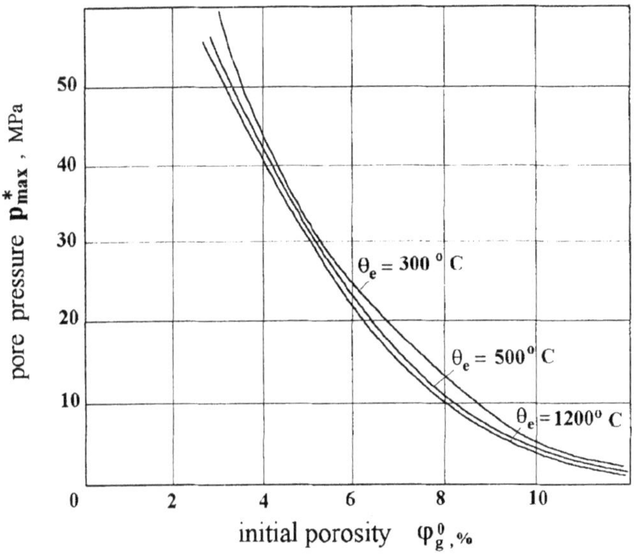

Figure 9.4 Dependence of stationary value of pore pressure p^*_{max} on initial porosity φ^0_g of the plate of composite for three values of temperature of overruning hot flow θ_e

p_{max} again; and at the fourth stage – it falls fast to the atmospheric value p_e. This fall is connected to the time when the peak of pore pressure p reaches the back cold side of the plate and then the 'pyrolytic power' of the composite material is exhausted. Maximum time t_{max} to the complete fall of pore pressure also decreases with growing φ^0_g. Maximum values of pore pressure p^*_{max} and time t_{max} depend on initial porosity of the composite material φ^0_g. The higher the value of initial porosity φ^0_g, the more intensive is filtration of gases into the surroundings and the lower are the values of p^*_{max} and t_{max} (Figures 9.4, 9.5).

Figure 9.4 presents the dependence of stationary value p_{max} on φ^0_g for different values of flow temperature θ_e. This figure shows that the temperature θ_e has a weak effect on pore pressure p^*_{max}: increase of θ_e from 300° to 1200° C leads to growth of p^*_{max} by approximately 1.4 times, for sufficiently high values of porosity $\varphi^0_g \sim 10$ %, and by 1.1 times for small values $\varphi^0_g \sim 3$ %.

Porosity φ^0_g has an essential effect on p^*_{max}: for $\varphi^0_g \sim 3$ % stationary value p^*_{max} is equal to ~ 50 MPa, but for $\varphi^0_g \sim 10$ % it is only 5 MPa.

Time t_{max} depends on φ^0_g and θ_e in another way. With increasing temperature θ_e the time t_{max} of pyrolysis completion decreases considerably: by 1.7 - 3 times. The effect of initial porosity φ^0_g is also considerable: e.g. by increasing φ^0_g from 3 to 12 % the time t_{max} decreases by approximately two times for $\theta_e = 500°$ C (Figure 9.5).

The presence of the pore pressure peak leads to the appearance of a local peak of tensile transverse stresses σ_{11} in the plate (Figure 9.6), which also moves with time towards the internal surface of the plate. These stresses are dangerous for a

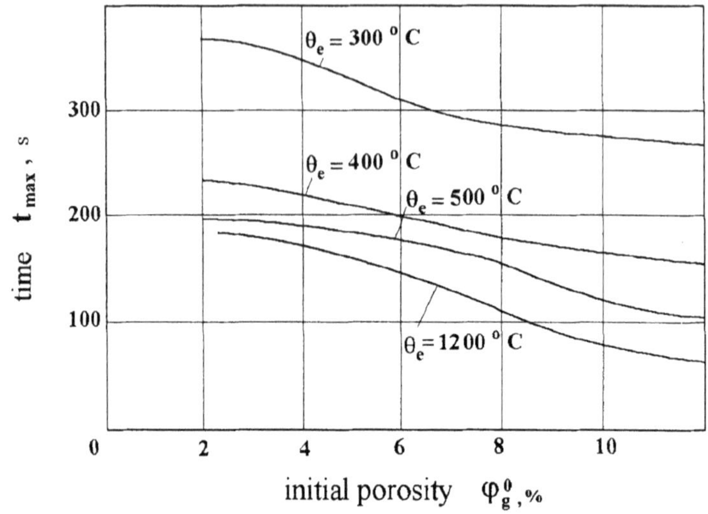

Figure 9.5 Dependence of time moment $t_{\max}$ on initial porosity φ_g^0 of the plate of composite for different values of temperature of overruning gas flow θ_e

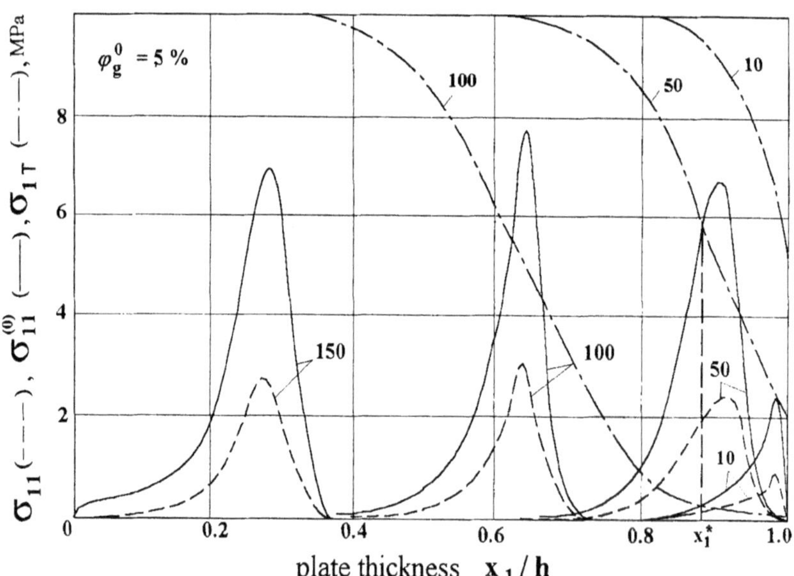

Figure 9.6 Distributions of stresses σ_{11}, $\sigma_{11}^{(0)}$ and strength σ_{1T} *versus* thickness of a composite plate under one-sided heating by a flow with temperature $\theta_e = 500\ ^\circ$C for different times t (symbols on curves are times, s)

composite material plate, as they can lead to its failure. As will be shown below, the peak character of transverse stresses σ_{11} is typical for thermodecomposition of the materials.

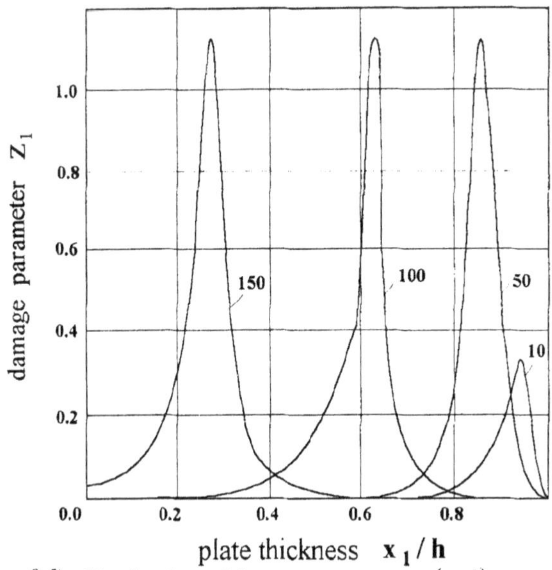

Figure 9.7 Distributions of damage parameter $z_1(x,t)$ *versus* thickness of a composite plate for different times t of one-sided heating (symbols on curves are times, s)

Actually, the strength of a composite is determined by the level of microstresses $\sigma_{kl}^{(0)}$ in the solid frame of its matrix; the microstresses are connected to macrostresses σ_{11} by relationships (4.28) of Chapter 4, and for this problem have the form

$$\sigma_{11}^{(0)} = (\sigma_{11} + \bar{N}^0 p)\frac{1}{h^0},$$

$$\sigma_{22}^{(0)} = \sigma_{33}^{(0)} = \frac{\bar{N}^0}{h^0}p. \tag{9.5}$$

The profile of microstresses $\sigma_{11}^{(0)}$ within the plate of composite material is similar to the profile of macrostresses (Figure 9.6), however the level of microstresses is considerably higher (by approximately 2.5 times) due to a more considerable effect of pore gas pressure p on $\sigma_{11}^{(0)}$.

Transverse strength of the composite σ_{1T} decreases in the heated domain of the plate (see Figure 9.6). With time, the domain of falling strength is displaced into the plate following the heat front, and at a certain time t^* microstresses $\sigma_{kl}^{(0)}$ reach their ultimate values at a certain point x_1^* (Figure 9.6) of the plate. This means that the condition $z_1(x_1^*, t^*) = 1$ is satisfied, and destruction of the composite plate occurs by the type (C), i.e. due to delamination.

Figure 9.7 shows that, as the plate is heated, the domain of delaminations of the composite is displaced into the plate depth following the heating front. This is typical for thermodecomposition of composite material under gradient heating.

Distributions of damage parameter $z_1(x_1, t)$ *versus* plate thickness for different times are shown in Figure 9.7. These distribuitons also have a peak form.

9.2. Plane Problems of Thermomechanics for Composites under High Temperatures

In paragraph 9.1 the one-dimensional problem of heating of a composite plate was considered. In actual plates of finite dimensions under gradient heating, there are bending stresses in addition to transverse stresses σ_{11}. To investigate the behaviour of a composite plate with finite dimensions, we consider a statement of a *plane problem* of composite thermomechanics at high temperatures.

Now, let us use the statements of problems 'A' and 'B' formulated in paragraph 3.6 to determine a stress-strain state in typical elements of structures made of composites. A large number of elements of structures (panels of inlets, aerofoil fragmenfs, etc.) can be modelled as plane elements.

Consider a textile composite, which can be regarded as orthotropic material, and orient the Ox_k axes of the Cartesian coordinate system so that the Ox_2 and Ox_3 axes are in the plane of the composite layers. Then for a plane-strain state in the Ox_1x_2 plane the following conditions are satisfied:

$$u_3 = 0, \quad u_1 = u_1(x_1, x_2), \quad u_2 = u_2(x_1, x_2), \tag{9.6}$$

$$\theta = \theta(x_1, x_2), \quad \rho_g = \rho_g(x_1, x_2), \quad \varphi_b = \varphi_b(x_1, x_2),$$

where u_k are the components of the displacement vector $\mathbf{u}$.

The heat-mass-transfer equations (3.41) - (3.44) in the coordinate system Ox_k take the form

$$\rho_b \frac{\partial \varphi_b}{\partial t} = -J, \tag{9.7}$$

$$\frac{\partial \varphi_g \rho_g}{\partial t} = \sum_{n=1}^{2} \frac{\partial}{\partial x_n}(RK_{nn} \frac{\partial \rho_g \theta}{\partial x_n}) + J\Gamma, \tag{9.8}$$

$$\rho c \frac{\partial \theta}{\partial t} = \sum_{n=1}^{2} \frac{\partial}{\partial x_n}(k_n \frac{\partial \theta}{\partial x_n}) + Rc_g \sum_{n=1}^{2} K_{nn} \frac{\partial \theta}{\partial x_n} \frac{\partial \rho_g \theta}{\partial x_n} - J\Delta e^0. \tag{9.9}$$

Equilibrium equations (3.45), Cauchy relations (3.46) and compatibility equations (3.60) are written as follows:

$$\sum_{l=1}^{2} \frac{\partial \sigma_{nl}}{\partial x_l} - \frac{\partial \varphi_g p}{\partial x_n} = 0, \quad \frac{\partial \sigma_{33} - \varphi_g p}{\partial x_3} = 0, \tag{9.10}$$

$$\varepsilon_{nl} = \frac{1}{2}(\frac{\partial u_n}{\partial x_l} + \frac{\partial u_l}{\partial x_n}), \tag{9.11}$$

$$\frac{\partial^2\varepsilon_{11}}{\partial x_2^2}+\frac{\partial^2\varepsilon_{22}}{\partial x_1^2}-2\frac{\partial^2\varepsilon_{12}}{\partial x_1\partial x_2}=0, \qquad n,l=1,2. \tag{9.12}$$

State equations (7.29) for orthotropic composite for the plane problem in the coordinate system Ox_k have the form

$$\varepsilon_{nn}=\overset{\circ}{\varepsilon}_n+\frac{1}{E_n}\Big(\sigma_{nn}+\widetilde{f}_n p-\nu_{nl}(\sigma_{ll}+\widetilde{f}_l p)-\nu_{nk}(\sigma_{kk}+\widetilde{f}_k p)\Big), \tag{9.13}$$

$$n=1,2, \qquad l,k=1,2,3, \qquad n\neq l\neq k;$$

$$\varepsilon_{12}=\frac{\sigma_{12}}{2G_{12}},$$

where we introduced coefficients of phase interaction $\widetilde{f}_n$ connected to f_n by the relationships

$$f_n=\frac{1}{E_n}\left(\widetilde{f}_n-\nu_{nl}\widetilde{f}_l-\nu_{nk}\widetilde{f}_k\right), \quad n,l,k=1,2,3, \quad n\neq l\neq k.$$

Elastic characteristics E_n, ν_{nl} and G_{12} depend on temperature θ and heating time t, and one can determine them from the formulae (7.31), which take the form

$$E_2=E_f^0(1-\zeta)a_{\theta 1}, \qquad E_3=E_f^0\zeta a_{\theta 1}, \tag{9.14}$$

$$E_1=E_b^0 a_{\theta 2}, \qquad G_{12}=G_b^0 a_{\theta 1},$$

$$\nu_{23}=\text{const}, \quad \nu_{13}=\text{const}, \quad \nu_{12}=\text{const}.$$

Function $a_{\theta 1}$ describes changing elastic properties of the composite in tension and shear in the fabric plane during heating; as does $a_{\theta 2}$ for tension in the transverse direction and interlayer shears. Expressions for functions $a_{\theta 1}$, $a_{\theta 2}$ are written in the form (7.32)

$$a_{\theta 1}=\frac{\varphi_f^0\widetilde{a}_2}{1+\vartheta_m^2\left((1-\varphi_f^0)\frac{\widetilde{a}_2}{m_E\widetilde{a}_1}-1\right)}, \qquad a_{\theta 2}=\widetilde{a}_1(1-\varphi_f^0)^{-1}, \tag{9.15}$$

where $\widetilde{a}_1$ and $\widetilde{a}_2$ are expressed by the formulae (7.33).

Expression for σ_{33} stress we can find from equations (9.13) and (2.6) as follows:

$$\sigma_{33}=p\widetilde{f}_3-\frac{\nu_{23}E_3}{E_2}(\sigma_{22}+\widetilde{f}_2 p)-\frac{\nu_{13}E_3}{E_1}(\sigma_{11}+\widetilde{f}_1 p)+E_3\overset{\circ}{\varepsilon}_3. \tag{9.16}$$

Relations (9.13) after substitution of the expression (9.16) are rewritten as follows:

$$\varepsilon_{nn}=\overset{\circ}{\widetilde{\varepsilon}}_{nn}+\sum_{l=1}^{2}\beta_{nl}(\sigma_{ll}+\widetilde{f}_l p), \quad n=1,2; \qquad \varepsilon_{12}=\frac{\beta_{66}}{2}\sigma_{12}, \tag{9.17}$$

where

$$\overset{\circ}{\tilde{\varepsilon}}_{11} = \overset{\circ}{\varepsilon}_1 - \frac{\nu_{13}E_3}{E_1}\overset{\circ}{\varepsilon}_3, \qquad \overset{\circ}{\tilde{\varepsilon}}_{22} = \overset{\circ}{\varepsilon}_2 - \frac{\nu_{23}E_3}{E_2}\overset{\circ}{\varepsilon}_3,$$

$$\beta_{11} = \frac{1}{E_1} - \frac{\nu_{13}^2 E_3}{E_1^2}, \qquad \beta_{22} = \frac{1}{E_2} - \frac{\nu_{23}^2 E_3}{E_2^2},$$

$$\beta_{12} = -\frac{\nu_{12}}{E_1} - \frac{\nu_{13}\nu_{23}E_3}{E_1 E_2}, \qquad \beta_{66} = \frac{1}{G_{12}}.$$

One can calculate the heat deformations $\overset{\circ}{\varepsilon}_n$ of textile ablative composite by formulae (7.39) as follows:

$$\overset{\circ}{\varepsilon}_n = (\alpha_f \varphi_f A_n + \alpha_b \varphi_b \Omega_n)(\theta - \theta_0) + \alpha_p \Omega_n \int_0^t (\theta(t) - \theta(\tau))\dot{\varphi}_p d\tau - \beta_p \Omega_n \varphi_p, \quad (9.18)$$

$$n = 1, 2, 3.$$

After introducing the *stress function* F according to (3.62)

$$\sigma_{11} = \frac{\partial^2 F}{\partial x_2^2} + \varphi_g p, \quad \sigma_{22} = \frac{\partial^2 F}{\partial x_1^2} + \varphi_g p, \quad \sigma_{12} = -\frac{\partial^2 F}{\partial x_1 \partial x_2}, \quad (9.19)$$

we satisfy the equilibrium equations (9.10) identically; the compability equation (9.12) takes the form

$$\frac{\partial^2}{\partial x_1^2}(\beta_{22}\frac{\partial^2 F}{\partial x_1^2} + \beta_{12}\frac{\partial^2 F}{\partial x_2^2}) + \frac{\partial^2}{\partial x_2^2}(\beta_{11}\frac{\partial^2 F}{\partial x_2^2} + \beta_{12}\frac{\partial^2 F}{\partial x_1^2}) +$$
$$+ \frac{\partial^2}{\partial x_1 \partial x_2}(\beta_{66}\frac{\partial^2 F}{\partial x_1 \partial x_2}) + \frac{\partial^2 \overset{\circ}{\tilde{\varepsilon}}_{22}}{\partial x_1^2} + \frac{\partial^2 \overset{\circ}{\tilde{\varepsilon}}_{11}}{\partial x_2^2} + \frac{\partial^2}{\partial x_1^2}(p(\beta_{22}\tilde{f}_2 +$$
$$+ \beta_{12}\tilde{f}_1)) + \frac{\partial^2}{\partial x_2^2}(p(\beta_{11}\tilde{f}_1 + \beta_{12}\tilde{f}_2)) = 0. \quad (9.20)$$

Boundary conditions on the contour $\partial\Sigma$ of the two-dimensional domain Σ in the Ox_1x_2 plane occupied by the composite have the form

$$\sum_{l=1}^{2} (\sigma_{kl} n_l - \varphi_g p n_k)_{\partial\Sigma} = S_k,$$

$$\sum_{l=1}^{2} K_{ll} n_l \frac{\partial \varphi_g p}{\partial x_l} = 0 \quad \text{at } \partial\Sigma_v;$$

$$p = p_e \quad \text{at } \partial\Sigma_p; \quad (9.21)$$

$$\sum_{l=1}^{2} k_{ll} n_l \frac{\partial \theta}{\partial x_l} = q_e - q_{bl} - q_w - q_{Rw} \quad \text{at } \partial\Sigma_q;$$

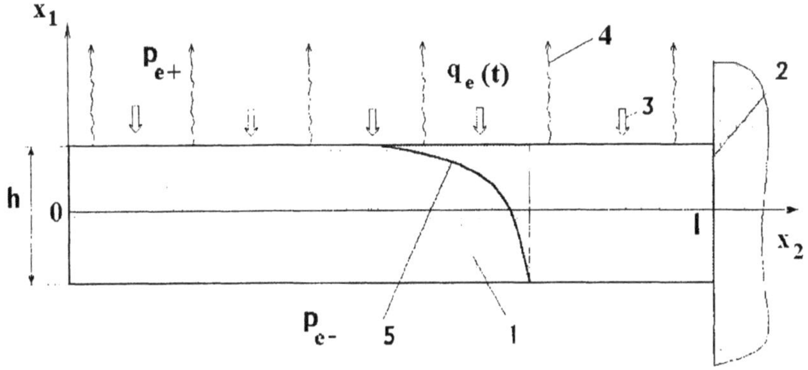

Figure 9.8. Schematic of action of convective heat flux q_e on a plate of ablative composite: 1 – composite, 2 – cantilever fastening, 3 – heat flux, 4 – gas outflow from composite, 5 – distribution of temperature $\theta(x_1)$ *versus* thickness of the plate

$$\theta = \theta_e \qquad \text{at } \partial\Sigma_\theta,$$

and

$$\partial\Sigma_v \cup \partial\Sigma_p = \partial\Sigma_q \cup \partial\Sigma_\theta = \partial\Sigma,$$

and initial conditions are

$$t = 0: \quad \rho_g = \rho_{g0}, \quad \varphi_b = \varphi_{b0}, \quad \theta = \theta_0, \tag{9.22}$$

where n_l are the components of the normal vector at contour $\partial\Sigma$.

Thus, the plane problem in the two-dimensional domain Σ consists of the set of four equations (9.7) - (9.9), (9.20) with boundary and initial conditions (9.21), (9.22), into which we have substituted the expressions (9.19). Its solution consists of functions φ_b, ρ_g, θ and F.

9.3. Heat Deformations, Stresses and Load-Bearing Capacity of a Composite Plate under Gradient Heating

9.3.1. PROBLEM STATEMENT

Within this formulation, we consider a problem of bending of a cantilever plate under the action of the pressure difference $p_{e+} - p_{e-} \geqslant 0$ uniformly distributed (where the pressures act on the surfaces $x_1 = \pm\frac{h}{2}$), and of heat flux $q_e(t)$ supplied to the surface $x_1 = \frac{h}{2}$ (Figure 9.8). The plate ends $x_2 = 0, l$ are assumed to be heat-insulated and gas-tight, so that the functions θ, φ_b, ρ_g depend only on coordinate x_1.

Integrating equlibrium equations (9.10) and taking account of the boundary conditions, we obtain

$$\int_{-h/2}^{h/2} \sigma_{12} dx_1 = (p_{e+} - p_{e-})x_2, \qquad \int_{-h/2}^{h/2} (\sigma_{22} - \varphi_g p) dx_1 = 0, \tag{9.23}$$

$$\int_{-h/2}^{h/2} (\sigma_{22} - \varphi_g p) x_1 dx_1 = (p_{e+} - p_{e-}) \frac{x_2^2}{2}, \qquad 0 \leqslant x_2 \leqslant l.$$

The conditions on the surfaces $x_1 = \pm \frac{h}{2}$ are

$$\sigma_{11} - \varphi_g p = -p_{e\pm}, \qquad \sigma_{12} = 0. \tag{9.24}$$

The solution of the compatibility equation (9.20) is sought in the form

$$F(x_1, x_2) = F_0(x_1) + F_1(x_1)x_2 + F_2(x_1)x_2^2. \tag{9.25}$$

Stresses are determined by formulae (9.19) as follows:

$$\sigma_{11} = 2F_2 + \varphi_g p; \qquad \sigma_{12} = -(F_1' + 2x_2 F_2'),$$

$$\sigma_{22} = (F_0'' + x_2 F_1'' + x_2^2 F_2'') + \varphi_g p, \tag{9.26}$$

where F_0' means $\partial F_0 / \partial x_1$.

The functions $\overset{\circ}{\tilde{\varepsilon}}_{nn}$, p, $\tilde{f}_n$ depend only on x_1; on substituting (9.25) into (9.20), we obtain

$$-x_2(\beta_{22} F_1'')'' - x_2^2(\beta_{22} F_2'')'' = (\beta_{22} F_0'')'' + 2(\beta_{12} F_2)'' + 2\beta_{12} F_2'' +$$

$$+ 2(\beta_{66} F_2')' + (\overset{\circ}{\tilde{\varepsilon}}_{22} + p(\beta_{22} \tilde{f}_2 + \beta_{12} \tilde{f}_1))''. \tag{9.27}$$

The right part of equation (9.27) is independent of x_2, and therefore the left part should be independent of x_2 too, and this is possible only when

$$\beta_{22} F_1'' = A x_1 + B, \qquad \beta_{22} F_2'' = C x_1 + D. \tag{9.28}$$

From these equations we can find the functions F_1 and F_2:

$$F_1 = A\psi_1 + B\psi_0 + G x_1 + M, \qquad F_2 = C\psi_1 + D\psi_0 + N x_1 + L, \tag{9.29}$$

where we have introduced the functions

$$\psi_n(x_1) = \int_{-h/2}^{x_1} \int_{-h/2}^{y} \frac{y^n dy}{\beta_{22}(y)}, \qquad n = 0, 1, 2;$$

$$\psi_3(x_1) = \int_{-h/2}^{x_1} \int_{-h/2}^{y} \frac{\beta_{12} y dy}{\beta_{22}}; \qquad \psi_4(x_1) = \int_{-h/2}^{x_1} \int_{-h/2}^{y} \frac{\beta_{12} dy}{\beta_{22}}. \tag{9.30}$$

Equating the right part of equation (9.27) to zero and substituting the expression (9.29) into equation (9.27), we derive the equation for $F_0(x_1)$. After integrating, we find

$$F_0'' = \frac{1}{\beta_{22}}\Big[Kx_1 + \bar{R} - p(\beta_{22} + \beta_{12}\widetilde{f}_1) - \\ - 2\beta_{12}(C\psi_1 + D\psi_0 + Nx_1 + L) - \overset{\circ}{\widetilde{\varepsilon}}_{22} - 2(\psi_3 C + D\psi_4) - \\ - 2\int_{-h/2}^{x_1} \beta_{66}(A\psi_1' + B\psi_0' + G)dx_1\Big], \tag{9.31}$$

where A, B, C, D, G, M, N, K, $\bar{R}$, L are constants.

On taking (9.29) and (9.31) into account, the stresses (9.26) take the form

$$\sigma_{11} = 2(C\psi_1 + D\psi_0 + Nx_1 + L) + \varphi_g p,$$

$$\sigma_{12} = -(A\psi_1' + B\psi_0' + G + 2x_2(C\psi_1' + D\psi_0' + N)), \tag{9.32}$$

$$\sigma_{22} = \frac{1}{\beta_{22}}(Kx_1 + \bar{R} - 2(C\psi_3 + D\psi_4) - 2\beta_{12}(C\psi_1 + D\psi_0 + Nx_1 + L) - \\ - 2\int_{-h/2}^{x_1} \beta_{66}(A\psi_1' + B\psi_0' + G)dx_1 - \overset{\circ}{\widetilde{\varepsilon}}_{22}) + \\ + (x_2(Ax_1'' + B\psi_0'') + x_2^2(C\psi_1'' + D\psi_0'')) - p(\frac{\beta_{12}}{\beta_{22}}\widetilde{f}_1 + \widetilde{f}_2 - \varphi_g).$$

On substituting expressions (9.32) into conditions (9.23) and (9.24), we obtain a set of equations to determine the constants A, B, C, D, G, M, N, K, $\bar{R}$, L as follows:

$$C\psi_1(\frac{h}{2}) + D\psi_0(\frac{h}{2}) = -\frac{p_{e+} - p_{e-}}{2}, \quad C\psi_1'(\frac{h}{2}) + D\psi_0'(\frac{h}{2}) = 0,$$

$$A\psi_1(\frac{h}{2}) + D\psi_0(\frac{h}{2}) = 0, \quad A\psi_1'(\frac{h}{2}) + B\psi_0'(\frac{h}{2}) = 0, \tag{9.33}$$

$$K\psi_1'(\frac{h}{2}) + \bar{R}\psi_0'(\frac{h}{2}) = \xi_1, \quad K\psi_2'(\frac{h}{2}) + \bar{R}\psi_1'(\frac{h}{2}) = \xi_2,$$

$$L = -\frac{p_{e-}}{2}, \quad G = 0, \quad N = 0,$$

where

$$\xi_n = \xi_{n0} + \xi_{n1}C + \xi_{n2}D + \xi_{n3}L + \xi_{n4}A + \xi_{n5}B, \quad n = 1, 2;$$

$$\xi_{n0} = \int_{-h/2}^{h/2} \frac{1}{\beta_{22}}\left[\overset{\circ}{\widetilde{\varepsilon}}_{22} + p(\beta_{22} + \beta_{12})\right] x_1^{n-1} dx_1;$$

$$\xi_{n1} = 2\int_{-h/2}^{h/2} \frac{1}{\beta_{22}}(\beta_{12}\psi_1 + \psi_3)x_1^{n-1} dx_1;$$

$$\xi_{n2} = 2\int_{-h/2}^{h/2} \frac{1}{\beta_{22}}(\beta_{12}\psi_0 + \psi_4)x_1^{n-1}dx_1;$$

$$\xi_{n3} = 2\int_{-h/2}^{h/2} \frac{\beta_{12}}{\beta_{22}}x_1^{n-1}dx_1; \qquad \xi_{n4} = 2\int_{-h/2}^{h/2} \frac{\beta_{66}}{\beta_{22}}\psi_0' x_1^{n-1}dx_1; \tag{9.34}$$

$$\xi_{n5} = 2\int_{-h/2}^{h/2} \frac{\beta_{66}}{\beta_{22}}\psi_1' x_1^{n-1}dx_1.$$

Resolving the system (9.33) one can find the constants C, K, D, $\bar{R}$, A and B as follows:

$$C = -\frac{p_{e+} - p_{e-}}{2}\psi_0'(\frac{h}{2})\frac{1}{\Delta_1}, \qquad K = \left[\xi_1\psi_1'(\frac{h}{2}) - \xi_2\psi_0'(\frac{h}{2})\right]\frac{1}{\Delta_2},$$

$$D = \frac{p_{e+} - p_{e-}}{2}\psi_1'(\frac{h}{2})\frac{1}{\Delta_1}, \tag{9.35}$$

$$\bar{R} = \left[\psi_1'(\frac{h}{2})\xi_2 - \xi_1\psi_2'(\frac{h}{2})\right]\frac{1}{\Delta_2}, \qquad A = 0, \qquad B = 0,$$

$$\Delta_1 = \psi_1(\frac{h}{2})\psi_0'(\frac{h}{2}) - \psi_0(\frac{h}{2})\psi_1'(\frac{h}{2}), \quad \Delta_2 = (\psi_1'(\frac{h}{2}))^2 - \psi_0'(\frac{h}{2})\psi_2'(\frac{h}{2}).$$

Heat-mass-transfer equations (9.7) - (9.9) for this problem have the form

$$\rho_b\dot{\varphi}_b = -J,$$

$$(\rho_g\varphi_g)^{\cdot} = (K_{11}R(\rho_g\theta)')' + J\Gamma, \tag{9.36}$$

$$\rho c\dot{\theta} = (k_{11}\theta')' + c_g RK_{11}(\rho_g\theta)'\theta' - J\Delta e^0.$$

Thus, the plane problem (9.7) - (9.9), (9.20) for the plate is reduced to solving the equation system (9.36) for functions φ_b, ρ_g, θ depending on x_1 and t.

Boundary and initial conditions for the system are

$$x_1 = \frac{h}{2}: \qquad k_{11}\theta' = q_e^0 - q_{bl} - q_w - q_{Rw}, \quad \rho_g\theta = \frac{p_{e+}}{R};$$

$$x_1 = -\frac{h}{2}: \qquad \theta' = 0, \quad \rho_g\theta = \frac{p_{e-}}{R}; \tag{9.37}$$

$$t = 0: \quad \varphi_b = \varphi_b^0, \quad \rho_g = \rho_g^0, \quad \theta = \theta_0,$$

where

$$q_e^0 = (\alpha/c_p)(\bar{I}_e - c\theta_0), \qquad q_{bl} = -\gamma_b(\bar{I}_e - c\theta)RK_{11}(\varphi_g\rho_g\theta)',$$

$$q_w = (\alpha/c_p)c(\theta_w - \theta_0), \qquad q_{Rw} = \varepsilon_w\sigma_{SB}\theta_w^4.$$

and $\theta_w = \theta(h/2)$ is the temperature of the heated surface of the plate, $\bar{I}_e = \bar{c}_g\theta_e$ is the enthalpy, θ_e is the temperature of the hot flow passing over the composite plate (see paragraph 1.1).

9.3.2. OTHER CASES OF LOADING A PLATE

Instead of the pressure difference $(p_{e+} - p_{e-})$ one can consider a bending moment M_0 applied to the plate ends, as the plate undergoes the action of pressure p_e uniformly distributed over each of the surfaces $x_1 = \pm\frac{h}{2}$, and heat flux $q_e(t)$ at surface $x_1 = \frac{h}{2}$. The plate ends $x_2 = 0, l$ and $x_3 = \pm a$ are assumed to be heat-insulated and gas-tight. Thus, φ_b, ρ_g, θ are functions only of coordinate x_1 and time t. The mechanical boundary conditions have the form

$$\sigma_{11} - \varphi_g p = -p_e, \quad \sigma_{12} = 0, \quad x_1 = \pm\frac{h}{2},$$

$$\int_{-h/2}^{h/2} (\sigma_{22} - \varphi_g p) dx_1 = 0, \tag{9.38}$$

$$\int_{-h/2}^{h/2} \sigma_{12} dx_1 = 0, \qquad \int_{-h/2}^{h/2} (\sigma_{22} - \varphi_g p) x_1 dx_1 = M_0.$$

Stresses σ_{11}, σ_{22} and σ_{12} at internal points of the plate for this problem have the same form as in equations (9.32). To determine constants A, B, C, D, N, K, $\bar{R}$ and L, we have relationships derived from equations (9.38) which are similar to formulae (9.35):

$$C = 0, \quad D = 0, \quad N = 0, \quad G = 0, \quad L = -\frac{p_e}{2}, \quad A = 0, \quad B = 0,$$

$$K = \left(\xi_1 \psi_1'(\frac{h}{2}) - (\xi_2 + M_0)\psi_0'(\frac{h}{2})\right) \frac{1}{\Delta_2}, \tag{9.39}$$

$$\bar{R} = \left((\xi_2 + M_0)\psi_1'(\frac{h}{2}) - \xi_1 \psi_2'(\frac{h}{2})\right) \frac{1}{\Delta_2}.$$

As a result, we obtain expressions for the stresses

$$\sigma_{11} = \varphi_g p - p_e, \qquad \sigma_{12} = 0, \tag{9.40}$$

$$\sigma_{22} = \frac{1}{\beta_{22}}(K x_1 + \bar{R} - \overset{\circ}{\varepsilon}_2) - p(\frac{\beta_{12}\tilde{f}_1}{\beta_{22}} + \tilde{f}_2 - \varphi_g) + \frac{\beta_{12} p_e}{\beta_{22}}.$$

The set of heat-mass-transfer equations (9.36) with conditions (9.37) takes place for this case too.

Now consider the combined action of tensile (compressive) force $b\sigma_0$ in the Ox_2 direction, pressure p_e on lateral surfaces $x_1 = \pm\frac{h}{2}$ and heat flux q_e at surface $x_1 = \frac{h}{2}$. Heat conditions at the plate ends $x_2 = 0, l$ are the same as for the problem of bending. Boundary conditions for this case are

$$\sigma_{11} - \varphi_g p = -p_e, \quad \sigma_{12} = 0, \quad \rho_g \theta = \frac{p_e}{R};$$

$$\pm k_{11}\theta' = q_e \pm \gamma_b(\bar{I}_e - c\theta)RK_{11}(\rho_g\theta)', \quad x_1 = \pm\frac{h}{2};$$

$$\int_{-h/2}^{h/2}(\sigma_{22} - \varphi_g p)dx_1 = b\sigma_0; \tag{9.41}$$

$$\int_{-h/2}^{h/2}(\sigma_{22} - \varphi_g p)x_1 dx_1 = 0; \qquad \int_{-h/2}^{h/2}\sigma_{12}dx_1 = 0;$$

then

$$A = 0, \quad B = 0, \quad C = 0, \quad D = 0, \quad N = 0, \quad G = 0, \quad L = -\frac{p_e}{2},$$

$$K = \left((\xi_1 + b\sigma_0)\psi_1(\frac{h}{2}) - \xi_2\psi_0'(\frac{h}{2})\right)\frac{1}{\Delta_2}, \tag{9.42}$$

$$\bar{R} = \left(\xi_2\psi_1'(\frac{h}{2}) - (\xi_1 + b\sigma_0)\psi_2'(\frac{h}{2})\right)\frac{1}{\Delta_2}.$$

9.3.3. COMPUTED RESULTS

To solve the problem (9.36), (9.37), the step-by-step method was applied using the implicit difference scheme and sweeping procedures [102].

Stresses

Computations were conducted for glass/epoxy-phenolic composite when $h = 10^{-2}$ m, $l = 10^{-1}$ m. Values of material constants of the composite used in computations are taken from Tables 4.1, 4.2, 5.1 and 5.2. It was also assumed that $\varphi_f^0 = 0.6$, $\zeta = 0.5$, $\Phi = 1°$, $\vartheta_{\max} = 10°$. The program of heating the surface $x_1 = h/2$ was the same as in paragraph 9.1: temperature θ_e of the overrunning flow was constant $\theta_e = 500$ °C, and $(\alpha/c_p) = 0.72$ kg/(m^2·s). At first, we consider the case of loading (9.38), when moment M_0 is absent: $M_0 = 0$.

Heat-mass-transfer problem (9.36), (9.37) for this case does not differ from (9.1), therefore fields of temperatures and intrapore pressure p within the plate at corresponding values of θ_e are similar (see Figure 9.2). The character of distribution of transverse stresses σ_{11} within the plate is similar too (Figure 9.6). However in the finite plate, besides σ_{11}, there appear stresses σ_{22} bending the plate. Distributions σ_{22} *versus* thickness of the glass/epoxy-phenolic plate for corresponding conditions of heating are shown in Figure 9.9.

A typical profile of stresses σ_{22} has the following peculiarities. At initial time of heating up to the beginning of thermodecomposition, the undersurface layer of the plate expands (deformation $\overset{\circ}{\varepsilon}_2 > 0$) and creates compressive stresses $\sigma_{22} < 0$ in this domain, as layers of the unheated composite material under gradient heating discourage a free expansion of the plate with finite dimensions. Further heating of the composite leads to its thermodecomposition and appearance of shrinkage deformations ($\overset{\circ}{\varepsilon}_2 < 0$) near its surface $x_1 = h/2$. A change of the sign of heat

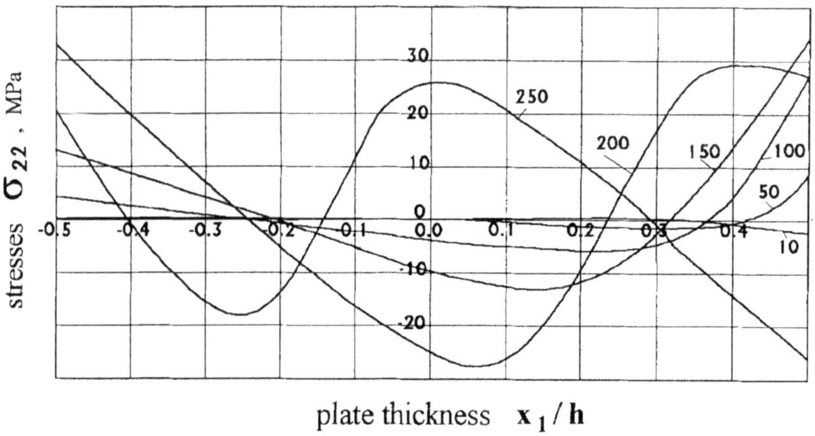

Figure 9.9. Distributions of stresses σ_{22} *versus* thickness of a glass/epoxy-phenolic plate under one-sided heating for different times t, s; $\theta_e = 500\ °C$

deformations $\overset{\circ}{\varepsilon}_2$ causes the appearance of tensile stresses σ_{22} within the thermodecomposition domain (Figure 9.9). While the heat front moves into the plate, the domain of shrinkage stresses also propagates towards the cold surface $x_1 = -h/2$. Simultaneously in the middle of the plate, where the level of temperatures is lower than the level of the onset of thermodecomposition, there is a domain of compressive stresses $\sigma_{22} < 0$, and, as follows from the equilibrium equation of the plate, at the cold surface $x_1 = -h/2$ there appears another domain of tensile stresses σ_{22} (Figure 9.9).

At sufficiently high temperatures θ_e and heating times (when $t \geqslant 220$ s), the heat deformation of the completely coked composite layer again becomes positive ($\overset{\circ}{\varepsilon}_2 > 0$) due to heat expansion of the coke, and therefore another domain of compression, where $\sigma_{22} < 0$, appears near the heated surface of the composite $x_1 = h/2$.

Depending on heating conditions and the thermostability of composite, the domains of tension-compression within the composite plate can be displaced with respect to the coordinate x_1 and time t; however, the character of longitudinal stresses σ_{22} described above is typical for ablative composites.

Deformations

Under gradient heating, an infinite plate expands or contracts uniformly along the Ox_1 axis of heating (see paragraph 9.1). But a finite plate under gradient heating buckles (Figure 9.10,*a*). Displacements u_n of the plate, which determine a change in geometry of the plate, can be found from the equation system (9.11), in which deformations ε_{nm} are known and satisfy the equation (9.12). In particular, for heating without the moment ($M_0 = 0$), we have $\varepsilon_{11} = \varepsilon_{11}(x_1)$, $\varepsilon_{22} = C_A x_1 + C_B$,

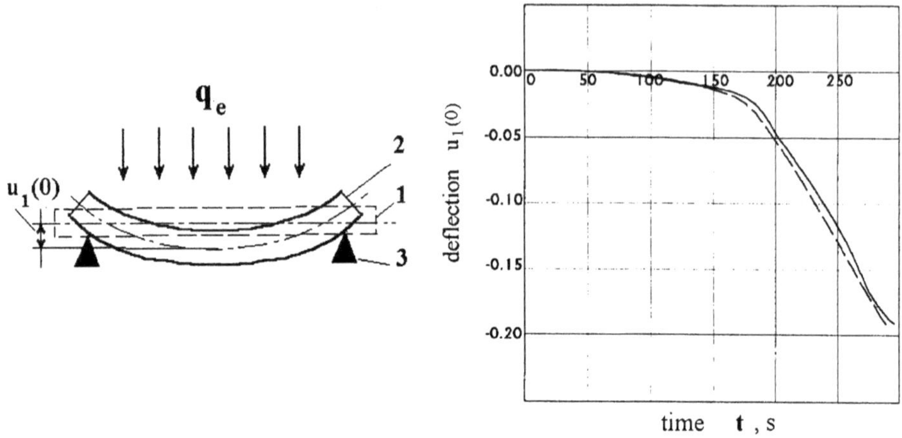

(a) (b)

Figure 9.10. a) Bending of a finite composite plate due to shrinkage under one-sided heating: 1 – the plate before heating, 2 – the plate under heating, 3 – supports; b) dependence of deflection $u_1(0)$ of the glass/epoxy-phenolic plate on time t of one-sided heating at $\theta_e = 500\ ^\circ C$; solid curve is computation by formula (9.46), dashed line – experimental data [116]

where C_A, C_B are the constants which, just as function the $\varepsilon_{11}(x_1)$, are determined from the equations (9.13) by known values of stresses σ_{nn} as follows:

$$u_1 = \int_{-h/2}^{x_1} \varepsilon_{11} dx_1 + C_A \frac{x_2^2}{2} + C_D x_2 + C_E, \tag{9.43}$$

$$u_2 = (C_A x_1 + C_B) - C_D x_1 + C_F.$$

To find the constants C_D, C_E and C_F, it is necessary to add conditions of fastening the plate

$$u_1\left(-\frac{h}{2};0\right) = u_1\left(-\frac{h}{2};l\right) = u_2\left(-\frac{h}{2};0\right) = 0 \tag{9.44}$$

to the conditions (9.38), whence one can find

$$C_E = 0, \quad C_D = -\frac{C_A l}{2}, \quad C_F = C_A \frac{lh}{4}. \tag{9.45}$$

The deflection of the plate, determined as a displacement of the symmetry center of the plate along the Ox_1 axis, is expressed as follows:

$$u_1(0) = \int_{-h/2}^{0} \varepsilon_{11} dx_1 - C_A \frac{l^2}{8}. \tag{9.46}$$

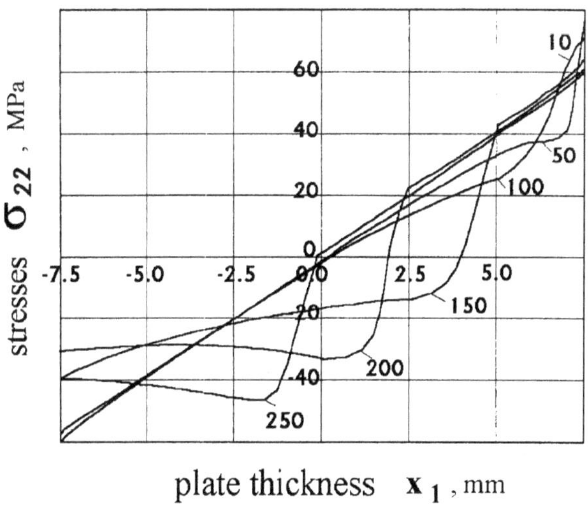

Figure 9.11. Distributions of longitudinal stresses σ_{22} *versus* thickness of the glass/phenolic plate under the combined action of bending moment and one-sided heating for different times t: symbols on curves are times, s; (heat flux $q_e^0 = 0.84$ MW/m^2, $M_0 = 20$ N·m)

Figure 9.10,*b* shows the dependence of the deflection $u_1(0)$ of the glass/epoxy-phenolic plate, calculated by formula (9.46), on duration t of one-sided heating. This figure also presents experimental values of $u_1(0)$ from [116]. The figure shows that, while the plate is heated, it bends in the direction away from the heat flux. This is caused by shrinkage of the heated surface of the composite plate at thermodecomposition (see Figure 9.10,*a*).

Load-Bearing Capacity of a Plate

Under combined gradient heating and mechanical loads (forces, moments), the concept of *load-bearing capacity* of a structure is usually used: this is a complex characteristic of the behaviour of the structure under thermomechanical loading.

Consider a composite plate under the action of one-sided heating and moment M_0 (9.38) bending the plate in the direction towards the heat flux. As a characteristic of the load-bearing capacity of the plate, the ultimate value of moment M_0^* was chosen, for which there occurs a failure of the plate for the given values of supplied heat flux q_e^0 and heating time t.

Computations were conducted for glass/phenolic composite with geometrical parameters: l = 80 mm, h =15 mm, b =10 mm and given values of the heat flux: q_e^0 = 0.861, 1.10 and 1.26 MW/m^2. Values of the material constants of the composite were taken from Tables 4.1, 4.2, 5.1 and 5.2.

Figure 9.11 shows distributions of longitudinal stresses σ_{22} *versus* thickness of the plate for different times of heating and for $q_e^0 = 0.84$ MW/m^2 and $M_0 = 20$ N·m. For this case, stress is a superposition of pure bending stress (positive for

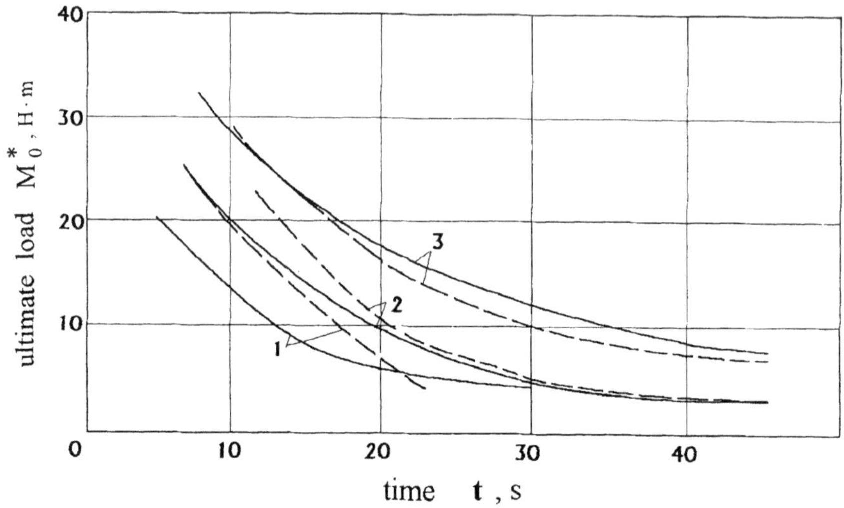

Figure 9.12. Dependence of ultimate load M_0^* of the composite plate on time t of the action of one-sided heating; 1 - $q_e = 0.84$ MW/m^2, 2 - $q_e =$ 1.1 MW/m^2, 3 - $q_e = 1.26$ MW/m^2; solid curves are computations by formula (9.47), dashed lines are experimental data [66]

$x_1 > 0$ and negative for $x_1 < 0$) and heat longitudinal stress, the profile of which has been considered above.

Failure of the plate for each value of M_0^* occurs after a certain time t of heating, when the damage parameter z_2 determined in accordance with (7.43) by the formula

$$z_2\left(\frac{h}{2}, t\right) = \frac{\sigma_{22}(h/2, t)}{\sigma_{2T}(t)} \tag{9.47}$$

reaches the value 1.

Figure 9.12 shows the dependence of the destructing moment M_0^* (load-bearing capacity) of the plate on time t of one-sided heating for different values of the heat flux q_e. This figure also presents experimental data of the load-bearing capacity [66].

As the heat flux q_e and heating time t increase, the load-bearing capacity M_0^* sharply decreases. For phenolic composites the dependence $M_0^*(t)$ comes to a horizontal asymptote at $t \longrightarrow +\infty$ (Figure 9.12), as phenolic plastics in a completely coked state usually hold relatively high mechanical properties.

LINEAR ABLATION OF COMPOSITES

For composites under gradient heating up to very high temperatures 1500 – 2000 °C, linear (surface) ablation must be considered in addition to volumetric ablation. The definitions and classification of linear ablation processes were given in paragraph 1.2. The purpose of the present chapter is to establish relationships for the calculation of the linear rate D for different types of surface ablation: combustion, sublimation, melting and thermomechanical erosion of composites.

Determination of combustion and sublimation rates of composites is based on the methods of mechanics for multicomponent media with chemical reactions. Details of these methods are developed in [83, 128, 107]. Applications of the theory of multicomponent media with chemical reactions for calculations of combustion and sublimation rates of composites [30 - 32, 71, 88, 95, 106] in high-speed gas flows are described below.

Derivation of expressions for rate of thermomechanical erosion of composites follows the logic of the methods of multiphase deformable media mechanics described earlier in the book. These results were obtained in [28, 30 - 32, 44]. Just as effective elasticity and heat conductivity moduli connect characteristics of a composite with properties of its separate phases (see Chapter 6 and 7) so, in the present chapter relationships are established between erosion rates of composites and these of fibres and matrices. and erosion rates of its fibres and matrix are established.

Determination of a linear rate of melting for composites under the action of high-speed hot gas flows is based on the analogy with thermomechanical erosion. Other methods based on analysis of flowing viscous liquids on solid surfaces are considered in [88, 95].

10.1. Main Types of Linear Ablation of Composites

Let $f_\Sigma(\mathbf{x}, t) = 0$ be the equation of the ablation surface $\Sigma(t)$ of a composite moving in three-dimensional space with Cartesian coordinates Ox_k and basis vectors $\mathbf{e}_k$. Provided that the surface is sufficiently smooth, the function $f_\Sigma(\mathbf{x}, t)$ satisfies the equation

$$\frac{\partial f_\Sigma}{\partial t} + D|\nabla_x f_\Sigma| + \mathbf{v}_\Sigma \cdot \nabla f_\Sigma = 0 \quad \mathbf{x} \in \Sigma(t). \tag{10.1}$$

Here ∇_x is the gradient with respect to coordinates x_k, $\mathbf{v}_\Sigma$ is the velocity of motion of material points of the composite on the surface Σ caused by mechanical forces or heat expansion, D is the linear rate of surface ablation, and

$$|\nabla_x f_\Sigma| = (\nabla_x f_\Sigma \cdot \nabla_x f_\Sigma)^{1/2}. \tag{10.2}$$

The rate $\mathbf{v}_\Sigma$ is determined by solving the deformation problem for a solid medium with a moving boundary. For small deformations of composite materials, $\mathbf{v}_\Sigma$ can be neglected compared to D, and equation (10.1) becomes (3.49).

The rate of surface ablation D is assumed to consist of a superposition of velocities of different independent ablation processes:

$$D = D_c + D_s + D_f + D_m, \tag{10.3}$$

where
D_c is the linear rate of chemical reaction of separate phases of the composite with the overrunning gas flow (the most important ablation of this type is combustion of carbon phases of the composite: charcoal and carbon fibre in oxygen contained in the gas flow);
D_s is the linear rate of evaporation (sublimation) of the composite;
D_m is the rate of thermomechanical erosion of the composite under the action of high-speed gas flow;
D_f is the linear rate of melting of the composite; this can occur due to melting of some phases, for example, of the crystalline l-phase of glass fibres.

We consider how to determine the rates.

10.2. Combustion Rate

10.2.1. GENERAL RELATIONS

Chemical interaction of a composite material with a hot oxygen-containing gas flow is *heterogeneous combustion* [21, 83, 108, 128], i.e. a process of the type

$$(\text{solid phase}) + (\text{gas}) \xrightarrow{\theta^\circ C} (\text{gas}).$$

The principal part of the chemical composition of solid phases of the composite taking part in the process is usually carbon C appearing in coking of polymer matrices, in thermodecomposition of organic fibres and also contained in carbon fibres. Therefore at a combustion surface Σ of composite we will consider a *multicomponent system* consisting of N chemical components ($j = 1, \ldots, N$), where $j = 1$ corresponds to a solid component (for example, carbon C), and the remaining $j = 2, \ldots, N$ correspond to components of a gas mixture of the overrunning flow and combustion products. All reactions of combustion will be assumed to occur only on the surface Σ of the composite.

In accordance with the main concept of the theory of multicomponent systems [83, 107], within each small volume ΔV of the gas mixture all N components are present (by contrast, in a multiphase system (see Chapter 2) each phase occupies a separate geometrical volume V_i), therefore instead of volumetric concentrations we should introduce mass concentrations y_i of chemical components $y_i = \rho_i/\rho$, where ρ_i is the specific mass (density) of the component, and ρ is the density of the mixture.

Our assumptions for the combustion surface Σ of the composite imply that on the one side (Σ^+) of the surface Σ there is only one component $j = 1$, and on the other side (Σ^-) there are all the remaining components $j = 2, \dots, N$, i.e.

$$\text{at } \Sigma^+: \; y_1 = 1, \quad y_j = 0 \; j = 2, \dots, N; \tag{10.4}$$

$$\text{at } \Sigma^-: \; y_1 = 0, \quad \sum_{j=2}^{N} y_j = 1.$$

We will assume below that in heterogeneous combustion of the composite, there are n chemical reactions, for each of which we can write the stoichiometric relationships

$$\sum_{j=1}^{N} \bar{\nu}_{kj} M_j = 0, \qquad k = 1, \dots, n, \tag{10.5}$$

where $\bar{\nu}_{kj}$ are the stoichiometric coefficients, M_j is the molecular mass of the j-th component; $\bar{\nu}_{kj} < 0$ if the component undergoes a reaction, and $\bar{\nu}_{kj} > 0$ if the component is formed in a reaction.

A change of concentrations y_j of components during the chemical reaction is determined by the *law of balance of component masses* [107] (this is an analog of relations (2.39))

$$\rho_w D_c(0 - y_1) + I_1 = \dot{m}_1, \tag{10.6}$$

$$\rho_w D_c(y_j - 0) + I_j = \dot{m}_j, \quad j = 2, \dots, N. \tag{10.7}$$

Here D_c is the linear rate of combustion of the composite, ρ_w is the density of the composite on the combustion surface, $\rho_w = (\rho_f \varphi_f + \rho_p \varphi_p)_\Sigma$, I_j is the diffusion flow of the j-th component from the reaction surface Σ, and $\dot{m}_j$ is the mass rate of formation of the j-th component.

Assumptions (10.4) on heterogenity of the combustion have been taken into account in equations (10.6), (10.7). The conditions give $y_1 = 1$ and $I_1 = 0$, as there is no diffusion of a solid component; then the relation (10.6) gives

$$D_c = -\frac{\dot{m}_1}{\rho_w}. \tag{10.8}$$

This formula allows us to determine a *linear rate of combustion* of the comopsite, if its mass rate $\dot{m}_1 < 0$ is known.

Mass rates of formation of components $\dot{m}_j$ are determined as follows:

$$\dot{m}_j = M_j \sum_{k=1}^{n} \bar{\nu}_{jk} U_k, \quad j = 1, \dots, N, \tag{10.9}$$

where U_k is the mass rate of the k-th chemical reaction calculated with the help of the *law of acting masses*:

$$U_k = U_k^0 \prod_{j=1}^{N} y_j^{\bar{\nu}_{kj}} \exp\left(-\frac{E_{Ak}M_j}{R\theta_w}\right), \qquad k = 1, \ldots, n. \tag{10.10}$$

Here $\prod_{j=1}^{N}$ denotes the product of concentrations y_j of components taking part in the k-th reaction; E_{Ak} is the activation energy of the chemical reaction. In addition, the *law of conservation of the mixture mass* [107] should be satisfied:

$$\sum_{j=1}^{N} \dot{m}_j = 0. \tag{10.11}$$

Under the action of a high-speed hot gas flow on composite material, diffusion fluxes I_j of the j-th component to and from the reaction surface are determined with the help of the so-called *Reynolds heat-mass-transfer analogy* [95] as follows:

$$I_j = \left(\frac{\alpha}{c_p}\right)(y_j - y_{je}), \qquad j = 2, \ldots, N. \tag{10.12}$$

Here (α/c_p) is the heat transfer coefficient (see Chapter 1), and y_{je} is the concentration of the j-th component at the external boundary layer of the overrunning flow.

By using equation (10.12), we can rewrite the relations (10.7) in the form

$$y_j = \frac{\dot{m}_j + (\alpha/c_p)y_{je}}{\rho_w D_c + (\alpha/c_p)}, \qquad j = 2, \ldots, N. \tag{10.13}$$

Equations (10.13) together with (10.9) are a system of $(2N-1)$ algebraic equations for the determination of $(2N-1)$ unknown functions $\dot{m}_1$, $\dot{m}_j$, y_j $j = 2, \ldots, N$. Solving the system, we can find the desired linear rate of combustion D_c from (10.8).

10.2.2. COMBUSTION RATE OF A COMPOSITE IN AIR FLOW

We consider now in more detail the chemical composition of the overrunning flow. Usually under very high temperatures (5000 K and higher), an overrunning air flow dissociates: besides 'heavy' components O_2, N_2 there are 'light' ones O, N. We assume that, at the external surface of the boundary layer of the overrunning flow, the gas is a five-component medium and contains O, O_2 N, N_2, NO (for the components $j = 2, 3, 4, 5, 6$, respectively). At the composite surface Σ^-, due to combustion reactions, the gas mixture contains CO, CO_2, CN and C_2N_2 ($j = 7, 8, 9, 10$) also. There are the following chemical reactions at the combustion surface Σ:

$$C + O = CO \qquad \text{reaction } k = 1;$$

$$C + O_2 = CO_2 \qquad \text{reaction } k = 2;$$
$$C + N = CN \qquad \text{reaction } k = 3; \tag{10.14}$$
$$2C + N_2 = C_2N_2 \qquad \text{reaction } k = 4.$$

The chemical composition of such a ten-component mixture (component $j = 1$ is carbon C of the solid phase) and the chemical reactions (10.14) allow us to find numerical values of the molecular masses M_j, the stoichiometric coefficients $\bar{\nu}_{kj}$ and also the concentrations y_{je} where

$$y_{je} = 0 \qquad j = 1, 7 \ldots 10. \tag{10.15}$$

The number n for such a mixture is $n = 4$. The mass rates of the chemical reactions U_k in (10.10) are determined as follows:

$$U_k = U_k^0 y_{k+1} \exp\left(-\frac{E_{Ak} M_{k+1}}{R\theta_w}\right) \qquad k = 1, \ldots, 4, \tag{10.16}$$

and the mass rate $\dot{m}_1$ of the disappearing carbon due to equations (10.9) and (10.16) takes the form

$$\dot{m}_1 = M_1 \sum_{k=1}^{4} \bar{\nu}_{1k} U_k^0 y_{k+1} \exp\left(-\frac{E_{Ak} M_{k+1}}{R\theta_w}\right). \tag{10.17}$$

Concentrations y_k contained in this equation are determined from formulae (10.13), into which we substitute the relations (10.8) and (10.16):

$$y_j = \frac{M_j \sum_{k=1}^{4} \bar{\nu}_{jk} y_{k+1} U_{k+1}^0 \exp\left(-\frac{E_{Ak} M_{k+1}}{R\theta_w}\right) + (\alpha/c_p) y_{je}}{M_1 \sum_{k=1}^{4} \bar{\nu}_{1k} y_{k+1} U_{k+1}^0 \exp\left(-\frac{E_{Ak+1} M_{k+1}}{R\theta_w}\right) + (\alpha/c_p)}; \tag{10.18}$$

$$j = 2, \ldots, 10.$$

The relations (10.18) are a set of nine non-linear algebraic equations for the determination of the nine concentrations of the components $y_j(\theta_w)$ in the form of functions of the surface temperature θ_w.

Expression for the linear rate of combustion D_c, as function of θ_w, is determined by formulae (10.8) and (10.17) after evaluation of concentrations $y_j(\theta)$ from equations (10.18).

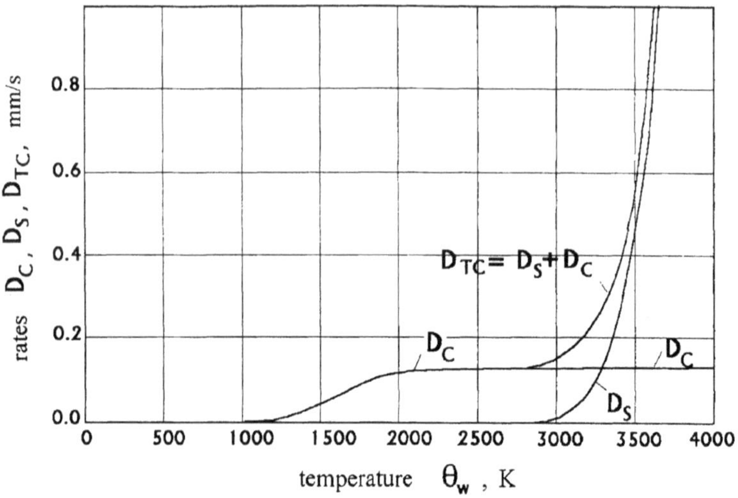

Figure 10.1. Dependences of combustion D_c and sublimation D_s rates of carbon/phenolic composite on surface temperature θ_w at pressure $p_e = 0.5$ MPa

10.2.3. COMPUTED RESULTS

Computations were conducted for the combustion rate D_c of a glass/phenolic composite. All the constants y_{je}, M_j, U_k^0 and E_{Ak} contained in equations (10.8), (10.15) - (10.18) are taken from [71, 108].

Figure 10.1 shows computed results for the linear rate of combustion D_c for a carbon/phenolic composite as a function of the surface temperature θ_w for the case of $p_e = 0.5$ MPa; the heat transfer coefficient was chosen as $(\alpha/c_p) = 2.2$ kg/(m$^2\cdot$ s). Figure 10.1 shows that the function $D_c(\theta_w)$ has the typical character for combustion of carbon materials [71]. One can observe two *combustion regimes*: *kinetic*, when the combustion rate D_c sharply increases (within the temperature interval from 1000 to 1700 K), and *diffusive*, when the combustion rate D_c is stabilized at a certain level (within the temperature interval from 1700 to 3000 K). In the diffusive regime, the combustion rate magnitude is determined, in the main, by rates of diffusion of oxidizers O_2 and O to the combustion surface Σ and of combustion products CO, CO_2 from the combustion surface Σ of the composite.

At higher temperatures θ_w of heating, the third *sublimational* combustion regime of the composite starts.

10.3. Sublimation Rate

At superhigh temperatures (higher than 3000 K) there occurs evaporation (sublimation) of solid carbon phases of the composite: carbon fibres and carbonized matrix

$$C(\text{solid}) \xrightarrow{\theta,{}^\circ C} C(\text{gas}). \tag{10.19}$$

For this case, there is a gaseous component $j = 1$ at the combustion surface Σ^-, and we can rewrite the equation of mass balance for component $j = 1$ (of the type (10.6)) as follows:

$$\rho_w D_s(y_1 - 1) + \left(\frac{\alpha}{c_p}\right)(y_1 - 0) = 0. \tag{10.20}$$

Here y_1 is the concentration of gaseous C at the surface Σ^-; at the external surface of the boundary layer, as before, the concentration of physico-chemical transformation products is zero; D_s is the *linear rate of sublimation*. Equation (10.20) gives

$$y_1 = \frac{\rho_w D_s}{\rho_w D_s + (\alpha/c_p)}. \tag{10.21}$$

For this case, we can assume that the remaining components of the gas mixture $j = 2, \dots, N$ at the surface Σ^- are inert with respect to the sublimation:

$$y_1 = 1 - \sum_{j=2}^{N} y_j.$$

The sublimation can be considered as a phase transformation. For phase transformations of the type (10.19), an expression for a rate of the phase transformation, the linear rate of sublimation D_s, is determined by the second law of thermodynamics (see Chapter 2) when there is no energy dissipation at the phase transformation surface. For this case, formula (2.107) shows that the rate D_s is determined by the eqation

$$\rho_w D_s = \left(\rho_w \frac{p_g^* \bar{S}_n - \sigma_n}{1 - \dfrac{\rho_w R\theta_w}{p_g^* \bar{S}_n M_1}}\right)^{1/2}, \tag{10.22}$$

$$\bar{S}_n = \exp\left(-\frac{E_{As} M_1}{R\theta_w}\right). \tag{10.23}$$

Here σ_n is the *partial pressure* of the gaseous carbon component in the gas mixture at the surface Σ^-,

$$\sigma_n = p_e y_1, \tag{10.24}$$

and p_g^* is the constant characterizing a pressure of 'initial state' of the generated gas phase for which the following relationship is valid:

$$(\rho_w R_a \theta_w)/(p_g^* S_n M_1) \ll 1.$$

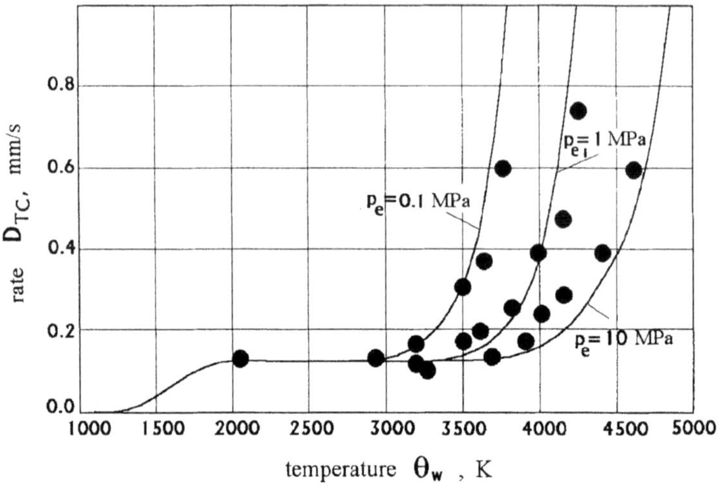

Figure 10.2. Dependence of thermomechanical ablation rate of carbon/phenolic composite on surface temperature θ_w for different values of local pressure $p_e = 0.1$, 1 and 10 MPa (lines are computations by formula (10.26), points are experimental data [71]

Equation (10.22) takes the form

$$\rho_w D_s = \left(\rho_w p_g^* \left(S_n - \left(\frac{p_e}{p_g^*}\right) y_1\right)\right)^{1/2} \tag{10.25}$$

and is called the *Langmuir-Knudsen dependence* [95].

The two equations (10.21) and (10.25) allow us to determine the sublimation rate D_s and concentration y_1 as functions of the surface temperature θ_w. Usually the linear rate of sublimation is relatively small, so that the relation

$$4p_g^* S_n(\alpha/c_p)^2/(\rho_w p_e^2) \ll 1$$

is satisfied; then the equation system (10.25), (10.21) admits an approximate solution in the form

$$\rho_w D_s = \left(\frac{\alpha}{c_p}\right)\left(\frac{p_g^*}{p_e}\right) \exp\left(-\frac{E_{As} M_1}{R\theta_w}\right). \tag{10.26}$$

The equation shows that the rate D_s is inversely proportional to pressure p_e of the surroundings. Expression (10.26) is assumed below as the sublimation rate of a composite material. Constants p_g^* and E_{As} in the expression (10.26) determined in experiments are

$$\frac{E_{As}}{RM_1} = 80 \cdot 10^3 \, \frac{1}{\mathrm{K}}, \qquad p_g^* = 10^9 \, \frac{\mathrm{N}}{\mathrm{m}^2}. \tag{10.27}$$

Figure 10.1 shows the dependence of function D_s on the surface temperature θ_w for carbon/phenolic composite as in equation (10.26). This figure also shows the dependence of the total *rate of thermochemical ablation*

$$D_{TC} = D_s + D_c \tag{10.28}$$

for carbon/phenolic composite on the temperature θ_w when $p_e = 0.5$ MPa. The figure shows that, under heating up to 3000 K, only kinetic and diffusive regimes of combustion of the composite in the air environment are realized; at higher temperatures (> 3000 K) a sublimational regime starts, for which the ablation rate D_{TC} sharply increases due to fast growth of the sublimation rate D_s.

The sublimation rate is well known to depend on local pressure p_e of a gas flow: with increasing pressure p_e the rate D_s decreases. Figure 10.2 shows results calculated by the formula (10.26) for the thermochemical ablation rate $D_{TC}(\theta_w)$ at three values of pressure $p_e = 0.1$, 1 and 10 MPa and corresponding experimental data [71]. The figure shows that the formula (10.26) describes the dependence of the rate D_s on pressure p_e not only qualitatively but also quantitatively.

10.4. Thermomechanical Erosion Rate

10.4.1. GENERAL RELATIONSHIPS

Thermomechanical erosion means the process of mechanical crumbling (erosion) of particles from a surface of a composite by a hot high-speed flow.

The principal causes of thermomechanical ablation are as follows:

- an overrunning external gas flow, a boundary layer of which moves along the ablation surface Σ of the composite with velocity $\mathbf{v}_e$ (see Figure 1.7),
- an internal flow of gaseous products of pyrolysis of the matrix which moves through pores towards the external surface Σ of the composite and 'breaks up' a thin weak coke layer of the composite which partially prevents gas filtration.

We can represent the rate of thermomechanical ablation D_m as a sum of rates D_e and D_i of these two processes:

$$D_m = D_e + D_i, \tag{10.29}$$

where D_e and D_i will be called respectively the *external and internal thermomechanical erosion rates.*

The definition of thermomechanical erosion means that rates D_e and D_i should depend on the characteristics of the resistance to the action of external and internal gas flows, i.e. on the strength of the composite. However, the composite strength depends, generally speaking, on the direction in which the mechanical forces act, and hence the linear rate of thermomechanical ablation also depends on the direction of motion of the gas flow with respect to symmetry axes of the composite properties. This distinguishes the rate D_m from the rate of thermochemical ablation D_{TC} which is independent of both the direction of flow and the orientation of symmetry axes of the composite.

Thus, the linear rate of thermomechanical ablation D_m should depend on some tensors $\mathbf{a}^{(\varkappa)}$ $\varkappa = 1, \dots, n$, characterizing the type of anisotropy of the composite material (isotropic, transversally isotropic, orthotropic etc.), and on the orientation of anisotropy axes with respect to the ablation surface Σ, i.e. on the normal vector $\vec{n}$ to the surface:

$$D_e = D_e\left(\mathbf{n}, \mathbf{a}^{(\varkappa)}, \mathbf{v}_e, \theta_w\right), \tag{10.30}$$

$$D_i = \left(\mathbf{n}, \mathbf{a}^{(\varkappa)}, \theta_w\right), \qquad \varkappa = 1, \dots, n. \tag{10.31}$$

In addition, the rate D_e depends on the surface temperature θ_w and also on both the rate and direction of motion of the hot gas flow $\mathbf{v}_e$ along the ablation surface Σ, as shown in formula (10.30).

Since the function D_m is a scalar, the arguments in expressions (10.30), (10.31) must also be scalar invariants of $\mathbf{n}$, $\mathbf{a}^{(\varkappa)}$, $\mathbf{v}_e$. We consider the following invariants:

$$a^{(\varkappa)} = \mathbf{n} \cdot \mathbf{a}^{(\varkappa)} \cdot \mathbf{n}, \tag{10.32}$$

$$p_\Sigma = \rho_e \mathbf{v}_e \cdot \mathbf{v}_e = \rho_e v_e^2, \qquad \varkappa = 1, \dots, n. \tag{10.33}$$

Here we assume that all tensors $\mathbf{a}^{(\varkappa)}$ are tensors of rank 2, and ρ_e is the gas density (scalar) of the overrunning flow at the external surface of the boundary layer.

As shown in [90], the number of invariants n cannot be more than 6 ($n \leqslant 6$).

We can present the rates D_e and D_i in the form of functions of two invariants $a^{(\varkappa)}$ and p_Σ as follows:

$$D_e = D_e\left(a^{(\varkappa)}, p_\Sigma, \theta_w\right), \tag{10.34}$$

$$D_i = D_i\left(a^{(\varkappa)}, \theta_w\right), \qquad \varkappa = 1, \dots, n. \tag{10.35}$$

Now let us assume that in equations (10.34), (10.35) the dependence on arguments $a^{(\varkappa)}$ is only linear

$$D_e = \sum_{\varkappa=1}^{n} D_{e\varkappa}\left(p_\Sigma, \theta_w\right) a^{(\varkappa)}, \tag{10.36}$$

$$D_i = \sum_{\varkappa=1}^{n} D_{i\varkappa}(\theta_w) a^{(\varkappa)}, \tag{10.37}$$

and investigate the effect of the type of anisotropy on the rates D_e and D_i of thermomechanical ablation.

10.4.2. ISOTROPIC COMPOSITES

For isotropic materials, for example, nonreinforced matrices, composites reinforced by dispersed particles or reinforced chaotically by short fibres, $n = 1$, the tensor $\mathbf{a}^{(1)}$ is the unit tensor $\mathbf{E}$, and the invariants (10.33) have the form

$$n = 1, \quad \mathbf{a}^{(1)} = \mathbf{E}, \quad a^{(1)} = \mathbf{n} \cdot \mathbf{E} \cdot \mathbf{n} = 1. \tag{10.38}$$

The expressions (10.36), (10.37) for D_e and D_i take the form

$$D_e = D_{Me} = D_{e1}(p_\Sigma, \theta_w), \tag{10.39}$$

$$D_i = D_{Mi} = D_{i1}(\theta_w). \tag{10.40}$$

As expected, the rates of thermomechanical ablation D_{Me} and D_{Mi} for isotropic media are independent of the direction of the normal $\mathbf{n}$ and are determined only by the surface temperature θ_w and the invariant p_Σ; the latter is the pressure head of the gas flow passing over the ablation surface Σ (see Chapter 1).

Formulae (10.39) and (10.40) establish only on what arguments the rates D_{Me} and D_{Mi} of matrix ablation depend. We now consider the determination of the functions $D_{Me}(p_\Sigma, \theta_w)$ and $D_{Mi}(\theta_w)$.

Rate D_{Mi} of Internal Erosion of a Matrix

The expression for the rate D_{Mi} is established on the basis of solving the system of one-dimensional equations of internal heat-mass-transfer (9.1), but for a non-reinforced matrix occupying a semi-infinite domain $x_1 \geqslant x_D(t)$, in the quasistatic approximation. Heat flux is supplied to the surface $x_1 = x_D(t)$, which moves due to thermomechanical erosion, and $\partial x_D/\partial t = D_{Mi}$.

We introduce a moving coordinate $\bar{x} = x_1 - x_D(t)$ (Figure 10.3), then due to quasistaticity of the heating, all derivatives with respect to time t in the system (9.1) are transformed as follows:

$$\frac{\partial \varphi_b}{\partial t} = -D_{Mi}\frac{\partial \varphi_b}{\partial \bar{x}}, \quad \frac{\partial \rho_g \varphi_g}{\partial t} = -D_{Mi}\frac{\partial \rho_g(1-\varphi_b)}{\partial \bar{x}}, \quad \frac{\partial \theta}{\partial t} = -D_{Mi}\frac{\partial \theta}{\partial \bar{x}}.$$

Here it is assumed for simplicity that $\Gamma \approx 1$ and $\varphi_g = 1 - \varphi_b$; all characteristics of the matrix remain constant during heating: $\rho_m \approx \rho_b$, $c = c_m = \text{const}$, $k_1 = k_m = \text{const}$, $K_{11} = K = \text{const}$, and the second and third terms at the right side of the heat conductivity equation (9.1) are neglected.

With these assumptions, we can rewrite the set (9.1) in the moving coordinates as follows:

$$\frac{\partial \varphi_b}{\partial \bar{x}} = \frac{J_m^0}{\rho_m D_{Mi}}\varphi_b \exp\Bigl(-\frac{E_{Am}}{R\theta}\Bigr), \qquad 0 \leqslant \bar{x} < +\infty;$$

$$\frac{\partial \theta}{\partial \bar{x}} + \frac{k_m}{\rho_m c_m D_{Mi}}\frac{\partial^2 \theta}{\partial \bar{x}^2} = 0;$$

$$\frac{\partial \rho_g(1-\varphi_b)}{\partial \bar{x}} + \frac{K}{D_{Mi}}\frac{\partial^2 p}{\partial \bar{x}^2} + \rho_m \frac{\partial \varphi_b}{\partial \bar{x}} = 0; \tag{10.41}$$

$$p = p_e, \quad \theta = \theta_w, \quad \bar{x} = 0, \qquad \theta = \theta_0, \quad \varphi_b = \varphi_b^0, \quad \bar{x} = +\infty;$$

$$\frac{\partial \rho_g(1-\varphi_b)}{\partial \bar{x}} = 0, \qquad \bar{x} = \bar{x}^*.$$

Here θ_0 and $\varphi_b^0 = 1$ is the temperature and concentration of the polymer phase within the unheated material, ρ_g is the density of the gas phase, J_m^0 is a pre-exponential multiplier, and E_{Am} is the activation energy of the thermodecomposition of the matrix.

A typical dependence of the pore pressure p on coordinate $\bar{x}$ for an ablative matrix has a local maximum (see Figure 10.3,*a*). The presence of the maximum is a typical peculiarity of ablative materials that was established by direct numerical solution of the equations of internal heat-mass-transfer (see paragraph 9.1). We can use the condition $\partial p/\partial \bar{x} = 0$ at the local maximum $x = \bar{x}^*$ as a boundary condition for the filtration equation of system (10.41).

Then the solution of the system (10.41) has the form

$$\frac{\partial p}{\partial \bar{x}} = \frac{D_{Mi}}{K}(1-\varphi_b)(\rho_m - \rho_g), \quad 0 \leqslant \bar{x} \leqslant \bar{x}^*;$$

$$\varphi_b = \exp\Big(-\frac{J_m^0}{\rho_m D_i} I(\bar{x})\Big), \quad 0 \leqslant \bar{x} < +\infty; \tag{10.42}$$

$$\theta - \theta_0 = \theta_w \exp\big(-\frac{\rho_m c_m D_{Mi}}{k_m}\bar{x}\big),$$

where

$$I(\bar{x}) \equiv \int_{\bar{x}}^{\infty} \exp\,\big(-\frac{E_{Am}}{R\theta(\bar{x}')}\big) d\bar{x}'. \tag{10.43}$$

Here we used the conditions $\rho_g \ll \rho_m$ and $1 - \varphi_b(\bar{x}^*) \ll 1$, that usually hold for composites.

To determine the integral $I(\bar{x})$ one can use the approximation

$$\exp\,\big(\frac{\rho_m c_m D_{Mi}\bar{x}}{k_m}\big) \longrightarrow \big(1 + \frac{\rho_m c_m D_{Mi}\bar{x}}{k_m}\big). \tag{10.44}$$

As the condition $\theta \gg \theta_0$ is usually satisfied when $\bar{x} \longrightarrow 0$, then

$$\frac{1}{\theta} \approx \frac{1}{\theta_w}\left(1 + \frac{\rho_m c_m D_{Mi}}{k_m}\bar{x}\right).$$

On substituting this value of θ into (10.43), we determine the integral $I(\bar{x})$; in particular, we obtain that

$$I(0) = \frac{k_m R\theta_w}{E_{Am}\rho_m c_m D_i}\exp\,\Big(-\frac{E_{Am}}{R\theta_w}\Big). \tag{10.45}$$

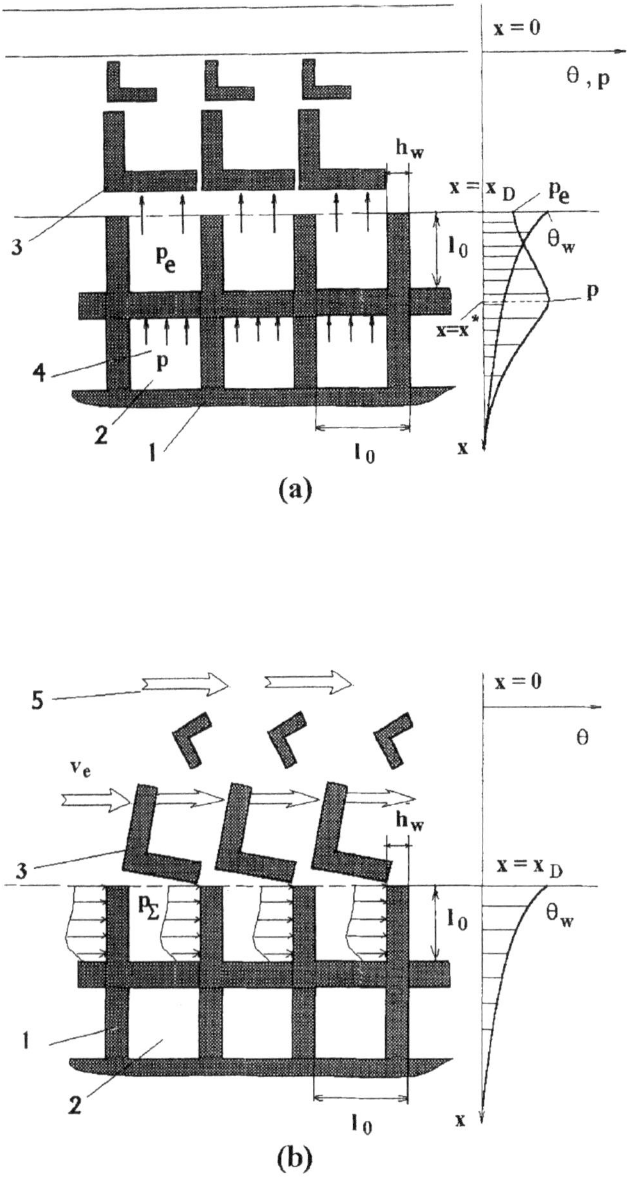

Figure 10.3. Models of thermomechanical erosion of ablative matrices:
a) model of internal erosion of the matrix for calculation of the rate D_{Mi},
b) model of external erosion of the matrix for calculation of the rate D_{Me}; (1 - solid frame of the matrix, 2 - pore, 3 - dispersed particles of the matrix in the gas flow, 4 - gas in pores, 5 - overrunning hot gas flow

Let the concentration of polymer phase of the matrix at the ablation surface $\bar{x} = 0$ be $\varphi_b = \varphi_w$. From the equation (10.42) at $\bar{x} = 0$ one can derive the expression for the rate D_{Mi}

$$D_{Mi} = -\frac{J_m^0 I(0)}{\rho_m \ln \varphi_w}. \tag{10.46}$$

In accordance with the considered model, the ablative porous matrix has a regular structure (see Figure 10.3) with a periodicity cell (PC) in the form of a hollow cube (see Chapter 4). The thickness h_w of the cube walls and the linear dimension of the pore l_0 are connected to φ_w by the relation

$$\left(1 + \frac{h_w}{l_0}\right)^{-3} = 1 - \varphi_w. \tag{10.47}$$

Pyrolysis causes the wall thickness h_w to decrease. However, the value $h_w = 0$ cannot be reached, as the walls of the PC fail at a certain value of h_w under the action of bending stresses caused by pore pressure p of pyrolysis gases (see Figure 10.3,a).

We can determine the stress state of the walls of each PC at the ablation surface Σ by modelling them as plates rigidly fixed at the edges and bending under the action of a uniformly distributed pressure difference $(p - p_e)$. When the bending stress in the wall exceeds the ultimate strength of the matrix in tension, the upper layer of the PC breaks off from its surface and the particles of the layer are carried away by the gas flow.

The considered model of erosion assumes that at each elementary destruction of the PC, the shape of a dispersed particle and the shape of the residual ablation surface remain self-similar (Figure 10.3).

The value of h_w for which there is a failure of a layer of PCs on the matrix surface is determined by the formula for bending stresses under the action of pressure difference $(p - p_e)$:

$$\left(\frac{h_w}{l_0}\right)^2 = \frac{6(p - p_e)}{\sigma_{mT}}, \tag{10.48}$$

where σ_{mT} is the ultimate stress in tension (strength) of the matrix at temperature θ_w.

We can find the expression for the pressure difference $(p - p_e)$ near the ablation surface with the help of the linearized formula (10.42):

$$p - p_e = \frac{D_{Mi}}{K}\rho_m(1 - \varphi_w)l_0, \qquad \bar{x}^* \approx l_0. \tag{10.49}$$

After substituting the formulae (10.49) into (10.48), and then (10.48) into (10.47), one can derive the equation for φ_w:

$$\varphi_w = 1 - \left(1 + \left(\frac{6D_{Mi}\rho_m l_0}{K\sigma_{mT}}(1 - \varphi_w)\right)^{1/2}\right)^{-3}. \tag{10.50}$$

As a rule, $(6D_{Mi}\rho_m l_0)/(K\sigma_{mT}) \gg 1$, and also $\varphi_w < 1$, so that one can write an approximate solution of the equation (10.50) in the form

$$-\ln\ \varphi_w = \left(\frac{K\sigma_{mT}}{6D_{Mi}\rho_m l_0}\right)^{3/5}. \tag{10.51}$$

After substitution of the formulae (10.51) and (10.45) into (10.46) we can derive the final expression for the erosion rate D_{Mi}:

$$D_{Mi} = \left(\frac{k_m J_m^0}{\rho_m c_m}\right)^{\omega_2} \left(\frac{6\rho_m l_0}{K\sigma_{mT}}\right)^{2\omega_2-1} \left(\frac{R\theta_w}{E_{Am}}\right)^{\omega_2} \exp\left(-\frac{E_{Am}\omega_2}{R\theta_w}\right), \tag{10.52}$$

$$\omega_2 = 5/7.$$

In the formula (10.52), characteristics of the matrix, in particular σ_{mT} and K, correspond to the ablation surface temperature θ_w.

Rate D_{Me} of External Erosion of a Matrix

Unlike internal erosion, when particles of material at its surface destruct under the action of intrapore pressure, erosion in the presence of external high-speed flow occurs due to local pressure p_Σ acting on bulging parts of the ablation surface (Figure 10.3,*b*).

The scheme of mechanical destruction of the matrix under the action of an external flow is similar to the scheme of internal erosion, therefore the formulae (10.42) - (10.48) also hold, except that instead of the pressure difference $p - p_e$ one should consider the pressure head p_Σ acting on the walls of the PC:

$$\left(\frac{h_w}{l_0}\right)^2 = \frac{6p_\Sigma}{\sigma_{mT}}. \tag{10.53}$$

As a result of substitution of the formulae (10.53) into (10.47), one can find the expression for the function φ_w:

$$-\ln\ \varphi_w = -\ln\Big(1 - \big(1 + \big(\frac{6p_\Sigma}{\sigma_{mT}}\big)^{1/2}\big)^{-3}\Big) \approx \left(\frac{6p_\Sigma}{\sigma_{mT}}\right)^{-\omega_1}, \quad \omega_1 = 3/2. \tag{10.54}$$

Then the equations (10.45), (10.46) and (10.54) give the final expression for D_{Me}:

$$D_{Me} = \frac{1}{\rho_m}\left(\frac{J_m^0 k_m}{c_m}\right)^{1/2} \left(\frac{6p_\Sigma}{\sigma_{mT}}\right)^{\omega_1} \left(\frac{R\theta_w}{E_{Am}}\right)^{1/2} \exp\left(-\frac{E_{Am}}{2R\theta_w}\right). \tag{10.55}$$

Erosion Rate of Particle-Reinforced Composites

As particle-reinforced composites (PRC) are isotropic composites, we can apply the previous theory to them. The rates of external and internal erosions D_e and D_i of PRC are determined by formulae (10.55) and (10.52), except that instead of the matrix characteristics ρ_m, k_m, c_m and σ_{mT}, we should use effective properties of the composite: ρ, k, c and σ_T calculated by the formulae of Chapter 8:

$$D_e = \frac{1}{\rho}\left(\frac{J_m^0 k}{c}\right)^{1/2}\left(\frac{6p_\Sigma}{\sigma_T}\right)^{\omega_1}\left(\frac{R\theta_w}{E_{Am}}\right)^{1/2}\exp\left(-\frac{E_{Am}}{2R\theta_w}\right), \tag{10.56}$$

$$D_i = \left(\frac{kJ_m^0}{\rho c}\right)^{\omega_2}\left(\frac{6\rho l_0}{k\sigma_T}\right)^{2\omega_2-1}\left(\frac{R\theta_w}{E_{Am}}\right)^{\omega_2}\exp\left(-\frac{E_{Am}\omega_2}{R\theta_w}\right).$$

10.4.3. TRANSVERSALLY ISOTROPIC COMPOSITES

For transversally isotropic media (for example, unidirectionally reinforced composites) which have only one axis of transverse isotropy characterized by unit vector $\mathbf{c} = c_k \mathbf{e}_k$, the number n is equal to 2, there are two tensors $\mathbf{a}^{(\varkappa)}$, and the invariants (10.32) have the form

$$n = 2, \quad \mathbf{a}^{(1)} = \mathbf{E} - \mathbf{c}\otimes\mathbf{c} = (\delta_{ks} - c_k c_s)\mathbf{e}_k\otimes\mathbf{e}_s,$$

$$\mathbf{a}^{(2)} = \mathbf{c}\otimes\mathbf{c} = c_k c_s \mathbf{e}_k \mathbf{e}_s, \tag{10.57}$$

$$a^{(1)} = 1 - n_k n_s c_k c_s, \qquad a^{(2)} = n_k n_s c_k c_s.$$

Expression (10.36) for the rate D_e takes the form

$$D_e = D_u(p_\Sigma, \theta_w) n_k n_s c_k c_s + D_u'(p_\Sigma, \theta_w)(1 - n_k n_s c_k c_s). \tag{10.58}$$

In particular, if the transverse isotropy axis is a coordinate axis, for example the Ox_3 axis, then $\mathbf{c} = \mathbf{e}_3$, $c_k = \delta_{k3}$, and formulae (10.56), (10.57) are rewritten in the form

$$a_{ks}^{(2)} = \delta_{k3}\delta_{s3}, \quad a_{ks}^{(1)} = \delta_{k1}\delta_{s1} + \delta_{k2}\delta_{s2},$$

$$a^{(2)} = n_3^2, \quad a^{(1)} = n_1^2 + n_2^2,$$

$$D_e = D_u(p_\Sigma, \theta_w) n_3^2 + D_u'(p_\Sigma, \theta_w)(n_1^2 + n_2^2). \tag{10.59}$$

Formulae (10.59) show that for transversally isotropic composites there exist two different functions: D_u is the rate of external erosion of the surface Σ orthogonal to the Ox_3 axis of transverse isotropy, and D_u' is the rate of external erosion of the composite surface Σ orthogonal to the axis Ox_1 or Ox_2.

Rate of Thermo-Mechanical Erosion of Fibres

Although fibres without a matrix are usually not construction material, theoretically we can consider separately the thermomechanical erosion of a hypothetic material, the properties of which are the same as the properties of monofibres. This hypothetic material, generally speaking, is transversally isotropic, as many fibres, in particular carbon fibres, are anisotropic: properties along and across the fibres can be considerably different. The rate of thermomechanical erosion of the fibres is determined as in the formula (10.59):

$$D_{ef} = D_{\perp}(p_{\Sigma}, \theta_w)(n_1^2 + n_2^2) + D_{\|}(p_{\Sigma}, \theta_w)n_3^2. \tag{10.60}$$

The concept of the hypothetic material of 'fibres' allows us to determine the rates $D_{\perp}$, $D_{\|}$ to use in the models developed for a matrix; the formulae for these rates have a form similar to (10.55):

$$D_{\perp}(p_{\Sigma}, \theta_w) = \frac{1}{\rho_f}\left(\frac{J_f^0 k_f}{c_f}\right)^{1/2}\left(\frac{6p_{\Sigma}}{\sigma_{f\perp}}\right)^{\omega_1}\left(\frac{R\theta_w}{E_{Af}}\right)^{1/2}\exp\left(-\frac{E_{Af}}{R\theta_w}\right),$$

$$D_{\|} = D_{\perp}\left(\frac{\sigma_{f\perp}}{\sigma_f}\right)^{\omega_1}. \tag{10.61}$$

Here σ_f and $\sigma_{f\perp}$ are the strengths of monofibres in the longitudinal and perpendicular directions, respectively; k_f, ρ_f, c_f are the thermal characteristics of the fibres; J_f^0 and E_{Af} are the pre-exponential multiplier and activation energy of the thermo-oxidative decomposition of the fibres.

Rates of External Erosion D_u' and D_u of Unidirectional Composites

The rate of external thermomechanical erosion D_u' in formula (10.59) is evaluated with the help of the model, shown in Figure 10.4, in the form of multilayer material consisting of periodically repeated layers of a matrix and hypothetic material of fibres. Let the thicknesses of the layers be h_m and h_f, and t_m, t_f be the times for which there occurs a complete removal of one layer of the matrix and fibre, respectively

$$t_m = h_m/D_{Me}, \qquad t_f = h_f/D_{\perp}, \tag{10.62}$$

and let us introduce the ablation rate D_u' of the whole multilayer material

$$D_u' = (h_m + h_f)/(t_m + t_f). \tag{10.63}$$

Then equations (10.62) and (10.63) give

$$D_u' = \left(\frac{1-\delta_m}{D_{\perp}} + \frac{\delta_m}{D_{Me}}\right)^{-1}, \tag{10.64}$$

where $\delta_m = h_m/(h_f + h_m)$ is the relative thickness of the matrix interlayer between the layers of fibres in the unidirectional composite. This is the formula that connects the ablation rates of matrix D_{me}, fibres $D_{\perp}$ and composite D_u'.

The rate of thermomechanical erosion D_u on the surface Σ, orthogonal to the Ox_3 axis, under the action of a flow across the direction of the fibres, can be determined with the help of the following model. We consider a multilayer material consisting of periodically repeated layers of matrix and fibres. Unlike the case

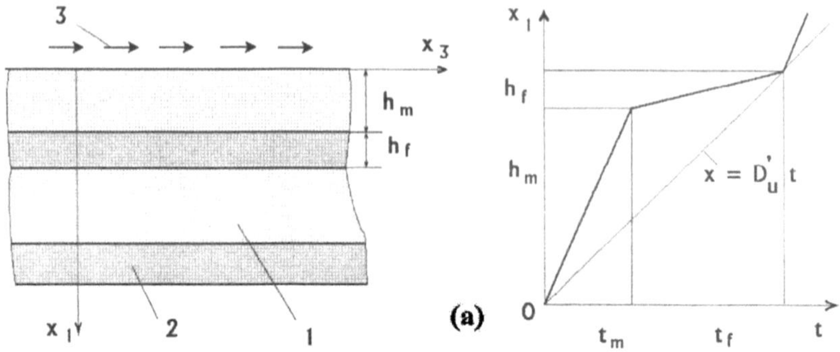

(a)

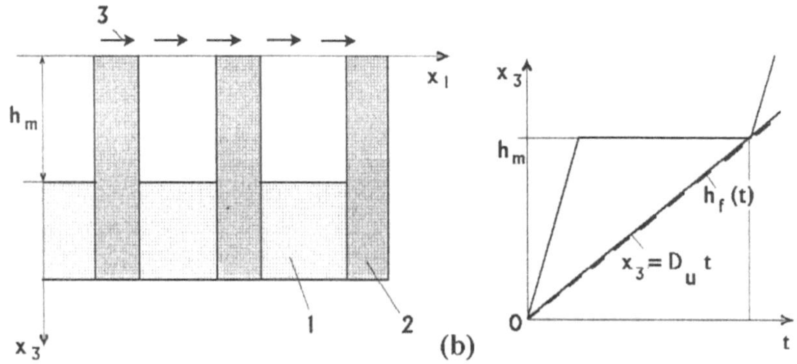

(b)

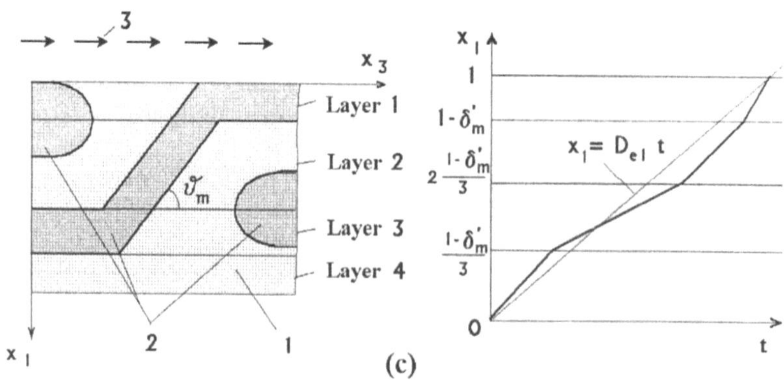

(c)

Figure 10.4. Models of external thermomechanical erosion of composite;
a) model of erosion of unidirectional composite for determination of the rate D'_u,
b) model of erosion of unidirectional composite for determination of the rate D_u,
c) model of erosion of textile composite for determination of the rate D_{e1} (1 - matrix, 2 - fibre, 3 - overrunning hot gas flow)

considered above, here the matrix and fibres layers are oriented across the direction of motion of the gas flow (Figure 10.4,*b*). Mechanical erosion of matrix and fibres in this multilayer material occurs simultaneously, but with different rates: $D_{Me} > D_{\|}$. Therefore for the same time t_f the thickness h_m of the matrix carried-off is more than h_f, where $h_m = D_{Me}t_f$; $h_f = D_{\|}t_f$. The ablation surface becomes rough: 'holes' forms in the places where the matrix was carried off. However, when the thickness h_m reaches a certain value, removal of the matrix comes to a halt as erosional action of the flow within the 'hole' eases towards its depth. The removal of the matrix is resumed when the thickness h_f of the fibre carried off reaches the value h_m (this model is shown in Figure 10.4,*a*). Hence, a total rate of removal of the multilayer material D_u will be close numerically to the value of removal rate of the fibre:

$$D_u = D_{\|}. \tag{10.65}$$

10.4.4. TEXTILE COMPOSITES

We can consider textile composites, properties of which in warp and weft directions are the same, as transversally isotropic media with a plane of isotropy being the plane of the fabric layer. Let the Ox_1 axis be normal to the plane. Then for the composite, number $n = 2$ and tensors $\mathbf{a}^{(\varkappa)}$ have the following form in the coordinate system Ox_i:

$$n = 2, \quad a_{ij}^{(2)} = \delta_{i2}\delta_{j2} + \delta_{i3}\delta_{j3}, \quad a_{ij}^{(1)} = \delta_{i1}\delta_{j1},$$

$$a^{(2)} = n_2^2 + n_3^2, \quad a^{(1)} = n_1^2, \tag{10.66}$$

and expressions (10.36) and (10.37) for the rates of external and internal erosions are

$$D_e = D_{e1}(p_\Sigma, \theta_w)n_1^2 + D_{e2}(p_\Sigma, \theta_w)(n_2^2 + n_3^2), \tag{10.67}$$

$$D_i = D_{i1}(\theta_w)n_1^2 + D_{i2}(\theta_w)(n_2^2 + n_3^2), \tag{10.68}$$

where D_{e1} is the rate of external erosion of the composite surface orthogonal to the Ox_1 axis under the action of gas flow on the surface; D_{e2} is the rate of external erosion on the composite surface Σ orthogonal to the axis Ox_2 or Ox_3 under the action of gas flow on the surface; D_u is the rate of internal erosion on the surface orthogonal to the Ox_1 axis, and D_{i2} – on the surfaces orthogonal to the Ox_2 and Ox_3 axes.

Rate of External Erosion

To determine the erosion rate D_{e1} let us consider a model of a multilayer material consisting of periodically repeated four layers (see Figure 10.4,*c*): the first and third layers are represented by unidirectional composites with perpendicular (with respect to the flow direction) fibre layers. Their removal rate is the same as D'_u; the fourth layer is an interlayer-matrix between the fabric layers, its erosion rate is D_{Me} and relative thickness is δ'_m, and the second layer is considered as unidirectional composite turned by angle ϑ_m with respect to the gas flow direction. Erosion rate of the second layer D_ϑ is evaluated with the help of formulae (10.59), where we should assume $n_2 = 0$, $n_3 = \vartheta_m$, $n_1 = (1 - \vartheta_m^2)^{1/2}$, $\vartheta_m = \sin\vartheta_{\max}$:

$$D_\vartheta = D_u \vartheta_m^2 + D_u'(1 - \vartheta_m^2). \tag{10.69}$$

The total removal rate of the four-layer material D_{e1} is determined by the method of summing up, as it was shown for the rate D_u' (10.64) of a unidirectional composite:

$$D_{e1} = \left(\frac{2(1-\delta_m')}{3D_u'} + \frac{1-\delta_m'}{3D_\vartheta} + \frac{\delta_m'}{D_{Me}} \right)^{-1}. \tag{10.70}$$

On the basis of conclusions made above for the rate D_u of a unidirectional composite and up to a certain accuracy, the erosion rate D_{e2} on the surface orthogonal to the Ox_3 axis under the action of flow along the Ox_1 axis can be considered as equal to the rate $D_{\|}$:

$$D_{e2} = D_u = D_{\|}. \tag{10.71}$$

Rate of Internal Erosion

We can determine the rates of internal erosion D_{i1} and D_{i2} as for ablative matrices, with the only difference that in place of the matrix strength σ_{mT} in (10.52) we should substitute the strengths of the textile composite in tension σ_{1T} and σ_{2T}, respectively:

$$D_{i1} = D_{Mi} \left(\frac{\sigma_{mT}}{\sigma_{1T}} \right)^{2\omega_2 - 1}, \qquad D_{i2} = D_{Mi} \left(\frac{\sigma_{mT}}{\sigma_{2T}} \right)^{2\omega_2 - 1}. \tag{10.72}$$

10.4.5. COMPUTED RESULTS

Figure 10.5 shows computed results for different components of thermomechanical erosion of textile carbon/phenolic composite: ablation rates of matrix D_M, unidirectional carbon/phenolic composite D_u and D_u' and also D_{e1}, D_{e2} and D_{i1} – as functions of the surface temperature θ_w. The figure shows that the rate of thermomechanical ablation of phenolic matrix D_M is considerably higher than the rates of ablation D_u and D_u', but rates of ablation of textile carbon/phenolic composite D_{e1} and D_{e2} are close to the ablation rates of the unidirectional composite.

The difference between the rates D_u and D_u' is approximately 30 % at $\theta_w = 3000$ K; the difference between the rates D_{e1} and D_{e2} of textile carbon/phenolic composite in different directions is higher, approximately 60 %. This is explained by noting that a matrix in a textile composite plays a greater role than one in a unidirectional composite.

Figure 10.5 shows that the internal erosion rate D_{i1} is also high: it is approximately 25 % of the external thermomechanical erosion rate D_{e1}.

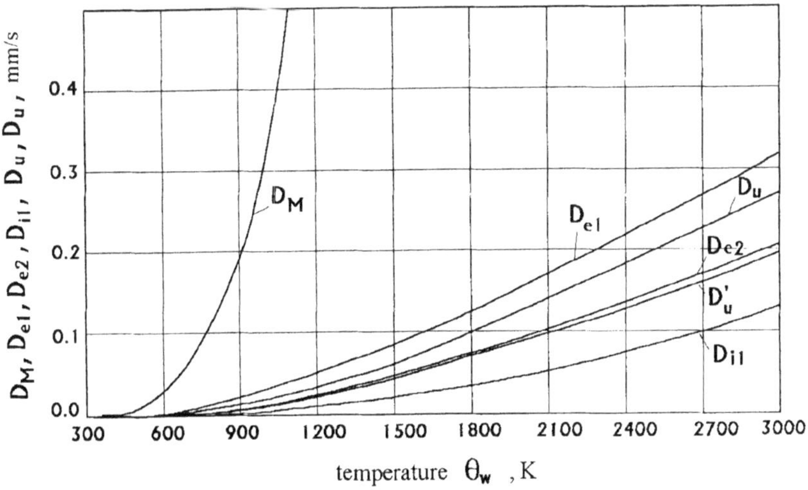

Figure 10.5. Dependence of rates of thermomechanical erosion of phenolic matrix (D_M), unidirectional composite (D_u) and (D'_u), textile carbon/phenolic composite (D_{e1}, D_{e2}, D_{i1}) on temperature θ_w at $p_\Sigma = 0.5$ MPa

10.5. Melting Rate

Some types of reinforcing fibres, mainly glass fibres based on SiO_2, or ceramic ones based on Al_2O_3 etc., under heating have a stage of melting before a stage of evaporation. The liquid phase of melted fibres expands and breaks up the weak carbonized matrix; as a result, between the solid surface of the composite and the boundary layer of the overrunning gas flow a thin film forms, this is composed of melted fibres mixed with dispersed particles of matrix coke.

Under the action of the flow, the liquid film spreads over the heated surface of the composite structure. The motion of the film is similar to motion of the overrunning gas flow: on the solid surface of the composite the film is retarded, and on the contact surface with the boundary gas layer its speed is maximum and close to the speed of the overrunning flow.

This scheme of melting of a composite allows us to apply the model of external thermomechanical erosion described in paragraph 10.4 to calculate the linear rate of melting. The linear melting rate within the scope of the model is considered as the speed of the front of a loss of mechanical strength in the undersurface layer under the action of the spread of the film of melted composite. Hence, for melting composites one should assume

$$D_f = D_e(p_\Sigma, \theta_w), \qquad D_m = D_i \tag{10.73}$$

in the formula (10.3), where D_e is determined by formulae (10.36) and (10.67) for textile composites. In the formulae, the high-temperature strength of fibres, σ_f,

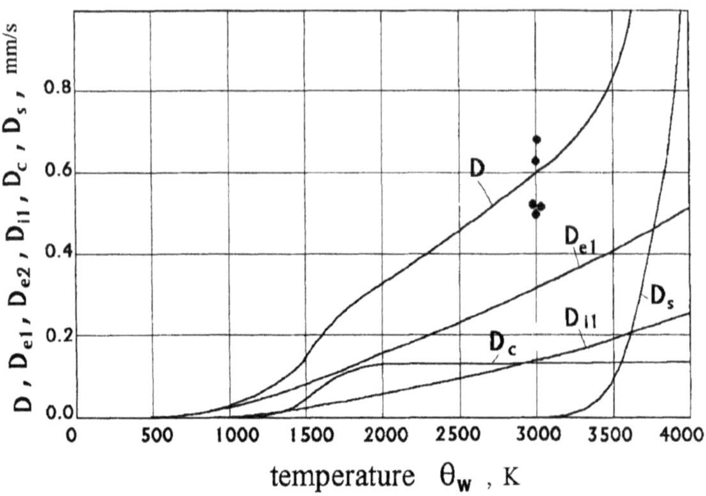

Figure 10.6. Dependences of rates of thermomechanical (D_{e1} and D_{i1}) and thermochemical (D_c and D_s) ablations and total ablation rate D for carbon/phenolic composite on surface temperature θ_w at $p_\Sigma = 0.5$ MPa. Solid curves are computations, points are experimental data for rate D [28]

is determined by the model developed in Chapter 5, and for high temperatures θ_w the strength is close to the limit of visco-plastic flow of fibre.

Other models for determining the rate of flow of a film over a surface of a melted composite are described in [95].

10.6. Comparison of Theoretical and Experimental Results

This paragraph presents results of computations by formula (10.3) for the linear rate of ablation D for different types of textile and PRC composites, and comparisons with corresponding experimental data.

Figure 10.6 shows the dependence of the total ablation rate $D = D_{e1} + D_{i1} + D_c + D_s$ of carbon/phenolic composite on the surface temperature θ_w and also dependence of its constituents on the temperature θ_w: combustion rate $D_c(\theta_w)$, sublimation rate D_s, external $D_{e1}(\theta_w)$ and internal $D_{i1}(\theta_w)$ thermomechanical erosion rates. The experimental data are taken from [28] and correspond to the pressure head $p_\Sigma = 0.5$ MPa; the surroundings is air. One can see that at temperatures $\theta_w \approx 3000$ K, experimental values of the rate D range between 0.5 and 0.67 mm/s; the theoretical magnitude $D = 0.6$ mm/s is quite acceptable.

Figure 10.6 also shows that the contribution of the thermomechanical erosion $D_m = D_{e1} + D_{i1}$ to the total rate D is considerable, although its value changes at different temperatures. At temperatures $\theta_w = 900 - 1200$ K, the contribution of thermomechanical erosion characterized by a *thermomechanical erosion coefficient* ϵ,

$$\epsilon = \frac{D_m}{D} \cdot 100 \ \%, \tag{10.74}$$

is approximately 90 %, as the combustion and sublimation rates of carbon/phenolic composite at these temperatures are small. At temperatures 1200 ... 2000 K, the contribution of the combustion rate D_c to the total value of D is a maximum and equal to 40... 60 % (thus, ϵ is 60...40 %, respectively). At temperatures 2000...3000 K, the rate of diffusive combustion D_c is stabilizes, and ϵ grows to 60...80 %, as the mechanical strength of carbon fibres falls with increasing temperature. Finally at temperatures 3000...4000 K, when sublimation starts, the contribution of thermomechanical erosion falls fast, so that ϵ is close to zero.

We will now consider how the type of matrix and reinforcing fibres affects the linear rate of ablation of textile composites, and also investigate the influence of pressure head p_Σ on the rate D.

10.6.1. EFFECT OF A MATRIX TYPE ON THE RATE OF LINEAR ABLATION OF COMPOSITES

To investigate the effect of the matrix type on the rate of linear ablation of composites, we computed the rate D from the formula (10.3) for textile carbon-plastics based on phenolic, silicon-organic and epoxy matrices. Values of all material constants for the matrices and carbon fibres used in the computations were taken from Tables 4.1, 4.2, 5.1 and 5.2; it was assumed that $\varphi_f = 0.7$, $\zeta = 0.5$, $\Phi = 1°$, $\vartheta_{\max} = 8°$. Figure 10.7 shows computed results in the form of functions $D(\theta_w)$ for the case of $p_\Sigma = 0.5$ MPa. Figure 10.7 also presents corresponding experimental data for rates of linear ablation of carbon-plastics [28].

Figure 10.7 shows that the effect of the matrix type on the ablation rate of carbon-plastics is considerable, namely, values of the rate D at temperature $\theta_w \approx 3000\ K$ are approximately 0.6, 0.8 and 1.1 mm/s for phenolic, silicon-organic and epoxy carbon-plastics, respectively. Computed values of the rate D are 0.55, 0.8 and 1 mm/s, respectively, i.e. they fall into the domain of spread in the experimental data (see Figure 10.7).

The effect of the matrix type on resistance of composites to linear ablation is determined, mainly, by two factors: heat-stability and initial porosity of the matrix. The heat-stability of the matrix ensures a high strength of the coked composite; high initial porosity maintains a low level of pore gas pressure during thermodecomposition.

Carbon/phenolic composite has the greatest high-temperature strength of coke residue for the three types of composites mentioned above, and has high values of initial porosity ($\varphi_g^0 \approx 15\ldots20$ %). Porosity in its undersurface layer during heating is $\varphi_{gw} \approx 30$ %; therefore pore gas pressure is low and, hence, also is the rate of internal thermal erosion D_{i1}.

Carbon/silicon-organic composite also has a high initial porosity $\varphi_g^0 \approx$ $\approx 13\ldots18$ %, high porosity within its undersurface layer $\varphi_{gw} \approx 27$ %, and

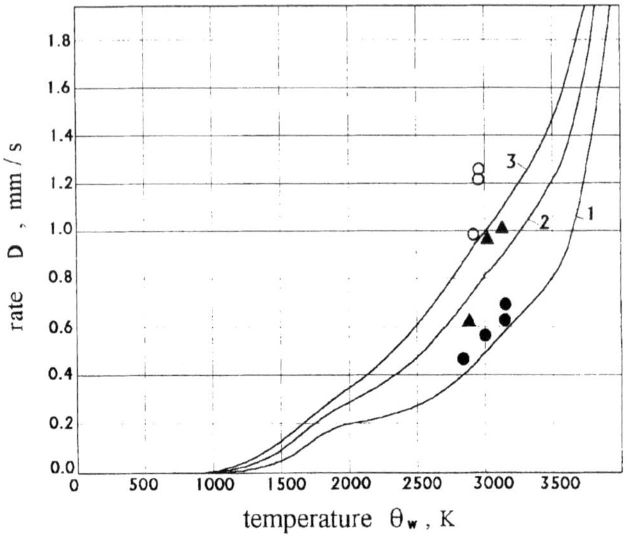

Figure 10.7. Dependence of ablation rate D of textile carbon-plastics on surface temperature θ_w and matrix type at $p_\Sigma = 0.5$ MPa in air. Matrices: 1(●) - phenolic, 2(▲) - silicon-organic, 3(○) - epoxy. Solid curves are computations, points are experimental data [28]

low pore gas pressure. However, the strength of the solid residue of thermodecomposition of its matrix is very small; therefore the rate of internal thermal erosion D_{i1} and, hence, the total rate of ablation D of the composite is higher than for carbon/phenolic composite.

For epoxy composite, the strength of solid residue of thermodecomposition of its matrix is higher than for silicon-organic, however, its initial porosity (and, hence, porosity in its undersurface layer) is lower: 5 and 12 %, respectively. This means that intrapore pressure of the gas phase in an epoxy composite is considerably higher than in a silicon-organic one, and, as a result, the rate of internal thermal erosion D_{i1} and the total ablation rate D are also higher.

We conclude that the matrix type affects mainly the rate of internal erosion D_{i1} of the composite; the remaining constituents of the ablation rate (D_{e1}, D_c and D_s) depend less on the matrix type.

10.6.2. EFFECT OF A TYPE OF FIBRES ON THE RATE OF LINEAR ABLATION OF COMPOSITES

We investigated the effect of reinforcing fibre type on the rate of linear ablation of composites for textile materials with carbon-, glass- and organo-reinforcement. The matrix for all the three types of composites was phenolic, and $p_\Sigma = 0.5$ MPa. Theoretical values of the rate D were determined by the formula (10.3), and for carbon- and organo-plastics $D_f = 0$, $D_m = D_{e1} + D_{i1}$, and for glass-plastic the model (10.72) was considered.

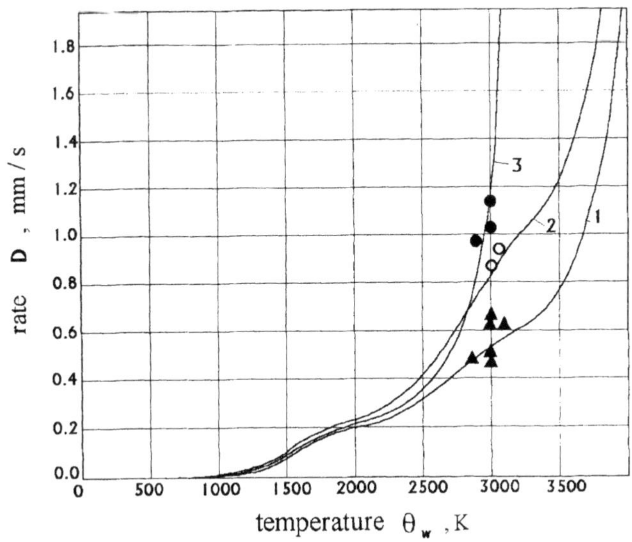

Figure 10.8. Effect of a type of fibres on ablation rate D of textile phenolic composites depending on the surface temperature θ_w at $p_\Sigma = 0.5$ MPa in air. Fibres: 1(▲) - carbon, 2(○) - organic, 3(●) - glass. Solid curves are computations, points are experimental data

Figure 10.8 shows computed results for function $D(\theta_w)$ and corresponding experimental data from [7, 28, 43].

The type of reinforcing fibre has an effect on all constituents of the linear ablation rate: D_{e1}, D_{i1}, D_c and D_s, as all the functions depend on thermal properties ρ_f, c_f, K_f, J_f^0, E_{Af} and mechanical properties of fibres. Glass fibres have higher density and smaller heat conductivity than carbon and organic fibres under high temperatures and do not burn; however, their thermal strength is lower. At temperatures θ_w up to 2000 K, these factors approximately compensate one another, so that the ablation rates D of the composites are close to each other (Figure 10.8).

Under higher temperatures, glass fibres lose practically all their mechanical strength (they melt and flow), therefore their ablation rate sharply grows. Organic fibres are carbonized at high temperatures and usually hold a certain mechanical strength, therefore their rate D is lower than for glass-reinforced plastics but higher than for carbon-plastics, the thermal strength of whose fibres is the highest. At temperature $\theta_w \approx 3000$ K, experimental values of the rate D are 0.5 - 0.65 mm/s for carbon-plastics, 0.8 - 0.9 mm/s for organo-plastics and 1.0 - 1.2 mm/s for glass-plastics. Theoretical magnitudes are 0.55, 0.83 and 1.15 mm/s, respectively, i.e. they are within the respective domains of spread in the experimental data.

10.6.3. EFFECT OF A PRESSURE HEAD ON THE RATE OF LINEAR ABLATION OF COMPOSITES

Increasing the pressure head p_Σ of flow about a composite leads to growing intensity of thermoerosion of the composite. Formula (10.3) allows us to take this important effect into account.

Figures 10.9 and 10.10 show results of comparison of computations for the ablation rate D and experimental data at three values of the pressure head $p_\Sigma =$ 0.1, 0.5 and 1.5 MPa for carbon/phenolic (Figure 10.9) and glass/phenolic (Figure 10.10) composites. One can see that increasing the pressure head p_Σ from 0.1 to 0.5 MPa leads to an increase of the rate D by almost four times for carbon-plastic (0.15 and 0.6 mm/s, respectively) and by almost three times for glass-plastic (0.3 and 1.0 mm/s, respectively) at temperature $\theta_w \approx 3000$ K.

The rate D_{e1} of external thermomechanical erosion is the main contributor to the increase of the ablation rate D with growing p_Σ for both the types of composites within the temperature range $\theta_w \approx 3000 \ldots 3500$ K. At temperatures $\theta_w \approx 4000$ K, when there is sublimation, increasing the pressure head p_Σ occurs to a considerably smaller degree.

10.6.4. PARTICLE-REINFORCED COMPOSITES

The total linear ablation rate of PRC is determined by the formula

$$D = D_e + D_i + D_c + D_s,$$

where D_e and D_i are calculated from formula (10.56). As for textile composites, linear ablation of PRC depends on a complex of effective characteristics of the composite: strength σ_T, density ρ, heat conductivity k, heat capacity etc. In order to estimate the contributions of the different characteristics to D we have conducted computations and experiments for two types of PRC: phenolic composite with carbon particles, and silicon-organic PRC with glass microspheres. Effective thermal and mechanical characteristics of both the composites were given in Chapter 8.

Figure 10.11 shows dependences of the ablation rate D on the surface temperature θ_w for both types of PRC at two values of the pressure head $p_\Sigma = 0.02$ and 0.1 MPa. Theoretical and experimental results agree well.

Figure 10.11 shows the qualitatively different characters of the ablation of PRC at small (0.02) and moderate (0.1) values of p_Σ. For the former, the ablation rate D of silicon-organic PRC is smaller than for phenolic PRC for all temperatures from 800 to 2500 °C. We can explain this fact by a considerably lower heat conductivity of silicon-organic PRC: $k = 0.06$ and 0.2 W/(m·K), respectively, at $\theta = 20$ °C.

For the latter, at p_Σ =0.1 MPa the linear rate D of silicon-organic PRC is smaller than that for phenolic PRC, but only for low temperatures $\theta_w \approx$ 500...700 °C. For temperatures $\theta_w \approx 1000$ °C and higher, the strength σ_T of silicon-organic PRC falls sharply practically to zero (we noted in Chapter 4, that silicon-organic matrices are usually weak at high temperatures), but the coke

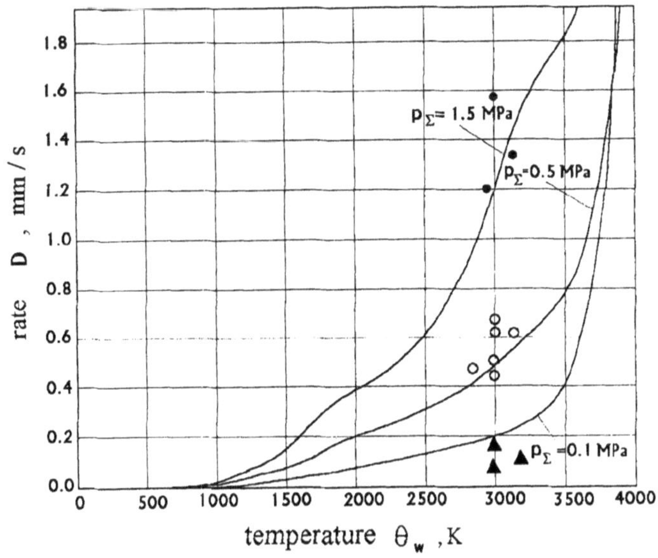

Figure 10.9. Effect of pressure head p_Σ of flow on ablation rate D of textile carbon/phenolic composite. Pressure head: (▲) - 0.1 MPa, (○) - 0.5 MPa, 3(●) - 1.5 MPa. Solid curves are computations, points are experimental data [28]

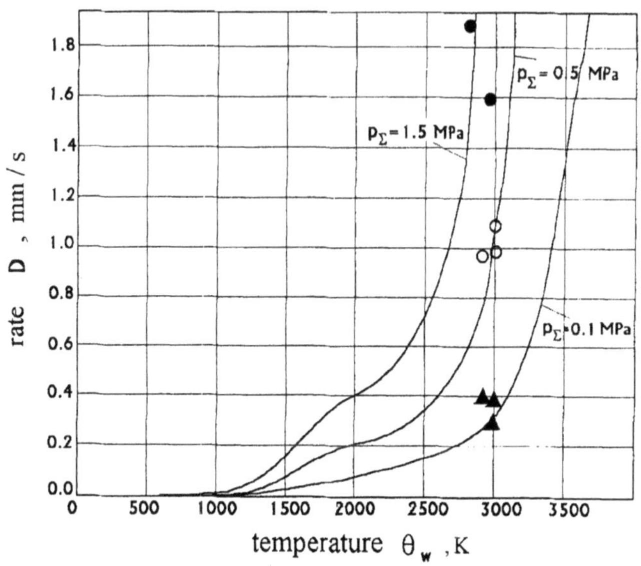

Figure 10.10. Effect of pressure head p_Σ of flow on ablation rate D of textile glass/phenolic composite. Pressure head: (▲) - 0.1 MPa, (○) - 0.5 MPa, 3(●) - 1.5 MPa. Solid curves are computations, points are experimental data

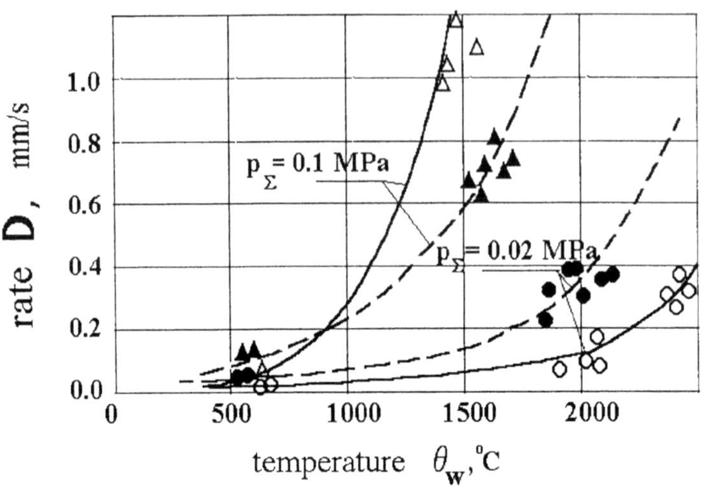

Figure 10.11. Dependence of the ablation rate D of PRCs on surface temperature θ_w at two values of $p_\Sigma = 0.02$ and 0.1 MPa. Curves are computations, points are experimental data: (o, △) and solid line – silicon-organic PRC; (•, ▲) and dashed line – phenolic PRC

residue of phenolic PRC remains strong. At elevated values of the pressure head p_Σ, the strength σ_t gives the principal contribution to the rate D; therefore phenolic composites have the smaller values of D at temperatures 1000...2000 °C.

We conclude that the linear ablation rate D depends in a complicated way on thermomechanical properties of the composite and on the heating conditions, i.e. on θ_w and p_Σ.

10.7. Heat Balance on Ablative Surface

To determine the true temperature θ_w of an ablative surface of a composite under the action of an overrunning hot gas flow, it is necessary to take account of processes of heat and mass transfer occurring on the surface.

Total heat flux q_Σ to the ablative surface of the composite, which is consumed in warming up the material, consists of the following components (see equation (3.56)):

$$q_\Sigma = q_e^0 + q_R - q_{Rw} - q_{\text{bl}} - q_w - q_{\text{sub}} + q_{cs} + q_{cg}, \tag{10.75}$$

where

$q_e^0 = (\alpha/c_p)(\bar{I}_e - I_0)$ – convective heat flux to the 'cold wall' supplied from the heated boundary layer to the composite surface Σ,

q_R – radiant heat flux supplied to the composite surface Σ, determined by formulae of paragraph 1.1,

$q_w = (\alpha/c_p)(I_w - I_0)$ – convective heat flux carried off into the surroundings from the surface Σ heated up to temperature θ_w,
$q_{Rw} = \varepsilon_w \sigma_{SB} \theta_w^4$ – radiational heat flux carried off from the surface Σ by radiation of the surface,
$q_{\text{bl}} = \gamma_{\text{bl}} (\rho_g v_g)_w (\bar{I}_e - I_w)$ – convective heat flux carried off from the heated surface Σ by injection of gaseous products of thermodecomposition of the composite into the boundary layer,
$q_{\text{sub}} = \rho_w D_s \Delta Q_{\text{sub}}$ – heat flux carried off from the composite surface Σ by sublimation of the composite,
$q_{cs} = \rho_w D_c y_1 \Delta Q_{\text{com}}$ – heat flux supplied to the composite surface Σ by exothermal process of combustion of solid phases of the composite in the oxidizing overrunning flow,
$q_{cg} = \gamma_{\text{bl}} (\rho_g v_g)_w D_c \sum_i \Delta y_i \Delta Q_i$ – heat flux supplied to the composite surface Σ by exothermal combustion of gaseous products of thermodecomposition of polymer binder injected into the boundary layer.

Here is some notation:
(α/c_p) – heat transfer coefficient between the composite surface Σ and boundary layer of hot gas;
$\bar{I}_e$ – enthalpy of the overrunning flow;
I_w – enthalpy of the composite surface heated;
I_0 – enthalpy of the 'cold wall';
σ_{SB} – Stefan-Boltzman constant;
ε_w – degree of blackness of the heated surface;
γ_{bl} – injection coefficient $(0 < \gamma_{\text{bl}} < 1)$;
ΔQ_{sub} – *heat of sublimation* of solid phases of the composite;
ΔQ_{com} – *heat of combustion* of solid phases of the composite (absolute value);
ΔQ_i – combustion heat of gaseous products of phenolic matrix thermodecomposition (absolute value);
y_1 – mass concentration of solid combustible component on the composite surface Σ;
Δy_i – changing mass concentrations of chemical components in the mixture of gaseous products of thermodecomposition of the matrix at combustion on the surface Σ;
$(\rho_g v_g)_w = -(K \nabla p)_w \cdot \mathbf{n}$ – mass rate of injection of gaseous products of composite thermodecomposition into the boundary layer,
$(\nabla p)_w$ – pore pressure gradient on the ablation surface Σ.

The heat flux q_l in (3.56) is connected to q_{sub}, q_{cs} and q_{cg} by the relationship

$$q_l = q_{\text{sub}} - q_{cs} - q_{cg}. \tag{10.76}$$

In numerical calculations, we use the following values of the constants corresponding to carbon/phenolic composite:

$$\varepsilon_w = 0.8, \qquad \gamma_{\text{bl}} = 0.4, \qquad \Delta Q_{\text{sub}} = 20 \text{ MJ/kg},$$

$$\Delta Q_{\text{com}} = 9.8 \text{ MJ/kg}, \qquad y_1 = 1.$$

10.8. Criteria of Efficiency of Composites

To find the most efficient material under ablation we introduce different criteria of efficiency.

Effective enthalpy I_{eff} of a material is determined as follows [7, 95]:

$$I_{\text{eff}} = Q_e/\Delta m, \tag{10.77}$$

where Q_e is the total heat flux falling onto the surface Σ of the composite for time interval $(0, t)$,

$$Q_e = \int_0^t \int_{\Sigma(\tau)} (q_e^0 + q_R) d\Sigma d\tau, \tag{10.78}$$

and Δm is the total carried-off mass of the composite and

$$\Delta m = m_0 - m(t), \qquad m_0 = \int_{V_0} \rho_0 dV, \qquad m(t) = \int_{V(t)} \rho dV. \tag{10.79}$$

Here V_0, ρ_0 are the initial volume and density of the composite, respectively; $V(t)$ and $\Sigma(t)$ are the volume and density of the ablative composite. Sometimes in equation (10.78) the heat flux $(q_e^0 + q_R - q_w - q_{Rw})$ is considered in place of $(q_e^0 + q_R)$.

The highest efficiency, in accordance with the criterion (10.77), corresponds to materials with the highest value of I_{eff}.

For constant heat flux q_e^0 and $q_R = 0$, the formula (10.77) takes the form

$$I_{\text{eff}} = q_e^0/\dot{m}, \tag{10.80}$$

where

$$\dot{m} = (m_0 - m(t))/|\Sigma| t \tag{10.81}$$

is the *mass rate of ablation* of the composite, and $|\Sigma|$ is the area of the surface carried-off. If only linear ablation occurs, then the formula is transformed to

$$\dot{m} = \rho_0 D. \tag{10.82}$$

Expression (10.80) is usually applied for experimental determination of I_{eff} by known values of q_e^0 and $\dot{m}$.

Dependence of the enthalpy I_{eff} on the the heat flux q_e^0 is usually non-linear:

$$I_{\text{eff}} = q_e^0/(\rho_0 D(p_e, p_\Sigma, \theta_w(q_e^0))). \tag{10.83}$$

Figure 10.12 presents dependences of effective enthalpy I_{eff} on q_e^0 for phenolic composites: glass-, organo- and carbon-plastics. The highest values of I_{eff} are for carbon-plastics, as they have the highest resistance to thermomechanical erosion. For values of the heat flux $q_e^0 \approx 2 \ldots 10$ MW/m^2, the effective enthalpies I_{eff} of

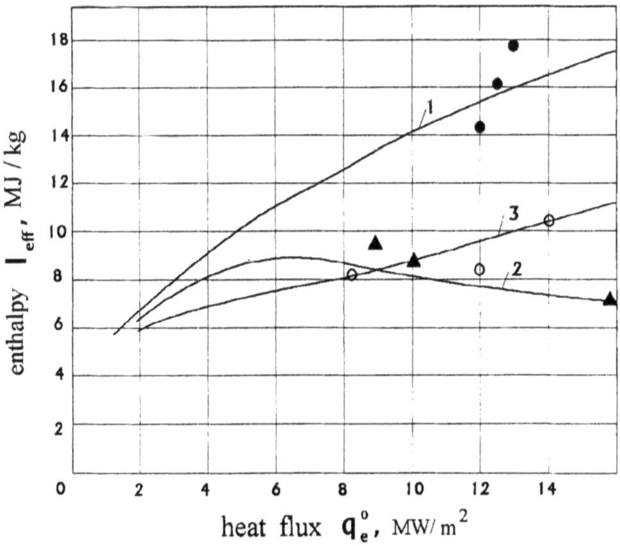

Figure 10.12. Effect of a type of fibres on effective enthalpy I_{eff} of phenolic composites at different values of heat flux. Fibres: 1 (•) – carbon, 2 (○) – organic, 3 (▲) – glass; pressure head $p_\Sigma = 0.5$ MPa. Solid curves are computations, points are experimental data

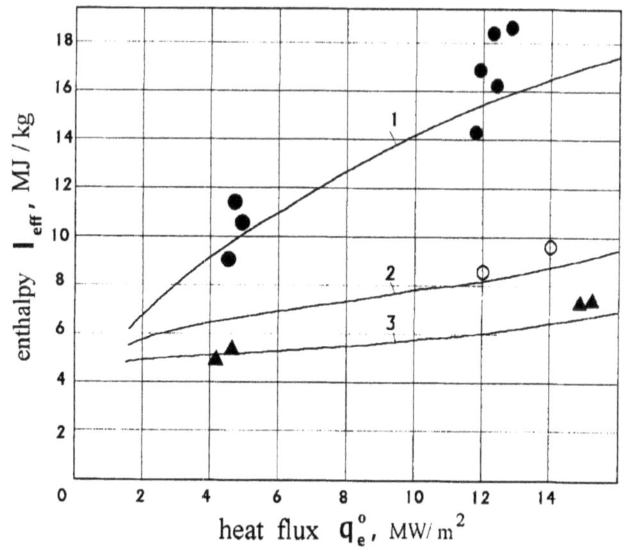

Figure 10.13. Effect of a type of matrix on effective enthalpy I_{eff} of textile carbon-plastics at different values of convective heat flux at $p_\Sigma = 0.5$ MPa in air. Matrices: 1 (•) – phenolic, 2 (○) – silicon-organic, 3 (▲) – epoxy; $(\alpha/c_p) = 2.0$. Solid curves are computations by (10.80), points are experimental data

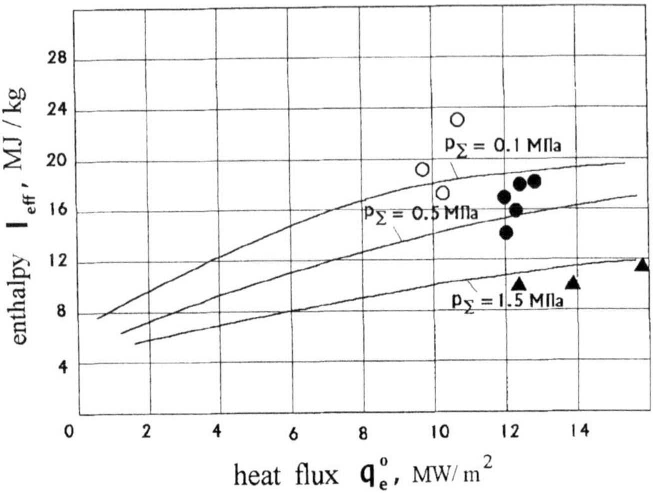

Figure 10.14. Dependence of effective enthalpy $I_{\rm eff}$ of carbon/phenolic composite on heat flux q_e^0 at different values $p_\Sigma = 0.1$ (○), 0.5 (●) and 1.5 (▲) MPa (corresponding values $(\alpha/c_p) = 0.7$, 2.0 and 4.0 kg/(m^2·s)). Solid curves are computations, points are experimental data

glass- and organo-plastics are practically the same, and only for magnitudes $q_e^0 \approx 12 \ldots 16$ MW/m^2 does the effective enthalpy of organo-plastics become higher.

Figure 10.13 shows dependences of effective enthalpy $I_{\rm eff}$ of carbon-plastics based on different matrices: phenolic, epoxy and silicon-organic. The matrix type has a considerable effect on $I_{\rm eff}$: the highest values of $I_{\rm eff}$ correspond to phenolic carbon-plastic, having highest thermal strength at high temperatures; the smallest ones relate to carbon/epoxy composite, thermal strength of which is negligible.

Figure 10.14 shows dependences of effective enthalpy of carbon/phenolic composite on pressure head p_Σ of a flow and convective heat flux q_e^0. Increasing the flux q_e^0 leads to growing values of $I_{\rm eff}$, while increasing the pressure head p_Σ from 0.1 to 1.5 MPa causes a considerable decrease of the enthalpy $I_{\rm eff}$ (by almost two times). This is connected to increase of the rate of thermomechanical erosion of carbon-plastics with increasing p_Σ.

To estimate the efficiency of ablative composites, we can also use the *dimensionless rate of ablation* [95]

$$\bar{G} = \dot{m}/(\alpha/c_p). \tag{10.84}$$

The rate $\bar{G}$ usually depends weakly on q_e^0 and p_e, therefore up to a certain accuracy we can consider the rate $\bar{G}$ as a constant of the composite at fixed values of p_Σ.

To estimate the efficiency of thermal-protective composites under ablation conditions, we may apply another dimensionless parameter [28]

$$\mathrm{Di} = Q_e(t)/(m_0 c \theta_s(t)), \tag{10.85}$$

where c is the heat capacity of the composite at temperature θ_0, and θ_s is the maximum temperature of the back composite surface protected.

CHAPTER 11

THERMAL STRESSES IN COMPOSITE STRUCTURES UNDER HIGH TEMPERATURES

11.1. Axisymmetric Problems of Thermomechanics for Composites under High Temperatures

11.1.1. BASIC EQUATIONS

Many stress-strain problems for elements of structures made of ablative composites are *axisymmetric*. We will use the statements of problems 'A' and 'B' formulated in paragraph 3.6.

Textile composite material is assumed to be a *cylindrically orthotropic medium* with orthotropy axis directed along the axis of the $Oz\vartheta r$ cylindrical coordinate system. Fabric layers are laid orthogonally to the Or axis. We consider the case of thermomechanical loading the composite, when its stress-strain state and fields of temperature and pore pressure are axisymmetric. For this case, all functions depend only on r and z, and equilibrium equations (3.45), Cauchy relations (3.46) and compatibility equation (3.60) for the cylindrical coordinate system have the form

$$\frac{\partial\sigma_r}{\partial r}+\frac{(\sigma_r-\sigma_\vartheta)}{r}+\frac{\partial\sigma_{rz}}{\partial z}-\frac{\partial\varphi_g p}{\partial r}=0,$$

$$\frac{\partial\sigma_{rz}}{\partial r}+\frac{\sigma_{rz}}{r}+\frac{\partial\sigma_z}{\partial z}-\frac{\partial\varphi_g p}{\partial z}=0, \tag{11.1}$$

$$\varepsilon_r=\frac{\partial u_r}{\partial r},\qquad \varepsilon_\vartheta=\frac{u_r}{r},\qquad \varepsilon_z=\frac{\partial u_z}{\partial z},\qquad \varepsilon_{rz}=\frac{1}{2}\Big(\frac{\partial u_r}{\partial z}+\frac{\partial u_z}{\partial r}\Big), \tag{11.2}$$

$$\frac{\partial^2\varepsilon_r}{\partial z^2}+\frac{\partial^2\varepsilon_z}{\partial r^2}-2\frac{\partial^2\varepsilon_{rz}}{\partial r\partial z}=0,\qquad \frac{\partial^2\varepsilon_\vartheta}{\partial z^2}+\frac{1}{r}\frac{\partial\varepsilon_z}{\partial r}-\frac{2}{r}\frac{\partial\varepsilon_{rz}}{\partial z}=0, \tag{11.3}$$

$$\frac{1}{r}\frac{\partial}{\partial r}\Big(r^2\frac{\partial\varepsilon_\vartheta}{\partial r}\Big)-\frac{\partial\varepsilon_r}{\partial r}=0,\qquad \frac{\partial}{\partial z}\Big(\varepsilon_r-\frac{\partial(r\varepsilon_\vartheta)}{\partial r}\Big)=0,$$

the remaining stresses, strains and displacements (i.e. $\sigma_{r\vartheta}$, $\sigma_{\vartheta z}$, $\varepsilon_{r\vartheta}$, $\varepsilon_{\vartheta z}$, u_ϑ) are zero.

Heat-mass-transfer equations (3.41) - (3.43) for the cylindric coordinate system are rewritten as follows:

$$\rho_f\frac{\partial\varphi_f}{\partial t}=-J_f\Gamma_f, \tag{11.4}$$

$$\rho_b \frac{\partial \varphi_b}{\partial t} = -J, \tag{11.5}$$

$$\frac{\partial \rho_g \varphi_g}{\partial t} = \frac{1}{r}\frac{\partial}{\partial r}(RK_{33} r \frac{\partial \rho_g \theta}{\partial r}) + \frac{\partial}{\partial z}(RK_{11}\frac{\partial \rho_g \theta}{\partial z}) + J\Gamma + J_f \Gamma_f, \tag{11.6}$$

$$\rho c \frac{\partial \theta}{\partial t} = \frac{1}{r}\frac{\partial}{\partial r}(k_{33} r \frac{\partial \theta}{\partial r}) + \frac{\partial}{\partial z}(k_{11}\frac{\partial \theta}{\partial z}) + c_g RK_{33}\frac{\partial \rho_g \theta}{\partial r}\frac{\partial \theta}{\partial r} +$$
$$+ c_g RK_{11}\frac{\partial \rho_g \theta}{\partial z}\frac{\partial \theta}{\partial z} - J\Delta e^0 - J_f \Delta e_f^0. \tag{11.7}$$

where the coordinate axes are numbered as follows: $1 \longrightarrow z,\ 2 \longrightarrow \vartheta,\ 3 \longrightarrow r$.

11.1.2. CONSTITUTIVE RELATIONS

For the axisymmetric problem, the constitutive relations for an orthotropic composite (7.29) have the form

$$\varepsilon_\alpha = \overset{\circ}{\varepsilon}_\alpha + \frac{1}{E_\alpha}\Big(\sigma_\alpha + \widetilde{f}_\alpha p - \nu_{\alpha\beta}(\sigma_\beta + \widetilde{f}_\beta p) - \nu_{\alpha\gamma}(\sigma_\gamma + \widetilde{f}_p)\Big),$$

$$\varepsilon_{rz} = \sigma_{rz}/(2G_{13}), \tag{11.8}$$

where functions $\widetilde{f}_\alpha$ are connected to f_α by the relationships

$$f_\alpha = \frac{1}{E_\alpha}(\widetilde{f}_\alpha - f_\beta \nu_{\alpha\beta} - f_\gamma \nu_{\alpha\gamma})), \quad \alpha,\beta,\gamma = 1,2,3,\ \alpha \neq \beta \neq \gamma, \tag{11.9}$$

here and below we introduce the following notation for stresses and strains: $\varepsilon_3 = \varepsilon_r$, $\varepsilon_2 = \varepsilon_\vartheta$, $\varepsilon_1 = \varepsilon_z$, $\sigma_3 = \sigma_r$, $\sigma_2 = \sigma_\vartheta$, $\sigma_1 = \sigma_z$.

The elastic characteristics E_α, G_{13} and $\nu_{\alpha\beta}$ of an orthotropic textile ablative composite depend on temperature θ and polymer phase concentrations φ_b and φ_f. As established in Chapter 7, the dependence is characterized by two functions $a_{\theta 1}$ and $a_{\theta 2}$ determined by formulae (7.31), (7.32) where one should perform the following substitution of indices: 1⟶3, 2⟶1, 3⟶2.

The inverse relations to (11.8) have the form

$$\sigma_{\alpha\alpha} = -\widetilde{f}_\alpha p + a_{\theta 1}\sum_{\beta=1}^{3} C_{\alpha\beta}(\varepsilon_\beta - \overset{\circ}{\varepsilon}_\beta), \qquad \alpha = 1,2; \tag{11.10}$$

$$\sigma_{33} = -p\widetilde{f}_3 + a_{\theta 1}\sum_{\beta=1}^{3} C_{3\beta}(\varepsilon_\beta - \overset{\circ}{\varepsilon}_\beta) + a_{\theta 2}C_{33}(\varepsilon_3 - \overset{\circ}{\varepsilon}_3); \quad \sigma_{rz} = a_{\theta 2}C_{44}\varepsilon_{rz}.$$

Elastic moduli $C_{\alpha\beta}$ of the orthotropic material are expressed as follows:

$$C_{11} = E_f^0 \frac{1 - \nu_{23}^2 m_E/\zeta}{\zeta\Delta}, \qquad C_{22} = E_f^0 \frac{1 - \nu_{13}^2 m_E/(1-\zeta)}{(1-\zeta)\Delta},$$

$$C_{12} = E_f^0 \frac{\nu_{12}\zeta + \nu_{13}\nu_{23}m_E}{\zeta(1-\zeta)\Delta}, \qquad C_{33} = E_b^0 \frac{1-\nu_{12}^2\zeta/(1-\zeta)}{\zeta(1-\zeta)\Delta}, \tag{11.11}$$

$$C_{23} = E_f^0 m_E \frac{\nu_{23} + \nu_{13}\nu_{12}\zeta/(1-\zeta)}{\zeta(1-\zeta)\Delta},$$

$$C_{13} = E_f^0 m_E \frac{\nu_{13} + \nu_{12}\nu_{23}}{\zeta(1-\zeta)\Delta}, \quad C_{44} = \frac{2G_b^0}{1-\zeta},$$

where we introduced the constant:

$$\Delta = \frac{1}{\zeta(1-\zeta)}\Big(1 - \nu_{12}^2\frac{\zeta}{1-\zeta} - m_E\big(\frac{\nu_{23}^2}{\zeta} + \frac{\nu_{13}^2}{1-\zeta} + \frac{2\nu_{13}\nu_{12}\nu_{23}}{1-\zeta}\big)\Big).$$

Pore gas pressure p is determined by the relation

$$p = R\rho_g\theta. \tag{11.12}$$

11.1.3. FUNCTIONS OF STRESSES

As known from [91], the compatibility equations (11.3) are reduced to two equations

$$\frac{\partial\varepsilon_z}{\partial z} + r\frac{\partial^2\varepsilon_\vartheta}{\partial z^2} = 2\frac{\partial\varepsilon_{rz}}{\partial z}, \quad r\frac{\partial\varepsilon_\vartheta}{\partial r} + \varepsilon_\vartheta - \varepsilon_r = 0. \tag{11.13}$$

There are only two stress functions F and χ for the axisymmetric problem, which, as follows from (3.62), are connected with stresses σ_r, σ_ϑ and σ_z by the relationships

$$\sigma_r = \frac{F}{r} + \varphi_g p, \quad \sigma_\vartheta = \big(\frac{\partial F}{\partial r} - \frac{\partial^2}{\partial z^2}(r\chi)\big) + \varphi_g p,$$

$$\sigma_z = \frac{1}{r}\frac{\partial}{\partial r}(r\chi) + \varphi_g p, \quad \sigma_{rz} = -\frac{\partial\chi}{\partial z}. \tag{11.14}$$

With relations (11.14), the equilibrium equations (11.1) are satisfied identically. To determine the functions F and χ we have two equations of compatibility (11.13) in which we should substitute formulae (11.8) and (11.14).

11.1.4. BOUNDARY CONDITIONS

The boundary conditions (3.34) - (3.37), (3.39) on the surface Σ, bounding the domain V occupied by the composite and consisting of four parts $\Sigma = \Sigma_u \cup \Sigma_\sigma \cup \Sigma_q \cup \Sigma_v$ (see paragraph 3.4), have the form

$$(\sigma_r - \varphi_g p)n_r + \sigma_{rz}n_z = S_r; \tag{11.15a}$$

$$\sigma_{rz}n_r + (\sigma_z - \varphi_g p)n_z = S_z, \quad \text{at } \Sigma_\sigma, \Sigma_q, \Sigma_v; \tag{11.15b}$$

$$u_r = u_r^0, \qquad u_\vartheta = u_\vartheta^0, \quad \text{at } \Sigma_u; \tag{11.15c}$$

$$\Big(K_{33} n_r \frac{\partial \rho_g \theta}{\partial r} + K_{11} n_z \frac{\partial \rho_g \theta}{\partial z}\Big) = 0, \quad \text{at } \Sigma_v, \Sigma_\sigma, \Sigma_u; \tag{11.15d}$$

$$\rho_g \theta = \frac{p_e}{R}, \quad \text{at } \Sigma_q; \tag{11.15e}$$

$$\Big(K_{33} n_r \frac{\partial \theta}{\partial r} + K_{11} n_z \frac{\partial \theta}{\partial z}\Big) = q_e^0 + q_R - q_{\text{bl}} - q_w - q_{Rw} - q_{\text{sub}} + q_{cs} + q_{cg}, \quad \text{at } \Sigma_q; \tag{11.15f}$$

$$\Big(K_{33} n_r \frac{\partial \theta}{\partial r} + K_{11} n_z \frac{\partial \theta}{\partial z}\Big) = 0, \quad \text{at } \Sigma_v, \Sigma_\sigma, \Sigma_v, \tag{11.15g}$$

where n_r and n_z are the components of the normal vector to the surface, S_r and S_z are the components of the force vector, and

$$S_r = -p_e n_r, \qquad S_z = -p_e n_z \quad \text{at } \Sigma_q \text{ and } \Sigma_v.$$

The convective part of the heat flux q_e^0, the heat flux q_{bl} carried off from Σ due to the ablation products flowing out into the surroundings, and the heat flux q_w carried off from the heated surface due to heat transfer have the form

$$q_e^0 = \Big(\frac{\alpha}{c_p}\Big)(\bar{I}_e - c\theta_0), \qquad q_w = \Big(\frac{\alpha}{c_p}\Big) c(\theta_w - \theta_0), \qquad q_{Rw} = \varepsilon_w \sigma_{SB} \theta_w^4, \tag{11.16}$$

$$q_{\text{bl}} = -\gamma_{\text{bl}} R \Big(K_{33} n_r \frac{\partial \rho_g \theta}{\partial r} + K_{11} n_z \frac{\partial \rho_g \theta}{\partial z}\Big)(\bar{I}_e - c\theta_w).$$

Expressions for the remaining parts q_R, q_{sub}, q_{cg} and q_{cs} are determined by the formulae of paragraph 10.7. For the case when there is neither linear ablation nor radiant flux from the outside, $q_R = q_{\text{sub}} = q_{cg} = q_{cs} = 0$.

Initial conditions for the set of equations (11.4), (11.6), (11.9) are written as follows:

$$t = 0: \qquad \rho_g = \rho_{g0}, \quad \varphi_f = \varphi_f^0, \quad \varphi_b = \varphi_b^0, \quad \theta = \theta_0. \tag{11.17}$$

11.1.5. A STATEMENT OF THE AXISYMMETRIC PROBLEM IN TERMS OF STRESSES

The axisymmetric thermomechanics problem in terms of stresses consists in solving the two equations of compatibility (11.13) and the four equations of heat-mass-transfer (11.4) - (11.7) in the domain V with boundary conditions (11.15) at the surface Σ (where $\Sigma_u = 0$) and initial conditions (11.17) for six functions φ_f, φ_b, ρ_g, θ, F and χ depending on three variables: r, z and t. After solving the equations, we determine stresses by formulae (11.14), and pore pressure p - by formula (11.12).

11.1.6. THE PROBLEM STATEMENT IN TERMS OF DISPLACEMENTS

For the axisymmetric problem in displacements, we substitute relations (11.2) and (11.10) into equation (11.1), and obtain the following equlibrium equations with respect to displacements u_r and u_z:

$$\Lambda_{11}u_r + \Lambda_{12}u_z + \Lambda_{13}u_r + \Lambda_{14}u_z + \Lambda_1 = 0,$$

$$\Lambda_{21}u_r + \Lambda_{22}u_z + \Lambda_{23}u_r + \Lambda_{24}u_z + \Lambda_2 = 0, \tag{11.18}$$

where there is following notation:

$$\Lambda_{11}u_r \equiv A_{11}\frac{\partial^2 u_r}{\partial r^2} + A_{14}\frac{\partial u_r}{\partial r} + A_{18}u_r, \quad \Lambda_{13}u_r \equiv A_{12}\frac{\partial^2 u_r}{\partial z^2} + A_{15}\frac{\partial u_r}{\partial z},$$

$$\Lambda_{14}u_z \equiv A_{17}\frac{\partial u_z}{\partial z}, \quad \Lambda_{12}u_z \equiv A_{16}\frac{\partial u_z}{\partial r}, \quad \Lambda_1 = A_{13}\frac{\partial^2 u_z}{\partial r \partial z} + A_{19},$$

$$\Lambda_{21}u_r \equiv A_{24}\frac{\partial u_r}{\partial r} + A_{28}u_r, \quad \Lambda_{22}u_z \equiv A_{12}\frac{\partial^2 u_z}{\partial r^2} + A_{26}\frac{\partial u_z}{\partial r},$$

$$\Lambda_{23}u_r \equiv A_{25}\frac{\partial u_r}{\partial z}, \quad \Lambda_{24}u_z \equiv A_{22}\frac{\partial^2 u_z}{\partial z^2} + A_{27}\frac{\partial u_z}{\partial z}, \tag{11.19}$$

$$\Lambda_2 = A_{13}\frac{\partial^2 u_r}{\partial r \partial z} + A_{29},$$

and coefficients

$$A_{11} = a_{\theta 2}C_{33}, \quad A_{12} = \frac{a_{\theta 2}}{2}C_{44}, \quad A_{13} = a_{\theta 2}\left(\frac{C_{44}}{2} + C_{23}\right)a_{\theta 1},$$

$$A_{14} = \frac{a_{\theta 2}C_{33}}{r} + \frac{\partial a_{\theta 2}C_{33}}{\partial r}, \quad A_{15} = \frac{1}{2}\frac{\partial a_{\theta 2}C_{44}}{\partial z}, \quad A_{16} = A_{15},$$

$$A_{17} = \frac{a_{\theta 1}}{r}(C_{13} - C_{12}) + \frac{\partial a_{\theta 1}C_{13}}{\partial r},$$

$$A_{18} = \frac{1}{r}\frac{\partial}{\partial r}(a_{\theta 1}C_{23}) - \frac{a_{\theta 1}C_{22}}{r^2}, \quad A_{22} = a_{\theta 1}C_{11}, \quad A_{24} = \frac{\partial}{\partial z}(a_{\theta 1}C_{13}), \tag{11.20}$$

$$A_{19} = -\frac{\partial \varphi_g p}{\partial r} - \frac{\partial}{\partial r}\left[a_{\theta 2}C_{33}\overset{\circ}{\varepsilon}_r + a_{\theta 1}(C_{23}\overset{\circ}{\varepsilon}_\vartheta + C_{13}\overset{\circ}{\varepsilon}_z)\right] - \frac{a_{\theta 1}}{r}\left[\left(\frac{a_{\theta 1}}{a_{\theta 2}}C_{33} - \right.\right.$$
$$\left.\left. - C_{23}\right)\overset{\circ}{\varepsilon}_r + (C_{23} - C_{22})\overset{\circ}{\varepsilon}_\vartheta + (C_{13} - C_{23})\overset{\circ}{\varepsilon}_z\right] - \frac{\partial \widetilde{f}_3 p}{\partial r} - \frac{\widetilde{f}_3 - \widetilde{f}_2}{r}p,$$

$$A_{25} = A_{26} + \frac{a_{\theta 1}C_{12}}{r}, \quad A_{26} = \frac{a_{\theta 2}C_{44}}{r} + \frac{\partial}{\partial r}(a_{\theta 2}C_{44}),$$

$$A_{27} = \frac{\partial}{\partial z}(a_{\theta 1}C_{11}), \quad A_{28} = \frac{1}{r}\frac{\partial}{\partial z}(a_{\theta 1}C_{12}),$$

$$A_{29} = -\frac{\partial \varphi_g p}{\partial z} - \frac{\partial}{\partial z}\left[a_{\theta 1}(C_{13}\overset{\circ}{\varepsilon}_r + C_{12}\overset{\circ}{\varepsilon}_\vartheta + a_{\theta 2}C_{11}\overset{\circ}{\varepsilon}_z\right] - \frac{\partial \widetilde{f}_1 p}{\partial z}.$$

The problem in terms of displacements consists in solving the equations (11.4) - (11.7) and (11.18) with conditions (11.15), (11.17) for functions u_r, u_ϑ, ρ_g, θ and φ_b, φ_f depending on r, z and t.

11.2. Thermal Stresses in Composite Structures of Heat-Energetical Systems

Typical elements of composite structures of heat-energetical systems classified among supersonic turbo-jet engines (STJE) are shown in Figures 1.11 and 1.12. Let us consider axisymmetric statements of problems for thermal stresses appearing in an external shell of an inlet of an STJE, and a shell of a central body made of textile composites under the action of an overrunning high-enthalpy gas flow. We must solve these problems to determine the optimal parameters of the heat-energetical systems corresponding to maximum structural strength.

11.2.1. EXTERNAL SHELL OF THE INLET OF STJE

The Problem Statement

As mentioned in 1.1.3, a hollow external shell of an inlet of an STJE undergoes two-sided heating by an overrunning hot gas flow, and simultaneously the action of excess internal pressure p_{e1}. The magnitude of p_{e1} is determined by equations (1.20) - (1.22). The outside of the shell experiences the pressure $p_{e2} = p_\infty \leqslant p_{e1}$.

The shell is assumed to be a hollow circular cylinder, with external and internal radii R_2 and R_1 (Figure 11.1).

Convective heat fluxes q_{e1}, q_{e2} and pressures p_{e1}, p_{e2} are uniformly distributed over its lateral surface. One end of the cylinder $z = 0$ is rigidly fastened; at the other end the pressure p_{e3} is given; the cylinder ends are assumed to be gas-tight and heat-insulated.

There is usually no linear ablation in the systems, and therefore the shell shape is constant and $D = 0$, $q_R = q_{\text{sub}} = q_{cs} = q_{cg} = 0$.

We use the statement in terms of stresses (see 11.1.5).

In the scope of these assumptions a solution of the problem (11.4) - (11.7) and (11.13) is sought in the form of functions depending only on r and t. Stresses are determined from (11.14) as follows:

$$\sigma_r = \frac{F}{r} + \varphi_g p, \quad \sigma_\vartheta = F' + \varphi_g p, \quad \sigma_{rz} = 0, \tag{11.21}$$

$$\sigma_z = E_1(C - \overset{\circ}{\varepsilon}_z) + \nu_{13}\sigma_r + \nu_{12}\sigma_\vartheta - p(\widetilde{f}_1 - \nu_{12}\widetilde{f}_2 - \nu_{13}\widetilde{f}_3),$$

where $C(t) = \varepsilon_z$ is the time-dependent function which must be found, and $F' \equiv \partial F/\partial r$.

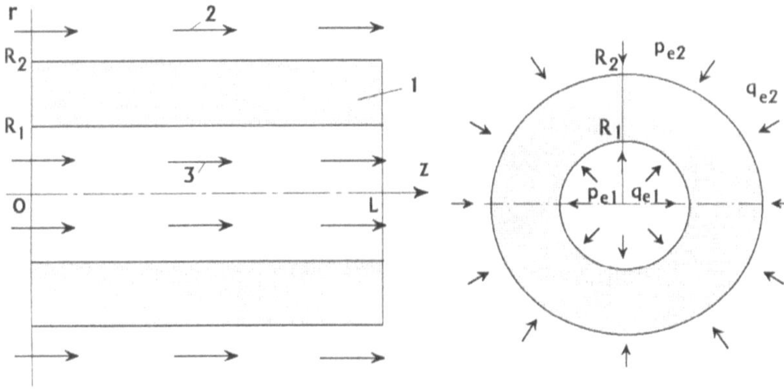

Figure 11.1. A scheme for determination of thermal stresses in the external shell of the STJE inlet under the action of high-enthalpy flow: 1 – composite cylindrical shell, 2 – outside gas flow, 3 – inside gas flow

Constitutive relations (11.8) take the form

$$\varepsilon_r = \overset{\circ}{\tilde{\varepsilon}}_r + \beta_{33}(\sigma_r + \tilde{f}_3 p) + \beta_{32}(\sigma_\vartheta + \tilde{f}_2 p), \qquad \varepsilon_{rz} = 0,$$

$$\varepsilon_\vartheta = \overset{\circ}{\tilde{\varepsilon}}_\vartheta + \beta_{32}(\sigma_r + \tilde{f}_3 p) + \beta_{22}(\sigma_\vartheta + \tilde{f}_2 p), \tag{11.22}$$

where

$$\overset{\circ}{\tilde{\varepsilon}}_r = \overset{\circ}{\varepsilon}_r - \nu_{13}(C - \overset{\circ}{\varepsilon}_z), \qquad \overset{\circ}{\tilde{\varepsilon}}_\vartheta = \overset{\circ}{\varepsilon}_\vartheta - \nu_{12}(C - \overset{\circ}{\varepsilon}_z), \tag{11.23}$$

and β_{ij} may be determined by the formulae

$$\beta_{33} = \frac{1 - \nu_{31}\nu_{13}}{E_3}, \qquad \beta_{22} = \frac{1 - \nu_{21}\nu_{12}}{E_2}, \qquad \beta_{32} = -\frac{\nu_{32} + \nu_{31}\nu_{12}}{E_3}. \tag{11.24}$$

After substituting equations (11.21), (11.22) into the second compatibility equation (11.13), we find that the last takes the form

$$\beta_{22}F'' + (\beta_{22}' + \frac{\beta_{22}}{r})F' + ((\frac{\beta_{32}}{r})' + \frac{\beta_{32} - \beta_{33}}{r^2})F + \overset{\circ}{\tilde{\varepsilon}}{}_\vartheta' + \frac{\overset{\circ}{\tilde{\varepsilon}}_\vartheta - \overset{\circ}{\tilde{\varepsilon}}_r}{r} +$$
$$+ (p(\tilde{f}_2\beta_{22} + \tilde{f}_3\beta_{32}))' + \frac{p}{r}(\beta_{22}\tilde{f}_2 - \beta_{33}\tilde{f}_3 + \beta_{32}(\tilde{f}_3 - \tilde{f}_2)) = 0. \tag{11.25}$$

The first compatibility equation is satisfied identically.

Heat-mass-transfer equations (11.4) – (11.7) take the form

$$\rho_f \dot{\varphi}_f = -J_f \Gamma_f, \qquad \rho_b \dot{\varphi}_b = -J,$$

$$(\rho_g \varphi_g)^{\bullet} = \frac{1}{r}(RK_{33} r(\varphi_g \rho_g \theta)')' + J\Gamma + J_f \Gamma_f, \tag{11.26}$$

$$\rho c\dot{\theta} = \frac{1}{r}(k_{33}r\theta')' + c_g R k_{33}(\varphi_g \rho_g \theta)'\theta' - J\Delta e^0 - J_f \Delta e_f^0.$$

Boundary and initial conditions for the set of equations (11.25), (11.26) are

$$r = R_{1,2}: \quad F = -R_{1,2}p_{e1,2}, \quad R\rho_g\theta = p_{e1,2},$$
$$k_{33}\theta' = q_{e1,2} - q_{\text{bl}1,2} - q_{w1,2} - q_{Rw1,2}; \tag{11.27}$$

$$t = 0: \quad \varphi_f = \varphi_f^0, \quad \varphi_b = \varphi_{b0}, \quad \rho_g = \rho_{g0}, \quad \theta = \theta_0. \tag{11.28}$$

where heat fluxes q_{ei}, q_{wi}, q_{Rwi} and $q_{\text{bl}i}$ have the form

$$q_{\text{bl}i} = (-1)^{(i+1)}\gamma_b R K_{11}(\rho_g\theta)'(\bar{I}_e - \theta), \quad q_{wi} = \left(\frac{\alpha}{c_p}\right)_i c(\theta - \theta_0),$$

$$q_{Rwi} = \varepsilon_w \sigma_{SB}\theta^4, \quad q_{ei} = \left(\frac{\alpha}{c_p}\right)(\bar{I}_e - c\theta_0), \quad i = 1, 2. \tag{11.29}$$

Values of the flow enthalpy $\bar{I}_e = \bar{c}_g\theta_e$, heat transfer coefficient $(\alpha/c_p)_i$ and pressures p_{ei} at the external and internal surfaces of the cylinder are assumed to be given, and are determined by the formulae (1.20), (1.23).

We can determine the function $C(t)$ from the condition at the end $z = L$. Neglecting Poisson's effects, one can find the approximate expression for C:

$$C = \overset{\circ}{\varepsilon}_z - \frac{p_{e3}(R_2^2 - R_1^2)}{2\int_{R_1}^{R_2} E_1 r dr}. \tag{11.30}$$

The problem of determination of the five functions F, φ_f, ρ_g and θ depending on r and t consists in solving the five equations (11.25) and (11.26) with conditions (11.27) - (11.29).

Condition of Failure

For the present problem the influence of axial stress σ_z as compared with tangential stress σ_ϑ is negligibly small, therefore only two types from the six possible destruction types of textile composite (see paragraph 7.7) can be realized: type (A) (breakage of textile fibres oriented in the ring direction) and type (C) (delamination of the composite).

With the help of the thermostrength criteria (7.43) and (7.51), the condition of composite detruction by type (A_2) (or (C)) can be presented as the condition when the corresponding damage parameter $z_2(r,t)$ or $z_1(r,t)$ reaches value 1 at time t_1^* or t_2^* in a certain domain of the cylinder with coordinate $r = r_1^*$ or $r = r_2^*$:

$$z_1(r_1^*, t_1^*) = 1 \qquad \text{type } (C); \tag{11.31}$$

$$z_2(r_2^*, t_2^*) = 1 \qquad \text{type } (A_2), \tag{11.32}$$

where z_1 and z_2 define the proximity of the stress state in the domain with coordinate r to the ultimate state, and given by the expressions

$$z_1(r,t) = (|\sigma_r + 3\bar{N}^0 p| + \sigma_r + 3\bar{N}^0 p)/2\sigma_{3T}, \tag{11.33}$$

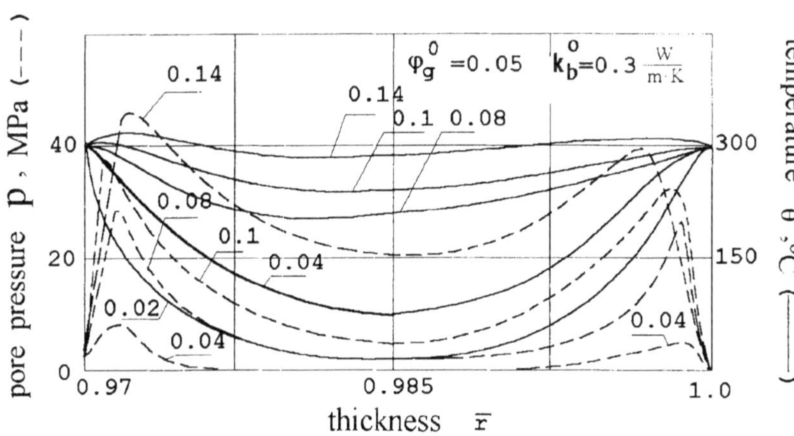

Figure 11.2. Distributions of temperature θ and pore pressure p *versus* the cylinder thickness for different times, symbols on curves are dimensionless times $\bar{t}$

$$z_2(r,t) = \frac{|\sigma_\vartheta(r,t)| + \sigma_\vartheta(r,t)}{2\sigma_{2T}(r,t)}. \tag{11.34}$$

Here σ_{3T} is the ultimate strength of the composite in tension across the fabric layers, and σ_{2T} - along the fabric layer in the warp direction, expressions for which are determined by formulae (7.44) and (7.52), where one should realize the substitution of indices: $1 \longrightarrow 3$, $2 \longrightarrow 1$, $3 \longrightarrow 2$. In (11.33) the function $\bar{N}^0 = N^0/B_{m1}$ is introduced, expression for N^0 is determined from (4.29), and B_{m1} - from formula (7.46).

Analysis of Numerical Solution

Solving the problem (11.25) - (11.29) for functions F, φ_f, φ_b, ρ_g and θ was carried out numerically with the help of a step-by-step method using an implicit difference scheme and matrix sweeping procedures. These methods are described in [102, 103].

Computations were conducted for textile glass/epoxy-phenolic composite, and the ratio of the external and internal radii was $R_1/R_2 = 0.97$. All material constants used in the computations of the composite are given in Tables 4.1, 4.2, 5.1 and 5.2, and it was assumed that $\zeta = 0.5$, $\varphi_f^0 = 0.6$, $\Phi = 1°$.

Distributions of temperatute θ, pore pressure p, volumetric concentrations φ_b and φ_p, stresses σ_r and σ_ϑ, damage parameters z_2 and z_1 *versus* the cylinder thickness are shown in Figures 11.2 - 11.6, indices at the curves correspond to dimensionless times $\bar{t} = t/t_0$, where t_0 is the characteristic time.

The programme of internal and external heating was the following: rapid growing the overrunning flow enthalpy $\bar{I}_e$ up to the level of $\bar{I}_e = 1$ MJ/kg for time $\bar{t} = 0.02$, and then this level remains for time $\bar{t} = 0.14$.

The heat transfer coefficient (α/c_p) at the stationary regime of heating, in accor-

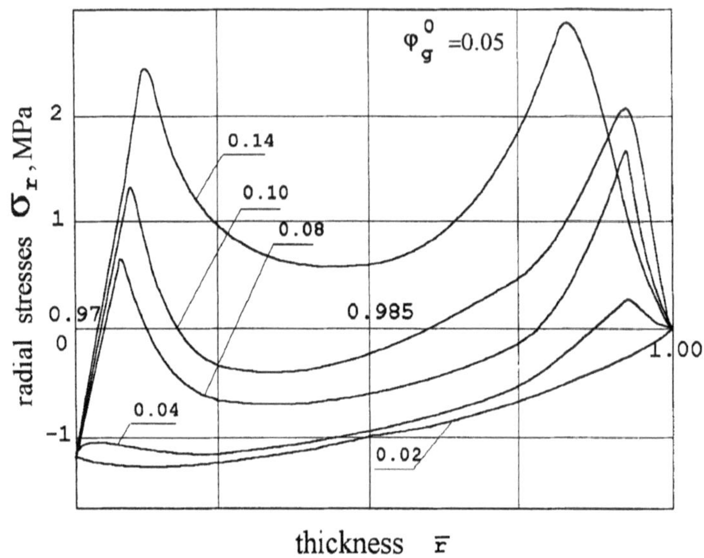

Figure 11.3. Distributions of radial stresses σ_r *versus* thickness of the cylinder of ablative composite for different times, notation is the same as in Figure 11.2

dance with Figure 1.13, was assumed to be (α/c_p)= 1.6 kg/(m^2·s); corresponding values of the heat fluxes were $q_{e1} = q_{e2} = 1$ MW/m^2, temperature $\theta_e \approx 500$ °C, internal pressure $p_{e1} = 1.3$ MPa, and external pressure p_{e2} =0.1 MPa.

In addition, some parameters were varied in different variants of calculations: the internal pressure p_{e1} = 0.13 MPa and 1.3 MPa, the initial porosity φ_g^0 from 0.05 to 0.15 and the polymer phase heat conductivity coefficient $k_b^0 = 0.2$ and 0.3 W/(m · K). A change of the initial porosity can be regulated by pressure and temperature regimes of solidification of the material. The binder heat conductivity can be changed, for example, by injection of hollow microspheres. The parameters substantially affect inner heat-mass-transfer processes in ablative composite materials.

Intrapore pressure p distribution within the cylinder under two–sided heating is characterized by the presence of two local maxima near the surfaces heated. As the material is being heated, these peaks grow and move towards the middle surface of the cylinder $\bar{r} = 0.975$, where $\bar{r} = r/R_2$ is the dimensionless radius. When $\bar{t} \geqslant 0.1$, confluence of the two fronts occurs, and the pore pressure increases abruptly at the middle surface. Equalizing the pressure over the cylinder thickness is obstructed only by failure of the composite at $\bar{t} = 0.14$ (Figure 11.2).

The tranverse stresses σ_r and damage parameter z_2 practically repeat the pore pressure p profile (see Figures 11.2, 11.3 and 11.5). Near the internal surface, the stress σ_r falls into the negative domain due to the presence of excess inner

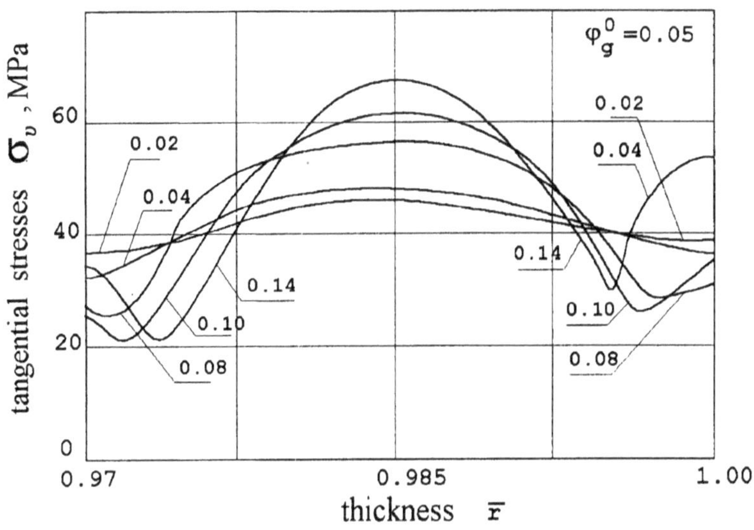

Figure 11.4. Distributions of tangential stresses σ_ϑ *versus* thickness of the cylinder for different times, notation is the same as in Figure 11.2

pressure $p_{e1} = 1.3$ MPa. With propagating heat into the material, the peaks of stresses σ_r and damage parameter z_2 move towards the middle surface of the cylinder and increase in their absolute values following the pore pressure p profile. At the beginning of heating, when $\bar{t}= 0.1$, the peak of stresses σ_r near the internal surface grows slower than near the external surface, and then, when $\bar{t} > 0.1$, their magnitudes become equal. Delamination of the composite occurs simultaneously both near the internal and external surfaces, where the measure z_1 reaches 1 at $\bar{t} = 0.14$ (Figure 11.5).

Volumetric ablation (pyrolysis) of glass–plastic substantially redistributes tangential stresses σ_ϑ within the cylinder. The pore pressure peaks lead to formation of local minima of stress σ_ϑ near the external and internal surfaces (Figure 11.4) and simultaneous formation of a local maximum at the middle surface of the cylinder. Shrinkage tensile stresses σ_ϑ appear at the surfaces $r = R_1, R_2$, where the coke phase content φ_p is the greatest (Figure 11.6). The maximum value of σ_ϑ increases persistently in the pyrolysis, and is equal to 65 MPa when $\bar{t} = 0.14$; while the classical shell theory without ablation effects gives the value $\sigma_\vartheta = p_{e1}/(1 - R_1/R_2) \approx$ 43 MPa for $p_{e1} = 13$ MPa.

Since, at the given pressure p_{e1}, the stress σ_ϑ is considerably smaller than the ultimate strength σ_{2T} and the damage parameter z_2 is considerably lower than z_1, failure of the cylinder occurs in accordance with the type (C) due to delamination.

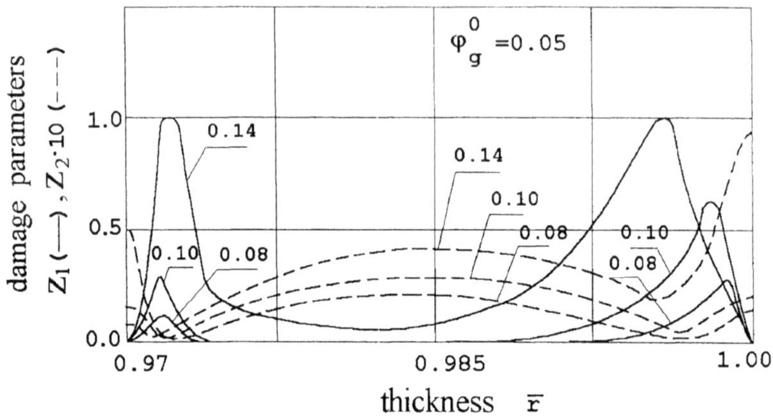

Figure 11.5. Distributions of damage parameters z_1 and z_2 *versus* thickness of the cylinder of ablative composite for different times, notation is the same as in Figure 11.2

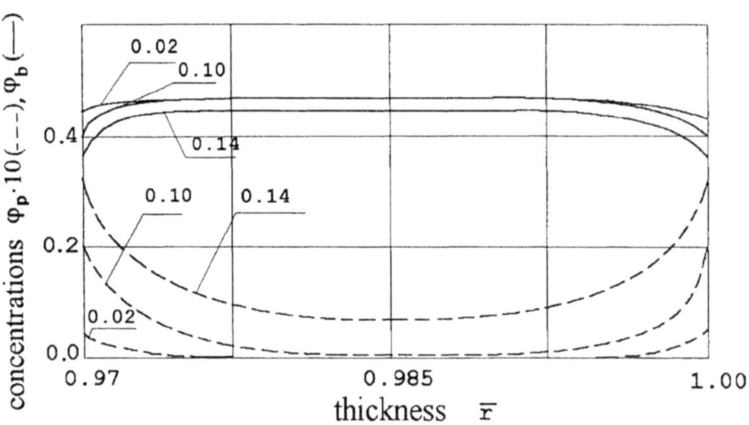

Figure 11.6. Distributions of volumetric concentrations of polymer phase φ_b and coke φ_p *versus* the cylinder thickness for different times, notation is the same as in Figure 11.2

The Effect of Composite Characteristics on Durability of the Structure

Figure 11.7 shows computed results for maximum pore pressure $p_{\max}$ in the cylinder for three values of initial porosity $\varphi_g^0 = 0.05$, 0.1 and 0.15, two values of inside pressure $p_{e1} = 0.13$ and 1.0 MPa and two values of the polymer phase heat conductivity coefficient $k_b^0 = 0.2$ and 0.3 W/(m · K).

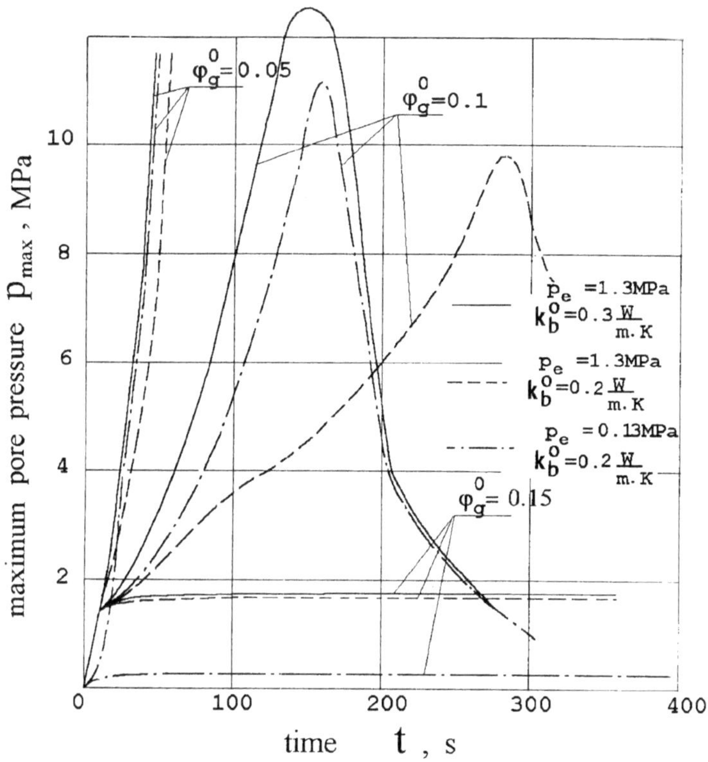

Figure 11.7. Distributions of maximum pore pressure $p_{\max}$ in the cylinder for different values of initial porosity φ_g^0, two values of inside pressure p_{e1} =1.3 Mpa and 0.3 MPa and two values of heat conductivity coefficient $k_b^0 = 0.2$ W/(m · K) and 0.3 W/(m · K)

As established in paragraph 9.1, the function $p_{\max}(t)$ has four typical stages, and initial porosity φ_g^0 qualitatively changes the picture of the pore pressure. Figure 11.7 shows that when $\varphi_g^0 = 0.05$ only the first stage of unsteady growth of $p_{\max}$ can be realized for all values of p_{e1} and k_b^0, as at time $t_1^* = 0.14\ t_0$ destruction occurs by type (C). When φ_g^0 =0.1, three stages are realized: a stage of rapid growth $0 < t < 10$ s, a stage of steady growth of pressure at $10 < t < 150 \ldots 300$ s and a stage of falling pressure for all p_{e1} and k_b^0. When $\varphi_g^0 = 0.15$ there are only two stages: a stage of initial growth, and a stationary regime for all values of p_{e1} and k_b^0. For $\varphi_g^0 \geqslant 0.1$, there is no failure of the composite.

Decreasing the heat conductivity coefficient k_b^0 leads to lowering the pore pressure intensity, this is noticeable especially for φ_g^0 =0.05 and φ_g^0 =0.1. When φ_g^0 =0.15, the level of pore pressure p is relatively small, and it is practically independent of the coefficient k_b^0. Diminishing the inside pressure p_{e1} also leads to lowering p (Figure 11.8), as obstacles to flow of the gaseous ablation products from

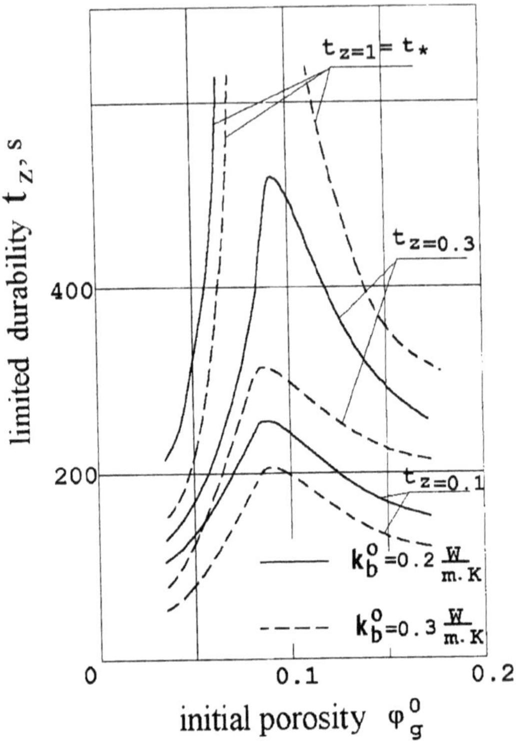

Figure 11.8. Dependence of limited durability t_z on initial porosity for an ablative cylinder under inside pressure p_{e1} =1.3 MPa and different values of k_b^0

pores into the surroundings decrease.

We introduce a concept of *limited durability* t_z; this is determined as the time t for which a maximum value of at least one of the damage parameters z_2 and z_1 reaches a certain value prescribed z. For example, the limited durability equal to $t_{0.1}$ means that one of the damage parameters z_1 or z_2 has reached the value equal to 0.1 for time $t = t_{0.1}$. The value of time $t_{1.0}$ of the limited durability corresponds to time t_* of complete failure of the structure.

Figure 11.8 shows the effect of porosity φ_g^0 and coefficient k_b^0 on the durability of the ablative cylinder under pressure p_{e1} =1.3 MPa. The function $t_z(\varphi_g^0)$ has a local maximum near the point where $\varphi_g^0 \approx 0.1$ for all values of z. The appearance of the maximum is connected to the fact that for $\varphi_g^0 \to 0$, $\varphi_g^0 < 0.1$, gas permeability of the composite decreases and the pore pressure level grows; this crushes the material. When $\varphi_g^0 \to 1$, $\varphi_g^0 > 0.1$, the strength σ_{2T} of glass-plastic falls abruptly; under the action of inside pressure the cylinder is destroyed by type (A_2).

Figure 11.8 shows that there exists a break of the function $t_z(\varphi_g^0)$ for $z = 1$ within the interval $0.08 > \varphi_g^0 > 0.12$; this means that the glass-plastic is not

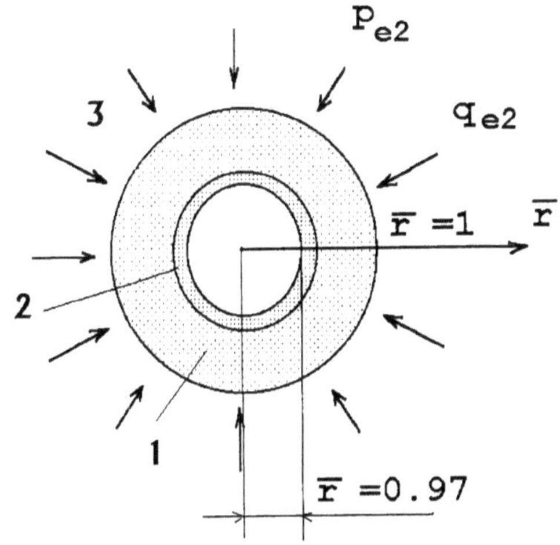

Figure 11.9. A scheme of the action of heat flux q_{e2} on a shell of the central body of an STJE inlet: 1 - load-bearing shell, 2 - composite shell, 3 - heat flux q_{e2}

destroyed at all under the given heat-force loads for the porosity values in this interval. Diminishing the heat conductivity coefficient k_b^0 increases the durability of the composite structure, displacing the curves $t_z(\varphi_g^0)$ into the domain of higher values of time t.

11.2.2. A SHELL OF A CENTRAL BODY OF STJE INLET

The Problem Statement

We now apply the statement of the axisymmetric problem in stresses to calculate thermal stresses in a shell of a central body of an STJE inlet (Figure 1.11). The shell is assumed to be a hollow circular cylinder fastened on the inside, together with a load-bearing shell and subjected to the action of heat flux q_{e2} and pressure p_{e2} uniformly distributed over its external surface $r = R_2$ (Figure 11.9).

The cylinder end $z = 0$ is supposed to be rigidly fastened and the deformation ε_z is given: $\varepsilon_z = \overset{\circ}{\varepsilon}_\Pi$, where $\overset{\circ}{\varepsilon}_\Pi = \alpha_\Pi(\theta_\Pi - \theta_0)$ is deformation in the z-direction, α_Π is the coefficient of heat expansion, $\theta_\Pi = \theta(R_1, t)$ is the temperature of the load-bearing shell, and R_2 and R_1 are the external and internal radii of the cylinder. Both ends are assumed to be gas-tight and heat-insulated. The material of the cylinder is textile ablative composite with orthotropy axes being the axes of the cylindrical coordinate system $Oz\vartheta r$.

Within the scope of our assumptions, stresses and characteristics of mass-transfer are determined from the solution of equations (11.25), (11.26). The only distinction is in that the function C is determined not by the formula (11.30), but in the following way:

$$C = \alpha_\Pi(\theta_\Pi - \theta_0), \tag{11.35}$$

and the boundary conditions are also modified:

$$r = R_1: \quad (\frac{F}{R_1} - \varphi_g p) = -p_{e1} +$$
$$+ \frac{E_\Pi h_\Pi}{(1-\nu_\Pi^2)R_1}(\overset{\circ}{\tilde{\varepsilon}}_\vartheta + \beta_{32}(\frac{F}{R_1} + \varphi_g p) + \beta_{22}(F' + \varphi_g p)),$$

$$(\rho_g \theta)' = 0, \qquad \theta' = 0; \tag{11.36}$$

$$r = R_2: \quad F = -R_2 p_{e2}, \quad \rho_g \theta = \frac{p_{e2}}{R}, \quad k_{33}\theta' = q_{e_2} - q_{\text{bl}} - q_{w2} - q_{Rw2},$$

$$q_{w2} = \left(\frac{\alpha}{c_p}\right) c(\theta - \theta_0), \qquad q_{\text{bl}} = -\gamma_b R K_{33}(\rho_g\theta)'(\bar{I}_e - c\theta), \tag{11.37}$$

$$q_{e_2} = \frac{\alpha}{c_p}(\bar{I}_e - c\theta_0), \qquad q_{Rw2} = \varepsilon_w \sigma_{SB} \theta^4.$$

Initial conditions for this problem have the form (11.28).

The first of equations (11.36) on the internal boundary is the condition of strengthening the cylinder by the load-bearing shell:

$$\sigma_r = \varphi_g p - p_{e1} + \frac{u_r E_\Pi h_\Pi}{(1-\nu_\Pi)^2 R_1^2}, \tag{11.38}$$

where E_Π, ν_Π and h_Π are the Young modulus, Poisson's ratio and thickness of the substrate, respectively, and p_{e1} is the internal pressure of the shell.

After the problem (11.25), (11.26) with conditions (11.35), (11.36), (11.37) and (11.29) has been solved, we determine the stresses by formulae (11.21).

Numerical Analysis of the Problem

Solving the formulated problem was realized numerically with the help of implicit difference schemes and sweeping procedures. The geometrical parameters were chosen as follows: $R_1/R_2 = 0.97$, $h_\Pi/R_2 = 0.003$. Material of the load-bearing substrate was aluminium alloy with the characteristics: $E_\Pi = 7 \cdot 10^4$ MPa, $\alpha_\Pi = 20 \cdot 10^{-6}\ K^{-1}$ and carbon-plastic with the characteristics $E_\Pi = 10^5$ MPa, $\alpha_\Pi = 2 \cdot 10^{-6}\ K^{-1}$. The main material of the cylinder was glass/epoxy-phenolic composite; its characteristics were calculated with the help of the material constants given in Tables 4.1, 4.2, 5.1 and 5.2.

The conditions of aerodynamical heating of the surface $r = R_2$ were the same as in 11.2.1: $\bar{I}_e = 1$ MJ/kg, $q_{e2} = 1$ MW/m^2, $(\alpha/c_p) = 1.6$ kg/(m^2·s).

For this problem, we investigate the effect of boundary conditions of the filtration equation on a stress-strain state and durability of the glass-plastic cylinder.

There is a common situation for actual glass-plastics: in manufacturing a thin layer of material near its external surface proves to have a higher density than one may expect, and substantially lower gas permeability. The thin layer hinders outflowing gaseous thermodecomposition products from the composite and

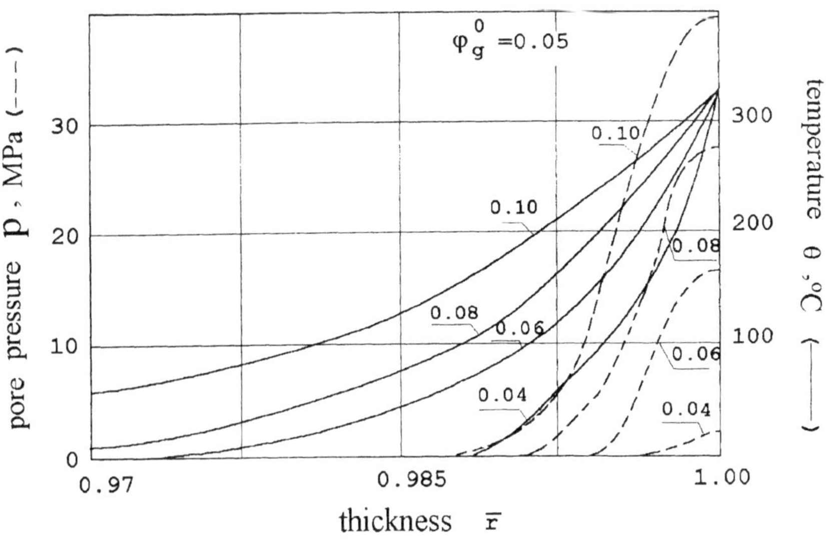

Figure 11.10. Distributions of temperature $\Delta\theta$ and pore pressure p *versus* thickness of a glass-plastic cylinder with strengthening made of Al alloy and with a gas-tight surface heated $\bar{r} = 1$, notation is the same as in Figure 11.2

hence increases the level of pore pressure. To investigate the effect, we compare computations for two situations:

1) with the given pressure $p_{e2} = 0.1$ MPa at the external surface $\bar{r} = 1$ and
2) with the gas-tight surface $\bar{r} = 1$: $(\rho_g\theta)' = 0$, where $\bar{r} = r/R_2$ is the dimensionless radius.

Figure 11.10 shows distributions of temperature θ and pore pressure p in the glass-plastic cylinder fastened together with the load-bearing substrate of aluminium alloy under one-sided heating of the surface $\bar{r} = 1$ when the surface is gas-tight. Here the pore pressure p profile differs from that considered in 11.2.1 for the nontight cylinder. The pressure maximum is not within the undersurface layer but on the surface $\bar{r} = 1$. For the initial porosity $\varphi_g^0 = 0.05$, maximum values of p are practically the same (see Figures 11.2 and 11.10). We can explain this fact: at low gas permeability of the composite, when $\varphi_g^0 = 0.05$, boundary conditions have practically no effect on distributions of the pore pressure.

Another situation is realized when $\varphi_g^0 \geqslant 0.1$. In nontight structures, due to high gas permeability of the material, gases hardly accumulate in the pores, and do not create high pore pressure; but for the structures with external and internal tight surfaces all gases generated in ablation accumulate near the material surface heated, creating a level of pore pressure p higher than that for the case of $\varphi_g^0 \leqslant 1$.

Because of the effect, tangential stresses σ_ϑ having the negative sign for $\bar{r} = 1$ (Figure 11.11) increase with growing φ_g^0 and thus diminish the structure durability. Curves of limited durability of the glass-plastic cylinder $t_z(\varphi_g^0)$ are presented in

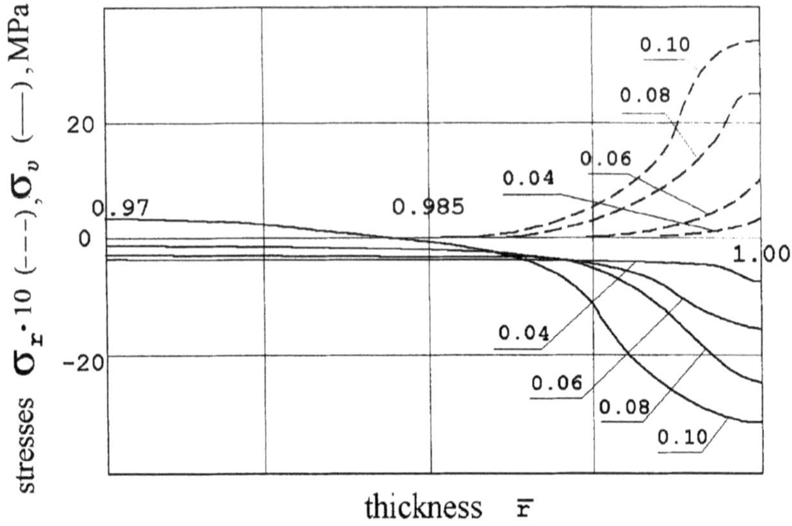

Figure 11.11. Distributions of radial σ_r and tangential σ_ϑ stresses *versus* thickness of a glass-plastic cylinder with strengthening and gas-tight surface heated $\bar{r} = 1$, notation is the same as in Figure 11.2

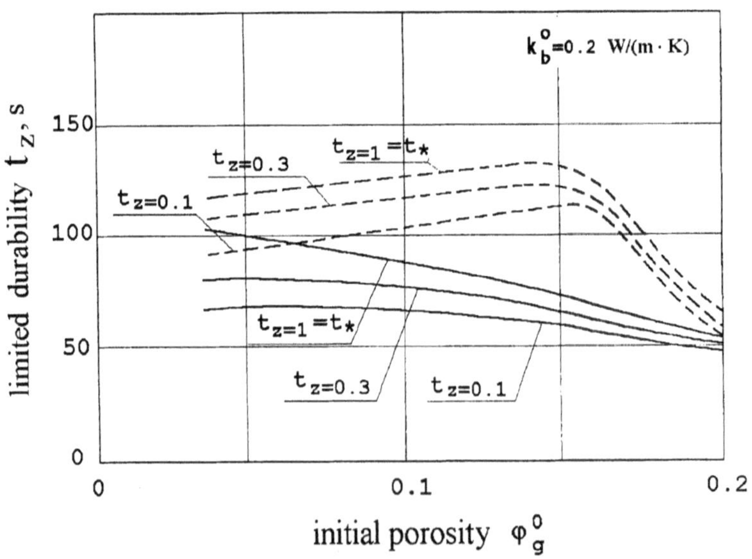

Figure 11.12. Effect of initial porosity φ_g^0 on limited durability t_z of a strengthened glass-plastic cylinder under one-sided heating for gas-tight (solid curves) and nontight (dashed lines) heated surfaces $\bar{r} = 1$

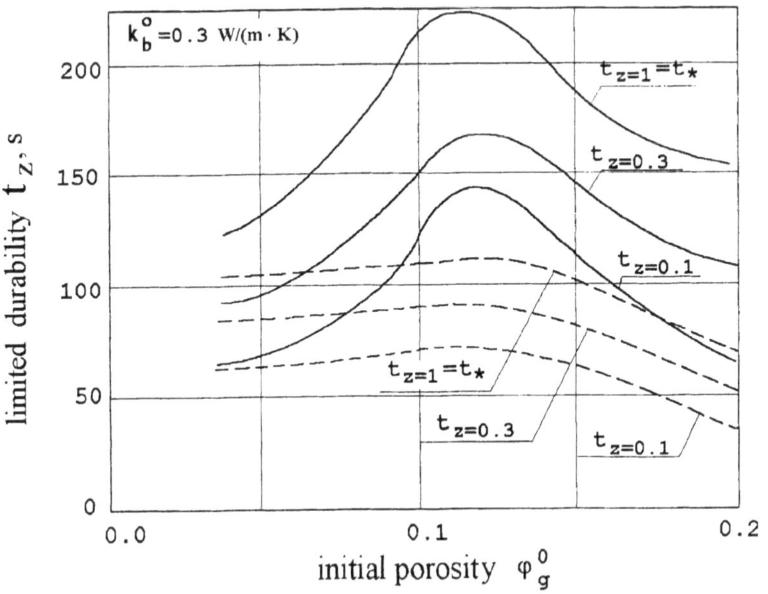

Figure 11.13. Effect of a type of strengthening on dependence of the limited durability t_z on initial porosity φ_g^0. Solid curves correspond to strengthening made of carbon-plastic, dashed lines - to strengthening made of Al alloy

Figure 11.12. The figure shows that for a tight structure the functions $t_z(\varphi_g^0)$ are monotonically decreasing. For the nontight cylinder, as in 11.2.1, the functions $t_z(\varphi_g^0)$ have a local maximum, as we explained earlier.

The effect of the load-bearing shell material on the function of limited durability $t_z(\varphi_g^0)$ is shown in Figure 11.13. The figure shows that the durability of glass-plastic material with carbon-plastic strengthening is higher than for the one strengthened with aluminium alloy. This effect is connected to the substantially lower coefficient of heat expansion for carbon-plastic; this leads to lower values of the axial stresses σ_z in the glass-plastic at the same times t when the heat front reaches the load-bearing shell.

11.3. Thermal Stresses in Thermoprotective Structures under Gasodynamical Heating

11.3.1. THE PROBLEM STATEMENT

To calculate thermal stresses in thermoprotective structures of nozzles of supersonic turbo-jet engines (STJE) (see Figure 1.10), we can apply the axisymmetric statement.

We consider a scheme of a nozzle insert structure (see Figure 1.10) in the form

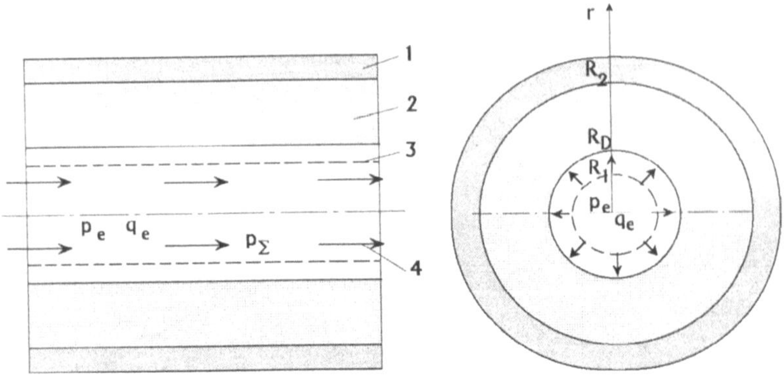

Figure 11.14. Scheme for calculations of thermal stresses in a thermoprotective structure of a nozzle of STJE: 1 – load-bearing shell, 2 – thermoprotective composite, 3 – initial location of a composite surface, 4 – hot gas flow

of a cylinder made of ablative textile composite fastened together with an external load-bearing shell. Location of the internal surface of the composite changes due to linear ablation caused by the action of high-temperature heat flux q_{e1} and pressure head p_Σ of hot gases on the material.

The problem of determination of thermal stresses and characteristics of internal heat-mass-transfer within the composite is investigated for the case of one-dimensional statement and reduced to solving the set of equations (11.25), (11.26) for functions F, φ_f, φ_b, ρ_g, θ depending on r and t, where $R_D \leqslant r < R_2$, $t \geqslant 0$. The internal surface $r = R_D(t)$ is moving. The equation of the surface motion is

$$\dot{R}_D = D, \qquad R_D(0) = R_1, \tag{11.39}$$

where D is the rate of linear ablation determined in accordance with the theory developed in Chapter 10 (see (10.3)):

$$D = D_s + D_e + D_i. \tag{11.40}$$

Rates of sublimation D_s and thermomechanical erosion D_e and D_i are determined by formulae (10.26), (10.70), (10.72):

$$D_s(p_{e1}, \theta_w) = \frac{1}{\rho}\left(\frac{\alpha}{c_p}\right)\left(\frac{p_g^*}{p_{e1}}\right)\exp\left(-\frac{E_{As}M_1}{R\theta_w}\right),$$

$$D_e(p_\Sigma, \theta_w) = D_{e1} =$$
$$= \frac{1}{\rho_f \gamma_D}\left(\frac{J_f^0 k_f}{c_f}\right)^{1/2}\left(\frac{6p_\Sigma}{\sigma_{f\perp}}\right)^{3/4}\left(\frac{R\theta_w}{E_{Af}}\right)^{1/2}\exp\left(-\frac{E_{Af}}{R\theta_w}\right), \tag{11.41}$$

$$D_i(p_\Sigma,\theta_w)=D_{i1}=$$
$$=\left(\frac{k_m J_m^0}{\rho_m c_m}\right)^{\omega_2}\left(\frac{6\rho_m l_0}{K\sigma_{1T}}\right)^{2\omega_2-1}\left(\frac{R\theta_w}{E_{Am}}\right)^{\omega_2}\exp\left(-\frac{E_{Am}\omega_2}{R\theta_w}\right).$$

We can express the coefficient γ_D in accordance with the relations derived in paragraph 10.4 as follows:

$$\gamma_D=\frac{(1-\delta_m)(1-\delta_m')}{3}\Big(2+\big(1-\vartheta_m^2+\vartheta_m^2(1-\delta_m)\big(\frac{\sigma_{f\|}}{\sigma_{f\perp}}\big)^{\omega_1}\big)^{-1}\Big).$$

Values of the material constants contained in the equations were given in Chapter 10, and $\theta_w=\theta(R_D,t)$ is the temperature of surface ablation.

Chemical composition of a hot gas flow moving in the nozzle zone is usually depleted of oxygen, therefore, at interaction of the flow with the composite, combustion of the material does not occur; we thus take $D_c=0$ in (11.40).

We introduce a moving coordinate y as follows:

$$y=\lambda(t)r+b(t), \tag{11.42}$$

$$\lambda(t)=\frac{R_2-R_1}{R_2-R_D(t)},\qquad b(t)=\frac{R_2(R_1-R_D(t)}{R_2-R_D(t)},$$

varying within the interval $R_1\leqslant y\leqslant R_2$ for $t\geqslant 0$. Then passing to variables (y,t) and introducing the function $\omega(y,t)$

$$\omega(y,t)=D\Big[\frac{R_2-R_1}{(R_2-R_D(t))^2}\Big(\frac{y-b(t)}{\lambda(t)}+R_2\Big)-\frac{R_2}{R_2-R_D(t)}\Big],$$

we can rewrite the system (11.25), (11.26) in the form

$$\beta_{22}F''+\big(\beta_{22}'+\frac{\beta_{22}}{y-b}\big)F'+\big((\frac{\beta_{12}}{y-b})'+\frac{\beta_{12}-\beta_{11}}{y-b}\big)F+$$
$$+\frac{1}{\lambda}[\overset{\circ}{\tilde{\varepsilon}}{}_\vartheta'+\frac{\overset{\circ}{\tilde{\varepsilon}}{}_\vartheta-\overset{\circ}{\tilde{\varepsilon}}{}_r}{y-b}+p(\beta_{22}+\beta_{12})'+\frac{p}{r}(\beta_{22}-\beta_{11})]=0, \tag{11.43}$$

$$\dot{\varphi}_b+\omega\varphi_b'=-\frac{J}{\rho_b^0}, \tag{11.44}$$

$$(\rho_g\varphi_g)^\bullet+\omega(\rho_g\varphi_g)'=\frac{\lambda^2}{y-b}(RK_{33}(y-b)(\rho_g\theta)')'+J\Gamma+J_f\Gamma_f, \tag{11.45}$$

$$\rho c\dot{\theta}+\rho c\omega\theta'=\frac{\lambda^2}{y-b}(k_{33}(y-b)\theta')'+c_g\lambda^2RK_{33}\theta'(\rho_g\theta)'-J\Delta\epsilon^0-J_f\Delta\epsilon_f^0. \tag{11.46}$$

Here $F'\equiv\partial F/\partial y$.

Boundary conditions at the surface $y = R_2$ are the condition of contact with the load-bearing shell assumed to be gas-tight and heat-insulated

$$y = R_2: \quad (\rho_g \theta)' = 0, \quad \theta' = 0, \tag{11.47}$$

$$\left(\frac{F}{R_2} - \varphi_g p\right) = -p_{e2} - \frac{E_\Pi h_\Pi}{(1-\nu_\Pi^2)R_2}\left(\overset{\circ}{\tilde{\varepsilon}}_\vartheta + \beta_{32}\left(\frac{F}{R_2} + \varphi_g p\right) + \beta_{22}(\lambda F' + \varphi_g p)\right).$$

Boundary conditions at the internal surface $y = R_1$ are

$$y = R_1: \quad F = -R_D p_{e1}, \quad \rho_g \theta = \frac{p_{e1}}{R},$$

$$-k_{33}\lambda\theta' = q_{e1} - q_{\mathrm{bl}} - q_w - q_{Rw} - q_s, \tag{11.48}$$

$$q_{\mathrm{bl}} = \gamma_b \lambda R K_{33}(\rho_g\theta)'(\bar{I}_e - c\theta_w), \qquad q_{Rw} = \varepsilon_w \sigma_{SB} \theta_w^4,$$

$$q_w = \left(\frac{\alpha}{c_p}\right) c(\theta_w - \theta_0), \quad q_{e1} = \frac{\alpha}{c_p}(\bar{I}_e - c\theta_w), \quad q_s = \rho D_s \Delta Q_{\mathrm{sub}}.$$

Initial conditions are

$$t = 0: \quad \varphi_f = \overset{0}{\varphi}_f, \quad \varphi_b = \varphi_{b0}, \quad \rho_g = \rho_{g0}, \quad \theta = \theta_0. \tag{11.49}$$

After determination of functions F, one can find stresses σ_r and σ_ϑ by the formulae

$$\sigma_r = \frac{\lambda F}{(y-b)} + \varphi_g p, \quad \sigma_\vartheta = \lambda F' + \varphi_g p. \tag{11.50}$$

Stresses σ_z and strains $\overset{\circ}{\tilde{\varepsilon}}_r$, $\overset{\circ}{\tilde{\varepsilon}}_\vartheta$ are determined by formulae (11.21) and (11.22).

The system of equations (11.43) - (11.46) with conditions (11.47) - (11.49) is the problem of determination of functions φ_f, φ_b, ρ_g, θ and F depending on arguments y and t.

11.3.2. NUMERICAL ANALYSIS OF THE PROBLEM

Solving the problem (11.43) – (11.49) was carried out numerically with the help of a step-by-step method using implicit difference schemes and sweeping procedures.

Figures 11.15 – 11.20 show computed results for the problem of thermoprotective glass/epoxy-phenolic composite rigidly fastened together with a load-bearing shell. Geometrical parameters were chosen as follows: $R_1/R_2 = 0.97$, $h_\Pi = 2 \cdot 10^{-3}$ m, $\nu_\Pi = 0.3$, $E_\Pi = 6 \cdot 10^4$ MPa. Conditions of thermo-force loading the structure were the following: for time $t = 0.2$ s temperature θ_e and pressure p_{e1} of gas flowing on the inside along the cylinder rapidly reach a stationary regime (see Figure 1.10). For the stationary regime, the parameters of gas flow were chosen as follows: $\bar{I}_e$ =7 MJ/kg, $(\alpha/c_p) = 2.2$ kg/(m^2·s), $p_{e1} = 6$ MPa, $q_{e1} = 14$ MW/m^2, $p_\Sigma = 4$ MPa, $\theta_e = 3500$ K, $c_g = 2$ kJ/kg.

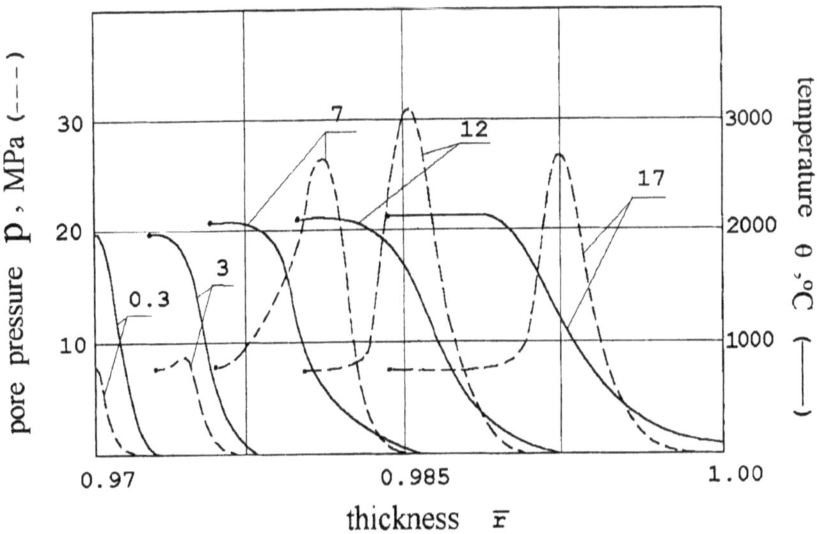

Figure 11.15. Distributions of temperature θ and pore pressure p *versus* thickness of the cylinder with taking account of surface ablation; symbols on curves are times t, s

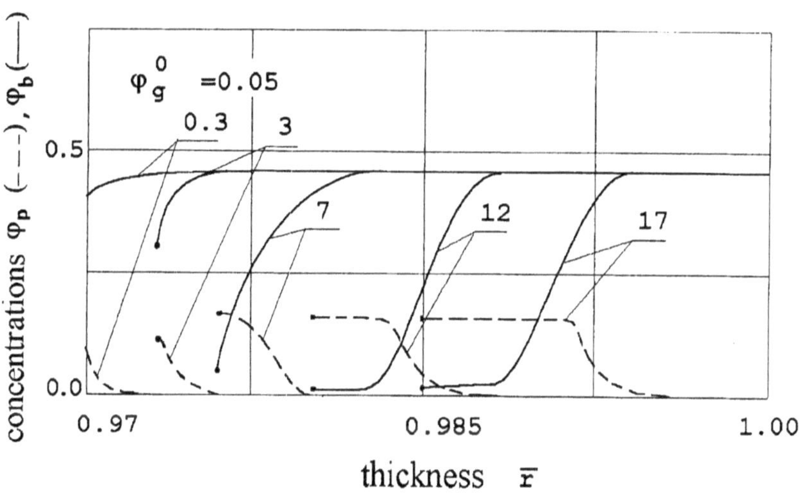

Figure 11.16. Distributions of volumetric phase concentrations φ_b φ_p *versus* thickness of the cylinder for different times; symbols at curves are times t, s

At the stationary stage of heating, the temperature θ_w of the material surface remains practically constant. Recession of the surface occurs almost by the linear

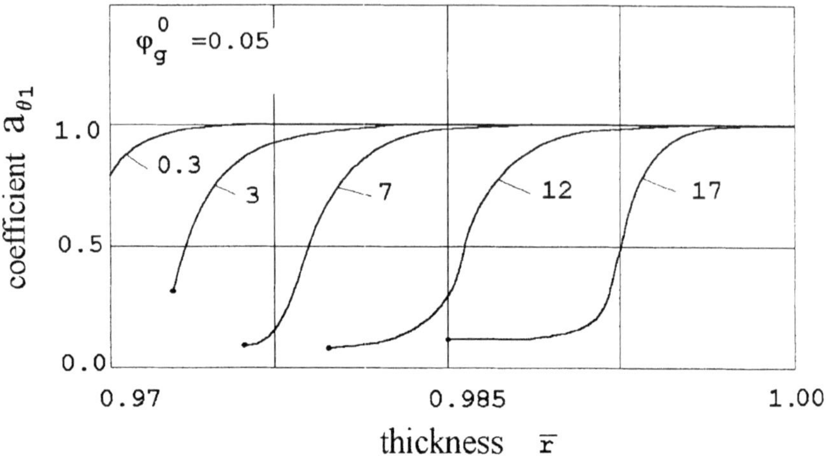

Figure 11.17. Distributions of elastic properties coefficient $a_{\theta 1}$ *versus* thickness of the cylinder for different times, symbols at curves are times t, s

law $\Delta R(t) \approx Dt$, where $\Delta R = R_D(t) - R_1$ is the total thickness of the composite which is carried away.

Let us note some peculiarities of the present problem. Due to the presence of surface ablation, there exists a solution only within the interval: $\bar{R}_D(t) \leqslant \bar{r} \leqslant 1$ (Figures 11.15 - 11.21).

Figure 11.15 shows that the pore pressure p is localized in a comparatively narrow zone corresponding to the temperature interval 300 - 1500 °C. It is connected to the fact that in a colder zone where the composite has no time for heating to thermodecomposition temperatures, the material gas permeability is very small, and gases do not reach the zone. On the contrary, in a coked zone with the porosity $\varphi_g \approx 0.26$ highly developed, high rates of gas filtration to the internal surface arise, leading to equalization of p to the inside pressure p_{e1}.

Computations were conducted for two values of the initial material porosity φ_g^0 =0.05 and 0.1. A considerable value of the inside pressure p_{e1}, which is typical for the problems of this type, leads to the appearance of compressive radial stress σ_r (Figure 11.18) having, at initial times $t < 3$ s, practically linear distribution *versus* the thickness. The appearance of peak stresses σ_r, when $t > 3$ s, is connected to a local growth of the pore pressure p in the zone of composite thermodecomposition. For material with initial porosity $\varphi_g = 0.05$, these peaks are considerably higher than for $\varphi_g = 0.1$; when $t > 12$ s, local tensile stresses arise in the composite.

A profile of the stress σ_ϑ at initial times $t < 10$ s is determined, in the main, by the presence of inside gas pressure p_{e1} and temperature gradient, therefore tensile stresses σ_ϑ are realized in the composite, and their minimum is on the internal surface (Figure 11.19). When $t > 10$ s, thermodecomposition of the material is inevitable; this leads to the appearance of two additional factors having an

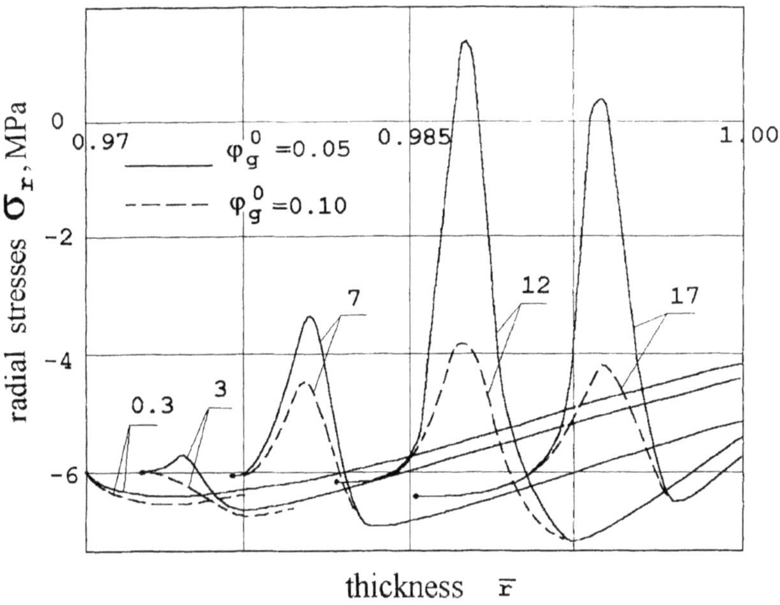

Figure 11.18. Distributions of radial stresses σ_r *versus* thickness of the ablative cylinder, symbols at curves are times t, s

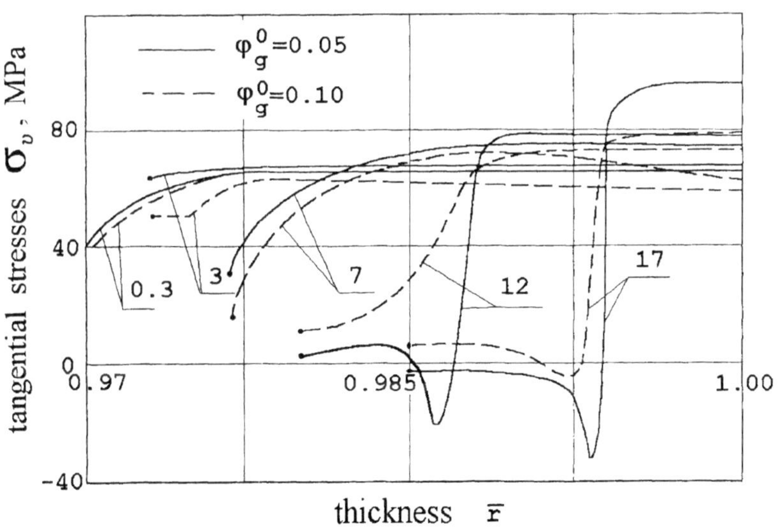

Figure 11.19. Distributions of tangential stresses σ_θ *versus* thickness of the ablative cylinder for different times; symbols at curves are times t, s

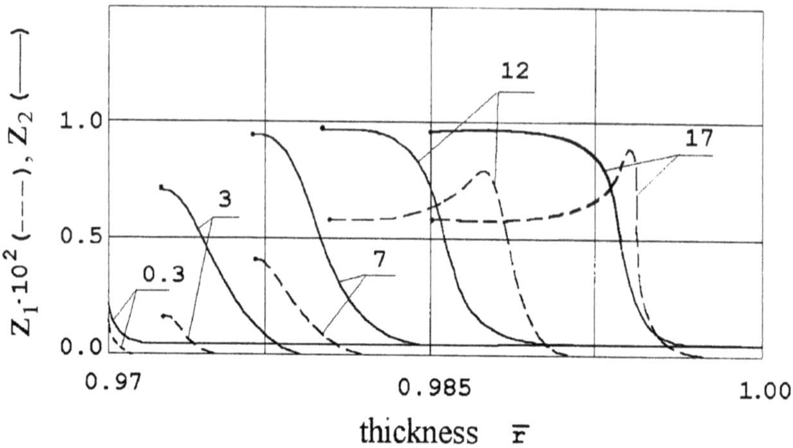

Figure 11.20. Distributions of damage parameters z_1 and z_2 in the ablative cylinder for different times, symbols at curves are times t, s

effect on the stress σ_ϑ, namely formation of local pore pressure p and shrinkage of the composite in coking. The local pore pressure causes the appearance of local minima of the stress σ_ϑ, but the shrinkage, on the contrary, leads to increasing stresses σ_ϑ at the internal surface of the cylinder. Decreasing the thickness of glass-plastic practically to the half–thickness due to surface ablation leads to raising the maximum of the stresses σ_ϑ from 60 MPa to 85 MPa.

Figure 11.20 shows that it is the tangential stresses σ_ϑ that are the most dangerous for these composite structures. Maximum value of the corresponding damage parameter z_2 determined by formula (11.34) at time $t = 17$ s is equal to about 0.9, while $z_1 < 0.01$; this is connected to the presence of high inner pressure p_{e1} displacing the stress σ_r into the domain of negative values. Thus, a characteristic type of failure of composites in structures of nozzle inserts of STJE is the type (A_2), which corresponds to the condition $z_2(t_*) = 1$.

The total rate of linear ablation D for the considered problem is equal to 0.35 mm/s. Formula (10.78) for the mass rate of ablation $\dot{m}$ is transformed as follows:

$$\dot{m} = \frac{1}{\lambda^2 t} \int_{R_1}^{R_2} (\rho_0 - \rho(t)) y dy. \tag{11.51}$$

Value of $\dot{m}$ for time t =17 s is $\dot{m}$ ≈0.73 kg/(m^2·s). Value of the effective enthalpy I_{eff} is determined by formula (10.77) and for the considered problem is $I_{\text{eff}} = q_e/\dot{m} = 19$ MJ/kg, the dimensionless rate of recession $\bar{G}$ determined by formula (10.81) is $\bar{G}$ =0.35, Value of parameter Di determined by formula (10.85) at time $t = 17$ s is Di = 42.

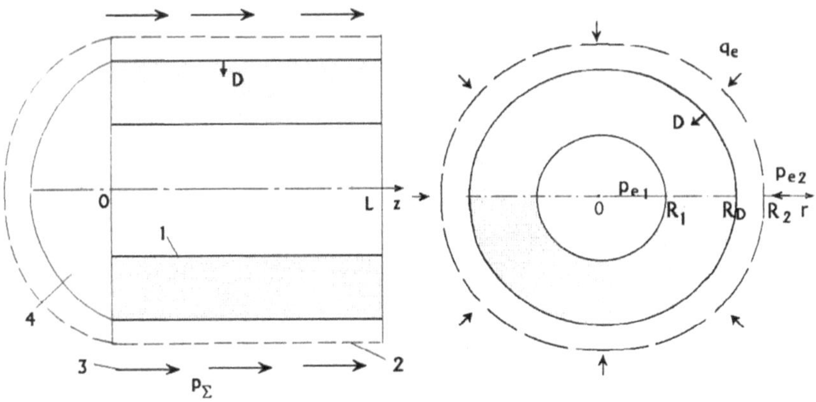

Figure 11.21. Model of a thermoprotective cylindrical shell of a re-entry vehicle under aerodynamical heating: 1 – thermoprotective composite, 2 – initial location of the external surface of the shell, 3 – overrunning gas flow, 4 – nose section of the re-entry vehicle

11.4. Thermal Stresses in Thermoprotective Structures under Aerodynamical Heating

11.4.1. THE PROBLEM STATEMENT

Typical elements of thermoprotective structures of re-entry vehicles are rotation shells (see Figure 1.4). By neglecting second order effects, we can consider many structures as cylindical, and apply the axisymmetric statement developed in paragraph 11.1 to determine thermal stresses.

We consider a hollow cylinder made of composite material subjected to intensive heating from the outside. Convective heat fluxes q_{e2}, pressure p_{e2} and pressure head p_Σ of the gas flow about the surface are assumed to be uniformly distributed over the lateral surface $r = R_D(t)$ which moves due to linear ablation of the composite. At the internal surface $r = R_1$ of the cylinder the pressure is considered to be equal to the static pressure p_{e1} (see Figure 11.21).

It is supposed that at both the ends of the cylindrical shell $z = 0$ and $z = L$ pressures $\sigma_z = -p_{e3}$ at $z = 0, L$ are given. The ends are assumed to be gas-tight and heat-insulated.

With the help of the axisymmetric problem statement we determine parameters of internal heat-mass-transfer within a structure made of ablative composite material and find stresses in the structure caused by external heating and ablation (surface and volumetric).

Within the framework of the assumptions mentioned above, solution of the problem is sought in the form of functions depending only on radius r and time t.

Stresses are determined as follows:

$$\sigma_r = \frac{F}{r} + \varphi_g p, \quad \sigma_\vartheta = \frac{\partial F}{\partial r} + \varphi_g p, \qquad \sigma_{rz} = 0, \tag{11.52}$$

$$\sigma_z = -p_{e3},$$

where F is the function of stresses determined from the equation of strain compatibility (11.23).

This equation is defined within the moving segment $[R_1, R_D(t)]$. In order to come to an immovable domain of definition, we introduce a moving coordinate y as follows:

$$y = \lambda(t)r + b(t), \tag{11.53}$$

where functions $\lambda(t)$ and $b(t)$ are determined by the formulae

$$\lambda(t) = \frac{R_2 - R_1}{R_D(t) - R_1}, \qquad b(t) = -\frac{R_1(R_2 - R_D(t))}{R_D(t) - R_1}. \tag{11.54}$$

Compatibility equations (11.25) and heat-mass-transfer equations (11.53) in the moving coordinate system take the form (11.43) - (11.46), with the only difference that the auxiliary function ω has another form, namely

$$\omega = D\frac{R_2 - R_1}{(R_D - R_1)^2}\left(\frac{y - b}{\lambda} - R_1\right). \tag{11.55}$$

Boundary and initial conditions in coordinates y, t have the form

$$y = R_1: \quad \lambda k_{33}\theta' = -q_{Rw1}, \quad F = -R_1 p_{e1}, \quad \rho_g\theta = \frac{p_{e1}}{R_1};$$

$$y = R_2: \quad \lambda k_{33}\theta' = q_{e2} - q_{\text{bl}} - q_w - q_{Rw} - q_{\text{sub}}, \tag{11.56}$$

$$F = -R_D p_e, \qquad \rho_g\theta = \frac{p_{e2}}{R_D},$$

where

$$q_{e2} = \left(\frac{\alpha}{c_p}\right)(\bar{I}_e - c\theta_w), \quad q_{\text{bl}} = \gamma_{\text{bl}}\lambda R K_{33}(\rho_g\theta)'(\bar{I}_e - c\theta_w),$$

$$q_w = \left(\frac{\alpha}{c_p}\right)c(\theta_w - \theta_0), \quad q_{Rw} = \varepsilon_w\sigma_{SB}\theta_w^4: \quad q_{\text{sub}} = \rho D_s\Delta Q_{\text{sub}}, \tag{11.57}$$

$$q_{Rw1} = \varepsilon_w\sigma_{SB}\theta_{w1}^4,$$

here θ_w and θ_{w1} are the temperatures of the surfaces R_D and R_1.

The equation system (11.43) - (11.46) with conditions (11.56), (11.49) is a problem for the determination of the functions F, ρ_g, φ_b, φ_f and θ depending on y and t. The linear ablation rate D is determined by the formulae (11.40), (11.41), however for thermoprotective structures of re-entry vehicles de-orbited in the Earth's atmosphere under aerodynamical heating, the rate D_c is usually different from zero, as the overrunning gas flow contains a considerable concentration of oxygen. Combustion rate D_c is determined by formulae paragraph 10.2. When only one chemical reaction from (10.14) is taken into account: $C + O_2 = CO_2$ (interaction of carbon components of the composite with oxygen of air), the formula (10.8) takes the form

$$D_c = \frac{1}{2\rho}\left(\left(B_w^2 + 4A_w C_w\right)^{1/2} - B_w\right),$$

where A_w, B_w and C_w are determined as follows:

$$A_w = M_1 U_1^0 \exp\left(-\frac{E_{A1}}{R\theta_w}\right),$$

$$B_w = \left(\frac{\dot{\alpha}}{c_p}\right) + \frac{M_2}{M_1}A_w, \qquad C_w = \left(\frac{\alpha}{c_p}\right) y_{2e}.$$

Here M_1 and M_2 are the molecular masses of carbon and oxygen, respectively, y_{2e} is the concentration of oxygen in the overrunning flow, and U_1^0, E_{A1} are the constants of the chemical reaction of oxidizing carbon from the work [71], $U_1^0 = 2\cdot 10^4$ mole/(m$^2\cdot$s), $E_{A1}/R = 2\cdot 10^4$ K $\cdot$ mole/kg.

Expressions for stresses in coordinates (y,t) are determined by formulae (11.50).

11.4.2. COMPUTED RESULTS

In accordance with the problem statement developed above, computations of thermal stresses were conducted for a thermoprotective cylindrical shell of a nose section of a re-entry vehicle (see Figure 11.21) in its de-orbiting in the Earth's atmosphere.

Characteristics of the composite correspond to carbon/phenolic material and all material constants are taken from Tables 4.1, 4.2, 5.1 and 5.2; it is assumed that $\varphi_f^0 = 0.6$, $\varphi_g^0 = 0.1$, $\Phi = 1°$, $\zeta = 0.5$, $\vartheta_{\max} = 10°$. Initial thickness of the shell is $h(0) = R_2 - R_1 = 11$ mm. Parameters of the overrunning gas flow $\bar{I}_e$, (α/c_p), p_{e2} and p_Σ during de-orbiting the re-entry vehicle are variables in time t and depend on the height $H(t)$ of the descent of the re-entry vehicle, the Mach number $M(t)$, the angle $\vartheta(t)$, which in turn are evaluated with the help of the system of ballistic equations (1.14). Ballistic constants for the re-entry vehicle have been taken from Table 1.2. The attack angle is zero; Figure 1.7 shows the flow about the nose section of the re-entry vehicle.

Figure 1.9 presents graphs of changing functions $H(t)$, $M(t)$ and also parameters $\bar{I}_e(t)$, $(\alpha/c_p)(t)$, $p_{e2}(t)$, $p_\Sigma(t)$ and $q_{e2}(t)$ of the flow acting on the lateral surface $r = R_D$ of the re-entry vehicle depending on time of de-orbiting the vehicle.

After evaluating the parameters $\bar{I}_e$, (α/c_p), p_{e2} and p_Σ of the overrunning flow, we determine the heat fluxes by formulae (11.57), and then solve the basic equation system (11.43) - (11.46). Solving the system has been conducted by the difference method, using implicit absolutely stable schemes and matrix sweeping procedure for the difference equations.

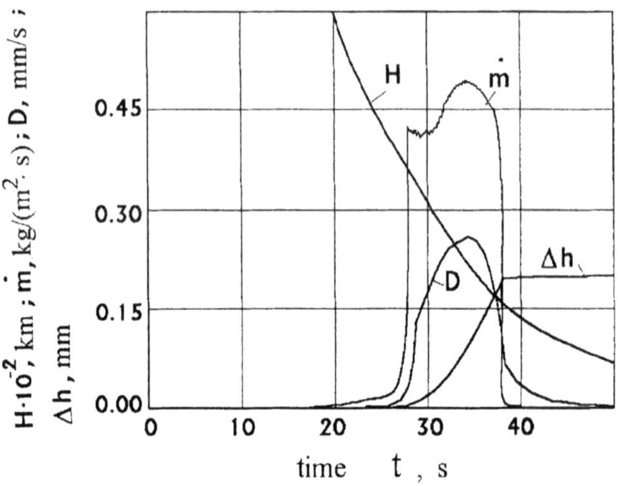

Figure 11.22. Dependence of linear D, mass $\dot{m}$ ablation rates and thickness Δh of carried-off carbon/phenolic composite on time t of de-orbiting the re-entry vehicle, H is the height of the descended re-entry vehicle

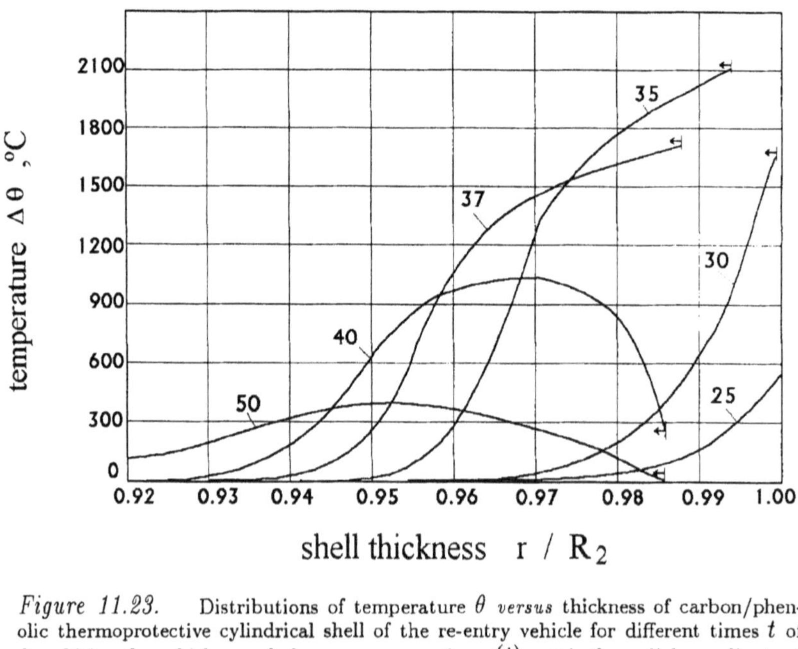

Figure 11.23. Distributions of temperature θ *versus* thickness of carbon/phenolic thermoprotective cylindrical shell of the re-entry vehicle for different times t of de-orbiting the vehicle: symbols on curves are times (t) s, r is the radial coordinate of the shell, arrows mean a location of the composite surface $r = R_D(t)$ carried-off

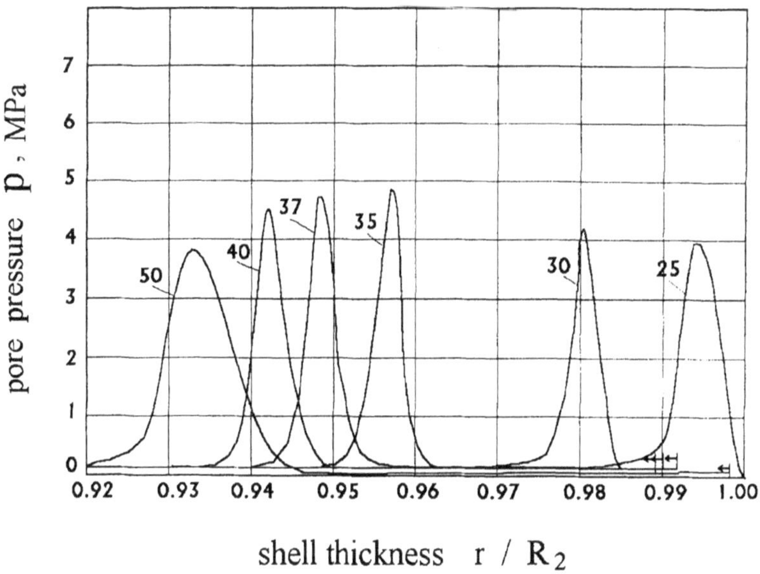

Figure 11.24. Distributions of pore gas pressure p *versus* thickness of carbon/phenolic thermoprotective cylindrical shell of the re-entry vehicle for different times t of de-orbiting the vehicle. Notations are the same as in Figure 11.23

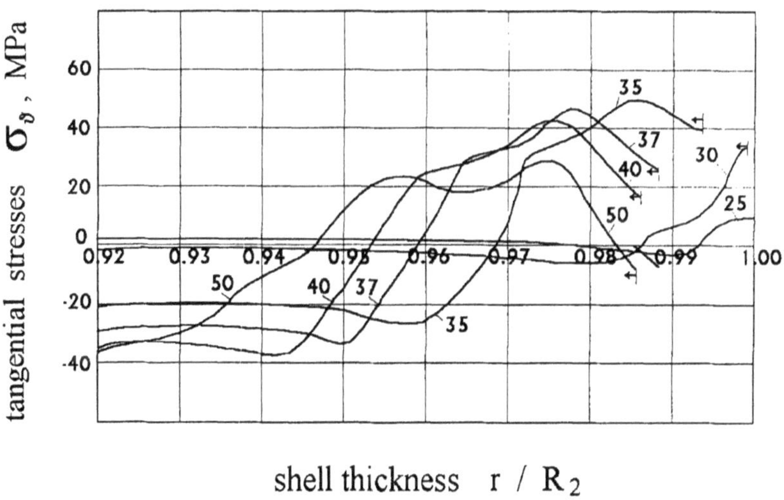

Figure 11.25. Distributions of tangential stresses σ_θ *versus* thickness of carbon/phenolic thermoprotective cylindrical shell of the re-entry vehicle for different times t of de-orbiting the vehicle. Notations are the same as in Figure 11.23

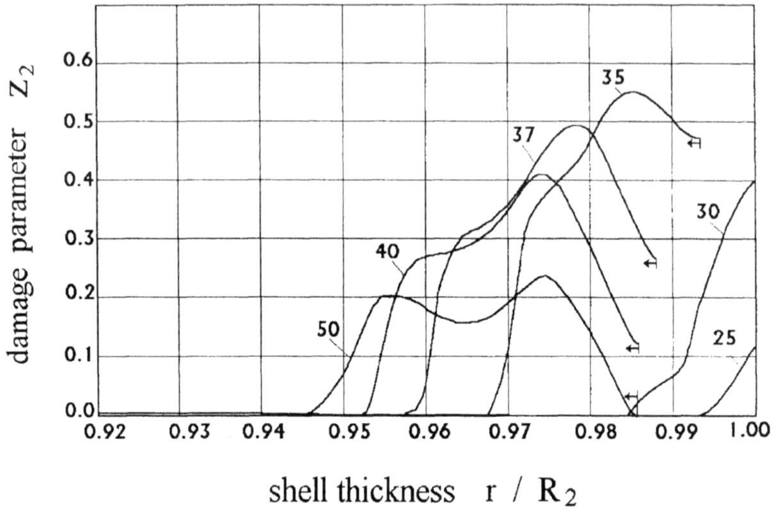

Figure 11.26. Distributions of damage parameter z_2 *versus* thickness of carbon/phenolic thermoprotective cylindrical shell of the re-entry vehicle for different times t of de-orbiting the vehicle. Notations are the same as in Figure 11.23

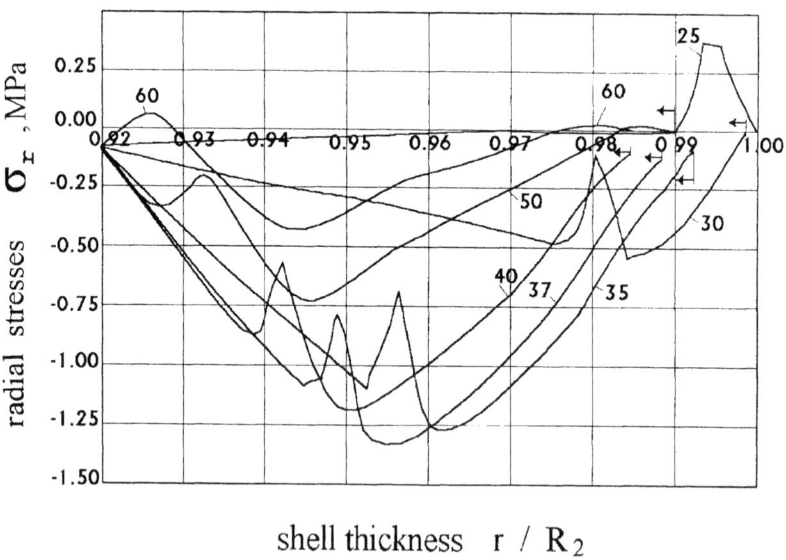

Figure 11.27. Distributions of radial stresses σ_r *versus* thickness of carbon/phenolic thermoprotective cylindrical shell of the re-entry vehicle for different times t of de-orbiting the vehicle. Notations are the same as in Figure 11.23

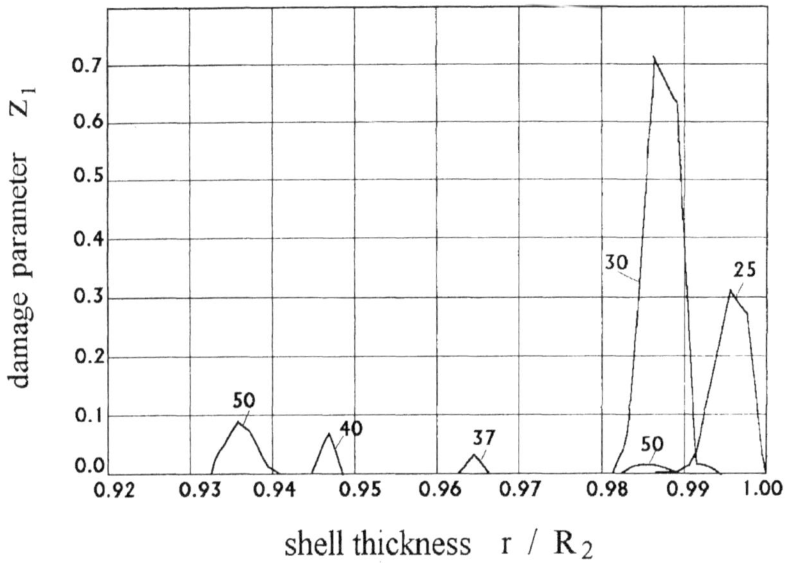

Figure 11.28. Distributions of damage parameter z_1 *versus* thickness of carbon/phenolic thermoprotective cylindrical shell of the re-entry vehicle for different times t of de-orbiting the vehicle. Notations are the same as in Figure 11.23

Figures 11.22 - 11.28 present results of computations. Figure 1.9 shows that intensive convective heat flux q_{e2} during de-orbiting a re-entry vehicle usually acts only for a limited time interval of order of 10 s (for the considered case, from t =27 s to t =37 s). Maximum values of parameters of the gas flow acting on the lateral surface $r = R_D$ are the following: heat flux $q_{e2} \approx 6$ MW/m^2, heat transfer coefficient $(\alpha/c_p) \approx 0.6$ kg/(m^2·s), pressure $p_{e2} \approx 0.12$ MPa, pressure head on the lateral surface of the shell $p_\Sigma \approx 1.6$ MPa. Functions $q_{e2}(t)$, $p_{e2}(t)$ and p_Σ reach their maxima not simultaneously (see Figure 1.9).

Under the action of the overrunning hot gas flow, there occurs a linear recession of the composite. Figure 11.22 shows dependences of the thickness $\Delta h(t) = R_2 - R_D(t)$ of the layer carried-off, the linear rate D and mass rate $\dot{m}$ of ablation on time t of de-orbiting the re-entry vehicle. Maximum magnitudes of D, $\dot{m}$ and Δh are 0.28 mm/s, 0.51 kg/(m^2·s) and 2.1 mm. Linear recession also occurs during the limited time interval t from 27 to 37 s.

Figure 11.23 shows distributions of temperature $\theta(r,t)$ through the shell thickness for different times t = 25, 30, 35, 37, 40 and 50 s. One can see that under heating, the external surface $r = R_D(t)$ of the shell and temperature profile $\theta(r,t)$ are displaced towards the internal surface $r = R_1$ with time.

The temperature θ_w of the ablative surface $r = R_D$ reaches its maximum value ≈ 2100 °C at the time of action of the maximum heat fluxes q_{e2}, then the temperature θ_w starts to decrease, and hence a total level of temperature $\theta(r,t)$ decreases. Maximum value of temperature θ_s of the 'cold' surface $r = R_1$ for the given thickness is $\theta_s \approx 120$ °C.

Figure 11.24 presents corresponding distributions of the pore gas pressure $p(r,t)$ through the shell thickness. The figure shows that the level of the pore gas pressure for the composite is high: $p_{\max} = 4 - 5$ MPa. Distribution of the pore pressure $p(r,t)$ *versus* the coordinate r has a typical peak, $p_{\max}$, which corresponds to the temperature value $\theta(r,t) \approx 350$ °C (see Figures 11.23 and 11.24). While the composite is heated, this peak is displaced towards the internal surface $r = R_1$ of the shell, and maximum value of the pore pressure $p_{\max}(t)$ increases with time from 4 MPa at $t = 25$ s to ≈ 5 MPa at $t = 35$ s, and then starts to diminish.

Figure 11.25 shows distributions of tangential stresses $\sigma_\vartheta(r,t)$ through the shell thickness for different times t of heating. During the initial time period of de-orbiting, when $t \leqslant 20$ s and ablation (volumetric and surface) of the composite has not yet started, the heat deformation $\overset{\circ}{\varepsilon}_\vartheta$ of the composite is positive; this leads to the appearance of compressive (negative) stresses σ_ϑ. As the composite is being heated, at $t \geqslant 20$ s near the external surface $r = R_D$ the volumetric ablation starts; this leads to the appearance of negative values of heat deformations $\overset{\circ}{\varepsilon}_\vartheta$, thus a shrinkage, which in turn causes the appearance of tensile (positive) stresses σ_ϑ. For this case, within the domain of relatively small temperatures, where $\theta(r,t) \leqslant 300$ °C, the stress σ_ϑ is compressive. During further heating, the profile of stresses $\sigma_\vartheta(r,t)$ is displaced towards the internal surface of the cylinder. The maximum level of the tensile stresses σ_ϑ increases up to 50 MPa. However, later on, near the external ablative surface, the stress σ_ϑ starts to fall (approximately from time $t = 33$ s) and becomes negative ($\sigma_\vartheta(R_D,t) = -8$ MPa at $t = 50$ s). This phenomenon is caused by peculiarities of the heat deformation $\overset{\circ}{\varepsilon}_\vartheta$ of composites of carbon/phenolic type under temperatures $\theta \geqslant 1800$ °C: shrinkage changes into heat expansion (see Figure 7.10), and leads to the appearance of negative values σ_ϑ within the domain immediately adjoining the ablative surface of the composite. Therefore at $t \geqslant 50$ s, the profile of tangential stresses $\sigma_\vartheta(r,t)$ has two domains of negative values and one domain of positive values. The appearance of such thermal stresses is a specific feature of these carbon/phenolic composites.

A degree of 'danger' of the tensile stresses σ_ϑ is characterized by the damage parameter z_2: the closer its value to 1, the closer the stress magnitude σ_ϑ is to the ultimate strength σ_{2T} of the composite in tension at the given temperature θ. Distributions of the damage parameter $z_2(r,t)$ through the shell thickness for different times t are shown in Figure 11.26. Functions $z_2(r,t)$ have a typical peak at $t \geqslant 30$ s, which moves towards the internal surface of the cylinder while the composite is being heated. The parameter z_2 reaches its absolute maximum equal to ≈ 0.55 at time $t \approx 35$ s, when $\overset{\circ}{\varepsilon}_\vartheta$ has a minimum at the ablative surface (the shrinkage reaches its maximum value). This time is the most dangerous for this composite from the point of view of possible destruction by type (A).

Besides the tangential stresses, there are also considerable radial thermal stresses σ_r in the cylindrical shell made of carbon/phenolic composite. Figure 11.27 shows distributions of stresses $\sigma_r(r,t)$ through the shell thickness for different times t. During the initial period of heating ($t \leqslant 20$ s), when the volumetric ablation has not yet started, heat deformation $\overset{\circ}{\varepsilon}_r$ is positive, and stresses σ_r are also positive (tensile). The volumetric ablation near the heated surface at time $t \geqslant 20$ s leads to the appearance of shrinkage deformations ($\overset{\circ}{\varepsilon}_r < 0$) and compressive stresses σ_r within the shell (see Figure 11.27). However inside the composite material heated (at $\theta \geqslant 300$ °C), as we mentioned earlier, a peak of pore pressure is formed. This peak causes the appearance of a local extremum of radial stresses σ_r. The magnitude of the stress σ_r at the peak can be positive or negative. Figure 11.27 shows that at time $t = 25$ s the extreme stress σ_r is positive, i.e. stresses σ_r are tensile ($\sigma_r \approx 0.4$ MPa), while within the interval $35 < t < 55$ s the peak stress σ_r is negative ($\sigma_r \approx$ -1.35 MPa). With heating, the peak stress σ_r is displaced towards the internal surface of the shell.

The danger of stresses σ_r, from the point of view of possible failure of the composite by type (C) (delamination), is determined by the damage parameter z_1. Figure 11.28 presents the distributions of the damage parameter $z_1(r,t)$ through the shell thickness. The figure shows that the parameter z_1 also has a peak. The peak is caused by the presence of pore gas pressure p. As follows from the computed results, maximum values of z_1 are very high and equal to approximately 0.7; this means that the composite state is close to failure by type (C) (due to delamination).

The parameter z_1 reaches its maximum value near the ablative surface at the depth $\approx$ 1.5 mm at time t =30 s. Hence, the appearance of delaminations in a composite structure of the considered type is possible under the external surface of the shell.

11.5. Thermal Stresses in Composites under Local Technological Heating

As we mentioned in paragraph 1.1, one of the sources of high-temperature actions is technological heating, in particular, laser cutting of composites, a peculiarity of which consists in the localization of the heat flux q_R supplied by radiation.

Under *local heating*, thermodecomposition and deforming processes of composites become essentially not one-dimensional. The purpose of the present paragraph is to investigate processes of internal heat-mass-transfer and the appearance of thermal stresses in composites under local heating by radiation, within the scope of an axisymmetric (two-dimensional) statement of the problem.

11.5.1. STATEMENT OF THE PROBLEM

Figure 1.14 shows a typical picture of the action of radiant heat flux q_R on a plate made of composite. Distribution of q_R on the plate surface is described by equation (1.25).

For this case, we can apply the axisymmetric statement in terms of displacements described in 11.1.6.

Boundary conditons (11.15) for this problem take the form

$$r = 0: \quad u_r = 0, \ \ \sigma_{rz} = 0, \ \ \frac{\partial \theta}{\partial r} = 0, \ \ \frac{\partial \rho_g \theta}{\partial r} = 0;$$

$$r = R_2: \quad \sigma_r - \varphi_g p = -p_{e3}, \ \ \sigma_{rz} = 0, \ \ \frac{\partial \theta}{\partial r} = 0, \ \ p = p_{e3};$$

$$z = 0: \quad \sigma_{rz} = 0, \ \ \sigma_z - \varphi_g p = -p_{e1}, \ \ p = p_{e1},$$
$$-k_{11}\frac{\partial \theta}{\partial z} = q_e^0 + q_R - q_{\text{bl}} - q_w - q_{Rw}; \tag{11.58}$$

$$z = L: \quad \sigma_{rz} = 0, \ \ \sigma_z - \varphi_g p = -p_{e2}, \ \ \frac{\partial \theta}{\partial r} = 0, \ \ p = p_{e2}.$$

The six equations (11.4) – (11.7), (11.18) for the six functions φ_g, φ_b, ρ_g, θ, u_r and u_z depending on two coordinates r, z and time t, with boundary conditions (11.58), should be supplemented by initial conditions (11.17).

The expressions for heat fluxes q_e^0, q_w and q_{Rw} are determined by formulae (11.16), and for q_{bl} we have

$$q_{\text{bl}} = \gamma_{\text{bl}} R K_{11} \frac{\partial \rho_g \theta}{\partial z}(\bar{I}_e - c\theta_w), \qquad \bar{I}_e = \bar{c}_g \theta_e. \tag{11.59}$$

We solved the two-dimensional problem numerically with the help of the iterative method of varying directions with the Chebyshev set of parameters [91, 92, 102] and the method of linearization of non-linear equations of heat-mass-transfer [92, 103]. The solution of the systems of linear algebraic equations has been realized by matrix sweeping [102]. Accuracy of the solution was verified for test problems.

11.5.2. COMPUTED RESULTS

For computations we used the material constants corresponding to glass/epoxy-phenolic composite, given in Tables 4.1, 4.2, 5.1 and 5.2; in addition, we assumed that

$$\Phi = 1^\circ, \quad \varphi_f^0 = 0.6, \quad \zeta = 0.5, \quad \vartheta_{\max} = 10^\circ, \quad \varphi_g^0 = 0.07,$$

$$\theta_e = \theta_0 = 22\ ^\circ\text{C}, \quad q_R^0 = \frac{E_0 P_\Sigma}{2\pi r_0^2} = 150\ \text{kW/m}^2,$$

$$R_2 = r_0 = 0.1\ \text{m}, \quad L = 10^{-2}\ \text{m}.$$

Figures 11.29 – 11.34 exhibit distributions (in the form of lines of equal level) of temperature $\theta(\bar{r}, \bar{z}, t)$, pore pressure $p(\bar{r}, \bar{z}, t)$, stresses $\sigma_z(\bar{r}, \bar{z}, t)$, $\sigma_r(\bar{r}, \bar{z}, t)$, $\sigma_\vartheta(\bar{r}, \bar{z}, t)$ and $\sigma_{rz}(\bar{r}, \bar{z}, t)$ in the two-dimensional domain $(\bar{r}, \bar{z}) \in (0, 1) \times (0.01)$ occupied by the plate under heating for three times $t = 5$, 15 and 30 s, where $\bar{r} = r/r_0$, $\bar{z} = z/r_0$.

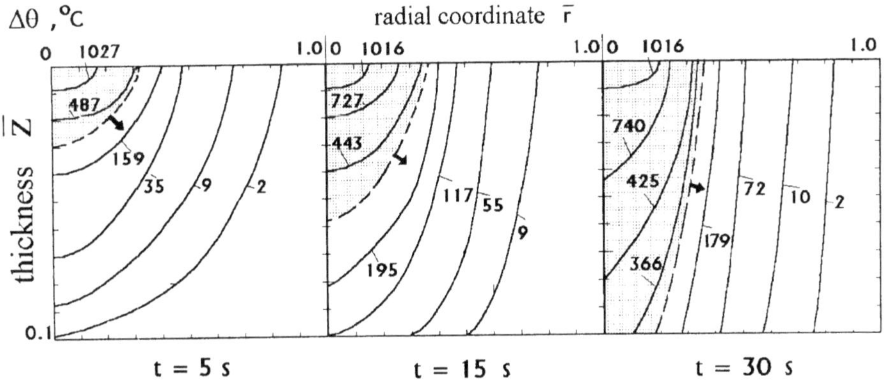

Figure 11.29. Distributions of temperature $\Delta\theta = \theta - \theta_0$, (°C), within a composite plate under local heating by radiation for three times t = 5, 15 and 30 s

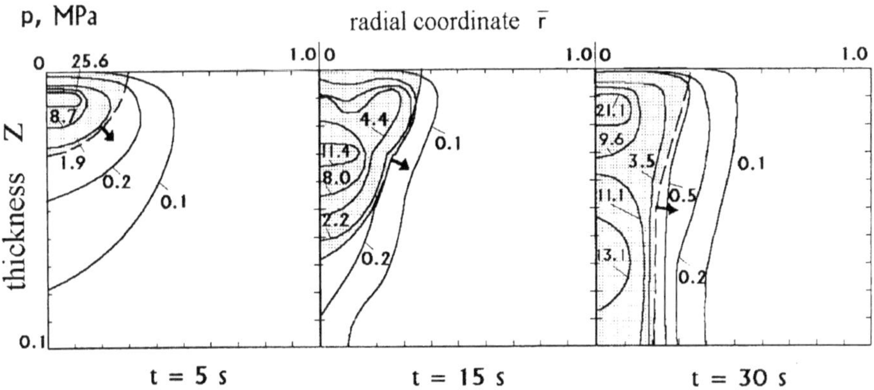

Figure 11.30. Distributions of pore pressure p, (MPa), within a composite plate under local heating by radiation for three times t = 5, 15 and 30 s

Figures 11.29 and 11.30 show shaded domains within the plate where there is thermodecomposition of the material at the considered time t; arrows indicate the direction of motion of the *thermodecomposition front*. Shaded domains in Figures 11.31 - 11.34 are the domains of positive values of stresses.

Since the maximum of heat flux q_R is in the centre of the plate $r = 0$, $z = 0$, the maximum temperature θ also occurs there, and the maximum rate of propagation of the heat front occurs along the Oz axis.

Considerable pore pressure appears in the domain of thermodecomposition, the maximum value of which, $p_{\max} \sim 25$ MPa, is localized on the Oz axis at a distance $z_0(t)$ from the heated surface. With heating, the thermodecomposition domain expands, and the peak pore pressure moves towards the back surface of the plate. At $t \approx 27$ s, when the thermodecomposition front comes to the back

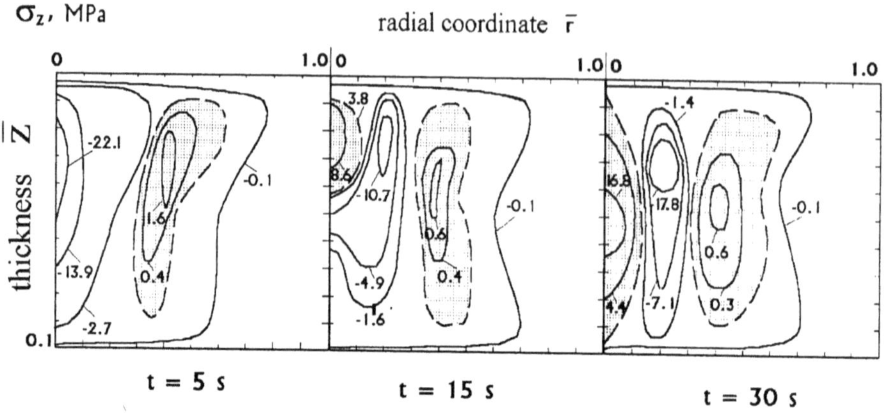

Figure 11.31. Distributions of transverse stresses σ_z, (MPa), within a composite plate under local heating by radiation for three times t = 5, 15 and 30 s

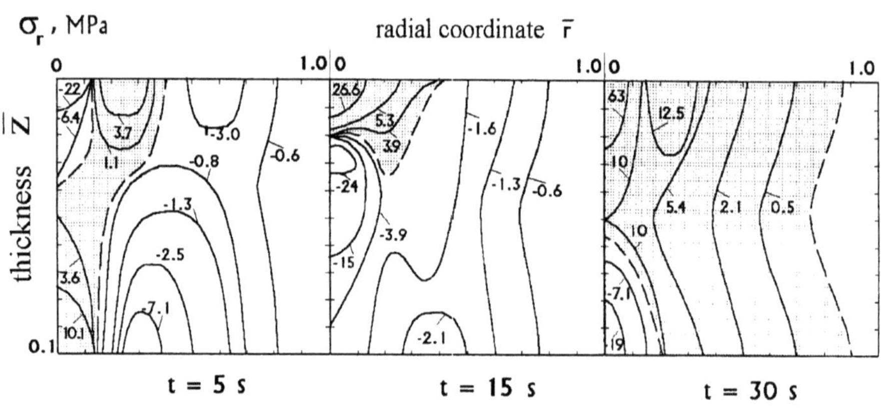

Figure 11.32. Distributions of radial stresses σ_r, (MPa), within a composite plate under local heating by radiation for three times t = 5, 15 and 30 s

surface, there occurs a reflection of the pore pressure wave: due to tightness of the surface $\bar{z} = 0.1$ pressure p sharply increases on the surface and simultaneously a peak $p_{\max}$ starts to move in the reverse direction towards the heated surface (see Figure 11.30).

Transverse stresses σ_z in the plate, just as for the one-dimensional processes (see paragraph 11.2), are determined by two factors: heat expansion and the presence of considerable pore pressure p. At initial time $t \lesssim 0.5$ s, a local tensile peak stress appears near the Oz axis; with heating, the peak moves into the peripheral zone towards the surface $\bar{r} = 1$ (see Figure 11.31), while a peak compressive stress $\sigma_z < 0$ forms on the Oz axis. When considerable pore pressure p is realized, within the thermodecomposition domain at the depth $z_0(t)$ along the Oz axis a peak tensile stress $\sigma_{z\max}$ reappears, and moves towards the back surface of the

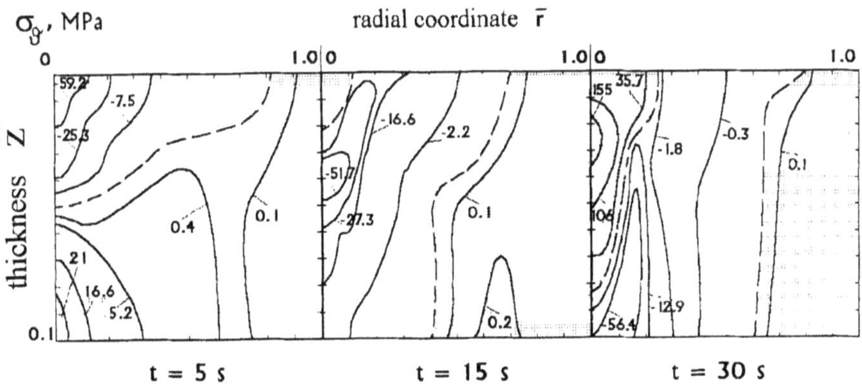

Figure 11.33. Distributions of tangential stresses σ_ϑ, (MPa), within a composite plate under local heating by radiation for three times t = 5, 15 and 30 s

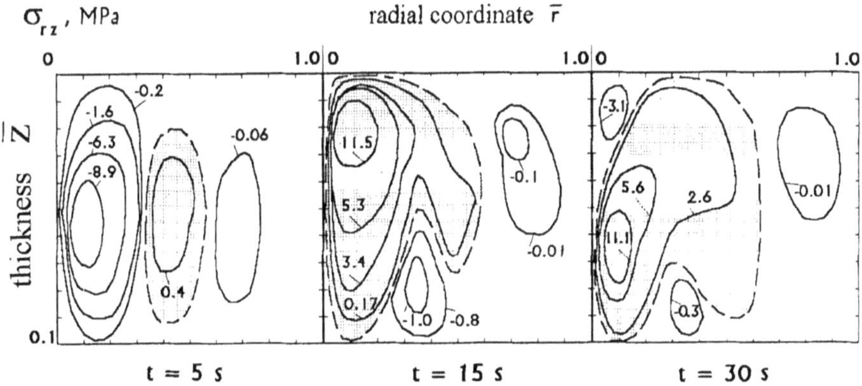

Figure 11.34. Distributions of shearing stresses σ_{rz}, (MPa), within a composite plate under local heating by radiation for three times t = 5, 15 and 30 s

plate following the peak $p_{\max}$. The magnitude $\sigma_{z\max}$ of the peak (≈16.8 MPa) is considerably higher than the first peak tensile stress σ_z appearing due to heat expansion (see Figure 11.31, t =30 s).

Distributions of radial σ_r and tangential σ_ϑ stresses within the plate are determined also by two factors: heat expansion and chemical shrinkage of the composite in thermodecomposition. At the initial time of heating $t \approx 5$ s, stresses σ_r and σ_ϑ have local minima at the point ($\bar{z} = 0$, $\bar{r} = 0$) (-22 and -60 MPa, respectively), and at the back surface at point ($\bar{z} = 0.1$, $\bar{r} = 0$) they have local maxima (10 and 21 MPa). With further heating, tensile stresses σ_r, σ_ϑ appear within the thermodecomposition domain (Figures 11.32, 11.33 at t = 15 s). The maxima of these stresses are localized at the point ($\bar{r} = 0, \bar{z} = 0$). As mentioned above, these stresses are caused by the change of sign of the heat deformations $\overset{\circ}{\varepsilon}_r$, $\overset{\circ}{\varepsilon}_\theta$, i.e.

by the appearance of chemical shrinkage. As the thermodecomposition domain expands, the shrinkage stresses σ_r and σ_ϑ become prevalent in the plate (Figures 11.32, 11.33 at $t = 30$ s), and their maxima reach values $\sigma_{r\max} \approx 63$ MPa and $\sigma_{\vartheta\max} \approx 155$ MPa, respectively.

Under local heating, in contrast to the one-dimensional case, shearing stresses σ_{rz} appear within the plate (Figure 11.34). During the initial time period $t \leqslant 5$ s, a local extremum is realized near the symmetry centre of the plate ($\bar{z} = 0.05, \bar{r} = 0$), however as the thermodecomposition domain expands, the local extremum of σ_{rz} is displaced into a zone of maximum pore pressure $p_{\max}$ and moves with the zone towards the back surface of the plate (Figure 11.34, $t = 15$ and 30 s). The value of $|\sigma_{rz}|_{\max}$ is high: ≈ 11 MPa.

Peak shrinkage stresses σ_r and σ_ϑ are the most dangerous, as well as peaks of transverse tensile σ_z and shearing σ_{rz} stresses, as they are realized within the thermodecomposition domain, where strength characteristics of the composite are minima. For the case considered above, the composite fails by both types (A) (breakage of fibres in the fabric plane) and (C) (delamination), as corresponding damage parameters z_1 and z_2 exceed 1.

CHAPTER 12

MECHANICS OF COMPOSITE THIN-WALLED SHELLS UNDER HIGH TEMPERATURES

Composite thin-walled shell structures are widely used now in design of high-speed air-space vehicles (ASVs) moving in dense layers of the atmosphere, for example, re-entry vehicles descending from the Earth's orbit. The typical aerospace composite structure is a three-layer shell (Figure 12.1): the external layer is a thermal-protective composite material based on thermostable resins (phenol, silicon-organic); the middle layer is a low-density thermoinsulative material, and the internal layer is a high strength composite material with epoxy, polyimide etc. resin base. These structures are much more effective than metallic ones. The behaviour of thin-walled composite structures under mechanical actions and normal temperatures (20–150 °C) has been well investigated (for example [3, 120, 121]). The behaviour of thin-walled composite structures under high temperatures, when there is ablation of composite materials, has specific features.

In previous chapters 1 - 3, we suggested a model for mechanical processes in ablative composites. We showed that the appearance of considerable thermal stresses in ablative composites is caused by two main effects:

1) nonlinear character of the heat deformation of the composite in heating: the usual thermal expansion is replaced by material shrinkage at ablation temperatures;
2) the interaction of internal mass-transfer and mechanical deformation at ablation temperatures.

These thermal stresses are not able to have an immediate effect on the stress state of the internal forced layer of the three-layer shell, but they can destroy the external thermal-protective layer. As shown in Chapter 9, 11, tensile tangential and transverse stresses are the most dangerous ones for laminated ablative composites. Therefore, we should expect that the external thermal-protective layer of the shell made of ablative composites would be destroyed by either cracking of the material due to breakage of reinforcing fibres, or delamination of the composite, i.e. separation of the fibre layers.

The aim of this chapter is to investigate thermal stresses caused by internal heat-mass-transfer processes in thin-walled multilayer shell structures of ablative composite materials under intensive heating.

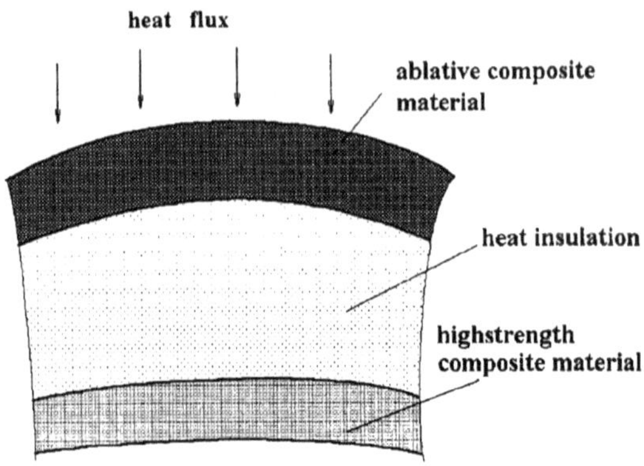

Figure 12.1. The scheme of a three-layer composite shell with an external ablative layer under intensive heat flux

12.1. General Equations for Thin-Walled Ablative Shells under High Temperatures

12.1.1. MODEL OF A MULTILAYER COMPOSITE SHELL

We consider a thin-walled shell consisting of N layers; each layer is assumed to be ablative composite material. Nonablative layers (such as the internal layer of the three-layer shell in Figure 12.1) also obey this model as a simple particular case.

According to the assumptions of Chapter 2, ablative composite material is a five-phase porous medium: a–phase is amorphic phase of reinforcing fibres, b–phase is a polymer phase of the matrix, p–phase is a solid pyrolytic phase of the matrix, g–phase is gaseous products of volumetric ablation in pores, l–phase is a crystalline phase of the fibres.

For different ablative composite layers of the shell, the solid phases can be different (different types of fibres, polymer matrices and pyrolysis residue), while the g–phase is assumed to be the same for all the layers.

The strains in all the layers of the shell are assumed to be small.

We consider the case when the ablative composite in each layer is an orthotropic laminated material, orthotropy axes of which are the axes of the chosen curvilinear orthogonal coordinate system Oq_α, $\alpha = 1, 2, 3$; the Oq_3 axis is orthogonal to the plane of the composite layers and to the shell surface. Let h_i be the thickness of the i-th layer ($i = 1, \ldots, N$), $q_3 = q_{3i-1}$ and $q_3 = q_{3i}$ the surfaces bounding the i-th layer, and h the total thickness of the shell: $q_{3N} - q_{30} = h$, $q_{3i} - q_{3i-1} = h_i$.

A general system of three-dimensional equations of internal heat-mass-transfer and deformation of a one-layer ablative composite material was given in paragraph 3.6 (the problem 'A' in displacements).

We write these equations for the i-th layer of the shell in the curvilinear coordinate system Oq_α. Mechanical equilibrium equations (3.45) have the form

$$\begin{aligned}&\frac{\partial}{\partial q_\alpha}(H_\beta H_\gamma \sigma^{(i)}_{\alpha\alpha}) + \frac{\partial}{\partial q_\beta}(H_\alpha H_\gamma \sigma^{(i)}_{\alpha\beta}) + \frac{\partial}{\partial q_\gamma}(H_\alpha H_\beta \sigma^{(i)}_{\alpha\gamma}) - \\ &\quad - \sigma^{(i)}_{\beta\beta} H_\gamma \frac{\partial H_\beta}{\partial q_\alpha} - \sigma^{(i)}_{\gamma\gamma} H_\beta \frac{\partial H_\gamma}{\partial q_\alpha} + \sigma^{(i)}_{\alpha\beta} H_\gamma \frac{\partial H_\alpha}{\partial q_\beta} + \\ &\quad + \sigma^{(i)}_{\alpha\gamma} H_\beta \frac{\partial H_\alpha}{\partial q_\gamma} - H_\beta H_\gamma \frac{\partial \varphi^{(i)}_g p^{(i)}}{\partial q_\alpha} = 0; \\ &\quad \alpha, \beta, \gamma = 1, 2, 3; \quad \alpha \neq \beta \neq \gamma; \quad i = 1, \dots, N;\end{aligned} \tag{12.1}$$

the equations of changing mass of the fibre, pyrolytic and polymer phases (3.41), (3.42) are

$$\rho^{(i)}_f \frac{\partial \varphi^{(i)}_f}{\partial t} = -J^{(i)}_f \Gamma^{(i)}_f, \quad \rho^{(i)}_p \frac{\partial \varphi^{(i)}_p}{\partial t} = (1 - \Gamma^{(i)}) J^{(i)}, \tag{12.2}$$

$$\rho^{(i)}_b \frac{\partial \varphi^{(i)}_b}{\partial t} = -J^{(i)}, \tag{12.3}$$

the equation of filtration of pyrolysis gaseous products in pores (3.43) is

$$\begin{aligned}\frac{\partial \rho^{(i)}_g \varphi^{(i)}_g}{\partial t} &= \frac{1}{H_1 H_2 H_3} \Big(\frac{\partial}{\partial q_1} \Big(\frac{H_2 H_3}{H_1} K^{(i)}_{11} R \frac{\partial \rho^{(i)}_g \theta^{(i)}}{\partial q_1} \Big) + \\ &+ \frac{\partial}{\partial q_2} \Big(\frac{H_1 H_3}{H_2} K^{(i)}_{22} R \frac{\partial \rho^{(i)}_g \theta^{(i)}}{\partial q_2} \Big) + \frac{\partial}{\partial q_3} \Big(\frac{H_1 H_2}{H_3} K^{(i)}_{33} R \frac{\partial \rho^{(i)}_g \theta^{(i)}}{\partial q_3} \Big) \Big) + \\ &+ J^{(i)} \Gamma^{(i)} + J^{(i)}_f \Gamma^{(i)}_f, \quad i = 1, \dots, N,\end{aligned} \tag{12.4}$$

and the equation of heat transfer (3.44) is

$$\begin{aligned}\rho^{(i)} c^{(i)} \frac{\partial \theta^{(i)}}{\partial t} &= \frac{1}{H_1 H_2 H_3} \Big(\frac{\partial}{\partial q_1} \Big(\frac{H_2 H_3}{H_1} k^{(i)}_{11} \frac{\partial \theta^{(i)}}{\partial q_1} \Big) + \\ &+ \frac{\partial}{\partial q_2} \Big(\frac{H_1 H_3}{H_2} k^{(i)}_{22} \frac{\partial \theta^{(i)}}{\partial q_2} \Big) + \frac{\partial}{\partial q_3} \Big(\frac{H_1 H_2}{H_3} k^{(i)}_{33} \frac{\partial \theta^{(i)}}{\partial q_3} \Big) \Big) + \frac{c_g K^{(i)}_{11} R}{H_1^2} \frac{\partial \rho^{(i)}_g \theta^{(i)}}{\partial q_1} \frac{\partial \theta^{(i)}}{\partial q_1} + \\ &+ \frac{c_g K^{(i)}_{22} R}{H_2^2} \frac{\partial \rho^{(i)}_g \theta^{(i)}}{\partial q_2} \frac{\partial \theta^{(i)}}{\partial q_2} + \frac{c_g K^{(i)}_{33} R}{H_3^2} \frac{\partial \rho^{(i)}_g \theta^{(i)}}{\partial q_3} \frac{\partial \theta^{(i)}}{\partial q_3} - \Delta e^{(i)}_0 J^{(i)} - \Delta e^{(i)}_f J^{(i)}_f, \\ &i = 1, \dots N.\end{aligned} \tag{12.5}$$

Here H_1, H_2, H_3 are *Lamé's parameters* of the shell; $k^{(i)}_{\alpha\alpha}$, $K^{(i)}_{\alpha\alpha}$ are the components of tensors of heat conductivity $\mathbf{k}^{(i)}$ and filtration $\mathbf{K}^{(i)}$ in the Oq_α coordinate system; $\sigma^{(i)}_{\alpha\beta}$ are the stresses; $\rho^{(i)}_b$ is the density of the polymer phase; $\rho^{(i)}_g$ is the gas density; $\theta^{(i)}$ is the temperature; $\varphi^{(i)}_f$, $\varphi^{(i)}_b$, $\varphi^{(i)}_p$, $\varphi^{(i)}_g$ are the volumetric concentrations of the fibre, polymer, pyrolytic and gas phases, ($\varphi^{(i)}_s = \sum_{j=1} \varphi^{(i)}_j$,

$\varphi_g^{(i)} = 1 - \varphi_s^{(i)}$); $p^{(i)}$ is the pore gas pressure; $J^{(i)}$ is the rate of volumetric ablation (pyrolysis); $\Gamma^{(i)}$ is the gasification coefficient of the polymer and $\Gamma_f^{(i)}$ is the gasification coefficient of the fibre, in the i-th layer.

The Cauchy relations connecting composite strains $\varepsilon_{\alpha\beta}^{(i)}$ with displacements $u_\alpha^{(i)}$ in the curvilinear coordinate system have the form

$$\varepsilon_{\alpha\alpha}^{(i)} = \frac{1}{H_\alpha}\frac{\partial u_\alpha^{(i)}}{\partial q_\alpha} + \frac{1}{H_\alpha H_\beta}\frac{\partial H_\alpha}{\partial q_\beta}u_\beta^{(i)} + \frac{1}{H_\alpha H_\gamma}\frac{\partial H_\alpha}{\partial q_\gamma}u_\gamma^{(i)}, \tag{12.6}$$

$$2\varepsilon_{\alpha\beta}^{(i)} = \frac{H_\alpha}{H_\beta}\frac{\partial}{\partial q_\beta}\left(\frac{u_\alpha^{(i)}}{H_\alpha}\right) + \frac{H_\beta}{H_\alpha}\frac{\partial}{\partial q_\alpha}\left(\frac{u_\beta^{(i)}}{H_\beta}\right),$$

$$\alpha, \beta, \gamma = 1, 2, 3, \quad \alpha \neq \beta \neq \gamma.$$

12.1.2. CONSTITUTIVE RELATIONS OF ABLATIVE COMPOSITES

Mechanical Characteristics

The ablative composite material of each layer is assumed to be thermoelastic, with a laminated textile structure of fibres, where the plane of textile layers within each layer of the shell is orthogonal to the Oq_3 axis. Composites with such structure can be considered as orthotropic ones.

Constitutive relations for orthotropic ablative composites have the form of (7.29)

$$\varepsilon_{\alpha\alpha}^{(i)} = \overset{\circ}{\varepsilon}{}_i^{(i)} + \frac{1}{E_\alpha^{(i)}}\left(\sigma_\alpha^{(i)} + \tilde{f}^{(i)}p^{(i)} - \nu_{\alpha\beta}^{(i)}(\sigma_\beta^{(i)} + \tilde{f}^{(i)}p^{(i)}) - \nu_{\alpha\gamma}^{(i)}(\sigma_\gamma^{(i)} + \tilde{f}^{(i)}p^{(i)})\right), \tag{12.7}$$

$$\varepsilon_{\alpha\beta}^{(i)} = \frac{\sigma_{\alpha\beta}^{(i)}}{2G_{\alpha\beta}^{(i)}}, \quad \alpha \neq \beta \neq \gamma, \quad \alpha, \beta, \gamma = 1, 2, 3, \quad i = 1, \dots, N,$$

where $E_\alpha^{(i)}$, $\nu_{\alpha\beta}^{(i)}$, $G_{\alpha\beta}^{(i)}$ are the elastic characteristics of the i-th layer of the shell. For ablative composites, the elastic characteristics depend considerably on temperature $\theta^{(i)}$ and heating time t, and, generally speaking, this dependence is different for each elastic modulus. However, as shown in Chapter 7, for textile composites we can introduce two functions $a_{\theta 1}^{(i)}$ and $a_{\theta 2}^{(i)}$ which adequately describe the temperature dependence of the elastic characteristics (as in relations (7.31), where one should make the substitution of indices $1 \longrightarrow 3$, $2 \longrightarrow 1$, $3 \longrightarrow 2$):

$$E_2^{(i)} = E_f^{(i)}\zeta^{(i)}a_{\theta 1}, \qquad G_{32}^{(i)} = G_b^{(i)}\frac{a_{\theta 2}^{(i)}}{1 - \zeta^{(i)}},$$

$$E_1^{(i)} = E_f^{(i)}(1 - \zeta^{(i)})a_{\theta 1}^{(i)}, \qquad G_{31}^{(i)} = G_b^{(i)}\frac{a_{\theta 2}^{(i)}}{\zeta^{(i)}}, \tag{12.8}$$

$$E_3^{(i)} = E_b^{(i)} a_{\theta 2}, \qquad G_{12}^{(i)} = G_f^{(i)} a_{\theta 1}^{(i)},$$

$$\nu_{23}^{(i)} = \text{const}, \quad \nu_{13}^{(i)} = \text{const}, \quad \nu_{12}^{(i)} = \text{const}, \quad i = 1 \dots N.$$

Here $E_f^{(i)}$, $E_b^{(i)}$, $G_f^{(i)}$, $G_b^{(i)}$ are the elastic and shear moduli of the amorphic and polymer phases, $\nu_{\alpha\beta}^{(i)}$ are Poisson's ratios at normal temperature, and $\zeta^{(i)}$ is the proportion of fibres in the Oq_2 direction, in the i-th layer.

The function $a_{\theta 1}^{(i)}$ describes the change in the elastic properties of the i-th layer in tension and shear in the textile plane under heating, and $a_{\theta 2}^{(i)}$ does this in tension in the transverse direction and interlayer shears. Expressions for functions $a_{\theta 1}^{(i)}$, $a_{\theta 2}^{(i)}$ are written in the form of (7.32)

$$a_{\theta 1}^{(i)} = \frac{\varphi_f^{(i)} \widetilde{a}_2^{(i)}}{1 + \vartheta_m^{(i)2} \left((1 - \varphi_f^{(i)}) \dfrac{\widetilde{a}_2^{(i)}}{m_E^{(i)} \widetilde{a}_1^{(i)}} - 1 \right)}, \tag{12.9}$$

$$a_{\theta 2}^{(i)} = \widetilde{a}_1^{(i)} (1 - \varphi_f^{(i)})^{-1}, \qquad i = 1, \dots, N,$$

where $\widetilde{a}_1^{(i)}$, $\widetilde{a}_2^{(i)}$ and $\widetilde{a}_f^{(i)}$ are expressed by the formulae (7.33) - (7.35). These formulae contain the following material constants of the layers:

$$E_f^{(i)}, \quad E_l^{(i)}, \quad E_b^{(i)}, \quad E_p^{(i)}, \quad \nu_m^{(i)}, \quad a^{(i)}, \quad a_f^{(i)}; \quad i = 1, \dots N,$$

and also the geometric constants:

$$\varphi_f^{(i)}, \qquad \zeta^{(i)}, \qquad \vartheta_{\max}^{(i)}, \qquad \phi^{(i)}.$$

Functions $\overset{\circ}{\varepsilon}{}_\gamma^{(i)}$ describe heat deformation of the i-th layer. For ablative composites, the dependence of $\overset{\circ}{\varepsilon}{}_\gamma^{(i)}$ on $\theta^{(i)}$ has the form (7.39):

$$\overset{\circ}{\varepsilon}{}_\gamma^{(i)} = \left(\alpha_f^{(i)} \varphi_f^{(i)} A_\gamma + \Omega_\gamma^{(i)} \alpha_b^{(i)} \varphi_b^{(i)} \right) (\theta^{(i)} - \theta_0) +$$
$$+ \alpha_p^{(i)} \Omega_\gamma^{(i)} \int_0^t (\theta^{(i)}(t) - \theta^{(i)}(\tau)) \dot{\varphi}_p^{(i)} d\tau - \beta_p^{(i)} \varphi_p^{(i)} \Omega_\gamma^{(i)}. \tag{12.10}$$

Here $\alpha_f^{(i)}$, $\alpha_b^{(i)}$, $\alpha_p^{(i)}$ are the coefficients of heat expansion of the amorphic, polymer and pyrolytic phases in the i-th layer, respectively, and $\beta_p^{(i)}$ is the coefficient of chemical shrinkage of the i-th layer. Constants $A_\gamma^{(i)}$ and $\Omega_\gamma^{(i)}$ are determined by formulae (7.42) for a composite layer.

Functions $f^{(i)}$ describe the influence of pore gas pressure on a stress state of the ablative composite and depend, in the main, on the value of porosity $\varphi_g^{(i)}$ of the i-th layer:

$$f_\alpha^{(i)} = \varphi_g^{(i)} \Omega_\alpha^{(i)}. \tag{12.11}$$

Functions $\widetilde{f}_\alpha^{(i)}$ are connected to $f_\alpha^{(i)}$ by formulae (11.9).

Relations inverse to (12.7) have the form

$$\sigma_{\alpha\alpha}^{(i)} = -\widetilde{f}_\alpha^{(i)} p^{(i)} + a_{\theta 1}^{(i)} \sum_{\beta=1}^{3} C_{\alpha\beta}^{(i)} (\varepsilon_{\beta\beta}^{(i)} - \overset{\circ}{\varepsilon}{}_\beta^{(i)}), \qquad \alpha = 1, 2;$$

$$\sigma_{33}^{(i)} = -\widetilde{f}_\alpha^{(i)} p^{(i)} + a_{\theta 1}^{(i)} \sum_{\beta=1}^{2} C_{3\beta}^{(i)} (\varepsilon_{\beta\beta}^{(i)} - \overset{\circ}{\varepsilon}{}_\beta^{(i)}) + a_{\theta 1}^{(i)} C_{33}^{(i)} (\varepsilon_{33}^{(i)} - \overset{\circ}{\varepsilon}{}_3^{(i)}), \tag{12.12}$$

$$\sigma_{12}^{(i)} = a_{\theta 1}^{(i)} C_{66}^{(i)} \varepsilon_{12}^{(i)}, \quad \sigma_{23}^{(i)} = a_{\theta 2}^{(i)} C_{55}^{(i)} \varepsilon_{23}^{(i)}, \quad \sigma_{13}^{(i)} = a_{\theta 2}^{(i)} C_{44}^{(i)} \varepsilon_{13}^{(i)},$$

where $C_{\alpha\beta}^{(i)}$ are the elastic moduli of orthotropic material which are expressed by the formulae similar to (11.11)

$$C_{11}^{(i)} = E_f^{(i)} \frac{1 - \nu_{23}^{(i)2} m_E^{(i)} / \varsigma^{(i)}}{\varsigma^{(i)} \Delta}, \qquad C_{22}^{(i)} = E_f^{(i)} \frac{1 - \nu_{13}^{(i)2} m_E^{(i)} / (1 - \varsigma^{(i)})}{(1 - \varsigma^{(i)}) \Delta},$$

$$C_{12}^{(i)} = E_f^{(i)} \frac{\nu_{12}^{(i)} \varsigma^{(i)} + \nu_{23}^{(i)} \nu_{13}^{(i)} m_E^{(i)}}{\varsigma^{(i)} (1 - \varsigma^{(i)}) \Delta}, \qquad C_{33}^{(i)} = E_b^{0(i)} \frac{1 - \nu_{12}^{(i)2} \varsigma^{(i)} / (1 - \varsigma^{(i)})}{m_E^{(i)} \varsigma^{(i)} (1 - \varsigma^{(i)}) \Delta},$$

$$C_{23}^{(i)} = E_f^{(i)} m_E^{(i)} \frac{\nu_{23}^{(i)} + \nu_{13}^{(i)} \nu_{12}^{(i)} \varsigma^{(i)} / (1 - \varsigma^{(i)})}{\varsigma (1 - \varsigma^{(i)}) \Delta},$$

$$C_{13}^{(i)} = E_f^{(i)} m_E^{(i)} \frac{\nu_{13}^{(i)} + \nu_{12}^{(i)} \nu_{23}^{(i)}}{\varsigma^{(i)} (1 - \varsigma^{(i)}) \Delta},$$

$$C_{66}^{(i)} = 2 G_b^{(i)}, \quad C_{44}^{(i)} = \frac{2 G_m^{(i)}}{1 - \varsigma^{(i)}}, \quad C_{55}^{(i)} = \frac{2 G_b^{(i)}}{\varsigma^{(i)}}, \tag{12.13}$$

where there are constants

$$m_E^{(i)} = E_b^{(i)} / E_f^{(i)},$$

$$\Delta = \frac{1}{\varsigma^{(i)} (1 - \varsigma^{(i)})} \left(1 - \nu_{12}^{(i)2} \frac{\varsigma^{(i)}}{1 - \varsigma^{(i)}} - m_E^{(i)} \left(\frac{\nu_{23}^{(i)2}}{\varsigma^{(i)}} + \frac{\nu_{13}^{(i)2}}{1 - \varsigma^{(i)}} + \frac{2 \nu_{13}^{(i)} \nu_{12}^{(i)} \nu_{23}^{(i)}}{1 - \varsigma^{(i)}} \right) \right).$$

Heat Characteristics

Gaseous products of pyrolysis in pores of all the layers of the shell are assumed to be ideal perfect gases:

$$p^{(i)} = R \rho_g^{(i)} \theta^{(i)}, \tag{12.14}$$

where R is the gas constant.

The coefficients of heat conductivity $k_{\alpha\alpha}^{(i)}$ and gas permeability $K_{\alpha\alpha}^{(i)}$ of ablative composite layers are determined as follows:

$$k_{33}^{(i)} = k_b^{0(i)} \varkappa_{\theta 2}^{(i)}, \quad k_{11}^{(i)} = k_f^{0(i)} (1 - \varsigma^{(i)}) \varkappa_{\theta 1}^{(i)}, \quad k_{22}^{(i)} = k_f^{0(i)} \varsigma^{(i)} \varkappa_{\theta 1}^{(i)}, \tag{12.15}$$

$$K_{\alpha\alpha}^{(i)} = K^0 \exp\,(n_k \varphi_g^{(i)}).$$

Constants $k_b^{0(i)}$, $k_f^{0(i)}$ can be obtained from experiments, and functions $\varkappa_{\theta 1}^{(i)}$, $\varkappa_{\theta 2}^{(i)}$ are defined by formulae (7.54).

The rate of volumetric ablation $J^{(i)}$ depends on temperature $\theta^{(i)}$ and is described by the equation of Arrhenius type:

$$J^{(i)} = J_0^{(i)} \exp\left(-E_A^{(i)}/R\theta^{(i)}\right), \tag{12.16}$$

where $J_0^{(i)}$ and $E_a^{(i)}$ are constants of the i-th layer.

12.1.3. BOUNDARY AND INITIAL CONDITIONS

Boundary conditions on separating surfaces of the layers have the form

$$\begin{gathered} q_3 = q_{3i}: \qquad \sigma_{3\alpha}^{(i+1)} = \sigma_{3\alpha}^{(i)}, \quad u_\alpha^{(i)} = u_\alpha^{(i+1)}, \quad \alpha = 1,2,3, \\ k_{33}^{(i)} \frac{\partial \theta^{(i)}}{\partial q_3} = k_{33}^{(i+1)} \frac{\partial \theta^{(i+1)}}{\partial q_3}, \quad \theta^{(i)} = \theta^{(i+1)}, \\ K_{33}^{(i)} \frac{\partial \rho_g^{(i)} \theta^{(i)}}{\partial q_3} = K_{33}^{(i+1)} \frac{\partial \rho_g^{(i+1)} \theta^{(i+1)}}{\partial q_3}, \qquad p^{(i)} = p^{(i+1)}, \\ i = 2, \ldots, N-1. \end{gathered} \tag{12.17}$$

Relations at the external and internal surfaces of the multilayer shell are

$$\begin{gathered} q_3 = q_{3N}: \qquad \sigma_{33}^{(N)} - \varphi_g^{(N)} p^{(N)} = -p_{eN}, \quad \sigma_{3\alpha}^{(N)} = 0, \quad \alpha = 1,2; \\ -k_{33}^{(N)} \frac{\partial \theta^{(N)}}{\partial q_3} = q_{eN} + q_{RN} - q_{\text{bl}N} - q_{wN} - q_{RwN}, \qquad p^{(N)} = p_{eN}; \\ q_3 = q_{30}: \qquad \sigma_{33}^{(1)} - \varphi_g^{(1)} p^{(1)} = -p_{e1}, \quad \sigma_{3\alpha}^{(1)} = 0, \quad \alpha = 1,2; \\ k_{33}^{(1)} \frac{\partial \theta^{(1)}}{\partial q_3} = q_{e0} + q_{R0} - q_{\text{bl}0} - q_{w0} - q_{Rw0}, \qquad p^{(1)} = p_{e0}, \end{gathered} \tag{12.18}$$

where q_{li}, q_{Ri}, $q_{\text{bl}i}$, q_{wi}, q_{Rwi}, p_{ei} $(i = 0, N)$ are the heat fluxes and gas pressure at the external and internal surfaces of the multilayer shell, respectively, and

$$q_{ei} = \left(\frac{\alpha}{c_p}\right)_i (\bar{I}_{ei} - c\theta_0), \qquad q_{\text{bl}i} = \gamma_{\text{bl}} R K_{33} \frac{\partial \rho_g^{(i)} \theta^{(i)}}{\partial q_3} (\bar{I}_{ei} - c\theta_0),$$

$$q_{wi} = \left(\frac{\alpha}{c_p}\right)_i c(\theta_\xi^{(i)} - \theta_0), \qquad q_{Rwi} = \varepsilon_w \sigma_{SB} \theta^{(i)4}, \qquad i = 0, N.$$

Here $\bar{I}_{ei} = c_g \theta_e^{(i)}$ is the enthalpy, $\theta_e^{(i)}$ is the temperature of the outside gas flow at the internal and external surfaces of the shell $i = 0, N$, and q_{Ri} is the radiant heat flux determined by (1.26).

Equation system (12.1) - (12.6) with constitutive relations (12.12), (12.14) and boundary conditions (12.17), (12.18) and initial conditions

$$t = 0: \quad \varphi_b^{(i)} = \varphi_{b0}^{(i)}, \quad \varphi_p^{(i)} = 0, \quad p^{(i)} = p_0, \quad \theta^{(i)} = \theta_0 \tag{12.19}$$

is the three-dimensional problem statement 'A' of thermomechanics for multilayer ablative composites *in curvilinear coordinates* Oq_α.

For the ablative shells we shall consider, these relations become two-dimensional.

12.2. Main Assumptions for Thin-Walled Ablative Shells

The ablative shell is assumed to be very *thin*. Then we can write the following expressions for Lamé's parameters:

$$H_3 = 1, \qquad H_\alpha = A_\alpha(1 + k_\alpha q_3) \approx A_\alpha, \qquad \frac{\partial H_\alpha}{\partial q_3} = A_\alpha k_\alpha, \tag{12.20}$$

where $A_\alpha(q)$ are the coefficients of the first quadratic form of the middle surface ($q_3 = 0$), k_α are its principal curvatures, and $k_\alpha q_3 \ll 1$.

There are many theories for thin-walled shells. For laminated composite materials, Timoshenko shell theory is simple and adequate. However, this theory and its analogs cannot be applied to ablative shells, as these theories do not include the stress σ_{33} from internal gas generation in the ablative laminated composite; this is the most dangerous delaminating stress. We must formulate a new theory.

We make the following assumptions.

(a) Distributions of displacements $u_\alpha^{(i)}$, $u_3^{(i)}$ through each layer are given by

$$u_\alpha^{(i)} = U_\alpha + q_3 \gamma_\alpha, \qquad u_3^{(i)} = W; \tag{12.21}$$

the five functions $U_\alpha(q_\beta)$, $W(q_\beta)$, $\gamma_\alpha(q_\beta)$ depend on the coordinates q_β, $\beta = 1, 2$.

(b) Distributions of shear $\sigma_{\alpha 3}^{(i)}$ and normal $\sigma_{33}^{(i)}$ stresses in the layers are chosen as quadratic functions of the coordinate q_3:

$$\sigma_{\alpha 3}^{(i)} = \eta^{(i)}(q_3)\psi_\alpha^{(i)} + \xi^{(i)}(q_3) g_\alpha^{(i)} + \sum_{j=1}^{i-1} g_\alpha^{(j)}, \tag{12.22}$$

$$\sigma_{33}^{(i)} = \eta^{(i)}(q_3)\psi_3^{(i)} + (p_{i-1} - p_i)\xi^{(i)}(q_3) + \varphi_g^{(i)} p^{(i)} - p_{i-1}, \tag{12.23}$$

$$\alpha = 1, 2, \qquad i = 1, \ldots, N,$$

where $p_i = p^{(i)}(q_{3i})$ is the pore pressure at the separation surface of the i-th and $(i-1)$-th layers, and $p_0 = p_{e0}$, $p_N = p_{eN}$.
Functions $\xi^{(i)}$ and $\eta^{(i)}$ of coordinate q_3 have the form

$$\eta^{(i)}(q_3) = \frac{1}{4} - \Big(\frac{q_3 - q_{3i-1}}{h_i} - \frac{1}{2}\Big)^2, \tag{12.24}$$

$$\xi^{(i)}(q_3) = \frac{q_3 - q_{3i-1}}{h_i}, \qquad i = 1, \ldots N;$$

and U_α, γ_α, W, $\psi_\alpha^{(i)}$, $\psi_3^{(i)}$, $g_\alpha^{(i)}$ are functions of t and q_α, $\alpha = 1, 2$.

(c) Instead of constitutive relations (the second, fourth and fifth equations of (12.12)), we consider the integral relations

$$\int_{q_{3i-1}}^{q_{3i}} \sigma_{33}^{(i)} dq_3 = -\int_{q_{3i-1}}^{q_{3i}} f_3^{(i)} p^{(i)} dq_3 + \sum_{\beta=1}^{2} C_{3\beta}^{(i)} \int_{q_{3i-1}}^{q_{3i}} a_{\theta 1}^{(i)} (\varepsilon_{\beta\beta}^{(i)} - \overset{\circ}{\varepsilon}{}_\beta^{(i)}) dq_3 +$$
$$+ C_{33}^{(i)} \int_{q_{3i-1}}^{q_{3i}} a_{\theta 2}^{(i)} (\varepsilon_{33}^{(i)} - \overset{\circ}{\varepsilon}{}_3^{(i)}) dq_3, \tag{12.25}$$

$$\int_{q_{3i-1}}^{q_{3i}} \sigma_{\alpha 3}^{(i)} dq_3 = C_{\alpha+3,\alpha+3}^{(i)} \int_{q_{3i-1}}^{q_{3i}} a_{\theta 2}^{(i)} \varepsilon_{\alpha 3}^{(i)} dq_3, \qquad \alpha = 1, 2. \tag{12.26}$$

(d) External heat fluxes q_{eN} and q_{e0} are considered to change slowly along the shell surface, and, thus, temperature $\theta^{(i)}$, pore pressure $p^{(i)}$, gas density $\rho_g^{(i)}$ and phase concentrations $\varphi_f^{(i)}$, $\varphi_b^{(i)}$, $\varphi_p^{(i)}$, $\varphi_g^{(i)}$ within each layer can be assumed to be functions only of coordinate q_3 and time t. Dependence on q_α is taken into account only parametrically, i.e. only by changing external conditions q_{ei}, p_{ei} at the surface.

12.3. Peculiarities of the Theory of Composite Shells under High Temperatures

12.3.1. MECHANICAL EQUATIONS FOR THIN-WALLED ABLATIVE SHELLS

Forces $T_{\alpha\beta}$, *moments* $M_{\alpha 1}$, M_{12}, *transverse forces* Q_α are introduced, as usual, for the multilayer shell:

$$T_{\alpha\beta} = \sum_{i=1}^{N} \int_{q_{3i-1}}^{q_{3i}} \sigma_{\alpha\beta}^{(i)} dq_3,$$

$$M_{\alpha\beta} = \sum_{i=1}^{N} \int_{q_{3i-1}}^{q_{3i}} \sigma_{\alpha\beta}^{(i)} q_3 dq_3, \qquad \alpha, \beta = 1, 2;$$

$$Q_\alpha = \sum_{i=1}^{N} \int_{q_{3i-1}}^{q_{3i}} \sigma_{\alpha 3}^{(i)} dq_3. \tag{12.27a}$$

In addition, for the ablative shell we introduce the *force* P_g and *moment* M_g *of pore pressure*:

$$M_g = \frac{1}{\bar{\varphi}_g} \sum_{i=1}^{N} \int_{q_{3i-1}}^{q_{3i}} \varphi_g^{(i)} p^{(i)} q_3 dq_3,$$

$$P_g = \frac{1}{\bar{\varphi}_g} \sum_{i=1}^{N} \int_{q_{3i-1}}^{q_{3i}} \varphi_g^{(i)} p^{(i)} dq_3, \tag{12.27b}$$

$$\bar{\varphi}_s = \frac{1}{h} \int_{q_{3i-1}}^{q_{3i}} \varphi_s^{(i)} dq_3, \qquad \bar{\varphi}_g = 1 - \bar{\varphi}_s.$$

Averaging the equilibrium equation (12.1) over the thickness, we obtain

$$\frac{\partial A_2 T_{11}}{\partial q_1} + \frac{\partial A_1 T_{21}}{\partial q_2} - \frac{\partial A_2}{\partial q_1} T_{22} + \frac{\partial A_1}{\partial q_2} T_{12} + A_1 A_2 k_1 Q_1 - A_2 \frac{\partial \bar{\varphi}_g P_g}{\partial q_1} = 0;$$

$$\frac{\partial A_1 T_{22}}{\partial q_2} + \frac{\partial A_2 T_{12}}{\partial q_1} - \frac{\partial A_1}{\partial q_2} T_{11} + \frac{\partial A_2}{\partial q_1} T_{21} + A_1 A_2 k_2 Q_2 - A_1 \frac{\partial \bar{\varphi}_g P_g}{\partial q_2} = 0; \tag{12.28}$$

$$-A_1 A_2 (k_1 T_{11} + k_2 T_{22}) + \frac{\partial A_2 Q_1}{\partial q_1} + \frac{\partial A_1 Q_2}{\partial q_2} - p_3 A_1 A_2 - (k_1 + k_2) A_2 A_1 \bar{\varphi}_g P_g = 0.$$

Multiplying the equation (12.1) for $\alpha = 1, 2$ by q_3 and integrating over the thickness, we find two equations for the moments

$$\frac{\partial A_2 M_{11}}{\partial q_1} + \frac{\partial A_1 M_{21}}{\partial q_2} + \frac{\partial A_1}{\partial q_2} M_{12} - \frac{\partial A_2}{\partial q_1} M_{22} - A_1 A_2 Q_1 - A_2 \frac{\partial \bar{\varphi}_g M_g}{\partial q_1} = 0; \tag{12.29}$$

$$\frac{\partial A_1 M_{22}}{\partial q_1} + \frac{\partial A_2 M_{12}}{\partial q_2} + \frac{\partial A_2}{\partial q_1} M_{21} - \frac{\partial A_1}{\partial q_2} M_{11} - A_1 A_2 Q_2 - A_1 \frac{\partial \bar{\varphi}_g M_g}{\partial q_2} = 0.$$

12.3.2. STRAINS AND STRESSES IN A MULTILAYER ABLATIVE SHELL

Distributions (12.21) - (12.23) satisfy the contact conditions (12.17) at separating surfaces of the shell layers, and the conditions (12.18) at the external and internal surfaces of the shell.

On substituting expressions (12.21) into kinematic relation (12.6) for each the layer, we derive the expression for strains $\varepsilon_{\alpha\beta}^{(i)}$ in layers:

$$\varepsilon_{\alpha\beta}^{(i)} = e_{\alpha\beta} + q_3 \varkappa_{\alpha\beta}, \qquad \varepsilon_{33}^{(i)} = 0, \qquad \alpha, \beta = 1, 2;$$

$$\varepsilon_{\alpha 3}^{(i)} = e_{\alpha 3}. \tag{12.30}$$

Strains $e\alpha\beta$, *curvatures* $\varkappa_{\alpha\beta}$ *of the middle surface* and shear strain $e_{\alpha 3}$ are determined by the formulae

$$e_{\alpha 3} = \frac{1}{A_\alpha}\frac{\partial W}{\partial q_\alpha} + \gamma_\alpha - k_\alpha U_\alpha,$$

$$e_{\alpha\alpha} = \frac{1}{A_\alpha}\frac{\partial U_\alpha}{\partial q_\alpha} + \frac{1}{A_1 A_2}\frac{\partial A_\alpha}{\partial q_\beta} U_\beta + k_\alpha W, \qquad \alpha = 1, 2; \tag{12.31}$$

$$2e_{12} = \frac{1}{2A_2}\left(\frac{\partial U_1}{\partial q_2}\right) + \frac{1}{2A_1}\left(\frac{\partial U_2}{\partial q_1}\right) - \frac{1}{2A_1 A_2}\left(\frac{\partial A_1}{\partial q_2}U_1 + \frac{\partial A_2}{\partial q_1}U_2\right),$$

$$\varkappa_{\alpha\alpha} = \frac{1}{A_\alpha}\frac{\partial \gamma_\alpha}{\partial q_\alpha} + \frac{1}{A_1 A_2}\frac{\partial A_\alpha}{\partial q_\beta}\gamma_\beta.$$

On substituting distributions (12.22) for stresses $\sigma_{\alpha 3}^{(i)}$ and (12.30) for strains $\varepsilon_{\alpha 3}^{(i)}$ into integral equations (12.26), we obtain relations connecting the functions $\psi_\alpha^{(i)}$, $g_\alpha^{(i)}$ in the layers to $e_{\alpha 3}$:

$$\frac{\psi_\alpha^{(i)}}{6} + \frac{g_\alpha^{(i)}}{2} + \sum_{j=1}^{i-1} g_\alpha^{(j)} = \frac{e_{\alpha 3}}{h_i} C_{\alpha+3,\alpha+3}^{(i)} a_{\theta 2}^{(i)(0)}, \tag{12.32}$$

$$\frac{\psi_\alpha^{(i)}}{12} + \frac{g_\alpha^{(i)}}{3} + \frac{1}{2}\sum_{j=1}^{i-1} g_\alpha^{(j)} = \frac{e_{\alpha 3}}{h_i} C_{\alpha+3,\alpha+3}^{(i)} \left(a_{\theta 2}^{(i)(1)} - q_{3i} a_{\theta 2}^{(i)(0)}\right), \quad i = 1, \ldots, N.$$

Here and below, the following notation for values averaged over the thickness of the i-th layer is used:

$$a_{\theta k}^{(i)(j)} = \int_{q_{3i-1}}^{q_{3i}} a_{\theta k}^{(i)} q_3^j dq_3 \qquad k = 1, 2, \qquad j = 0, 1, 2;$$

$$\overset{\circ}{\varepsilon}{}_\beta^{(i)(j)} = \int_{q_{3i-1}}^{q_{3i}} a_{\theta 1}^{(i)} \overset{\circ}{\varepsilon}{}_\beta^{(i)} q_3^j dq_3, \qquad \beta = 1, 2; \tag{12.33}$$

$$\overset{\circ}{\varepsilon}{}_3^{(i)(j)} = \int_{q_{3i-1}}^{q_{3i}} a_{\theta 2}^{(i)} \overset{\circ}{\varepsilon}{}_3^{(i)} q_3^j dq_3, \quad i = 1, \ldots N;$$

$$P_{g\alpha}^{(i)} = \int_{q_{3i-1}}^{q_{3i}} p^{(i)} f_\alpha^{(i)} dq_3, \quad M_{g\alpha}^{(i)} = \int_{q_{3i-1}}^{q_{3i}} f_\alpha^{(i)} p^{(i)} q_3 dq_3, \qquad \alpha = 1, 2, 3;$$

$$P_g^{(i)} = \int_{q_{3i-1}}^{q_{3i}} \varphi_g^{(i)} p^{(i)} dq_3.$$

From equations (12.32) we find

$$\frac{g_\alpha^{(i)}}{6} = \frac{e_{\alpha 3}}{h_i^2} C_{\alpha+3,\alpha+3}^{(i)} \left(2a_{\theta 2}^{(i)(1)} - (2q_{3i} + h_i) a_{\theta 2}^{(i)(0)}\right), \tag{12.34}$$

$$\frac{1}{6}\psi_\alpha^{(i)} = 2\frac{e_{\alpha 3}}{h_i^2} C_{\alpha+3,\alpha+3}^{(i)}\big((3q_{3i}+2h_i)a_{\theta 2}^{(i)(0)} - 3a_{\theta 2}^{(i)(1)}\big) - \sum_{j=1}^{i-1} g_\alpha^{(j)},$$

$$i = 1,\ldots,N, \qquad \alpha = 1,2.$$

As before, on substituting distribution (12.23) for stress $\sigma_{33}^{(i)}$ and (12.30) for strains $\varepsilon_{\alpha\alpha}^{(i)}$ into the integral equation (12.25), we obtain relationships connecting functions $\psi_3^{(i)}$ to $e_{\beta\beta}$ and $\varkappa_{\beta\beta}$:

$$\frac{1}{6}\psi_3^{(i)} = \frac{p_i + p_{i-1}}{2} + \frac{P_g^{(i)} + P_{g3}^{(i)}}{h_i} + \frac{e_{\beta\beta}}{h_i}\sum_{\beta=1}^{2} C_{3\beta}^{(i)} a_{\theta 1}^{(i)(0)} +$$
$$+ \frac{\varkappa_{\beta\beta}}{h_i}\sum_{\beta=1}^{2} C_{3\beta}^{(i)} a_{\theta 1}^{(i)(1)} - \frac{1}{h_i}\sum_{\beta=1}^{3} C_{3\beta}^{(i)} \overset{\circ}{\varepsilon}{}_\beta^{(i)(0)}, \qquad i = 1,\ldots,N. \quad (12.35)$$

If the functions γ_α, U_α and W are known, then all the functions $\psi_\alpha^{(i)}$, $g_\alpha^{(i)}$ and $\psi_3^{(i)}$ can be found from formulae (12.34) and (12.35), and, hence, stresses $\sigma_{\alpha 3}^{(i)}$, $\sigma_{33}^{(i)}$ can be determined by formulae (12.22) and (12.23).

The stresses $\sigma_{\alpha\alpha}^{(i)}$ in the shell layers are determined by the constitutive relations (12.12), into which we should substitute the strains (12.30):

$$\sigma_{\alpha\alpha}^{(i)} = -p^{(i)} f^{(i)} + a_{\theta 1}^{(i)} \sum_{\beta=1}^{2} C_{\alpha\beta}^{(i)}(e_{\beta\beta} + q_3\varkappa_{\beta\beta}) - \sum_{\beta=1}^{3} C_{\alpha\beta}^{(i)} \overset{\circ}{\varepsilon}{}_\beta^{(i)}, \qquad \alpha = 1,2;$$

$$\sigma_{12}^{(i)} = a_{\theta 1}^{(i)} C_{66}^{(i)}(e_{12} + q_3\varkappa_{12}). \qquad (12.36)$$

12.3.3. CONSTITUTIVE RELATIONS FOR MULTILAYER ABLATIVE SHELLS

On substituting expressions (12.36) into the force and moment equations (12.27), we obtain

$$T_{\alpha\alpha} = -P_{g\alpha} + \sum_{\beta=1}^{2} \bar{C}_{\alpha\beta}(e_{\beta\beta} + N_{\alpha\beta}\varkappa_{\beta\beta}) - \overset{\circ}{T}_\alpha,$$

$$T_{12} = \bar{C}_{66} e_{12} + N_{66}\varkappa_{12},$$

$$M_{\alpha\alpha} = -M_{g\alpha} + \sum_{\beta=1}^{2} N_{\alpha\beta}(e_{\beta\beta} + D_{\alpha\beta}\varkappa_{\beta\beta}) - \overset{\circ}{M}_\alpha, \qquad (12.37)$$

$$M_{12} = N_{66} e_{12} + D_{66}\varkappa_{12},$$

$$Q_\alpha = \bar{C}_{\alpha+3,\alpha+3} e_{\alpha 3}, \qquad \alpha = 1,2.$$

Equations (12.37) are the *constitutive relations for the multilayer ablative shell.* Here we use the following notation for the *membrane* $\bar{C}_{\alpha\beta}$, *combined* $N_{\alpha\beta}$ and *bending* $D_{\alpha\beta}$ *stiffnesses* of a multilayer shell:

$$\bar{C}_{\alpha\beta} = \sum_{i=1}^{N} C_{\alpha\beta}^{(i)} a_{\theta 1}^{(i)(0)}, \qquad N_{\alpha\beta} = \sum_{i=1}^{N} C_{\alpha\beta}^{(i)} a_{\theta 1}^{(i)(1)},$$

$$D_{\alpha\beta} = \sum_{i=1}^{N} C_{\alpha\beta}^{(i)} a_{\theta 1}^{(i)(2)}, \quad \alpha, \beta = 1, 2; \tag{12.38}$$

$$\bar{C}_{\alpha+3,\alpha+3} = \sum_{i=1}^{N} C_{\alpha+3,\alpha+3}^{(i)} a_{\theta 2}^{(i)(0)},$$

for *forces of thermal stresses* $\overset{\circ}{T}_\alpha$ and *moments of thermal stresses* $\overset{\circ}{M}_\alpha$:

$$\overset{\circ}{T}_\alpha = \sum_{\beta=1}^{3} \sum_{i=1}^{N} C_{\alpha\beta}^{(i)} \overset{\circ}{\varepsilon}{}_{\beta}^{(i)(0)}, \qquad \overset{\circ}{M}_\alpha = \sum_{\beta=1}^{3} \sum_{i=1}^{N} C_{\alpha\beta}^{(i)} \overset{\circ}{\varepsilon}{}_{\beta}^{(i)(1)}, \tag{12.39}$$

and for *forces of phase interaction* $P_{g\alpha}$ and *moments of phase interaction* $M_{g\alpha}$ of a multilayer shell:

$$P_{g\alpha} = \sum_{i=1}^{N} P_{g\alpha}^{(i)}, \qquad M_{g\alpha} = \sum_{i=1}^{N} M_{g\alpha}^{(i)}. \tag{12.40}$$

12.3.4. A PROBLEM STATEMENT OF HEAT-MASS-TRANSFER AND DEFORMING FOR A MULTILAYER ABLATIVE SHELL

With the assumption (d) made in paragraph 12.2, the heat-mass-transfer equations for a multilayer ablative shell become one-dimensional, but contain $3N$ unknown functions:

$$\rho_f^{(i)} \frac{\partial \varphi_f^{(i)}}{\partial t} = -J_f^{(i)} \Gamma_f^{(i)}, \qquad \rho_b^{(i)} \frac{\partial \varphi_b^{(i)}}{\partial t} = -J^{(i)},$$

$$\frac{\partial \rho_g^{(i)}}{\partial t} = \frac{1}{A_1 A_2} \frac{\partial}{\partial q_3} (A_1 A_2 K_{33}^{(i)} R \frac{\partial \rho_g^{(i)} \theta^{(i)}}{\partial q_3}) + \Gamma^{(i)} J^{(i)} + \Gamma_f^{(i)} J_f^{(i)}, \tag{12.41}$$

$$\begin{aligned} \rho^{(i)} c^{(i)} \frac{\partial \theta}{\partial t} = & \frac{1}{A_1 A_2} \frac{\partial}{\partial q_3} (A_1 A_2 k_{33}^{(i)} \frac{\partial \theta^{(i)}}{\partial q_3}) + \\ & + c_g^{(i)} R K_{33}^{(i)} \frac{\partial \rho_g^{(i)} \theta^{(i)}}{\partial q_3} - \Delta e^{0(i)} J^{(i)} - \Delta e_f^{(i)} J_f^{(i)}, \end{aligned}$$

$$i = 1, \ldots, N.$$

Averaging the equations (12.41) of heat-mass-transfer over a thickness of a multilayer shell would lead to a very rough approximation, as temperature and pore pressure for ablative thermal-protective shells are distributed, as a rule, very nonuniformly over the thickness (see Chapters 9, 11).

The contour L bounding the multilayer shell, according to the assumption (d), should be air-tight and heat-insulated, i.e. at the contour (q_α = const) the following conditions should be satisfied:

$$\frac{\partial \theta^{(i)}}{\partial q_\alpha} = 0, \qquad \frac{\partial \rho_g^{(i)} \theta^{(i)}}{\partial q_\alpha} = 0. \tag{12.42}$$

Thermal boundary conditions for the problem (12.41) are set at the surfaces $q = q_{3i}$, $i = 1, \dots N$ in the form of relations (12.17) and (12.18).

Initial conditions for the system (12.41) are

$$t = 0: \quad \varphi_f^{(i)} = \varphi_f^{(i)0}, \quad \varphi_b^{(i)} = \varphi_{b0}^{(i)}, \quad \rho_g^{(i)} = \frac{p_0}{R\theta_0}, \quad \theta^{(i)} = \theta_0. \tag{12.43}$$

Mechanical conditions for the multilayer shell are set as for the one-layer one, and, for example, at the contour L determined by the equation q_α = const they are given by five values (i.e. by one value from each pair)

$$(T_{11} - \bar{\varphi}_g P_g, u_\alpha), \ (T_{12}, u_\beta), \ (Q_\alpha, W), \ (M_{11} - \bar{\varphi}_g M_g, \gamma_\alpha), (M_{12}, \gamma_\beta). \tag{12.44}$$

Five equilibrium equations (12.28), (12.29), into which the constitutive relations (12.37) and kinematic relations (12.31) have been substituted, and also $4N$ heat-mass-transfer equations (12.41) with boundary conditions (12.44), (12.17), (12.8) and initial conditions (12.43) give the *coupled problem of heat-mass-transfer and deforming of multilayer ablative shells*, being the closed system of $5+4N$ equations to determine $5 + 4N$ functions

$$U_\alpha, \ \gamma_\alpha, \ W, \ \varphi_f^{(i)}, \ \varphi_b^{(i)}, \ \rho_g^{(i)}, \ \theta^{(i)}, \quad i = 1, \dots N; \tag{12.45}$$

U_α, γ_α and W depend on q_β and t, ($\beta = 1, 2$), and the remaining functions depend on q_3, q_β and t.

12.4. Cylindrical Composite Shells under High Temperatures

12.4.1. BASIC EQUATIONS

Now we consider the special case of a *cylindrical* ablative shell that can be used for a wide variety of applied problems.

The coordinate surface $q_3 = 0$ is assumed to be the middle surface of the shell; the meridional arc s and the azimuth angle Φ counted off from a certain point M_0

are chosen as coordinates q_1 and q_2, and $q_3 = r - R_2 + h/2$, where r is the radius in the cylindrical coordinate system $Oz\Phi r$, and R_2 is the radius of the external surface of the shell $q_3 = q_{3N} = h/2$.

For the case of cylindrical shells, the principal curvatures k_1 and k_2 are constants:

$$k_1 = 0, \quad k_2 = \frac{1}{R_2}, \quad A_1 = 1, \quad A_2 = R_2. \tag{12.46}$$

Now we can simplify equilibrium equations (12.28), (12.29) as follows:

$$\begin{gathered} \frac{\partial}{\partial s}(T_{11} - \bar{\varphi}_g P_g) = 0, \\ \frac{\partial Q_1}{\partial s} - \frac{T_{22} - \bar{\varphi}_g P_g}{R} + (p_{eN} - p_{e0}) = 0, \\ \frac{\partial}{\partial s}(M_{11} - \bar{\varphi}_g M_g) - Q_1 = 0. \end{gathered} \tag{12.47}$$

Kinematic relations (12.31) take the form

$$\begin{gathered} e_{11} = \frac{\partial U_1}{\partial s}, \quad e_{22} = \frac{W}{R}, \quad e_{12} = 0, \\ \varkappa_{11} = \frac{\partial \gamma_1}{\partial s}, \quad \varkappa_{22} = 0, \quad \varkappa_{12} = 0. \end{gathered} \tag{12.48}$$

Constitutive relations (12.37) for a cylindrical shell have the form

$$\begin{gathered} T_{\alpha\alpha} = -P_{g1} + \sum_{\beta=1}^{2} C_{\alpha\beta}(h e_{\beta\beta} - \overset{\circ}{\varepsilon}{}_{\beta}^{(0)}) - C_{\alpha 3}\overset{\circ}{\varepsilon}{}_{3}^{(0)}, \quad \alpha = 1, 2; \\ M_{11} = -M_{g1} + C_{11}\varkappa_{11}\frac{h^3}{12} - \sum_{\beta=1}^{3} C_{1\beta}\overset{\circ}{\varepsilon}{}_{\beta}^{(1)}. \end{gathered} \tag{12.49}$$

Consider the system (12.47), (12.48) with the boundary conditions

$$\begin{gathered} s = s_+ : \quad T_{11} - \bar{\varphi}_g P_g = T_0, \quad \gamma_1 = 0, \\ s = s_- : \quad U_1 = 0, \quad \gamma_1 = 0, \end{gathered} \tag{12.50}$$

where $s = s_{+,-}$ are the coordinates of the shell ends, and T_0 is the given axial force.

Under these conditions, the system (12.47), (12.48) has the solution

$$\begin{gathered} T_{11} = \bar{\varphi}_g P_g + T_0, \quad T_{22} = (p_{eN} - p_{e0})R_2 + \bar{\varphi}_g P_g, \\ Q_1 = 0, \quad M_{11} = \Big(-M_{g1} - \sum_{\beta=1}^{3} \bar{C}_{1\beta}\overset{\circ}{\varepsilon}{}_{\beta}^{(1)}\Big), \quad \gamma_1 = 0, \quad \varkappa_{11} = 0. \end{gathered} \tag{12.51}$$

On substituting formulae (12.51) into (12.49), one can find strains $e_{\beta\beta}$:

$$e_{11} = \frac{1}{\Delta_1}\Big(\bar{C}_{22}(T_{11} + P_{g1} + \overset{\circ}{T}_{11}) - \bar{C}_{12}(T_{22} + P_{g2} + \overset{\circ}{T}_{22})\Big),$$

$$e_{22} = \frac{1}{\Delta_1}\Big(\bar{C}_{11}(T_{22} + P_{g2} + \overset{\circ}{T}_{22}) - \bar{C}_{12}(T_{11} + P_{g1} + \overset{\circ}{T}_{11})\Big).$$

We can determine the stresses in the cylindrical shell by the formulae

$$\sigma^{(i)}_{\alpha 3} = 0, \qquad \sigma^{(i)}_{12} = 0,$$

$$\sigma^{(i)}_{\alpha\alpha} = -p^{(i)} f^{(i)} + a^{(i)}_{\theta 1} \sum_{\beta=1}^{2} C^{(i)}_{\alpha\beta} e_{\beta\beta} - \sum_{\beta=1}^{3} C^{(i)}_{\alpha\beta} \overset{\circ}{\varepsilon}{}^{(i)(0)}_{\beta}, \qquad \alpha = 1, 2, \qquad (12.52)$$

where

$$\Delta_1 = \bar{C}_{11}\bar{C}_{22} - \bar{C}^2_{12}.$$

Heat and mass transfer equations (12.41) are solved numerically.

12.4.2. COMPUTED RESULTS

As an example, we consider a one-layer cylindrical shell of ablative composite subjected to external heating on its surface $r = R_2$ ($q_3 = h/2$). For this case superscripts $^{(i)}$ on all functions are omitted.

The system (12.41) was solved by using a numerical method with a step-by-step method and matrix sweeping. Solution of the system is given by functions $\rho_g(r,t)$, $\theta(r,t)$, $\varphi_b(r,t)$ and $\varphi_f(r,t)$, with the help of which one can evaluate the pore gas pressure

$$p(r,t) = R\rho_g(r,t)\theta(r,t).$$

All computations were conducted for a composite shell with the following geometric parameters: $R_2 = 0.3$ m, $h = 10^{-2}$ m, and with thermal characteristics corresponding to material with epoxy resin matrix and glass-textile reinforcement. Material constants of the glass-plastic were taken from Tables 4.1, 4.2, 5.1 and 5.2.

External and internal pressures are considered to be atmospheric: $p_{eN} = p_{e0} = 0.1$ MPa, the end surfaces of the glass/epoxy shell are assumed to be heat-insulated. At the external surface $q_3 = h/2$ of the shell, only a radiant heat flux has been given, which is determined by (1.26) and is equal to $q_{RN} = 12$ kW/m^2 and $q_{eN} = 0$. At the internal surface, the conditions of heat insulation have been given: $q_{R0} = q_{b10} = q_{w0} = q_{Rw0} = q_{e0} = 0$, the remaining heat fluxes can be determined by (12.18a). These boundary conditions simulate the behaviour of a composite shell lying in a zone of flame action (see paragraph 1.1), when a heat flux from the source is supplied by radiation without immediate contact with the flame.

Figure 12.2,*a* shows the temperature $\theta(r,t)$ distribution *versus* the shell thickness for different times $t = 50$, 100, 150, 200 and 250 s. Figure 12.2,*b* shows the

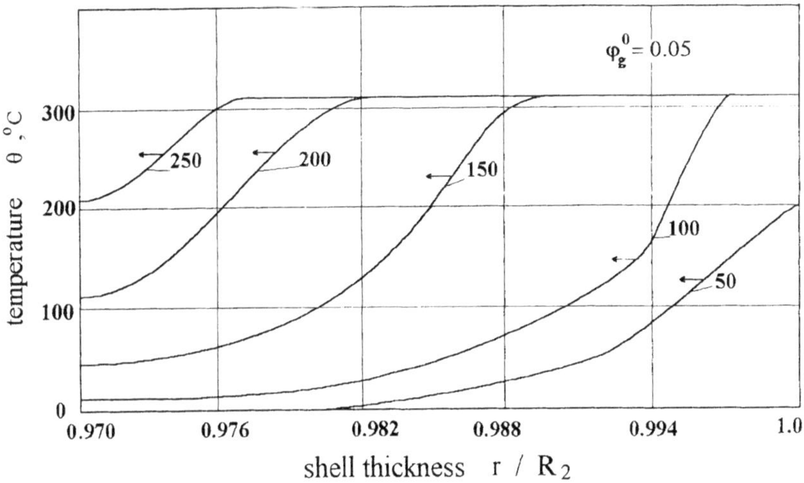

(a)

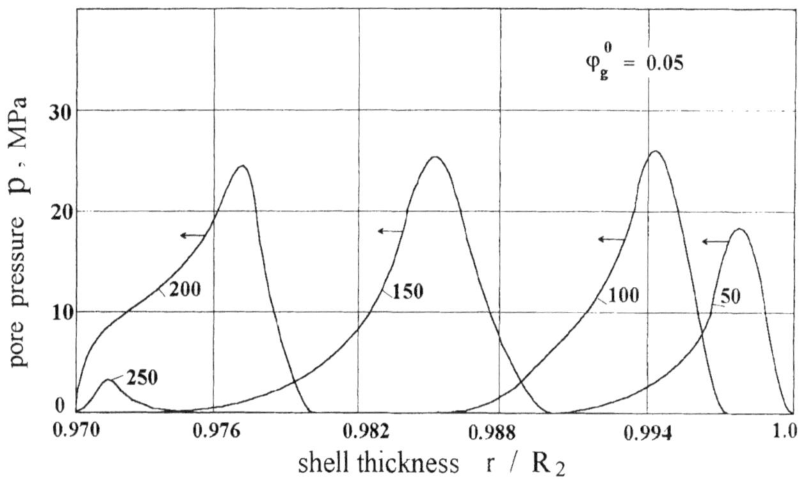

(b)

Figure 12.2. Distributions of temperature θ (a) and pore pressure p (b) *versus* thickness of a composite shell for different times t. Initial porosity of the composite φ_g^0 =0.05, symbols at the curves are times (s). Arrows indicate the direction of motion of heating and pore pressure fronts

distribution of pore gas pressure $p(r, t)$ through the shell thickness for the same times.

For the one-layer shell, equations (12.23) give the expression for transverse

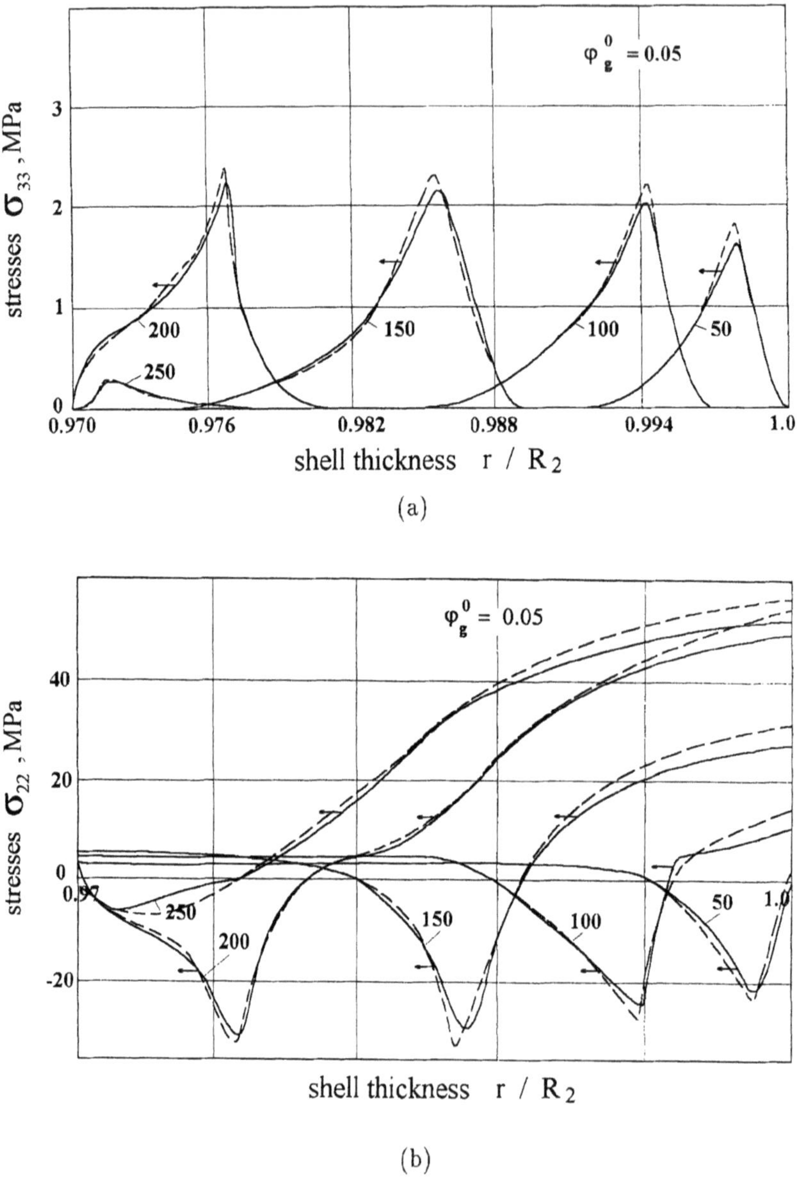

Figure 12.3. Distributions of transverse stresses σ_{33} (a) and tangential stresses σ_{22} (b) *versus* thickness of a composite shell for different times t. Initial porosity of the composite φ_g^0 =0.05, symbols at the curves are times (s), solid curves are solutions determined by the exact equations, dashed lines - by the shell theory

stresses σ_{33}

$$\sigma_{33} = \varphi_g p - p_{e0} - (p_{eN} - p_{e0})(\frac{q_3}{h} + \frac{1}{2}) + 6(\frac{1}{4} - \frac{q_3^2}{h^2})((p_{eN} - p_{e0})\xi_2 - \\ - (P_{g1} + \bar{\varphi}_g P_g)\xi_1 - C_{33}\overset{\circ}{\varepsilon}{}^{(0)}_{33}), \qquad (12.53)$$

where

$$\xi_1 = \frac{1}{h}(1 - \frac{1}{\Delta}(C_{31}(C_{22} - C_{12}) + C_{32}(C_{11} - C_{12})),$$

$$\xi_2 = \frac{1}{2} + \frac{R_2}{h\Delta}(C_{32}C_{11} - C_{31}C_{12}).$$

Figures 12.3,*a* and 12.3,*b* show distributions of radial σ_{33} and tangential σ_{22} stresses in the shell for different times.

As mentioned in Chapters 9 and 11, a peculiarity of the behaviour of composite shells under high temperatures is the presence of a specific profile of intrapore pressure p (Figure 12.2,*b*). Intensive pyrolysis of the polymer matrix leads to growth of the pore gas pressure. Only gases near the external suface $q_3 = h/2$ of the shell, where porosity and gas permeablity of the composite are high, have time to filter out into the surroundings. Near the internal surface of the shell $q_3 = -h/2$, where porosity and gas permeability are low, gases accumulate to form a local peak of pore pressure (Figure 12.2,*b*). Over time, this peak moves towards the internal cold surface of the shell.

The peak of pore pressure leads to the appearance of a local peak of tensile normal transverse (through-thickness) stresses σ_{33} in the shell (Figure 12.3,*a*) that also moves with time towards the internal surface of the shell. These stresses are the most dangerous for laminated textile composite shells, as there appears a chance of their delamination.

Tangential stresses σ_{22} in the cylindrical composite shell under high temperatures have two local extrema: one in the domain of compressive stresses and another in the domain of tensile stresses (Figure 12.3,*b*). The peak of the compressive stresses is connected with the presence of a peak of pore pressure, that also moves with time towards the cold internal surface. The peak of tensile tangential stresses is caused by shrinkage of the composite at matrix pyrolysis temperatures. Equations (12.52) give the tensile shrinkage stresses

$$\sigma_{22} = -C_{22}\overset{\circ}{\varepsilon}_2 \approx E_2\beta_{sh}\Omega_2.$$

These stresses are localized near the most heated external surface $q_3 = h/2$ of the shell. If the level of these stresses is sufficiently high and exceeds the strength limit in tension σ_{2T}, then longitudinal cracks can appear in the cylindrical shell, they are directed along the Oq_1 axis of the shell (see Figure 12.7,*a*).

To estimate the accuracy of the solution, we conduct a comparison between the stresses determined by formulae (12.52), (12.53) and those evaluated by equations (11.21) with the help of solution of the compatibility equation (11.25). Figure 12.3

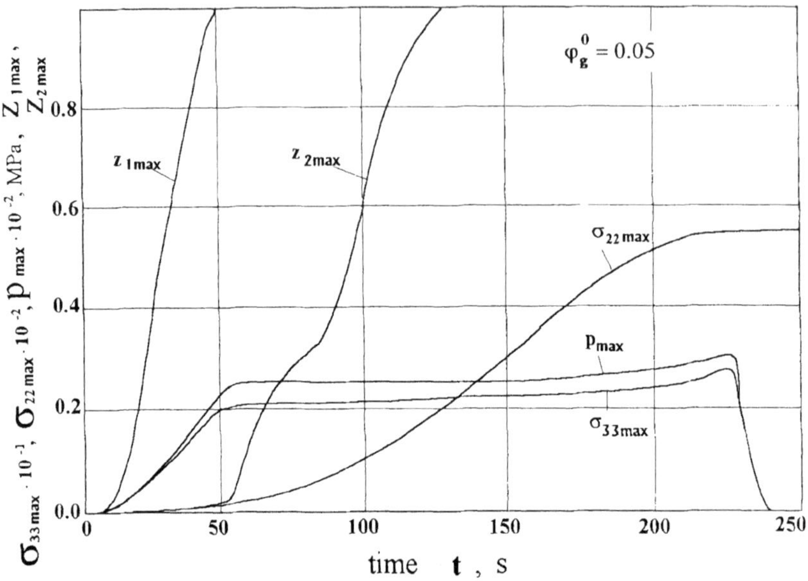

Figure 12.4. Dependences of maximum values of functions in a composite shell: transverse stresses σ_{33max}, tangential stresses σ_{22max}, pore pressure p_{max}, damage parameters z_{2max} and z_{1max} on heating time t. Initial porosity of the composite φ_g^0 =0.05

shows that there is not only qualitative but also quantitative agreement between the results. The error due to the use of the shell theory does not exceed 10 %.

The maximum values of the pore pressure p and also of the transverse σ_{33} and tangential σ_{22} stresses depend on the initial porosity φ_g^0 of the composite. The higher the initial porosity φ_g^0, the more intensive is the filtration of gases into the surroundings, and the lower are the values of p_{max} and σ_{33max}.

Figures 12.4 - 12.6 show dependences of p_{max}, σ_{22max} and σ_{33max} on heating time t for three values of $\varphi_g^0 = 0.05$, 0.07 and 0.1.

These figures show that the kinetics of growth of functions p_{max} and σ_{33max} for all three values of φ_g^0 are the same, and correspond to the type of function p_{max} in ablative composites established in paragraph 9.1. First there is a sharp growth (approximate duration 50 s), then a stabilization stage at time t = 100 - 180 s, and finally a sharp reduction that is connected with completion of the intensive stage of matrix pyrolysis. Tangential stresses $\sigma_{22max}(t)$ at all stages and for all three values of φ_g^0 grow monotonically, that is connected with increasing the content of coke phase φ_p near the heated surface and, hence, with increasing shrinkage stresses σ_{22}.

However, as the initial porosity φ_g^0 decreases, the absolute values of p_{max} and σ_{33max} considerably increase (Figures 12.4 - 12.6): for $\varphi_g^0 = 0.1$ the highest values of p_{max} and σ_{33max} are 3 and 0.2 MPa; for $\varphi_g^0 = 0.07$ they are 20 and 1.5 MPa;

while for $\varphi_g^0 = 0.05$ they are 25 and 2.3 MPa, respectively. The absolute values of $\sigma_{22\max}$ show smaller variation: they are 15, 26 and 50 MPa, respectively; they are determined only by the degree of charring of the composite matrix φ_p and are independent of the level of pore pressure p.

12.5. Failure of Composite Structures under High Temperatures

12.5.1. CONDITIONS OF THE APPEARANCE OF FAILURE

As shown above, in composite shells under intensive nonstationary heating to the pyrolysis temperatures of the material, considerable thermal stresses appear: both normal through-thickness σ_{33} and tangential σ_{22} occur even without external mechanical loads. To estimate the danger of these thermal stresses (when they reach the ultimate strength of the composite and the composite structure is destroyed) it is convenient to introduce parameters z_α called damage parameters. The parameters z_α are determined as ratios of the thermal stresses $\sigma_{\alpha\alpha}$ and the corresponding limits of strength $\sigma_{\alpha T}$ of the composite in tension (see equations (11.33), (11.34)):

$$z_1(q_3,t) = \frac{|\sigma_{33} + 3\bar{N}^0 p| + \sigma_{33} + 3\bar{N}^0 p}{2\sigma_{3T}}, \qquad z_2(q_3,t) = \frac{|\sigma_{22}| + \sigma_{22}}{2\sigma_{2T}}, \tag{12.54}$$

where $|\sigma_{\alpha\alpha}|$ is the modulus of stresses $\sigma_{\alpha\alpha}$.

Ultimate strengths $\sigma_{\alpha T}$ depend on temperature θ and are determined by formulae (7.44), (7.52) where one should substitute indices: $3 \longrightarrow 2$, $1 \longrightarrow 3$.

If at a certain time t_α^* within a zone of the shell with coordinate q_3 the following condition is satisfied:

$$\max_{-(h/2)\leqslant q_3 \leqslant (h/2)} z_\alpha(q_3, t_\alpha^*) = 1, \qquad \alpha = 1,2, \tag{12.55}$$

then failure of the shell occurs. This means that the corresponding stresses $\sigma_{\alpha\alpha}$ reach the ultimate strength $\sigma_{\alpha T}$ of the composite in tension.

As was shown above (see Figure 12.3,*b*), the maximum tensile stresses σ_{22} are always near the external surface of the shell $r = R_2$ (i.e. $q_3 = h/2$). Therefore, if at the certain time t_2^*, the condition (12.55) for $\alpha = 2$ is satisfied, then during destruction of the shell longitudinal cracks appear on the external surface $q_3 = h/2$ (Figure 12.7,*a*), as the destruction for this case occurs by the type (B) - due to breakage of fibres by tangential stresses σ_{22} at the surface $q_3 = h/2$ of the shell.

Figure 12.3,*a* shows that stresses σ_{33} can reach their maximum values at any surface with coordinate q_3^*: $-h/2 \leqslant q_3^* \leqslant h/2$. Therefore if at the certain time t_1^* the condition (12.55) for $\alpha = 1$ is satisfied, then the shell is destroyed by the type (C) - due to delamination (Figure 12.7,*b*), and the delamination appears within the shell at a surface with coordinate $q_3 = q_3^*$.

If condition (12.55) is satisfied for both $\alpha = 1$ and $\alpha = 2$, then both types of failure of the shell take place.

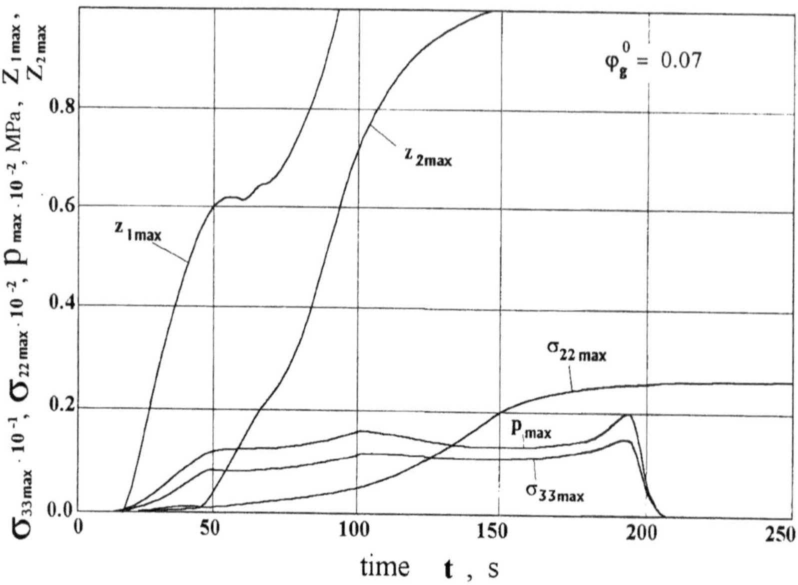

Figure 12.5. Dependences of maximum values of functions in a composite shell: transverse stresses $\sigma_{33\max}$, tangential stresses $\sigma_{22\max}$, pore pressure $p_{\max}$, damage parameters $z_{1\max}$ and $z_{2\max}$ on heating time t. Initial porosity of the composite φ_g^0 =0.07

Thus, the damage parameter z_α is a convenient indicator of the danger of thermal stresses appearing within the shell during heating: the closer the value of z_1 or z_2 is to 1, the higher the level of thermal stresses and the greater the danger of destruction of the shell by type (B) (longitudinal cracking) or by type (C) (delamination) failure, respectively.

Below, calculated damage parameters z_α will be given for a composite cylindrical shell for the considered case of external heating.

Figures 12.4 - 12.6 show the kinetics of the maximum values of damage parameters, max $z_\alpha(t)$, in the composite shell for three values of initial porosity $\varphi_g^0 =$ 0.05, 0.07 and 0.1. The function $z_{2\max}(t)$ does not change considerably with φ_g^0, but $z_{1\max}(t)$ changes dramatically. This can be explained by the fact that z_2 is determined mainly by shrinkage of the composite during pyrolysis, which is independent of porosity φ_g^0, while parameter z_1, on the other hand, is defined, in general, by pore pressure which depends significantly on the porosity of the composite. For $\varphi_g^0 = 0.05$ and 0.07, the condition $z_{1\max} = 1$ is satisfied earlier than $z_{2\max} = 1$, this means that delamination of the shell occurs earlier under high temperatures. For $\varphi_g^0 = 0.1$, the condition $z_1 = 1$ is not realized at all, i.e. the composite shell is destroyed only due to cracking on its external heated surface.

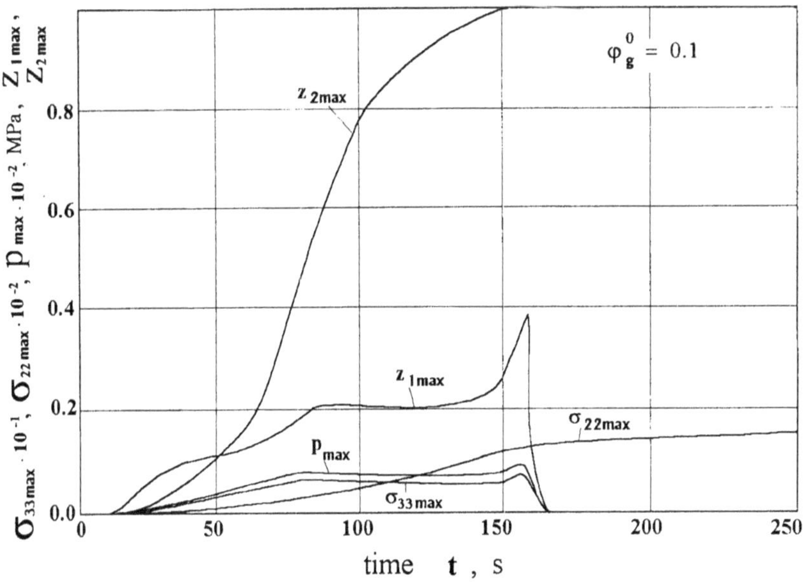

Figure 12.6. Dependences of maximum values of functions in a composite shell: transverse stresses $\sigma_{33\max}$, tangential stresses $\sigma_{22\max}$, pore pressure $p_{\max}$, damage parameters $z_{1\max}$ and $z_{2\max}$ on heating time t. Initial porosity of the composite $\varphi_g^0 = 0.1$

12.5.2. BEHAVIOUR OF A COMPOSITE SHELL AFTER THE APPEARANCE OF FAILURE

Composite structures are usually examined only for the time interval $0 \leqslant t \leqslant t_\alpha^*$, till the time t_α^* when failure first appears, for example, until composite delamination appears due to accumulating intrapore pressure of gaseous pyrolysis products. This restriction is justified for many heat-shield structures because, after delamination has occurred in a piece of hardware having (for example) a plane shape (plates, unclosed shells, panels), the piece fails and completely loses its thermal protective properties. However, for some types of structure there are conditions under which the hardware still performs its functional protective purpose for some time interval $t_1^* \leqslant t \leqslant t_{**}$ after delamination has taken place.

This situation occurs under external heating of a shell with a closed contour made of textile composite material, e.g. under the action of high temperatures on a cylindrical fuel container made of glass-reinforced plastic (see Figure 1.16). The action of a temperature field uniformly distributed over the surface on the cylindrical shell of the fuel container leads to the formation of ring delaminations having a closed contour. Because the cylindrical shell has a closed contour, it does not fail after the first delamination, and continues to perform its functions. Upon further heating, new ring delaminations appear, the formation of which is directed from the outer surface of the shell to the inner surface. Pore pressure is accumulated in each ring crack so that the pressure difference $p_d(r_i,t) - p_d(r_{i-1},t)$ (where i=1,...N, r_i is the radius of i-th delamination, $r_i = R_2 - h/2 + q_{3i}^*$, where q_{3i}^* is the coordinate of a surface of the i-th delamination), uniformly distributed, acts on each cylindrical layer stripped off. The other non-delaminated part of the cylindrical shell of the fuel container with thickness $h_d(t)$ proves to be the most loaded, as it is subjected to the action of the maximum pressure difference $p_d(r_N,t) - p_{e0}$, where $h_d = r_N - R_2 + h$.

The delamination continues until time t_{**} when, at a certain critical thickness $h_d(t_{**})$, the inner part of the shell loses its stability under the action of the external pressure difference $p_d(r_N,t_{**}) - p_{e0} > 0$. After that the structure stops performing its functions and the fuel container fails completely. However, the time interval $t_{**} - t_1^*$ from the appearance of first delamination till the loss of stability of the fuel container is long: $t_{**} - t_1^* \gg t_*$, therefore the method of stresses and heat-mass-transfer calculations developed above for $t < t_1^*$ should be continued for the time interval $t_1^* < t \leqslant t_{**}$.

The condition of the appearance of delamination in the composite under high temperatures has the form (12.55) for $\alpha = 1$. We will assume that the appearance of delamination does not change the picture of heat-mass transfer in the shell.

The critical external pressure p^*, for which a *loss of stability* of the non-delaminated shell section occurs, was obtained in [122] and, for the considered problem, is written as follows:

$$p^*(t) = 0.92E_f^0 a_{\theta 1}(t)\left(\frac{h_d(t)}{R_2}\right)^{2.5} \cdot \frac{R_2}{L}, \tag{12.56}$$

where L is the shell length, and the stability condition has the form

$$p_d(r_N,t) - p_{e0} \leqslant p^*(t), \qquad t_1^* \leqslant t \leqslant t_{**}. \tag{12.57}$$

Complete destruction of the shell due to the loss of stability is called *destruction by type (E)*. To determine the pressure p_d of gaseous pyrolysis products accumulated within the ring crack, the state equation of a perfect gas is used:

$$p_d = R\rho_{gd}\theta, \tag{12.58}$$

where R is the gas constant. In order to determine the density ρ_{gd} of the gas within the crack, we apply the equation of gas mass conservation before and after formation of the crack:

$$\rho_{gd}V_d = \rho_g V_g, \tag{12.59}$$

where ρ_g is the gas density in the pores, V_g is the gas volume in the pores coming to the surface of the ring crack:

$$V_g = 2\pi R_2 L l_g \varphi_g, \tag{12.60}$$

l_g is the characteristic diameter of the pores ($10^{-4}\ldots10^{-3}$ m), V_d is the crack volume opened by action of pressure p_d:

$$V_d = \pi L(R_{\text{def}}^2 - R^2) = 2\pi L\frac{p_d R_2^3}{h_d E_f^0 a_{\theta 1}}, \tag{12.61}$$

here R_2 and R_{def} are the radii of the delamination surface in the non-opened and opened (by deformation) states, respectively, and

$$R_{\text{def}} = R_2(1 + \varepsilon_\theta) = R_2\Big(1 + \frac{p_d R_2}{h_d E_f^0 a_{\theta 1}}\Big).$$

Here $\varepsilon_\theta = \sigma_\theta / E_f^0 a_{\theta 1}$ is the tangential deformation of the shell, σ_θ is the tangential stress of the shell connected to the pressure p_d in the crack by the relation $\sigma_\theta = p_d R_2 / h_d$.

Having substituted formula (12.59) and (12.60) into (12.58), we can find the gas density within the crack, ρ_{gd}. Then equation (12.58) gives the expression for the pressure of the pyrolysis gases p_d within the crack

$$p_d(r_N, t) = \Big(\frac{h_d}{R_2^2} \cdot p E_f^0 a_{\theta 1} l_g \varphi_g\Big)^{1/2}, \tag{12.62}$$

where p is determined from the solution of heat-mass-transfer equations (12.41).

On substituting (12.56) and (12.62)into (12.57), we obtain the condition of the absence of destruction in the composite cylindrical shell by stability loss under high temperatures. If at time $t = t_{**}$ there is equality in (12.62), then there occurs a loss of stability of the internal layer of the composite shell.

12.5.3. COMPUTED RESULTS

Figure 12.8 shows the graph of the function $(h_d(t)/h)$ for a glass/epoxy cylindrical shell with the characteristics given before for four values of $\varphi_g^0 = 0.05$, 0.07, 0.08 and 0.1. For $\varphi_g^0 = 0.05$, during the first stage of heating $t \leqslant 50$ s there is no delamination in the shell and $(h_d/h) = 1$.

At time $t = t_1^* \approx 50$ s the condition (12.55) is realized for the first time and the first delamination occurs near the external surface of the shell. Figure 12.7,c presents the picture for the appearance of new delaminations; this graph is characterized by the thickness $h_d(t)$ of the non-delaminated material. Figure 12.8 shows that the dependence $(h_d(t)/h)$ is practically linear. At time $t = t_{**} \approx 160$ s, when h_d/h reaches the value 0.4, the non-delaminated shell thickness becomes so small that the pore pressure p_d of gas accumulated in delamination cavities exceeds the stability limit (12.57) and the internal layer of the composite shell loses its stability; the cylindrical composite shell fails completely.

These computations show that the time interval $t_{**} - t_1^*$ is approximately 110 s; it is more than twice t_1^*. Thus, the calculation of internal heat-mass-transfer and stresses is necessary up to time t_{**} as the shell still performs its designated purpose at these times.

For a composite shell with initial porosity $\varphi_g^0 = 0.07$, the first delamination appears at $t_1^* \approx 90$ s, and delaminations continue to arise up to $t = t_{**} = 170$ s, when a loss of stability of the internal layer of the shell occurs (Figure 12.8). For $\varphi_g^0 = 0.1$, delaminations do not appear at all, and $(h_d/h) = 1$ during the whole

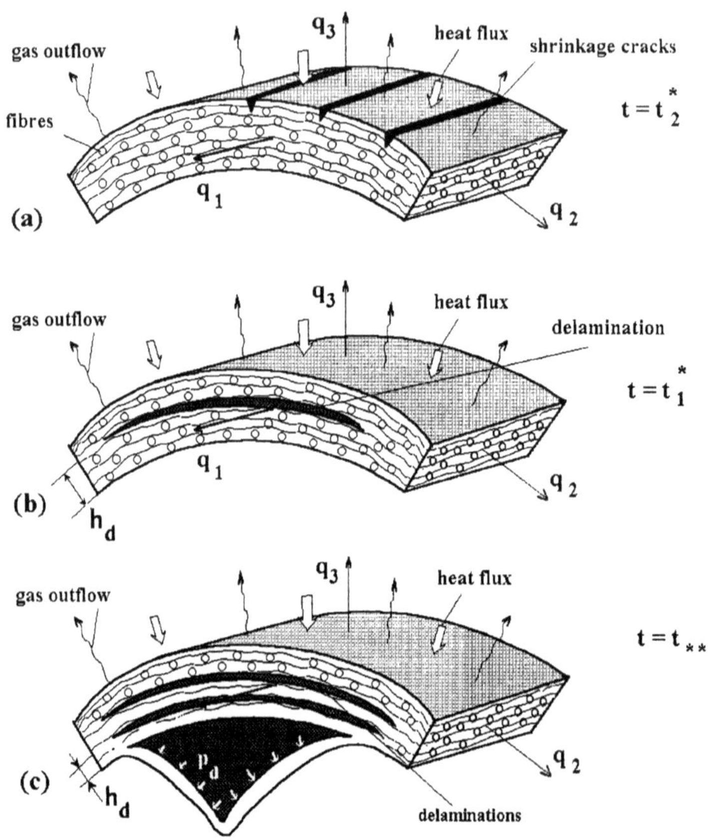

Figure 12.7. A scheme of three types of failure of a composite shell under high-temperature heating:
(a) cracking of the heated surface of the composite at $t = t_2^*$ (type (A));
(b) delamination of the composite at $t = t_1^*$ (type (C));
(c) appearance of several delaminations and loss of their stability at $t = t_{**}$ (type (E))

heating time due to the low level of pore gas pressure p; hence, there is a low level of normal transverse stresses σ_{33}.

Figure 12.9 shows the dependences of the times t_1^*, t_2^* and t_{**} for appearance of the three types ((A), (C) and (E)) of failure on the initial porosity of the composite φ_g^0. For $\varphi_g^0 \geqslant 0.08$, there are no delaminations and no loss of stability of the shell during heating (as t_1^* and $t_{**} \longrightarrow \infty$), and there is only cracking due to chemical shrinkage at time $t = t_2^* \approx 150$ s. For $\varphi_g^0 \leqslant 0.08$, delamination of the composite shell (type (A)) occurs first at $t = t_1^* \approx 50 - 100$ s, then there is shrinkage cracking (type (C)) of the shell surface at $t = t_2^* \approx 100 - 130$ s, and finally a loss of stability of the delaminated internal part (type (E)) of the shell at $t = t_{**} \approx 160$ - 200 s.

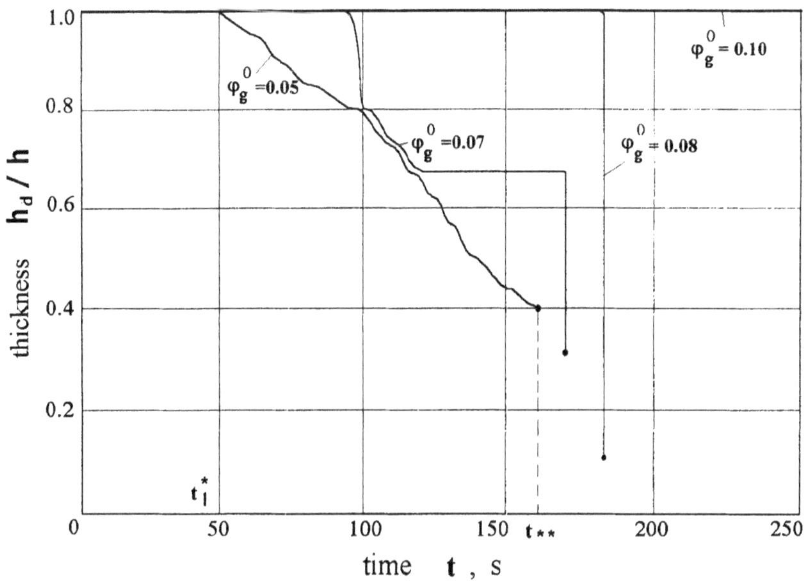

Figure 12.8. Distributions of nondelaminated thickness h_d of a composite shell *versus* heating time t for different values of initial porosity of composite φ_g^0 =0.05, 0.07, 0.08 and 0.1

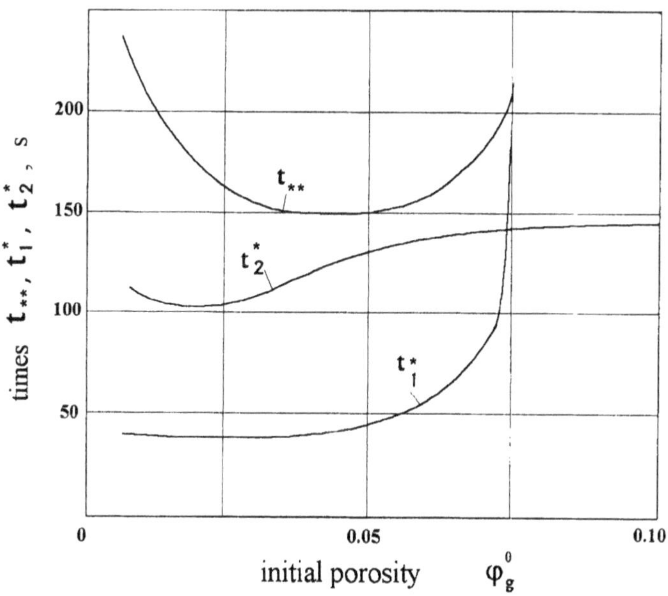

Figure 12.9. Times of the appearance of three types of failure t_2^*, t_1^* and t_{**} *versus* initial porosity φ_g^0 of the composite

12.5.4. EXPERIMENTAL RESULTS

To test the accuracy of the model, heat testing was conducted for cylindrical glass/epoxy composite shells with the following geometrical parameters: $L = 0.8m$, $h/R = 0.03$; the results were compared with the corresponding computed ones. Heating was performed with the aid of infra-red heaters so that the heat flux q_+ to the composite shell and the surface temperature $\theta(R,t)$ would be the same as the values used in the computations, namely 10 $\mathrm{kW \cdot m^{-2}}$ and 300° C. The initial porosity φ_g^0 of the textile glass/epoxy composite shell which was manufactured by pressing in a press mould, was approximately 0.04 - 0.05. During testing, acoustic emission was used to determine the time of delamination onset t_1^* in the shell; it was detected easily by the peak of intensity in the acoustic signal and also by the sharp growth of gas generation from the shell, which was noticeable with the naked eye. The time t_{**} of stability loss of the internal layer of the shell was also detected easily by the peak of the acoustic signal and also by a characteristic audible click. After testing, we measured the thickness h_d of the shell part that lost its stability.

Three composite shells with the same parameters were tested. The points in Figure 12.8 show experimental values of the times t_1^* for the beginning of delamination and t_{**} for the stability loss, and also the corresponding experimental values of thickness h_d at t_1^* and t_{**}. The agreement between experimental and computed results is sufficient to allow us to confirm the adequacy of the model developed for thermomechanical behaviour of composite structures under high temperatures.

APPENDIX

METHODS OF EXPERIMENTAL INVESTIGATION OF HIGH-TEMPERATURE PROPERTIES OF COMPOSITE MATERIALS

This appendix is devoted to a brief description of methods of experimental determination of composite material properties under high temperatures. The following parameters were considered:

- density $\rho(\theta)$ under heating;
- heat conductivity $k_\alpha(\theta)$ under heating;
- gas-permeability K under heating;
- heat deformations $\overset{\circ}{\varepsilon}_\alpha(\theta)$ under heating;
- elastic moduli $E_\alpha(\theta)$ under heating;
- strength $\sigma_{\alpha T}(\theta)$ under heating;
- linear ablation rate D of composites under nonuniforn heating.

The data obtained by experiment are given in Chapters 4 – 12 of the book.

A1. Determination of Density under Heating

Density measurement for composites and their separate components, matrix and fibre, is usually realized with the help of thermogravimetric analysis (TGA) [69, 75], the essence of which consists in measurement by an electronic balance of the changing mass m of a tested specimen of investigated material under uniform heating in an electric stove with a constant rate of temperature growth: $\dot{\theta}_0 =$ const. The specimen is usually a fine-grained powder with mass $m_0 \approx 2$ g. As a result, we obtain the experimental dependence $\rho(\theta)/\rho_0 = m(\theta)/m_0$ of the relative density on temperature θ_0. The general form of the dependence $\rho(\theta)/\rho_0$ was shown in Figures 4.4, 4.5 and others.

If the test is conducted for a non-reinforced matrix (or for fibres), then by substituting the experimental values of the function $\rho_m(\theta)/\rho_m^0$ into the theoretical equation (4.6), we get mathematical relations to determine the following constants of thermodecomposition: J_0, Γ and E_A (or J_f^0, E_{fA}^0 for fibres).

We can readily determine the gasification coefficient as follows:

$$\Gamma = 1 - \frac{\rho_m(\theta_{\max})}{\rho_m^0}, \qquad (A1)$$

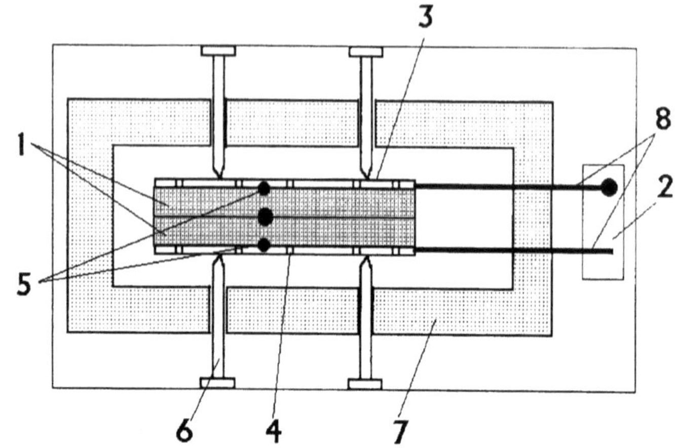

Figure A1. A scheme of the device for determination of thermal properties of composites under high temperatures [108]: 1 – two plates of investigated composite material, 2 – source of electric current, 3 – heated metallic plates, 4 – holes in metallic plates, 5 – places of location of thermocouples, 6 – holders of specimens, 7 – heat insulation of heaters of the chamber, 8 – current-suppliers

where $(\rho_m(\theta_{\max})/\rho_m^0)$ is the density at the temperature $\theta_{\max}$ for which the function ρ_m/ρ_m^0 comes to a horizontal asymptote (see Figure 4.4). After that, we can find the constants J_0 and E_A in different ways [69, 75]: the simplest way is to find the numerical values of J_0 and E_A which give the best fit between the theoretical values of $\rho_m(\theta)/\rho_m^0$ determined by equation (4.6), and experimental values, in the temperature range $\theta_0 \leqslant \theta \leqslant \theta_{\max}$). TGA can be conducted in either inert environment or air.

A2. Determination of Thermal Characteristics under Heating

A2.1. EXPERIMENTAL DEVICE

Methods and experimental devices for the determination of the heat conductivity and heat capacity of ablative composite materials differ from those used for metals. Composites differ from metals in the following ways:

- they have considerably lower heat conductivity and thermoconductivity;
- there is gas generation in composites;
- their thermal properties change nonmonotonically;
- the thermal characteristics of composites depend on rate of heating (or, generally speaking, on 'history' of heating).

Figure A1 shows a scheme of the experimental device for determination of the thermal characteristics of composites at high temperatures [108]. Heat source in the device is joule heat released by electric current. To regulate the heating rate and uniformity of the temperature field, one should apply metallic heated plates from both sides to the two composite plates investigated. The metallic plates have holes for gas outflow from the composite. In the test, we measure the heat q used for heating the composite plates, and the temperatures θ_w and θ_s on the external (contacting with the metallic plate) and internal surfaces of the plates; then we determine the heating rate $b = (\partial/\partial t)\theta_w(t)$. From the experimental data, one can find the heat conductivity coefficient k, heat capacity c and thermoconductivity $a = k/(\rho c)$ which are determined with the help of approximate analytical solution of two special problems of heat conduction.

A2.2. DETERMINATION OF THERMOCONDUCTIVITY

We consider a one-dimensional heat conduction problem with a boundary condition of the first type, i.e. at one end of the plate, the temperature $\theta = \theta_w$ is given and varies linearly with time:

$$\frac{\partial \theta}{\partial t} = a\frac{\partial^2 \theta}{\partial x^2}, \qquad 0 < x < h, \qquad 0 < t < t_{\max}; \tag{A2}$$

$$x = 0: \qquad \frac{\partial \theta}{\partial x} = 0;$$

$$x = h: \qquad \theta = \theta_w = \theta_0 + bt;$$

$$t = 0: \qquad \theta = \theta_0.$$

After introducing the dimensionless values

$$T = \frac{\theta - \theta_0}{b\, t_{\max}}, \qquad \tau = \frac{t}{b\, t_{\max}}, \qquad y = \frac{x}{h}, \qquad \mathrm{Fo} = \frac{a\, t_{\max}}{h^2}, \tag{A3}$$

one can rewrite the problem (2.2) in the form

$$\frac{\partial T}{\partial \tau} = \mathrm{Fo}\frac{\partial^2 T}{\partial y^2}, \qquad 0 < y < 1, \qquad 0 < \tau < 1; \tag{A4}$$

$$y = 0: \qquad \frac{\partial \theta}{\partial y} = 0;$$

$$y = 1: \qquad T = \tau;$$

$$\tau = 0: \qquad T = 0,$$

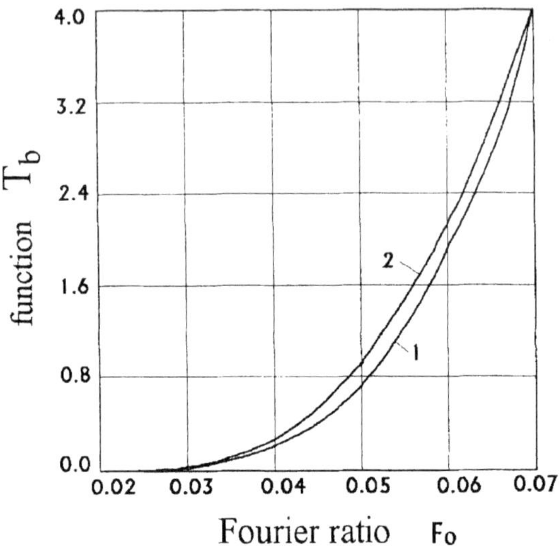

Figure A2. Dependence of function T_b on the Fourier ratio Fo (1 – exact numerical solution, 2 – approximate solution)

where $t_{\max}$ is the maximum time of heating in testing, Fo is the Fourier ratio.

The problem (A4) contains only one parameter, the dimensionless Fourier ratio Fo; so that the solution $T(y,t)$ depends parametrically only on Fo. We need only the value $T(0,1)$ (dimensionless temperature on the heat-insulated surface at the final time $t_{\max}$. This magnitude depends parametrically also only on value of Fo:

$$T(0,1) = T_b(\mathrm{Fo}), \tag{A5}$$

where $T_b(\mathrm{Fo})$ is the certain function of Fo. We can determine numerical values of this function by numerical solution of the heat conduction problem (A4), using the difference method of sweeping [102]. Figure A2 shows graphs of function $T_b(\mathrm{Fo})$ for Fo from 0.02 to 0.07. Within this interval there are characteristic values of Fo for typical glass/ and carbon/phenolic textile composites.

We can approximate the function $T_b(\tau_0)$ up to a sufficient accuracy by the exponential dependence

$$T_b(\mathrm{Fo}) = T_b^0 \exp\left(-\frac{M_b}{\mathrm{Fo}}\right), \tag{A6}$$

where T_b^0 and M_b are numbers which depend only on the interval in which ratio Fo varies. For $\mathrm{Fo} \in [0.01, 0.1]$, from the condition (A6) we find $T_b^0 = 0.34$, $M_b = 0.55$.

For the Fourier ratio $\mathrm{Fo} \geqslant 0.1$, the function $T_b(\mathrm{Fo})$ is adequately approximated by the exponential dependence (A6) with the constants: $T_b^0 = 0.89$, $M_b = 0.49$.

After substituting (A6) into (A5) and then into (A3), one can derive the expression for temperature $\theta_s = \theta(h, t_{\max})$ on the heat-insulated surface:

$$\theta_s = \theta_0 + bt_{\max} T_b^0 \exp\left(-\frac{M_b}{\mathrm{Fo}}\right). \tag{A7}$$

The heat conduction problem (A2) describes heat propagation in the composite plates investigated in the experimental device shown in Figure A1. If the temperature of the substrate $\theta_s(t_{\max})$ and heating rate b are determined in experiments then from formula (A7) one can find thermoconductivity $a = \mathrm{Fo}h^2/t_{\max}$ of the composite as follows:

$$a = \frac{M_b h^2 b}{(\theta_w - \theta_0)\ln\left((\theta_w - \theta_0)T_b^0/(\theta_s - \theta_0)\right)}. \tag{A8}$$

Values of thickness of specimens are usually chosen so that the temperature difference $\Delta\theta = \theta_w(t_{\max}) - \theta_s(t_{\max})$ is not very large. Formula (A7) shows that for linear growth of surface temperature θ_w, the value of $\Delta\theta$ can reach its minimum $\Delta\theta = (1 - T_b^0)(\theta_w - \theta_0)$.

Formula (A8) allows us to determine experimental dependence of a on temperature θ_w with the help of the device shown in Figure A1.

A2.3. DETERMINATION OF HEAT CONDUCTIVITY

If we measure the heat flux q supplied by contact to the surface of the composite plate, as well as the temperatures θ_w and θ_s, then we can determine not only a but also the heat conductivity k of the composite. For this, in the problem (A2) in place of the boundary condition $\theta = \theta_w$ one should use the condition of the second type:

$$k\frac{\partial\theta}{\partial x} = q, \qquad x = h. \tag{A9}$$

We introduce the dimensionless temperature T given by

$$T = \frac{k(\theta - \theta_0)}{qh}, \tag{A10}$$

and rewrite the heat conduction problem in dimensionless form similar to (A4) but with the boundary condition:

$$y = 1: \qquad \frac{\partial\theta}{\partial y} = 1. \tag{A11}$$

The particular solution of the problem is the dimensionless temperature $T(1,1)$ on the heated surface $y = 1$ at time $\tau = 1$, which depends only on the Fourier ratio $T(1,1) = T_q(\mathrm{Fo})$, and this dependence is approximated by the power function

$$T(1,1) = T_q(\mathrm{Fo}) = T_q^0\sqrt{\mathrm{Fo}}, \tag{A12}$$

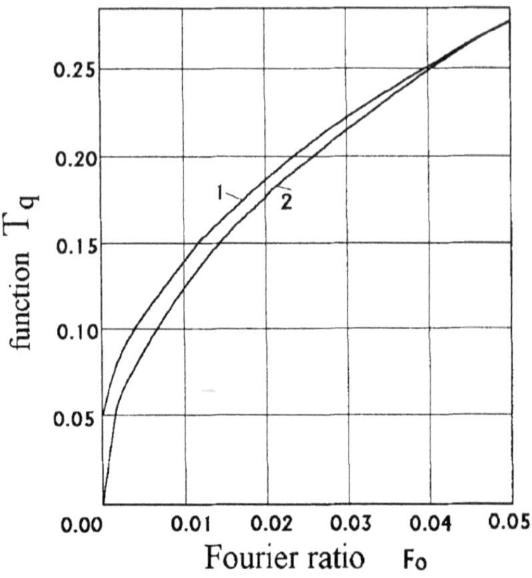

Figure A3. Dependence of function T_q on the Fourier ratio Fo(1 – exact numerical solution, 2 – approximate solution)

where T_q^0 is a constant. Figure A3 shows a graph of function $T_q(\mathrm{Fo})$ obtained from numerical solution of (A4), (A11) and approximated by the power function (A12); its accuracy is adequate.

From (A10) and (A12) one can find the expression connecting temperature θ_w and heat flux q:

$$\theta_w = \theta_0 + \frac{qh}{k} T_q^0 \sqrt{\mathrm{Fo}}, \tag{A13}$$

whence

$$k = \frac{q}{(\theta_w - \theta_0)} T_q^0 \sqrt{a t_{\max}}, \tag{A14}$$

permitting us to determine the heat conductivity coefficient k as a function of temperature θ_w; we can find a from formula (A8).

If functions $\rho(\theta)$, $k(\theta)$, $a(\theta)$ are determined experimentally, then the heat capacity $c(\theta)$ of the material depending on heating temperature is evaluated by the formula

$$c = k/(\rho a).$$

The methods mentioned in A2.2 and A2.3 were used for determination of thermal constants of composites at high temperatures.

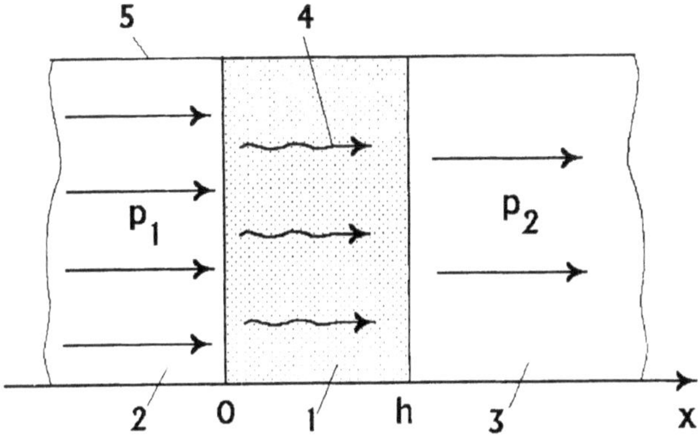

Figure A4. A scheme of the device for determination of gas-permeability of composites [7]: 1 – specimen, 2 – gas pressure p_1 at the entrance, 3 – gas pressure p_2 at the outlet, 4 – gas filtration through the specimen, 5 – gas-tight walls of the device

A3. Determination of Gas Permeability

Experimental values of gas-permeability coefficient K of composites are usually determined by Darcy's law, i.e. by the filtration rate $\dot{m}$ (kg/s) of standard gas (nitrogen, helium, CO_2, air, gas mixture etc.) through the standard plate with thickness h at the given value of the gas pressure difference $(p_1 - p_2)$ on the plate surfaces [7]:

$$K = \frac{\dot{m}h}{S(p_1 - p_2)}. \qquad (A15)$$

Figure A4 shows the testing scheme. Mass rate of filtration is connected to consumption of filtering gas by the relation $\dot{m} = \rho_g v_g S$, where S is the area of the specimen crosssection.

Testing is conducted under constant temperature $\theta_0 = 200$ °C; changing gas permeability due to volumetric ablation is determined after first heating the specimen up to high temperatures and then cooling it down to temperature θ_0 (Figure A4).

A4. Determination of Heat Deformations under Heating

Determination of the heat deformation $\overset{\circ}{\varepsilon}$ of composites under high temperatures is conducted by using dilatometric devices [116], which differ one from another by the determination method of specimen extension $\Delta l = l - l_0$, and by the

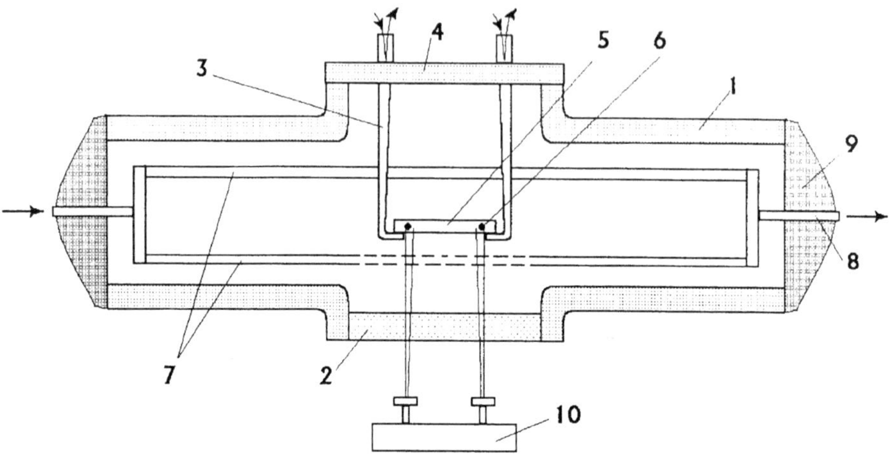

Figure A5. A scheme of the device for determination of heat deformations of composites under high temperatures [116]: 1 – frame of a gas-tight chamber heated, 2 – port-holes, 3 – water-cooled holders of the specimen, 4 – cover, 5 – composite specimen, 6 – markers on specimens, 7 – ceramic heaters, 8 – water-cooled current-suppliers, 8 – hermetic seal, 10 – optical watching system for measurement of displacements

heater structure. Figure A5 exhibits a device with an optical method of specimen extension measurement [116]. Heating in the device is realized by ceramic heaters; the specimen in the chamber is held by cooled supports. Optical observation of displacements Δl was conducted by a special optical watching system through a port-hole in the heated chamber. The port-hole is usually made of quartzy glass. Watching is realized on two markers on the specimen placed at the distance l_0 one from another and drawn by contrast heat-resistant substances.

Air or argon is usually supplied into the heated chamber. The device allows one to determine heat deformation $\overset{\circ}{\varepsilon} = \Delta l / l_0$ up to temperatures 1500...1800 °C with different rates of heating.

A5. Determination of Strength and Elastic Modulus of Composites under High Temperatures

Determination of high-temperature physico-mechanical characteristics of composites (elastic moduli, strength) is conducted by using breaking testing devices with a heating chamber. The structure of the device is shown schematically in Figure A6. The greatest complication in such devices consists in problems of determination of displacements of a specimen in tension, and in methods of keeping the specimen in the holder under heating up to high temperatures. To overcome these difficulties,

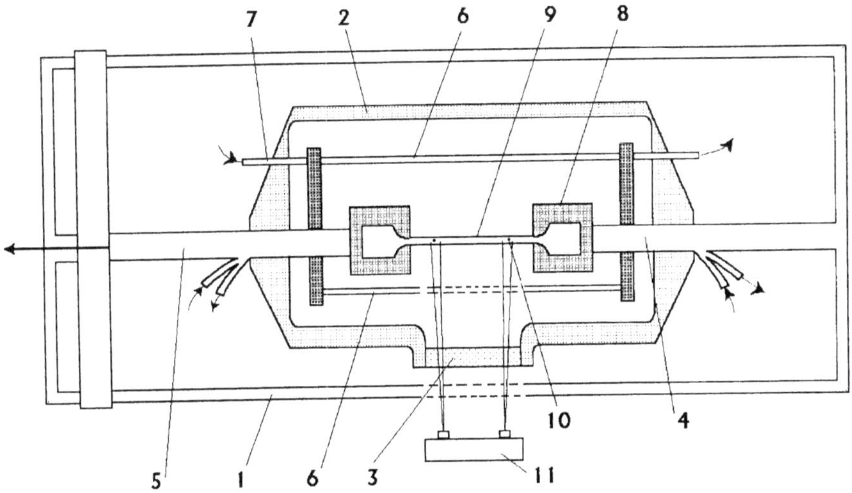

Figure A6. A scheme of the device for determination of thermostrength and elastic moduli of composites under high temperatures: 1 – device for testing of breakage, 2 – frame of a gas-tight chamber heated, 3 – port-hole, 4 – stationary water-cooled holder, 5 – moving water-cooled holder, 6 – ceramic heater, 7 – water-cooled current-suppliers, 8 – fasteners of the specimen and holders, 9 – composite specimen, 10 – markers on the specimen, 11 - optical system for measurement of deformations

one usually applies holders of the water-cooled type made of heat-resistant alloy, or holders made of carbon or ceramic materials. Determination of deformations is realized, as a rule, by the optical method shown in Figure A6 (similarly to that described in A4) or by the displacement of the holders at the outlet of the heated chamber. For the second method, additional testing should be conducted with the holders connected to one another and without the specimen; the values of displacements so obtained are subtracted from the displacements obtained in testing with the specimen. Thus, heat and elastic deformations of the testing system itself are eliminated.

To determine strength of composites in tension in the fabric plane, one can apply a specimen in the form of blades (Figure A7) with a working part thinned. For this, strength $\sigma_{\alpha T}$ (orientation of axes Ox_i is shown in Figure A7) and elastic moduli are determined as follows:

$$\sigma_{\alpha T} = \frac{F_{\alpha T}}{S}, \qquad E_{\alpha} = \frac{\Delta F_{\alpha T}}{S} \frac{l_0}{\Delta l_{\alpha}}, \qquad \alpha = 1, 2, \qquad (A16)$$

where $F_{\alpha T}$ are the forces tensile the specimen cut out in the Ox_{α} axis direction, l_0 is the length of the working part of the specimen, Δl_{α} is the displacement of the working part at changing force by value ΔF_{α}, S is the area of the specimen crosssection.

If the specimen is cut out from a composite plate with an angle $\vartheta = 45°$ to the orthotropy axes Ox_{α} (Figure A7,b) and testing in breakage is conducted as before,

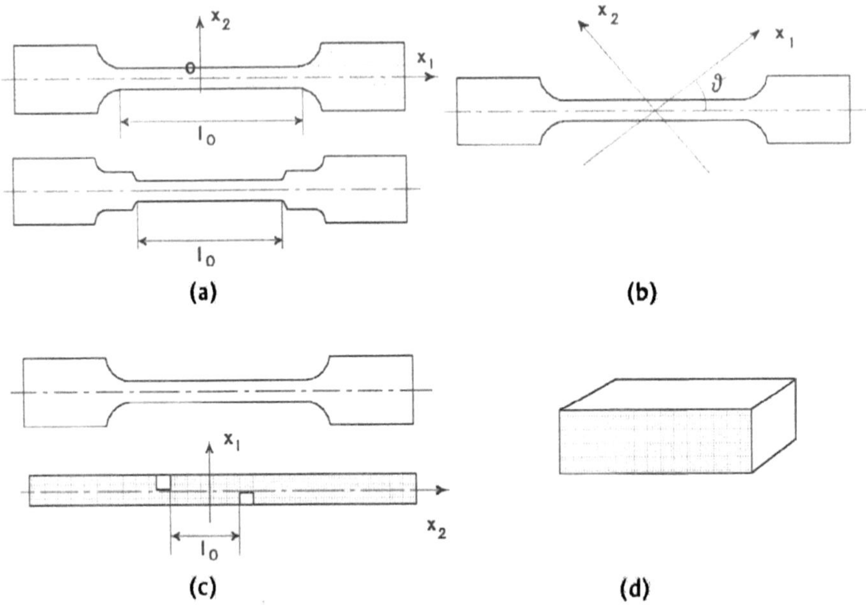

Figure A7. Typical forms of specimens of textile composites for testing:
a) in tension (to determine σ_{1T}, σ_{2T}, E_1, E_2),
b) in shear in the plane of fabric (to determine σ_{12S}, G_{12}),
c) in interlayer shear (to determine σ_{13S}, σ_{23S}, G_{13}, G_{23}),
d) in compression (to determine $\sigma_{\alpha C}$, E_3)

then one can determine shear modulus G_{12} in the fabric plane:

$$G_{12} = \left(\frac{4}{E_{45}} - \frac{1}{E_1} - \frac{1}{E_2} \right)^{-1}, \tag{A17}$$

where

$$E_{45} = \frac{\Delta F_{45}}{S} \frac{l_0}{\Delta l_{45}}, \tag{A18}$$

here ΔF_{45} and Δl_{45} are the force increment and corresponding extension of the specimen cut out with the angle ϑ.

To determine the elastic modulus G_{12} and strength σ_{12S} in shear in the fabric plane, the device for testing in torsion is usually used, as shown in Figure A6, but in place of the holder realizing tension-compression, one should use a rotary holder creating a torque M. For this case, shear modulus G_{12} and strength σ_{12S} are determined as follows:

$$G_{12} = \frac{\Delta M}{2\pi R^3 h \Delta\gamma_{12}},$$

$$\sigma_{12S} = \frac{M}{2\pi R^2 h}, \tag{A19}$$

where M is the ultimate value of the torque for which the specimen fails, ΔM is the increment of the torque to which the increment of the speciment twisting angle $\Delta\gamma_{12}$ corresponds, R is the radius, h is the thickness of the tubular specimen.

Measurement of the twisting angle γ_{12} of the specimen under high temperatures is realized either optically or by changing location of markers on the specimen, or by twisting the holders outside the heated chamber. For the second case, as in tension, auxiliary experiments are conducted to eliminate the twist in the measuring device itself from calculations.

To determine strength in compression $\sigma_{\alpha C}$, a specimen in the form of a prism is used (Figure A7,d); the specimen cut out from a composite plate in the corresponding direction Ox_α is placed between plane holders in the device shown in Figure A6 and is compressed to failure. Strength is determined by the formula

$$\sigma_{\alpha C} = \frac{F_{\alpha C}}{S}, \qquad \alpha = 1, 2, 3, \tag{A20}$$

where $F_{\alpha C}$ is the compressing force of the testing device.

Determination of strength $\sigma_{\alpha 3S}$ and modulus $G_{\alpha 3}$ in interlayer shear can also be realized by using the device for tension shown in Figure A6. For these purposes a specimen with two cuts is used (Figure A7); the cuts are made across the plane of fibres lying along the Ox_3 axis; their depth is usually equal to a half of the specimen thickness h. For this case, failure of the specimen in tension occurs due to interlayer shear between the cuts. Shear modulus G_{13} (or G_{12} if the specimen is cut out in the Ox_2 axis direction) is determined by the formula

$$G_{\alpha 3} = \frac{\Delta F_{\alpha 3}}{b l_0 \Delta\gamma_{\alpha 3}}, \qquad \alpha = 1, 2, \tag{A21}$$

where ΔF_{23} is the increment of tensile load, b is the specimen width, l_0 is the length of the working part between the cuts, $\Delta\gamma_{23}$ is the increment of the shear angle calculated by the formula

$$\Delta\gamma_{\alpha 3} = \frac{\Delta l_{\alpha 3} - \Delta\bar{l}_{\alpha 3}}{h_m}, \qquad \Delta\bar{l}_{\alpha 3} = \frac{\Delta F_{\alpha 3}}{b h E_\alpha}. \tag{A22}$$

Here Δl_{23} is the increment of extension of the working part of the specimen determined in experiments, $\Delta\bar{l}_{23}$ is the part of the extension defined by tension of the specimen, and h_m is the thickness of one layer of the matrix between the fabric layers.

Strength in interlayer shear is determined by the formula

$$\sigma_{\alpha 3S} = \frac{F_{\alpha 3}}{l_0 b}\xi \operatorname{cth}\xi, \qquad \xi = \frac{2l_0}{h}\sqrt{\frac{G_{\alpha 3}}{2E_\alpha}}, \qquad \alpha = 1, 2, \tag{A23}$$

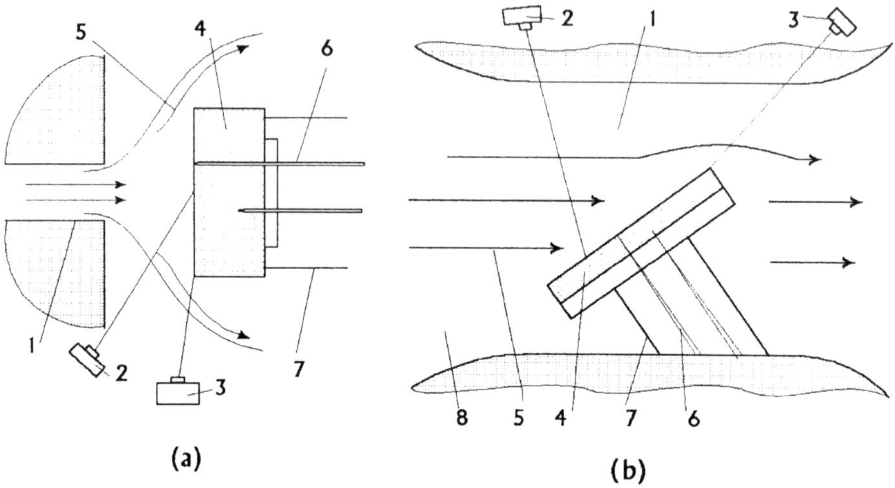

Figure A8. A scheme of the device for gas-dynamical testing of composites in a subsonic jet of hot gas (a) simulation of conditions of the critical point, b) simulation of the lateral surface): 1 – nozzle, 2 – pyrometer for measurement of the surface temperature, 3 – watching camera, 4 – specimen, 5 – overrunning hot gas flow, 6 – thermocouples, 7 – protective water-cooled causing, 8 – burning chamber

where $F_{\alpha 3}$ is the ultimate tensile load for which there occurs a failure of the specimen. Values of shear modulus $G_{\alpha 3}$ and elastic modulus E_α under given temperature θ of testing are determined by formulae (A16) and (A21).

A6. Gas-Dynamical Testing of Composites

To determine the ablation rate of composites under the action of high-speed hot gas flows, special, so-called gas-dynamical, testing is conducted in different devices. For this, the following devices are usually used: electro-arc heating of air and other gases (plasmatron), chemical heating (usually with combustion gaseous or liquid fuels), a shock tube where heating is realized due to deceleration of a flow etc. [94]. Both subsonic and supersonic devices can be applied. Figure A8 shows typical subsonic devices for testing of composites applied in structures of nose parts of re-entry vehicles (stagnation point) (Figure A8,*a*) and lateral surfaces of aerospace vehicles (Figure A8,*b*).

As mentioned in Chapter 1, main parameters defining heat-mechanical action of high-speed hot gas flow on composites are

- heat-transfer coefficient (α/c_p),
- flow enthalpy I_e and
- pressure head p_Σ of the ambient flow.

In simulation of characteristics of overruning flow in gas-dynamical devices, one usually strives for fine representation of these parameters. Other parameters such as flow temperature θ_∞, flow speed M_∞, static pressure of flow p_∞, Reynolds number Re etc. need not correspond to actual values of the overrunning flow in modeling testing.

Considerable efforts are undertaken in development of special methods for fine representation of the whole complex of parameters (α/c_p), I_e and p_Σ. Many works [59, 89, 94, 95 etc.] are devoted to the question.

As mentioned in Chapter 10, the main characteristics of the behaviour of composites under the action of high-speed hot gas flow are the following: linear ablation rate D, mass rate of ablation $\dot{m}$ and temperature of ablating surface θ_w of a composite. Besides that, in testing one often determines a temperature of a 'cold' surface θ_s and sometimes the temperature θ_i at the certain depth x_i from the ablating surface, with the help of which the accuracy of prediction of thermal characteristics of the material is verified.

The temperature of the ablating surface θ_w is usually determined either optically with the help of pyrometers, or with the help of a thermocouple placed under the heated surface of the composite. In the optical method, the so-called brilliant temperature θ_{wc} is determined; this is connected to the true temperature θ_w by the relation

$$\theta_w = \left(\theta_{wc}^{-1} + \frac{\lambda}{c_\lambda^2} \lg \varepsilon_{\lambda w}\right)^{-1},$$

where λ is the wavelength of the radiation at which the pyrometer works (usually within the interval from 0.3 to 10 mcm), $c_\lambda = 14.38 \cdot 10^3$ mcm·K is the pyrometric constant, $\varepsilon_{\lambda w}$ is the spectral degree of blackness of the composite surface corresponding to the wavelength λ and determined with the help of special additional testing [95]. However, for most ablative composites, values of $\varepsilon_{\lambda w}$ range from 0.8 to 0.9; for this case at $\lambda = 0.3 \ldots 0.5$ mcm, the difference between θ_w and θ_{wc} does not exceed 3...5 %.

To measure temperature θ_w one can also apply tungsten-rhenium, platinum-rhodium, chromel-aluminel and other thermocouples. As the ablating surface moves into the depth of material during the test, the end of the thermocouple is initially placed at a certain depth, the value of which is chosen in experiments.

The linear ablation rate D of composite is determined either optically (watching optical device) or by video-survey, and also by measuring the thickness Δh of the material carried away after testing:

$$D = \Delta h/t,$$

where t is the duration of the stationary regime of testing (when all the parameters correspond to calculated values). The value of the rate of mass removal is often determined only after completion of testing by measuring the change in specimen mass Δm:

$$\dot{m} = \frac{\Delta m}{|\Sigma| t},$$

where $|\Sigma|$ is the area of the ablating surface of the composite.

From these data on the composite behaviour, one determines the following experimental functions: $D(\theta_w)$ and $\dot{m}(\theta_w)$, and also $J_{\text{eff}}(\theta_w)$ or $I_{\text{eff}}(q_e)$, the effective enthalpy defined by formula (10.77).

REFERENCES

1. *Ablative Materials*, Handbook, U.S. Polymeric Inc. Santa Ana. CA, 1964.
2. Adams, M.K., Power, W.E. and Georgiev, S.: An experimental and theoretical study of quarter ablation at the stagnation point, *JASS*, **7** (1960), 535-547.
3. Alfutov, N.A., Zinoviev, P.A. and Popov, B.G.: *Calculation of Multilayer Plates and Shells of Composite Materials*, Moscow, Machinostroenie, 1984 (in Russian).
4. Arai, N. and Karashima, K.: Transient thermal response of ablating bodies, *AIAA Journal*, **17** (1979), 191-195.
5. Arai, N., Tani, T. and Sato, K.: Transient ablation of blunt bodies at angles attack, *Journal of Spacecraft and Rockets*, **21** (1984), N 3, 241-245.
6. Bachtel, J.H.: Heating of solid targets with laser pulses, *Journal of Applied Physics*, **46** (1975), 1585-1593.
7. Baer, E. Ed.: *Engineering Design for Plastics*, Reinhold Publishing Corporation, Chapmen & Hall LTD, London, 1966.
8. Bakhvalov, N.S. and Panasenko, G.P.: *Homogenization: Averaging Processes in Periodic Media (Mathematical Problems in Mechanics of Composite Materials)*, Kluwer Academic Publishers, Dordrecht, 1989.
9. Balabuh, L.I., Kolesnikov, K.S., Zarubin, V.S., Alfutov, N.A., et al.: *Basis of Designing Rocket Mechanics*, Moscow, Vyischaya schola, 1969 (in Russian).
10. Berdichevskiy, V.L.: *Variational Principles of Continuum Mechanics*, Moscow, Nauka, 1983 (in Russian).
11. Bondaryuk, N.M. and Ilyashenko, S.M.: *Straight-Flow Air-Rocket Engines*, Moscow, Oborongiz, 1958 (in Russian).
12. Broutman, L.J. and Krock, R.H. Eds: *Composite Materials*, **2**, *Mechanics of Composite Materials*, Academic Press, New York, 1974.
13. Broutman, L.J. and Krock, R.H. Eds: *Composite Materials*, **5**, *Fracture and Fatigue*, Academic Press, New York, 1974.
14. Chang, C.I., Griffis, C.A., Stonesifer, F.R. and Nemes, J.A.: Thermomechanical effects of intense heating on materials/structures, *Journal of Thermophysics and Heat Transfer*, **1** (1987), 175-181.
15. Chen, J.K., Sun, C.T. and Chang, C.I.: Failure analysis of graphite/epoxy composite subjected to combined thermal and mechanical loading, *J.Composite Mater.*, **19** (1985), 408-423.
16. Chen, J.K., Perea, A. and Allahdadi, F.A.: Laser effects on the dynamic response of laminated composites, *Composite Engineering*, **5** (1995), 1135-1147.

17. Chernyih, I.F: *Introduction to Anisotropic Elasticity*, Moscow, Nauka, 1988 (in Russian).
18. Chin, J.: *Thermoanalysis of Fibers and Fiber-Forming Polymers*, Interscience Publishers, New York, 1966.
19. Chou, Tsu-Wei and Ko, F.K. Eds: *Textile Structural Composites*, Elsevier Science, 1989.
20. Christensen, R.M.: *Mechanics of Composite Materials*, Wiley-Interscience Publication John Wiley & Sons, New York, 1980.
21. D'Alelio, G.F. and Parker, J.A. Eds: *Ablative Plastics*. Marcel Dekker Inc., New York, 1971.
22. Day, W.A.: *The Thermodynamics of Simple Materials with Fading Memory*, Springer-Verlag, Berlin-Heidelberg-New York, 1972.
23. Dimitrienko, Yu.I.: Coupled thermo-viscoelasticity problems for composites with periodic structure, *Mechanics of Composite Materials* (Riga), **6** (1985), 996-1004 (Translated in English by Consulting Bureau, New York).
24. Dimitrienko, Yu.I.: Technological stresses in carbon-carbon composite materials, *Mechanics of Composite Materials* (Riga), **6** (1991), 1030-1042 (Translated in English by Consulting Bureau, New York).
25. Dimitrienko, Yu.I.: Failure of composite materials at high temperatures and finite strains, *Mechanics of Composite Materials* (Riga), **1** (1992), 43-55 (Translated in English by Consulting Bureau, New York).
26. Dimitrienko, Yu.I.: Mathematical modelling of ceramic composite processing based on combustion, *Journ. Math. Comput. Modelling*, **21** (1995), N 8, 69-83.
27. Dimitrienko, Yu.I.: Thermal stresses and heat-mass-transfer in ablating composite materials, *Journal of Heat Mass Transfer*, **38** (1995), N 1, 139-146.
28. Dimitrienko, Yu.I., Efremov, G.A. and Chernyavsky, S.A.: Optimum design of erosion-stable heat-shield composite materials, *Applied Composite Materials*, **4** (1997), 35-52.
29. Dimitrienko, Yu.I.: Modelling of mechanical properties of composite materials under high temperatures. Part 1. Matrix and fibres, Part 2. Properties of unidirectional composites, *Applied Composite Materials*, **4** (1997), 219-261.
30. Dimitrienko, Yu.I.: Experimental and theoretical investigation of thermomechanical ablation processes of multiphase materials under high enthalpy heat flux effects, *Proceedings of 4th World Conference on Experimental Heat Transfer, Fluid Mechanics and Thermodynamics*, Brussels, **1** (1997), 401 - 408.
31. Dimitrienko, Yu.I.: Modelling of erosion combustion of energetic materials in high-enthalpy flows, *Combustion and Flame*, **111** (1997), 161-174.
32. Dimitrienko, Yu.I.: Effect of microstructure on erosive combustion of energetic materials, *Proc. of 27th Int. Conference of ICT 'Energetic Materials - Technology, Manufacturing and Processing'*, Karlsruhe, Germany, 1996, 63.1 - 63.14.

33. Dimitrienko, Yu.I.: Mathematical modelling of composite materials manufacturing processes, *Proc. of the IMACS Symposium on Mathematical Modelling*, Austria, **5** (1994), 878-881.
34. Dimitrienko, Yu.I.: Ultra-light thermal-protective materials, *Proceedings of Int. Conference on Composite Engineering (ICCE/2)*, New Orleans, USA, 1995, 189-190.
35. Dimitrienko, Yu.I.: Mathematical modelling of ablating materials, *Proceedings of the Second ECCOMAS Conference on Numerical Methods in Engineering*, Paris, France, 1996, 7 p.
36. Dimitrienko, Yu.I.: Effect of finite deformations on internal heat-mass-transfer in elastomer ablating materials, *Int. Journal of Heat Mass Transfer*, **40** (1997), N 3, 699-709.
37. Dimitrienko, Yu.I.: Internal heat-mass-transfer and stresses in thin-walled structures of ablating materials, *Int. Journal of Heat Mass Transfer*, **40** (1997), N 7, 1701-1711.
38. Dimitrienko, Yu.I.: Thermal stresses in ablative composite thin-walled structures under intensive heat flows, *Int. Journal of Engineering Science*, **35** (1997), N 1, 15-31.
39. Dimitrienko, Yu.I.: Heat-mass-transport and thermal stresses in porous charring materials, *Transport in Porous Media*, **27** (1997), N 2, 143-170.
40. Dimitrienko, Yu.I.: Thermomechanical behaviour of composite materials and structures under high temperatures. 1. Materials, 2. Structures, *Composites, Part A: Applied Science and Manufacturing*, **28A** (1997), 453 - 471.
41. Dimitrienko, Yu.I.: Mechanics of porous media with phase transformations and periodical structure. 1. Method of asymptotic averaging, 2. Solutions of local and global problems, *European Jounal of Mechanics (A: Solids)* **17** (1998), N 2, 305-337.
42. Dimitrienko, Yu.I. and Dimitrienko, I.P.: Long-duration strength of reinforced plastics, *Mechanics of Composite Materials* (Riga), **1** (1989), 16-22 (Translated in English by Consulting Bureau, New York).
43. Dimitrienko, Yu.I., Efremov, G.A. and Epifanovsky, I.S.: Reusable re-entry vehicles with reclaimable ablating thermal protection, *Preprints of 19th Int.Symp. on Space Technology and Science*, 1994, Yokohama, Japan, 94-b-26.
44. Dimitrienko, Yu.I., Efremov, G.A., Chernyavsky, S.A. and Medvedev, Yu.V.: Theory and synthesis of advanced thermal-protective composite materials, *Applied Composite Materials*, **2** (1995), 367-384.
45. Dimitrienko, Yu.I. and Epifanovsky, I.S.: Deforming and strength of destructing thermo-protective materials at high temperatures, *Mechanics of Composite Materials* (Riga), **3** (1990), 460-468 (Translated in English by Consulting Bureau, New York).
46. Dimitrienko, Yu.I. and Epifanovsky, I.S.: Investigation of high temperature deformation of composites on inorganic matrix, *Proceedings of the Moscow Conference on Composites*, Elsevier Science, 1991.

47. Dimitrienko, Yu.I. and Limonov, V.A.: Effect of fiber orientation on dissipative heating and deformability of viscoelastic composites under cyclic loading, *Mechanics of Composite Materials* (Riga), **5** (1988), 797-805 (Translated in English by Consulting Bureau, New York).
48. Favaloro, M.: Ablative materials, in *Encyclopedia of Chemical Technology*, 4-th ed., J. Wiley & Sons, **1**, 1992.
49. Gladwell, G.M.L., Barber, J.R. and Olesiak, Z.: Thermal problems with radiation boundary conditions, *Quart.J.Mech. and Appl. Math.*, **36** (1983), N 3, 387-401.
50. Gladwell, G.M.L. and Barber, J.R.: Thermoelastic contact problems with radiation boundary conditions, *Quart.J.Mech. and Appl. Math.*, **36** (1983), N 3, 403-417.
51. Green, A.E. and Zerna, W.: *Theoretical Elasticity*, McGraw-Hill, New York, 1956.
52. Greenwood, T.F., Lee, Y.C., Bender, R.L. and Carter, R.E.: Space shuttle base heating, *J. Spacecraft and Rockets*, **21** (1984), N 4, 339-345.
53. Griffis, C.A., Nemes, J.A., Stonesifer, F.R. and Chang, C.I.: Degradation in strength of laminated composites subjected to intense heating and mechanical loading, *J.Composite Mater.*, **20** (1986), 216-235.
54. Grigoriev, I.S. and Meylihov, E.Z. Eds: *Physical Magnitudes.* Handbook, Moscow, Energoatomizdat, 1991 (in Russian).
55. Grigoryantz, A.G.: *Basis of Laser Processing of Materials*, Moscow, Machinostroenie, 1989 (in Russian).
56. Grinfeld, M.A.: *Thermodynamic Methods in the Theory of Heterogeneous System*, Longman, 1991.
57. Gruntfest, I.J. and Shanker, L.H.: Behavior of reinforced plastics at very high temperatures, *Soc. Plastics. Industr. 13th Annual Conference*, Reinforced Plastics Division, 1958.
58. Gularte, R.C., Nemes, J.A., Stonesifer, F.R. and Chang, C.I.: Failure of mechanically loaded laminated composites subjected to intense localized heating, *Int. J. Numer. Meth. Eng.*, **25** (1988), N 2, 561-570.
59. Hankey, W.L.: *Re-Entry Aerodynamics*, AIAA textbook, 1988.
60. Heidug, W. and Lehner, F.K.: Thermodynamics of coherent phase transformation in nonhydrostatically stressed solids, *Pure Appl. Geophys.*, **123** (1985), N 1, 91-98.
61. Herakovich, C.T. and Tarnopol'skii, Yu.M. Eds: *Handbook of Composites*, **2**, *Structures & Design*, Amsterdam, North Holland, 1988.
62. Jones, R.M.: *Mechanics of Composite Materials*, McGraw-Hill, New York, 1975.
63. Ilyushin, A.A.: *Plasticity*, Moscow, Science Academy of USSR, 1963 (in Russian).
64. Ilyushin, A.A.: *Continuum Mechanics*, Moscow, Moscow State University Publ., 1978 (in Russian).
65. Ilyushin, A.A. and Pobedria, B.E.: *Basis of Mathematical Theory of Thermoviscoelasticity*, Moscow, Nauka, 1970 (in Russian).

66. Isahanov, G.V.: *Strength of Nonmetallic Materials under Non-uniform Heating*, Kiev, Naukova Dumka, 1971 (in Russian).
67. Kondaurov, V.I.: On conservation laws and symmetrisation of equations of non-linear theory of thermoelasticity, *Dokladi Academii Nauk*, **256** (1981), 819-823 (in Russian).
68. Kondaurov, V.I. and Nikitin, L.V.: Phase transformations of the first type in elastoviscoplastic medium, *Solid Body Mechanics*, **4** (1986), 130-139.
69. Korshak, V.V.: *Chemical Structure and Temperature Characteristics of Polymers*, Moscow, Nauka, 1970 (in Russian).
70. Krasnov, N.F.: *Aerodynamics*, Moscow, Vyischaya schola, 1971 (in Russian).
71. Laub, B.: Thermochemical ablation of tantalum carbide loaded carbon-carbons, *AIAA Paper*, (1980), N 1476.
72. Lehnitski, S.G.: *Elasticity Theory for Anisotropic Bodies*, Moscow, Nauka, 1977 (in Russian).
73. Lions, J.L.: Remarks on nonlocal phenomena in composite materials and in perforated materials, *Proceedings of the IUTAM Symposium*, Nothwestern Univ., North Holland, 1979.
74. Luryie, A.I.: *Nonlinear Elasticity Theory*, Moscow, Nauka, 1980 (in Russian).
75. Madorsky, S.L.: *Thermal Degradation of Organic Polymers*, Interscience Publishers, New York - London - Sydney, 1964.
76. Malmeyster, A.K., Tamuzh, V.P. and Teters, G.A.: *Resistance of Polymer and Composite Materials*, Riga, Zinatne, 1980 (in Russian).
77. McHenry, M.R. and Laub, B.: Ablative radome material thermal-ablation and erosion modelling, *13th Intersociety Conf. on Environ. Systems*, San Francisco, 1983.
78. McManus, H.N. and Springer, G.S.: High temperature thermomechanical behavior of carbon-phenolic and carbon-carbon composites: I. Analysis, II. Results, *J.Composite Mater.*, **26** (1992), 206-255.
79. Mobilia, J.J. and Rosato, D.V.: Reinforced plastics in missiles and rockets, *S.P.E.J.*, (1960), 969-980.
80. Molchanov, Yu.M. Ed.: *Physical and Mechanical Properties of Glass-Plastics*. Handbook, Riga, Zinatne, 1969 (in Russian).
81. Morland, L.W.: Generation of thermoelastic stress waves by impulse electromagnetic radiation, *AIAA Journal*, **6** (1985), 5123-5126.
82. Moskvitin, V.V.: *Resistance of Viscoelastic Materials*, Moscow, Nauka, 1972 (in Russian).
83. Nigmatulin, R.I.: *Dynamics of Multiphase Media*, Part I, II. Elsevier Science, New York, 1987.
84. Oleynik, O.A., Iosifyan, G.A. and Shamaev, A.S.: *Mathematical Problems of Strongly Inhomogeneous Elastic Media Theory*, Moscow, Moscow State University Publ., 1990 (in Russian).
85. Omelchenko, K.G., Savelov, N.V. and Timoshenko, V.P.: On investigation of heat mass transfer in decomposing porous materials, *Heat Physics of High Temperatures*, **12** (1974), 761-768.

86. Ovchinski, A.S., Kopyiev, I.M., Saharova, E.N. and Moskvitin, V.V.: Redistribution of stresses at breakage of brittle fibres in metallic composite materials, *Polymer Mechanics*, **1** (1977), 19-29.
87. Pagano, N.J. Ed.: *Interlaminar Response of Composite Materials*, Elsevier Science, 1989.
88. Pankratov, B.M., Polezhaev, Yu.V. and Rud'ko, A.K.: *Interaction of Materials with Gas Flows*, Moscow, Machinostroenie, 1976 (in Russian).
89. Park, C., Lundell, J.H., Green, M.J., Winovich, W. and Covington M.A.: Ablation of carbonaceous materials in a hydrogen-helium arcjet flow, *AIAA Journal*, **22** (1984), N 10, 1491-1498.
90. Pobedria, B.E.: *Lectures on Tensor Analysis*, Moscow, Moscow State University Pbl., 1986 (in Russian).
91. Pobedria, B.E.: *Mechanics of Composite Materials*, Moscow, Moscow State University Publ., 1984 (in Russian).
92. Pobedria, B.E.: *Numerical Methods in Elasticity and Plasticity theories*, Moscow, Moscow State University Publ., 1981 (in Russian).
93. Pobedria, B.E. and Dimitrienko, Yu.I.: Coupled problems of linear thermomechanics of deformable solid body, *Progress in Mechanics*, **10** (1987), 97-137.
94. Polezhaev, Yu.V. and Shishkov, A.A.: *Gasodynamical Testing of Thermal Protection*, Moscow, Promedek, 1992 (in Russian).
95. Polezhaev, Yu.V. and Yurevich, F.B.: *Thermal Protection*, Moscow, Energia, 1976 (in Russian).
96. Pozdeev, A.A., Trusov, P.V. and Nyashin, Yu.I.: *Finite Elastoplastic Deformations*, Moscow, Nauka, 1986 (in Russian).
97. Regan, F.J. and Anandakrishnan, S.M.: *Dynamics of Atmospheric Re-Entry*, AIAA textbook, 1993.
98. Richard, T.G.: The mechanical behavior of a solid microsphere filled composite, *Journal of Composite Materials*, **9** (1975), 108-116.
99. Rosato, D.V. and Schwartz, R.T. Eds: *Environmental Effects on Polymeric Materials*, Wiley-Interscience, New York, 1968.
100. Rosen, B.W.: Tensile failure of fibrous composites, *AIAA Journ.*, **2** (1964), 1985-1994.
101. Sallivan, J.M. and Kobayashi, W.S.: Spallation modeling in the charring material thermal response and ablation computer program, *AIAA Pap.*, N 1516 (1987), 1-6.
102. Samarskiy, A.A.: *Difference Scheme Theory*, Moscow, Nauka, 1983 (in Russian).
103. Samarskiy, A.A. and Popov, Yu.P.: *Difference Schemes of Gas Dynamics*, Moscow, Nauka, 1975 (in Russian).
104. Sanchez-Palencia, E.: *Non-Homogeneous Media and Vibration Theory*, Springer-Verlag, New York, 1980.
105. Scala, S.M.: *Jet. Propulsion*, **28** (1958), N 5, 340-351.
106. Scala, S.M. and Hilbert, L.M.: Heat destruction of charring plastic at hypersonic flights, *Rocket Technika*, **6** (1962), 72-83.

107. Sedov, L.I.: *Continuum Mechanics*, Moscow, Nauka, 1976 (in Russian).

108. Shlensky, O.F., Shashkov, A.V. and Aksenov, L.N.: *Thermal Decomposition of Materials*, Amsterdam, Elsevier Science, 1991.

109. Sinyukov, A.M., Volkov, L.I., Lyvov, A.I. and Shishkevich, A.M.: *Ballistic Rocket on Solid Propellant*, Moscow, Voenizdat, 1972 (in Russian).

110. Skvortsova, N.V.: Experimental investigations of the effect of heating rate and environment on Young modulus of carbon-plastics in a wide interval of temperatures, *Mechanics of Composite Materials* (Riga), **1** (1986), 23-27 (Translated in English by Consulting Bureau, New York).

111. Sobolev, N.V., Kavun, T.N., Kiselev, B.A., et al.: Varying properties of glass- and carbon-filled polymers during pyrolysis, in *Composite Materials*, Moscow, Nauka, 1981, 244-247 (in Russian).

112. Tamuzh, V.P. and Kuksenko, V.S.: *Micromechanics of Fracture of Polymer Materials*, Riga, Zinatne, 1978 (in Russian).

113. Tamuzh, V.P. and Protasov, V.D. Eds: *Destruction of Composite Material Constructions*, Riga, Zinatne, 1986 (in Russian).

114. Tarnopolsky, Yu.N., Zjigune, I.G. and Polyakov, V.A.: *Space-Reinforced Composite Materials*, Moscow, Machinostroenie, 1987 (in Russian).

115. Thornton, E.A.: *Thermal Structures and Materials for High-Speed Flight*, Progress in Astronautics and Aeronautics, 1992.

116. Tretyachenko, G.N. and Gracheva L.N.: *Thermal Deforming of Non-Metallic Destructing Materials*, Kiev, Naukova Dumka, 1983 (in Russian).

117. Trostyanskaya, E.B. Ed.: *Thermal Resistance of Plastics for Constructional Purposes*, Moscow, Chimia, 1980 (in Russian).

118. Tsai, S.W. and Hahn, H.T.: *Introduction to Composite Materials*, Technomic Pub.Co., Lancaster, PA, 1980.

119. Vanin, G.A.: *Micromechanics of Composite Materials*, Kiev, Naukova Dumka, 1985 (in Russian).

120. Vasilyev, V.V.: *Mechanics of Composite Constructions*, Moscow, Machinostroenie, 1988 (in Russian).

121. Vasilyev, V.V. and Tarnopolsky, Yu.M. Eds: *Composite materials*, Handbook, Moscow, Machinostroenie, 1989 (in Russian).

122. Volmir, A.S.: *Stability of Deformable Systems*, Moscow, Nauka, 1967 (in Russian).

123. Wu, E.M.: Phenomenological anisotropic failure criterion, in: *Composite Materials*, edited by L.J.Broutman and R.H.Krock, **2**, edited by G.P.Sendeckyj, Academic Press, New York, 1974.

124. Young, R.W.: Sensitivity of thermomechanical response to thermal boundary conditions and material constants, *Int. J. Sol. Str.*, **15** (1979), N 7, 513-517.

125. Zarubin, V.S.: *Applied Thermostrength Problems for Structural Elements*, Moscow, Machinostroenie, 1985 (in Russian).

126. Ziegler, H.: *An Introduction to Thermodynamics*, Amsterdam, North-Holland, 1977.

127. Ziering, M.B.: Thermochemical ablation of ceramic heat shields, *AIAA Journ.*, **13** (1975), 610-616.
128. Zinchenko, V.I. and Yakimenko, A.S.: Regimes of thermochemical destruction of carbon/phenolic composite material under the action of heat flux, *Physika Gorenia i Vzryiva*, **24** (1988), 141-149 (in Russian).

SUBJECT INDEX

Ablation 19
–, surface (linear) 19
–, volumetric (thermodecomposition) 19
Adhesion conditions 49
Angle of bending of a thread 157
Asymptotic expansion 59
Basis, dyadic 28
–, local, coupling 28
–, –, general 28
Chemical shrinkage 89
Coefficient, chemical shrinkage 47
–, gasification 55
–, heat expansion 47
–, heat transfer 7
–, injection 65
–, Lamé's 47
– of varying a strength 96
–, stress concentration 134
–, thermomechanical erosion 245
Combustion regime, diffusive 228
– –, kinetic 228
– –, sublimational 228
Component-layer 159
Composite materials, ablative 19
– –, charring 21
– –, reinforced by particles 191
– –, textile 157
– –, unidirectional 127
Configuration, actual 27
–, initial 27
–, natural 34
Conservation equations for continuum mechanics 30
Constitutive relations, effective 64
– – for a multilayer ablative shell 307
– – for phases 37
Coordinates, global 55
–, local 55
–, material 27
Cubic model 74
Curvatures of a middle surface 305
Dependence, Arrhenius 52
–, Langmuir-Knudsen 230
–, Weibull 120
Derivative, Gateaux's 36
–, total with respect to time 33
Divergent form of conservation laws 31
Dyad 28
Effective characteristics 69
Enthalpy, effective 252
– of a gas flow 7
Entropy 31
Equation, kinematic 31
– of continuity 31
– of energy 31
– of entropy balance 31
– of heat conductivity 48
– of kinetic energy 33
– of momentum 31
– of strain compatibility 31, 71
Evaporation (sublimation) 21
Failure criterion for a matrix 94
– – for textile composite 175
– – for unidirectional composite 155
Fire action 16
Force 303
– of phase interaction 307
– of pore pressure 304
– of thermal stresses 307
–, transverse 303
Function, energy dissipation 31
–, quasiperiodic 56
–, stress 212
Gas constant 48
Gas phase, linear-viscous 37
– –, perfect 47
General thermodynamical identity 33
Gradient, phase transformation 34
–, strain 30
Hamilton nabla-operator 29
Heat of combustion 251
– of sublimation 251
Heat capacities of phases 46
Heat flux, radiant 17
– –, convective 2
– – to the 'cold wall' 8
Heating, aerodynamical 1
–, gas-dynamical 9
–, gradient 203
–, local 289
–, technological 15
Helmholtz free energy 33
Heterogeneous combustion 224
Hierarchy of internal structural levels 128
Infinitesimal strains 44
Internal energy 31
Jump of a function 32
Law, Darcy's 64
–, Fourier's 33
– of acting masses 226
– of balance of component masses 225
– of conservation of a mixture mass 226
Limited durability 268

Load-bearing capacity 221
Loss of stability of a shell 318
Macrostresses 93
Medium, cylindrically orthotropic 255
–, homogeneous 29
–, isotropic 47
–, linear-elastic 46
–, orthotropic 165
–, transversally isotropic 130
Melting 21
Microcomposite 130
Microstresses 92
–, thermal 150
Misalignment of fibres 128
Moment 303
– of phase interaction 307
– of pore pressure 304
– of thermal stresses 307
Monofibre 117
Multicomponent system 224
Non-effective length 124, 140
Parameters, damage 178
–, Lamé's 297
Periodic structure 53
Periodicity cell 53
Phase 29
–, 'amorphic' 25
–, 'crystalline' 25
–, gas 25
–, polymer 25
–, pyrolytic 25
Phase interface, homothermal 40
– –, nondissipative 39
– – with ideal contact 40
– – with slip 40
Phase transformation 30
Poisson's ratio 47
Pressure, local 3
–, partial 229
–, pore gas 77
Pressure head 3
Problem, averaged 65
–, axisymmetric 255
–, coupled 71
–, – of heat and mass transfer 71
–, – of heat-mass-transfer anf deforming of multilayer ablative shells 308
–, deformation 71
–, heat transfer 71
– in terms of displacements ('A') 70
– – – – – in curvilinear coordinates 302
– in terms of stresses ('B') 72
–, local over the periodicity cell, first-level, gas dynamics 62
–, - - - - -, –, heat conduction 62
–, - - - - -, –, solid deformable body mechanics 62
–, - - - - -, zero-level, gas dynamics 60
–, - - - - -, –, heat conduction 60
–, - - - - -, –, solid deformable body mechanics 60
–, mass transfer 71
–, plane 210
–, thermomechanics for ablative composites 59
Product, scalar 28
–, tensor 29
–, vector 29
Pyrolysis (thermodecomposition) 20
Pyrolytic residue 19, 21
Quasistatic process 48
Rate of ablation, dimensionless 254
– – –, mass 252
– of combustion, linear 225
– of external thermomechanical erosion 231
– of formation of components 225
– of internal thermomechanical erosion 231
– of phase transformation 32
– – – –, mass 39
– of sublimation, linear 229
– of thermochemical ablation 231
Realization of elastic properties of a monofibre 138, 170
Reynolds heat-mass-transfer analogy 226
Section of a trajectory, active (AST) 3
– – –, passive (PST) 3
Shell, thin 302
–, –, cyindrical 308
Speed of a phase interface 32
Stiffness, bending 307
–, combined 307
–, membrane 307
Surface, gas-tight 65
–, heat-insulated 65
–, strength 99
– of a strong break 31
– of phase transformation 40
Symbol, Kronecker 28
–, Levi-Civita 29
Temperature 31
– of heat-resistance 178
– of thermostability 87, 187
Tensor, inverse 29
–, inverse-transpose 29
–, metric (unit) 28
– of chemical potential, Bowen's 42
– of chemical shrinkage 45
– of elastic moduli 46
– of elastic pliabilities 47, 71
– of elastic stresses 33
– of finite strains, Almancy's 33
– of gas permeability 64
– of heat conductivity 32
– of heat deformations 46
– of heat expansion 36
– of kinetic energy of phase transformation 42
– of phase interaction 69
– of stress functions 71
– of stresses, Cauchy's 31
– of viscous stresses 33
–, transpose 29

Tensor components, combined 28
– –, contravariant 28
– –, covariant 28
Tensor invariants, first 47, 94
– –, second 94
– –, sign-constant 94
Tensor-invariant relationship 134
Thermochemical reaction 22
Thermodecomposition (TD) 20
Thermodecomposition front 291
Thermogravimetric analysis 77
Thermomechanical erosion 22
– –, external 231
– –, internal 231
Thermo-oxidative decomposition (TOD) 21
Thread 117
Triad 29
Types of destruction of a unidirectional composite: (1), (2), (3), (4) 138
– – – of a textile composite: (A_α), (B_α), (C), (D) 175, (E) 318
Ultimate strength 94, 196
Vector, displacement 30
–, heat flux 31
Warp 174
Weft 174

Mechanics

SOLID MECHANICS AND ITS APPLICATIONS

Series Editor: G.M.L. Gladwell

Aims and Scope of the Series

The fundamental questions arising in mechanics are: *Why?*, *How?*, and *How much?* The aim of this series is to provide lucid accounts written by authoritative researchers giving vision and insight in answering these questions on the subject of mechanics as it relates to solids. The scope of the series covers the entire spectrum of solid mechanics. Thus it includes the foundation of mechanics; variational formulations; computational mechanics; statics, kinematics and dynamics of rigid and elastic bodies; vibrations of solids and structures; dynamical systems and chaos; the theories of elasticity, plasticity and viscoelasticity; composite materials; rods, beams, shells and membranes; structural control and stability; soils, rocks and geomechanics; fracture; tribology; experimental mechanics; biomechanics and machine design.

1. R.T. Haftka, Z. Gürdal and M.P. Kamat: *Elements of Structural Optimization*. 2nd rev.ed., 1990 ISBN 0-7923-0608-2
2. J.J. Kalker: *Three-Dimensional Elastic Bodies in Rolling Contact*. 1990 ISBN 0-7923-0712-7
3. P. Karasudhi: *Foundations of Solid Mechanics*. 1991 ISBN 0-7923-0772-0
4. *Not published*
5. *Not published.*
6. J.F. Doyle: *Static and Dynamic Analysis of Structures*. With an Emphasis on Mechanics and Computer Matrix Methods. 1991 ISBN 0-7923-1124-8; Pb 0-7923-1208-2
7. O.O. Ochoa and J.N. Reddy: *Finite Element Analysis of Composite Laminates*. ISBN 0-7923-1125-6
8. M.H. Aliabadi and D.P. Rooke: *Numerical Fracture Mechanics*. ISBN 0-7923-1175-2
9. J. Angeles and C.S. López-Cajún: *Optimization of Cam Mechanisms*. 1991 ISBN 0-7923-1355-0
10. D.E. Grierson, A. Franchi and P. Riva (eds.): *Progress in Structural Engineering*. 1991 ISBN 0-7923-1396-8
11. R.T. Haftka and Z. Gürdal: *Elements of Structural Optimization*. 3rd rev. and exp. ed. 1992 ISBN 0-7923-1504-9; Pb 0-7923-1505-7
12. J.R. Barber: *Elasticity*. 1992 ISBN 0-7923-1609-6; Pb 0-7923-1610-X
13. H.S. Tzou and G.L. Anderson (eds.): *Intelligent Structural Systems*. 1992 ISBN 0-7923-1920-6
14. E.E. Gdoutos: *Fracture Mechanics*. An Introduction. 1993 ISBN 0-7923-1932-X
15. J.P. Ward: *Solid Mechanics*. An Introduction. 1992 ISBN 0-7923-1949-4
16. M. Farshad: *Design and Analysis of Shell Structures*. 1992 ISBN 0-7923-1950-8
17. H.S. Tzou and T. Fukuda (eds.): *Precision Sensors, Actuators and Systems*. 1992 ISBN 0-7923-2015-8
18. J.R. Vinson: *The Behavior of Shells Composed of Isotropic and Composite Materials*. 1993 ISBN 0-7923-2113-8
19. H.S. Tzou: *Piezoelectric Shells*. Distributed Sensing and Control of Continua. 1993 ISBN 0-7923-2186-3

Kluwer Academic Publishers – Dordrecht / Boston / London

Mechanics

SOLID MECHANICS AND ITS APPLICATIONS

Series Editor: G.M.L. Gladwell

20. W. Schiehlen (ed.): *Advanced Multibody System Dynamics.* Simulation and Software Tools. 1993 ISBN 0-7923-2192-8
21. C.-W. Lee: *Vibration Analysis of Rotors.* 1993 ISBN 0-7923-2300-9
22. D.R. Smith: *An Introduction to Continuum Mechanics.* 1993 ISBN 0-7923-2454-4
23. G.M.L. Gladwell: *Inverse Problems in Scattering.* An Introduction. 1993 ISBN 0-7923-2478-1
24. G. Prathap: *The Finite Element Method in Structural Mechanics.* 1993 ISBN 0-7923-2492-7
25. J. Herskovits (ed.): *Advances in Structural Optimization.* 1995 ISBN 0-7923-2510-9
26. M.A. González-Palacios and J. Angeles: *Cam Synthesis.* 1993 ISBN 0-7923-2536-2
27. W.S. Hall: *The Boundary Element Method.* 1993 ISBN 0-7923-2580-X
28. J. Angeles, G. Hommel and P. Kovács (eds.): *Computational Kinematics.* 1993 ISBN 0-7923-2585-0
29. A. Curnier: *Computational Methods in Solid Mechanics.* 1994 ISBN 0-7923-2761-6
30. D.A. Hills and D. Nowell: *Mechanics of Fretting Fatigue.* 1994 ISBN 0-7923-2866-3
31. B. Tabarrok and F.P.J. Rimrott: *Variational Methods and Complementary Formulations in Dynamics.* 1994 ISBN 0-7923-2923-6
32. E.H. Dowell (ed.), E.F. Crawley, H.C. Curtiss Jr., D.A. Peters, R. H. Scanlan and F. Sisto: *A Modern Course in Aeroelasticity.* Third Revised and Enlarged Edition. 1995 ISBN 0-7923-2788-8; Pb: 0-7923-2789-6
33. A. Preumont: *Random Vibration and Spectral Analysis.* 1994 ISBN 0-7923-3036-6
34. J.N. Reddy (ed.): *Mechanics of Composite Materials.* Selected works of Nicholas J. Pagano. 1994 ISBN 0-7923-3041-2
35. A.P.S. Selvadurai (ed.): *Mechanics of Poroelastic Media.* 1996 ISBN 0-7923-3329-2
36. Z. Mróz, D. Weichert, S. Dorosz (eds.): *Inelastic Behaviour of Structures under Variable Loads.* 1995 ISBN 0-7923-3397-7
37. R. Pyrz (ed.): *IUTAM Symposium on Microstructure-Property Interactions in Composite Materials.* Proceedings of the IUTAM Symposium held in Aalborg, Denmark. 1995 ISBN 0-7923-3427-2
38. M.I. Friswell and J.E. Mottershead: *Finite Element Model Updating in Structural Dynamics.* 1995 ISBN 0-7923-3431-0
39. D.F. Parker and A.H. England (eds.): *IUTAM Symposium on Anisotropy, Inhomogeneity and Nonlinearity in Solid Mechanics.* Proceedings of the IUTAM Symposium held in Nottingham, U.K. 1995 ISBN 0-7923-3594-5
40. J.-P. Merlet and B. Ravani (eds.): *Computational Kinematics '95.* 1995 ISBN 0-7923-3673-9
41. L.P. Lebedev, I.I. Vorovich and G.M.L. Gladwell: *Functional Analysis.* Applications in Mechanics and Inverse Problems. 1996 ISBN 0-7923-3849-9
42. J. Menčik: *Mechanics of Components with Treated or Coated Surfaces.* 1996 ISBN 0-7923-3700-X
43. D. Bestle and W. Schiehlen (eds.): *IUTAM Symposium on Optimization of Mechanical Systems.* Proceedings of the IUTAM Symposium held in Stuttgart, Germany. 1996 ISBN 0-7923-3830-8

Kluwer Academic Publishers – Dordrecht / Boston / London

Mechanics

SOLID MECHANICS AND ITS APPLICATIONS

Series Editor: G.M.L. Gladwell

44. D.A. Hills, P.A. Kelly, D.N. Dai and A.M. Korsunsky: *Solution of Crack Problems.* The Distributed Dislocation Technique. 1996 ISBN 0-7923-3848-0
45. V.A. Squire, R.J. Hosking, A.D. Kerr and P.J. Langhorne: *Moving Loads on Ice Plates.* 1996 ISBN 0-7923-3953-3
46. A. Pineau and A. Zaoui (eds.): *IUTAM Symposium on Micromechanics of Plasticity and Damage of Multiphase Materials.* Proceedings of the IUTAM Symposium held in Sèvres, Paris, France. 1996 ISBN 0-7923-4188-0
47. A. Naess and S. Krenk (eds.): *IUTAM Symposium on Advances in Nonlinear Stochastic Mechanics.* Proceedings of the IUTAM Symposium held in Trondheim, Norway. 1996 ISBN 0-7923-4193-7
48. D. Ieşan and A. Scalia: *Thermoelastic Deformations.* 1996 ISBN 0-7923-4230-5
49. J. R. Willis (ed.): *IUTAM Symposium on Nonlinear Analysis of Fracture.* Proceedings of the IUTAM Symposium held in Cambridge, U.K. 1997 ISBN 0-7923-4378-6
50. A. Preumont: *Vibration Control of Active Structures.* An Introduction. 1997 ISBN 0-7923-4392-1
51. G.P. Cherepanov: *Methods of Fracture Mechanics: Solid Matter Physics.* 1997 ISBN 0-7923-4408-1
52. D.H. van Campen (ed.): *IUTAM Symposium on Interaction between Dynamics and Control in Advanced Mechanical Systems.* Proceedings of the IUTAM Symposium held in Eindhoven, The Netherlands. 1997 ISBN 0-7923-4429-4
53. N.A. Fleck and A.C.F. Cocks (eds.): *IUTAM Symposium on Mechanics of Granular and Porous Materials.* Proceedings of the IUTAM Symposium held in Cambridge, U.K. 1997 ISBN 0-7923-4553-3
54. J. Roorda and N.K. Srivastava (eds.): *Trends in Structural Mechanics.* Theory, Practice, Education. 1997 ISBN 0-7923-4603-3
55. Yu. A. Mitropolskii and N. Van Dao: *Applied Asymptotic Methods in Nonlinear Oscillations.* 1997 ISBN 0-7923-4605-X
56. C. Guedes Soares (ed.): *Probabilistic Methods for Structural Design.* 1997 ISBN 0-7923-4670-X
57. D. François, A. Pineau and A. Zaoui: *Mechanical Behaviour of Materials.* Volume I: Elasticity and Plasticity. 1998 ISBN 0-7923-4894-X
58. D. François, A. Pineau and A. Zaoui: *Mechanical Behaviour of Materials.* Volume II: Viscoplasticity, Damage, Fracture and Contact Mechanics. 1998 ISBN 0-7923-4895-8
59. L. T. Tenek and J. Argyris: *Finite Element Analysis for Composite Structures.* 1998 ISBN 0-7923-4899-0
60. Y.A. Bahei-El-Din and G.J. Dvorak (eds.): *IUTAM Symposium on Transformation Problems in Composite and Active Materials.* Proceedings of the IUTAM Symposium held in Cairo, Egypt. 1998 ISBN 0-7923-5122-3
61. I.G. Goryacheva: *Contact Mechanics in Tribology.* 1998 ISBN 0-7923-5257-2
62. O.T. Bruhns and E. Stein (eds.): *IUTAM Symposium on Micro- and Macrostructural Aspects of Thermoplasticity.* Proceedings of the IUTAM Symposium held in Bochum, Germany. 1999 ISBN 0-7923-5265-3

Kluwer Academic Publishers – Dordrecht / Boston / London

Mechanics

SOLID MECHANICS AND ITS APPLICATIONS

Series Editor: G.M.L. Gladwell

63. F.C. Moon: *IUTAM Symposium on New Applications of Nonlinear and Chaotic Dynamics in Mechanics.* Proceedings of the IUTAM Symposium held in Ithaca, NY, USA. 1998
ISBN 0-7923-5276-9
64. R. Wang: *IUTAM Symposium on Rheology of Bodies with Defects.* Proceedings of the IUTAM Symposium held in Beijing. China. 1999 ISBN 0-7923-5297-1
65. 'Yu. I. Dimitrienko: *Thermomechanics of Composites under High Temperatures.* 1999
ISBN 0-7923-4899-0

Kluwer Academic Publishers – Dordrecht / Boston / London

MIX
Papier aus verantwortungsvollen Quellen
Paper from responsible sources
FSC® C105338
FSC
www.fsc.org

If you have any concerns about our products,
you can contact us on
ProductSafety@springernature.com

In case Publisher is established outside the EU,
the EU authorized representative is:
Springer Nature Customer Service Center GmbH
Europaplatz 3, 69115 Heidelberg, Germany

Printed by Libri Plureos GmbH
in Hamburg, Germany